Statistical Mechanics in a Nutshell

Luca Peliti

Translated by Mark Epstein

PRINCETON UNIVERSITY PRESS • PRINCETON AND OXFORD

Originally published in Italian as *Appunti di meccanica statistica*,

Published by Princeton University Press, 41 William Street,
Princeton, New Jersey 08540

In the United Kingdom: Princeton University Press,
6 Oxford Street, Woodstock, Oxfordshire OX20 1TW

press.princeton.edu

Library of Congress Cataloging-in-Publication Data

Peliti, L. (Luca), 1948–
[Appunti di meccanica statistica. English]
Statistical mechanics in a nutshell / Luca Peliti.
p. cm. — (In a nutshell)
Includes bibliographical references and index.
ISBN 978-0-691-14529-7 (hardcover : alk. paper) 1. Statistical mechanics. I. Title.
QC174.8.P4513 2011
530.13—dc22

2010046591

British Library Cataloging-in-Publication Data is available

This book has been composed in Scala

Printed on acid-free paper. ∞

Printed in the United States of America

10 9 8 7 6 5 4

Contents

Preface to the English Edition

Make things as simple as possible, but not simpler.
—*attributed* to Albert Einstein

The aim of the present book is to provide a concise introduction to statistical mechanics: a field of theoretical physics that has experienced tumultuous development in recent years. In the early 1970s, it was recognized that quantum field theory—the most widely accepted paradigm for the description of elementary particles—can be interpreted in a statistical mechanics language. This has allowed for extensive interchange of researchers and results between the two fields, leading to the impressive achievements of the modern theory of phase transitions. More recently, the intensive investigation of the statistical physics of disordered systems has led to the establishment of new methods and concepts, which have found surprising applications in a number of fields, ranging from economy (under the label of *econophysics*) to social sciences (*sociophysics*), to information processing. Meanwhile, within statistical physics proper, there has been a continuous interest in complex systems, especially of biological interest, and a new understanding of soft matter and of physical biology has arisen, thanks to newly developed theoretical tools and to the amazing advances in experimental techniques.

I am unable to provide a thorough introduction to these exciting developments in a short book kept at an elementary level. What I hope I have succeeded in doing is to introduce the reader to the basic concepts, to the essentials of the more recent developments, leaving her or him on the threshold of the different fields in which current research is most active. In view of this goal, the book is organized so that the first five chapters, which should be read sequentially, deal with the basic concepts and techniques of modern statistical mechanics, while the last five chapters, which can be read independently, each deal at an introductory level with a different aspect of interest in current research. I strived to keep the discussion at the most elementary level the subject could afford, keeping in mind the quotation in the epigraph.

"Every book is a collaborative effort, even if there is only one author on the title page" (M. Srednicki) [Sred06]. This book is no exception. A number of friends and colleagues have helped me collect my notes for their publication in the Italian edition of this book:

Stanislas Leibler, whose lecture notes were an important source of inspiration; Mauro Sellitto and Mario Nicodemi, who helped me with the first draft; and Antonio Coniglio, Franco Di Liberto, Rodolfo Figari, Jean-Baptiste Fournier, Silvio Franz, Giancarlo Franzese, Giuseppe Gaeta, Marc Mézard, Fulvio Peruggi, and Mohammed Saber, who provided a number of important suggestions and corrections. Further observations and comments on the published book were made to me by Jeferson Arenzon, Alberto Imparato, Andrea Longobardo, and a number of students of my courses. I am also grateful to Alberto Clarizia, Giancarlo D'Ambrosio, Pietro Santorelli, Maurizio Serva, and Angelo Vulpiani for discussions that have led me to reconsider some points of my exposition. Giuseppe Trautteur encouraged me to publish my lecture notes in book form and connected me with Alfredo Salsano at Bollati Boringhieri. I am grateful to Tony Zee for suggesting to publish the book in English and putting me in contact with Princeton University Press. A reader at the Press suggested adding the discussion of a few relevant points, which I did with great pleasure. I received from Natalie Baan, Ingrid Gnerlich, and Jennifer Harris all the help I could dream of during the production process. Mark Epstein was brilliantly able, not without effort, to transform my involved Italian into fluid English prose. Sarah Stengle designed the attractive jacket, based on an illustration of turbulent flow that was kindly provided by Guido Boffetta.

I renew the dedication of this book to Giovanni Paladin. We are still reaping the fruit of his scientific activity, so untimely and tragically interrupted.

Preface

> I learned more from my pupils than from all others.
>
> —Lorenzo Da Ponte

This work is a result of the notes I have gathered while preparing lessons in statistical mechanics—a course I have been teaching for 15 years at the Federico II University in Naples (Italy). The course's aim is to be both a beginner's and an advanced course. The first section therefore contains an introduction to the problems and methods of statistical mechanics, devised for students without any prior knowledge of the subject, while the second section contains an introduction to some topics of current interest. The work is structured accordingly, with a first group of chapters (1 to 5) that are meant to be studied sequentially, corresponding to the first section, while the following chapters can also be studied independently of one another.

This arrangement has led to a drastic shift in the presentation of the subject matter as compared to more traditional manuals. I thought it necessary to begin with a modern exposition of thermodynamics, using the geometric method introduced by J. W. Gibbs and reworked by L. Tisza and H. Callen. This work is therefore not merely another manual about thermodynamics, but an exposition that is quite innovative for texts at this level. It has significant conceptual and mnemonic advantages when compared to more traditional presentations. I have chosen to introduce the fundamental postulates of statistical mechanics on the basis of the expression of entropy in terms of the volume of the accessible phase space, due to Boltzmann. In this, I follow the point of view of L. D. Landau and S.-K. Ma. Although this point of view is not commonly represented in traditional texts, it allows one to do without the ergodic hypothesis, thus avoiding a series of delicate discussions that are ultimately not very relevant from a practical standpoint. In this manner, the basic techniques of statistical mechanics can be explained in their general outlines without taking too many detours. I have drastically simplified both the terminology and formalisms used in the definition of the fundamental ensembles, relying at every step on parallels with the formulation of thermodynamics I had previously introduced. I have experienced that this setting allows the teacher to reach in few lectures a point where the basic techniques can be used by the students to solve problems in elementary statistical mechanics (interaction-free systems). It is then possible to apply these techniques to simple physically interesting systems like quantum gases and paramagnets, and even to less traditional problems like the elasticity of polymers.

Working knowledge of methods for solving interaction-free systems is the foundation needed to introduce the most important method to explore the behavior of systems with interaction—in other words, *mean-field theory*. This theory is rarely mentioned in introductory statistical mechanics textbooks, and in my opinion, the space allocated to it in advanced textbooks is rarely sufficient. In my experience, it is not only fundamental, but it can be presented and understood at a beginner's level. It allows the teacher to introduce the problems connected to phase transitions and their fluctuations following a natural sequence; the phenomenological theory concerning these issues (*scaling*) is the final topic of the introductory section.

The second part of this work includes topics that can be studied independently from each other. The *renormalization group theory* of phase transitions represented a true conceptual revolution for our discipline. Explaining it is a notoriously difficult task, and I therefore attempt, in chapter 6, to present it in a manner that can be productive on a variety of levels. On the one hand, I explain its basic structure by introducing the concept of a Kadanoff transformation independent of its specific realizations, showing what the existence of this transformation entails for the system's critical behavior. I then introduce two of the transformation's concrete realizations: one elementary, but not systematic (decimation in the Ising model), and another that is more systematic—in the sense that it can lead to a series of successive approximations—even though it requires more complex calculations. In this case, it is important to transmit the method's principle, since its actual implementation is a more advanced topic, one that is also discussed in several other texts.

I devote one chapter to the modern theory of classical fluids, which has seen remarkable progress in the 1970s and 1980s. I believe that I have found a simple way to present the principles of diagrammatic development of the fluids' properties, as well as the derivation of integral equations. This is a good starting point to introduce more advanced methods, such as the density functional method.

Given the importance of numerical simulations in modern statistical mechanics, I devoted a chapter to this topic: I spend more time on basic problems than on the actual writing of simulation programs. This is a deliberate choice because discussing these programs would require the explicit introduction of a specific programming language, the analysis of a certain number of technical aspects related to its implementation, and so on—all items that would require too much space and distract the reader. Most often the students have already met some of these topics in other courses, and if this is not the case, discussing them would transform this chapter into a necessarily unsatisfactory introduction to the use of computers. On the other hand, I believe that discussions of the foundations of the Monte Carlo method, of the treatment of statistical errors, and above all of the extrapolation of the thermodynamic limit are of great relevance in the context of our discipline but are usually neglected in the textbooks of this level I have been able to examine. I have also insisted on the contribution that numerical simulation provides in clearing up problems in the foundations of statistical mechanics—a topic that as far as I know has been treated only in the text by S.-K. Ma [Ma85].

Another chapter is devoted to dynamic equilibrium phenomena, especially the theory of linear response and the fluctuation–dissipation theorem (and its consequences, such as the Onsager reciprocity relations). These topics are usually not treated in introductory textbooks at this level. My experience teaching the course has shown me that, starting with the theory of Brownian motion and making use of the concept of generalized Brownian motion, one can rapidly and easily derive these fundamental properties. Last, chapter 10 contains an introduction to the study of complex systems in statistical mechanics: it discusses the theory of linear polymers and then the theory of percolation as an example of a system with frozen disorder. It then contains a short introduction to the more general theory of disordered systems and allows a glimpse into one of its interdisciplinary applications—namely, the statistical theory of associative memories. In this fashion, the student can perceive the general nature of the methods of statistical mechanics, which the more traditional mechanistic approach instead tends to obscure.

In order to keep the size of the work within reasonable boundaries, I decided to omit the discussion of several interesting topics that are normally discussed in other similar textbooks. More specifically, I include only a brief introduction to the kinetic theory of gases (and I do not even mention Boltzmann's equation), I do not discuss the so-called ergodic problem, and I generally avoid discussions about the foundations of statistical mechanics. Indeed, the debate about these foundations is still ongoing but appears to me to bear little relevance for the concrete results that can be obtained within a more pragmatic approach. I also had to omit any reference to quantum statistical mechanics, except for those few cases that lead to calculations that are basically classical in nature. The problem is that, except for some special circumstances, this subject cannot be treated with the elementary methods I use in this textbook. This allows me to discuss the problems that lie at the heart of contemporary statistical mechanics, especially the theory of phase transitions and of disordered systems, with relative ease.

This book could not have been written without the help of many who helped and encouraged me in the course of these years. I especially thank Mauro Sellitto and Mario Nicodemi, who valiantly contributed to the completion of a first draft, as well as Stanislas Leibler, whose lessons inspired me when I was giving my course its shape. Many students in my statistical mechanics course contributed both to making certain proofs clearer and to reducing the number of errors. Although I cannot thank them all here one by one, I want to express my gratitude to all of them. I also thank Antonio Coniglio, Franco Di Liberto, Rodolfo Figari, Jean-Baptiste Fournier, Silvio Franz, Giancarlo Franzese, Giuseppe Gaeta, Marc Mézard, Fulvio Peruggi, and Mohammed Saber for their observations and suggestions, as well as Giuseppe Trautteur for his constant encouragement and for having suggested I publish my notes in book format.

I am grateful to the Naples Unit of the Istituto Nazionale di Fisica della Materia for the continuous aid they provided for my research during these years. I also thank the Naples Section of the Istituto Nazionale di Fisica Nucleare for the support they have provided.

This book is dedicated to the memory of Giovanni Paladin, whose curious and sharp mind promised great achievements for statistical mechanics, but one that a tragic fate has prematurely taken from us.

Statistical Mechanics in a Nutshell

1 Introduction

> Lies, damned lies, and statistics.
>
> —Disraeli

1.1 The Subject Matter of Statistical Mechanics

The goal of statistical mechanics is to predict the macroscopic properties of bodies, most especially their thermodynamic properties, on the basis of their microscopic structure.

The macroscopic properties of greatest interest to statistical mechanics are those relating to *thermodynamic equilibrium*. As a consequence, the concept of *thermodynamic equilibrium* occupies a central position in the field. It is for this reason that we will first review some elements of thermodynamics, which will allow us to make the study of statistical mechanics clearer once we begin it. The examination of *nonequilibrium* states in statistical mechanics is a fairly recent development (except in the case of gases) and is currently the focus of intense research. We will omit it in this course, even though we will deal with properties that are time-dependent (but always related to thermodynamic equilibrium) in the chapter on dynamics.

The microscopic structure of systems examined by statistical mechanics can be described by means of mechanical models: for example, gases can be represented as systems of particles that interact by means of a phenomenologically determined potential. Other examples of mechanical models are those that represent polymers as a chain of interconnected particles, or the classical model of crystalline systems, in which particles are arranged in space according to a regular pattern, and oscillate around the minimum of the potential energy due to their mutual interaction. The models to be examined can be, and recently increasingly are, more abstract, however, and exhibit only a faint resemblance to the basic mechanical description (more specifically, to the quantum nature of matter). The explanation of the success of such abstract models is itself the topic of one of the more interesting chapters of statistical mechanics: the *theory of universality* and its foundation in the renormalization group.

The models of systems dealt with by statistical mechanics have some common characteristics. We are in any case dealing with systems with a large number of degrees of freedom: the reason lies in the corpuscular (atomic) nature of matter. Avogadro's constant, $N_A = 6.02\ 10^{23}$ mol^{-1}—in other words, the number of molecules contained in a gram-mole (mole)—provides us with an order of magnitude of the degrees of freedom contained in a thermodynamic system. The degrees of freedom that one considers should have more or less comparable effects on the global behavior of the system. This state of affairs excludes the application of the methods of statistical mechanics to cases in which a restricted number of degrees of freedom "dominates" the others—for example, in celestial mechanics, although the number of degrees of freedom of the planetary system is immense, an approximation in which each planet is considered as a particle is a good start. In this case, we can state that the *translational degrees of freedom* (three per planet)—possibly with the addition of the *rotational degrees of freedom*, also a finite number—dominate all others. These considerations also make attempts to apply statistical concepts to the human sciences problematic because, for instance, it is clear that, even if the behavior of a nation's political system includes a very high number of degrees of freedom, it is possible to identify some degrees of freedom that are disproportionately important compared to the rest. On the other hand, statistical methods can also be applied to systems that are not strictly speaking mechanical—for example, neural networks (understood as models of the brain's components), urban thoroughfares (traffic models), or problems of a geometric nature (percolation).

The simplest statistical mechanical model is that of a large number of identical particles, free of mutual interaction, inside a container with impenetrable and perfectly elastic walls. This is the model of the **ideal** gas, which describes the behavior of real gases quite well at low densities, and more specifically allows one to derive the well-known equation of state.

The introduction of pair interactions between the particles of the ideal gas allows us to obtain the standard model for **simple fluids**. Generally speaking, this model cannot be resolved exactly and is studied by means of perturbation or numerical techniques. It allows one to describe the behavior of real gases (especially noble gases), and the liquid–vapor transition (boiling and condensation).

The preceding models are of a classical (nonquantum) nature and can be applied only when the temperatures are not too low. The quantum effects that follow from the inability to distinguish particles are very important for phenomenology, and they can be dealt with at the introductory level if one omits interactions between particles. In this fashion, we obtain models for **quantum gases**, further divided into **fermions** or **bosons**, depending on the nature of the particles.

The model of **noninteracting fermions** describes the behavior of conduction electrons in metals fairly well (apart from the need to redefine certain parameters). Its thermodynamic properties are governed by the Pauli exclusion principle.

The model of **noninteracting bosons** has two important applications: radiating energy in a cavity (also known as **black body**) can be conceived as a set of particles (**photons**) that are bosonic in nature; moreover, helium (whose most common isotope, ^{4}He, is bosonic in nature) exhibits, at low temperatures, a remarkable series of properties that can be interpreted on the basis of the noninteracting boson model. Actually, the transition of ^{4}He to a superfluid state, also referred to as **λ transition**, is connected to the **Einstein condensation**, which occurs in a gas of noninteracting bosons at high densities. Obviously, interactions between helium atoms are not negligible, but their effect can be studied by means of analytic methods such as perturbation theory.

In many of the statistical models we will describe, however, the system's fundamental elements will not be "particles," and the fundamental degrees of freedom will not be mechanical (position and velocity or impulse). If we want to understand the origin of ferromagnetism, for example, we should isolate only those degrees of freedom that are relevant to the phenomenon being examined (the orientation of the electrons' magnetic moment) from all those that are otherwise pertinent to the material in question. Given this moment's quantum nature, it can assume a finite number of values. The simplest case is that in which there are only two values—in this fashion, we obtain a simple model of ferromagnetism, known as the **Ising model**, which is by far the most studied model in statistical mechanics. The ferromagnetic solid is therefore represented as a regular lattice in space, each point of which is associated with a degree of freedom, called **spin**, which can assume the values $+1$ and -1. This model allows one to describe the paramagnet—ferromagnet transition, as well as other similar transitions.

1.2 Statistical Postulates

The behavior of a mechanical system is determined not only by its structure, represented by motion equations, but also by its initial conditions. The *laws* of mechanics are therefore not sufficient by themselves to define the behavior of a mechanical system that contains a large number of degrees of freedom, in the absence of hypotheses about the relevant initial conditions. It is therefore necessary to complete the description of the system with some additional postulates—the **statistical postulates** in the strict sense of the word—that concern these initial conditions.

The path to arrive at the formulation of statistical postulates is fairly twisted. In the following section, we will discuss the relatively simple case of an ideal gas. We will formulate some statistical hypotheses on the distribution of the positions and velocities of particles in an ideal gas, following Maxwell's reasoning in a famous article [Maxw60], and we will see how from these hypotheses and the laws of mechanics it is possible to derive the equation of state of the ideal gas. What this argument does not prove is that the hypotheses made about the distribution of positions and velocities are compatible with the equations of motion—in other words, that if they are valid at a certain instant in time, they remain valid, following the system's natural evolution, also at every successive instant. One of Boltzmann's greatest contributions is to have asked this question in a clear way and to have made a bold attempt to answer it.

1.3 An Example: The Ideal Gas

1.3.1 Definition of the System

In the model of an **ideal gas**, one considers N point-like bodies (or particles, with mass equal to m, identical to one another), free of mutual interaction, inside a container of volume V whose walls are perfectly reflecting. The system's mechanical state is known when, for each particle, the position vector $\boldsymbol{r}$ and the velocity vector $\boldsymbol{v}$ are known. These vectors evolve according to the laws of mechanics.

1.3.2 *Maxwell's Postulates*

The assumption is that the vectors are distributed "randomly," and more specifically that:

1. The vectors pertaining to different particles are *independent* from one another. This hypothesis certainly does not apply, among other examples, to particles that are very close to each other, because the position of two particles that are very close is undoubtedly influenced by the forces that act between them. One can however expect that if the gas is very diluted, deviations from this hypothesis will have negligible consequences. If one accepts the hypothesis of independent particles, the system's state is determined when the number of particles $\mathrm{d}N$ whose position is located within a box, with sides $\mathrm{d}\boldsymbol{r} = (\mathrm{d}x, \mathrm{d}y, \mathrm{d}z)$ placed around a point defined by $\boldsymbol{r} = (x, y, z)$, and that are simultaneously driven by a velocity whose vector lies in a box defined by sides $\mathrm{d}\boldsymbol{v} = (\mathrm{d}v_x, \mathrm{d}v_y, \mathrm{d}v_z)$ around the vector $\boldsymbol{v} = (v_x, v_y, v_z)$: $\mathrm{d}N = f(\boldsymbol{r}, \boldsymbol{v})\, \mathrm{d}\boldsymbol{r}\, \mathrm{d}\boldsymbol{v}$ is known. This defines the **single-particle distribution** $f(\boldsymbol{r}, \boldsymbol{v})$.
2. Position is independent of velocity (in the sense given by probability theory), and therefore the probability distribution $f(\boldsymbol{r}, \boldsymbol{v})$ is factorized: $f(\boldsymbol{r}, \boldsymbol{v}) = f_r(\boldsymbol{r}) f_v(\boldsymbol{v})$.
3. Density is uniform in the space occupied by the gas, and therefore $f_r(\boldsymbol{r}) = N/V = \rho = \text{const.}$ if $\boldsymbol{r}$ is inside the container, and equal to zero otherwise.
4. The velocity components are mutually independent, and therefore $f_v(\boldsymbol{v}) = f_x(v_x) f_y(v_y) f_z(v_z)$.
5. The distribution $f_v(\boldsymbol{v})$ is isotropic in velocity space, so that $f_v(\boldsymbol{v})$ will in actual fact depend only on the modulus $v = |\boldsymbol{v}|$ of $\boldsymbol{v}$.

Exercise 1.1 Prove that the only distribution that satisfies postulates 4 and 5 is Gaussian:

$$f_v(\boldsymbol{v}) \propto \exp(-\lambda v^2), \tag{1.1}$$

where λ is a constant, related to the average quadratic velocity of the particles:

$$\langle v^2 \rangle = \frac{3}{2\lambda}, \tag{1.2}$$

and where therefore the average kinetic energy is given by

$$\left\langle \frac{1}{2} m v^2 \right\rangle = \frac{3m}{4\lambda}. \tag{1.3}$$

1.3.3 *Equation of State*

We will now prove that Maxwell's postulates allow us to derive the equation of state for ideal gases and provide a microscopic interpretation of absolute temperature in terms of kinetic energy.

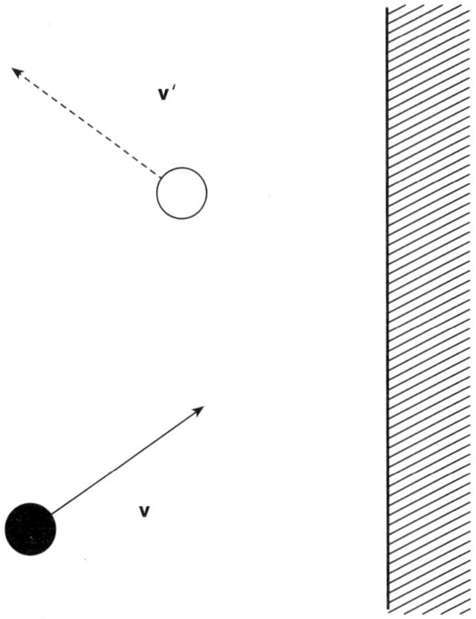

FIGURE I.1. Derivation of the relation between average velocity and pressure. The particle driven by velocity $\boldsymbol{v}$ is about to hit the wall. After the impact, it will have velocity $\boldsymbol{v}'$, in which the sign of the x component (normal to the wall) has changed.

Let us consider a particle of velocity $\boldsymbol{v} = (v_x, v_y, v_z)$ which, coming from the left, hits a wall, parallel to the plane (yz) (see figure 1.1). After the impact, it will be driven by velocity $\boldsymbol{v}' = (-v_x, v_y, v_z)$. The change $\Delta\boldsymbol{p}$ in its momentum $\boldsymbol{p}$ is given by $\Delta\boldsymbol{p} = \boldsymbol{p}' - \boldsymbol{p} = m(\mathrm{v}' - \mathrm{v}) = m(-2v_x, 0, 0)$. The number of impacts of this type that occur in a time interval Δt on a certain region of the wall of area S is equal to the number of particles driven by velocity $\boldsymbol{v}$ that are contained in a box of base equal to S and of height equal to $v_x\Delta t$. The volume of this box is equal to $Sv_x\Delta t$, and the number of these particles is equal to $\rho f_v(\mathbf{v})v_x S\Delta t$.

The total momentum $\Delta\boldsymbol{P}$ transmitted from the wall to the gas, during the time interval Δt, is therefore

$$\Delta\boldsymbol{P} = \int_0^\infty dv_x \int_{-\infty}^{+\infty} dv_y \int_{-\infty}^{+\infty} dv_z f_v(\boldsymbol{v})\,\rho S\Delta t(-2m)\,v_x^2\boldsymbol{i}. \tag{1.4}$$

where $\boldsymbol{i} = (1, 0, 0)$ is the versor of axis x. In this expression, the integral over v_x runs only on the region $v_x > 0$ because only those particles that are moving toward the right contribute to pressure on the wall we are examining.

The total force that the wall exercises on the gas is given by $\boldsymbol{F} = -\Delta\boldsymbol{P}/\Delta t$, and therefore the pressure p is given by

$$p = \frac{|\boldsymbol{F}|}{S} = 2m\rho \cdot \frac{1}{2}\langle v_x^2\rangle = \frac{\rho m}{2\lambda}. \tag{1.5}$$

In this equation, the factor $1/2$ comes from the integration over v_x, which runs only on the region $v_x > 0$. It is well known that the equation of state for perfect gases takes the form

$$pV = nRT, \tag{1.6}$$

where $n = N/N_A$ is the number of moles, T is the absolute temperature, and $R = 8.31$ JK^{-1} mol^{-1} is the constant of the gas. By introducing the **Boltzmann constant**

$$k_B = \frac{R}{N_A} = 1.38\ 10^{-23}\ \mathrm{JK}^{-1}, \tag{1.7}$$

and the particle density

$$\rho = \frac{N}{V}, \tag{1.8}$$

it can be written as

$$p = \rho\, k_B T. \tag{1.9}$$

If we compare this expression with equation (1.5), we obtain the constant λ:

$$\lambda = \frac{m}{2k_B T}. \tag{1.10}$$

The Gaussian velocity distribution implies the following distribution of the magnitude v of the velocity, known as the **Maxwell distribution**:

$$\phi(v) = \int \mathrm{d}\mathbf{v}\, \delta(v - |\mathbf{v}|) f_v(\mathbf{v}) = \mathcal{N}\left(\frac{1}{2\pi m k_B T}\right)^{3/2} 4\pi v^2 \exp\left(-\frac{mv^2}{2k_B T}\right), \tag{1.11}$$

where $\mathcal{N}$ is a normalization factor.

1.3.4 *The Marcus and McFee Experiment*

The Maxwell distribution can be measured directly by means of experiments on the molecular beams. We will follow the work by Marcus and McFee [Marc59]. A diagram of the experiment is given in figure 1.2. Potassium atoms are heated to a fairly high temperature (several hundred degrees Celsius) in an oven. The oven is equipped with a small opening

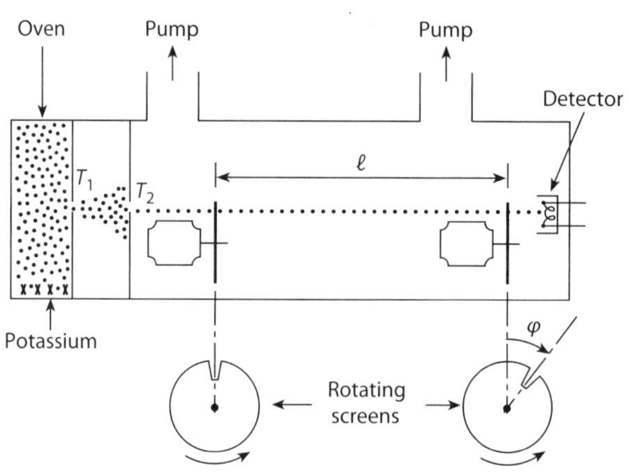

FIGURE 1.2. The Marcus and McFee experiment. Adapted from [Marc59].

from which the molecular beam departs. In the region traversed by the beam, a vacuum is maintained by a pump. Two rotating screens, set at a distance ℓ from each other, act as velocity selectors. Each is endowed with a narrow gap, and they rotate in solidarity with angular velocity ω. The two gaps are out of phase by an angle ϕ. Therefore, only those particles driven by a velocity $v = \ell\omega/\phi$ will be able to pass through both gaps, hit the detector, and be counted. If we denote the total beam intensity by j_0, and the solid angle by which the detector is seen from the opening by $\mathrm{d}\Omega$, the number of particles driven by a velocity between v and $v + \mathrm{d}v$ that hit the detector in a given unit of time, is given by

$$j\,\mathrm{d}v\,\mathrm{d}\Omega = \mathcal{N} j_0 v\,\phi(v)\,\mathrm{d}v\,\mathrm{d}\Omega = \frac{j_0}{2\pi\,(2mk_BT)^2}\,v^3\exp\left(-\frac{mv^2}{2k_BT}\right)\mathrm{d}v\,\mathrm{d}\Omega, \qquad (1.12)$$

where $\mathcal{N}$ is a normalization factor.

By varying ϕ or ω, one can measure the particle flow at various velocities v. We can introduce the variable $\eta = v\sqrt{m/k_BT}$, thus obtaining a law that is independent of both m and T:

$$\frac{j\mathrm{d}v}{j_0 d\Omega} = \frac{\eta^3}{2}\exp\left(-\frac{\eta^2}{2}\right)\mathrm{d}\eta. \qquad (1.13)$$

This law is well suited to experimental proof, which is shown in figure 1.3. In order to arrive at a quantitative agreement with the results of the experiment, it is necessary to take into account the fact that the disks do not instantaneously interrupt the particle flow and that velocity selection is therefore not ideally precise.

1.4 Conclusions

We have therefore been able to show that, if one formulates some statistical hypotheses about the distribution of the particles' velocities, the equation of state of ideal gases is

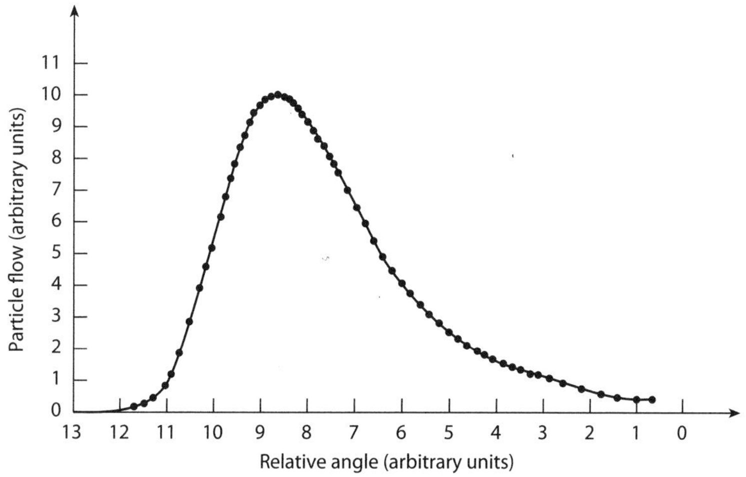

FIGURE 1.3. Distribution of the particle flux in the Marcus and McFee experiment. Adapted from [Marc59].

compatible with the ideal gas model. In the argument we have laid out, there is, however, no proof that the statistical hypotheses concerning the distribution of position and velocity of particles are compatible with the laws of mechanics. In order to prove this compatibility, we will need to establish some hypotheses about initial conditions. We will therefore be satisfied if we manage to prove that for "almost all" initial conditions that satisfy certain criteria, the statistical properties of the relevant distributions are "almost always" valid. These considerations will be incomplete, however, since it is not clear how hypotheses that are valid for almost all initial conditions can be relevant to explain the result of a specific experiment, carried out with specific initial conditions.

The statistical postulates are ultimately founded in thermodynamics. As S.-K. Ma [Ma85] has stressed, they can be summed up in the relation (discovered by Boltzmann) between thermodynamic entropy and a mechanical quantity: of accessible phase space. In order to clarify what entropy is and what we mean by volume of accessible phase space, it will be necessary to briefly discuss the principles of thermodynamics, which is what I will do in the next chapter.

Recommended Reading

The kinetic theory of gases is treated in greater detail in a large number of works. A brief but careful introduction can be found in K. Huang, *Statistical Mechanics,* New York: Wiley, 1987. A classic reference work is R. C. Tolman, *The Principles of Statistical Mechanics,* Oxford, UK: Oxford University Press, 1938 (reprinted New York: Dover, 1979). The (often surprising) history of the kinetic theory has been retraced by S. G. Brush, *The Kind of Motion We Call Heat,* Amsterdam: North-Holland, 1976.

2 Thermodynamics

> The method of "postulating" what we want has many advantages; they are the same as the advantages of theft over honest toil.
>
> —Lord Russell

2.1 Thermodynamic Systems

In this chapter, I will present a brief review of thermodynamics, following an order that is the reverse of the most common way by which it is introduced. The usual course is inductive: one starts with certain regularities that can be observed in thermal phenomena in order to present them as general principles (the **principles of thermodynamics**), one then proceeds to define entropy and its properties, and one finally applies these properties to the study of the equilibrium of thermodynamic systems. We will instead use a deductive approach: we will postulate the properties of thermodynamic equilibrium—specifically, the existence of the entropy function—in order to then deduce the properties of thermodynamic equilibrium. This approach, which was established by Gibbs in the 1870s [Gibb73] [Gibb75], and then developed by various authors (especially L. Tisza in the 1960s [Tisz66]), is the one followed by the most modern textbooks, especially Callen's [Call85], which I shall follow fairly closely.

A **thermodynamic system** is a *macroscopic system* whose behavior is identified thanks to a *small and finite* number of quantities—the **thermodynamic properties**.

The most common example is a given quantity of a **simple fluid**—in other words of a physical phase that is *homogeneous and isotropic*—formed by particles that belong to only one chemical species. In this case, the thermodynamic properties are the volume V, the number of particles N, and the internal energy E. In the case of nonsimple fluids (composed of mixtures of particles belonging to several chemical species), one will have to consider the numbers of particles N_1, N_2, and so on, of the relevant chemical species.

One can also examine more general thermodynamic properties—for example, in magnetic systems, one considers the total magnetic dipole of the sample $\mathbf{M}$, usually called

magnetization; in crystals, there are several thermodynamic quantities that characterize its structure and therefore allow one to distinguish the various crystalline phases.

Thermodynamics is a "miracle": the system has $\sim 10^{23}$ degrees of freedom, and yet it can be well described by a much smaller number of variables. It is clear that we will not be able to foresee *all* the properties of our system, but for our purposes, those few parameters will be sufficient. There is a certain degree of circularity in the definition of thermodynamic parameters, which is resolved by experiment. One considers only a restricted set of manipulations on thermodynamic systems. In practice, one allows them to be put in contact with one another, or one acts upon them by changing a few macroscopic properties such as their volume or the electric or magnetic field in which they are immersed. One then identifies a number of properties such that, if they are known before the manipulation, their values after the manipulation can be predicted. The smallest set of properties that allows one to successfully perform such a prediction can be selected as the basis for a thermodynamic description of the system. Unfortunately, we are not able, at least for the time being, to define a priori, in general for any given system, what the thermodynamic properties that characterize it will be. In order to do this, we must resort to experiments.

If the state of a thermodynamic system can be fully characterized by the values of the thermodynamic variables, and if these values are invariant over time, one says that it is in a state of **thermodynamic equilibrium**. Thermodynamic equilibrium, as R. Feynman [Feyn72] says, occurs when all *fast* processes have already occurred, while the *slow* ones have yet to take place. Clearly the distinction between fast and slow processes is dependent on the observation time τ that is being considered. This property has been emphasized especially by S.-K. Ma [Ma85, p. 3]:

> If we pour some boiling water from a thermos flask into a tea cup, after a few seconds the water in the cup becomes an unchanging state at rest, and this is an equilibrium state. The volume of the water in the cup can be measured from the height of the water and the cross-section area of the cup. The temperature of the water can be measured by a thermometer. Within one or two seconds, these measured quantities will not change significantly, and thus during this observation time the water is in equilibrium. If the time is too long, the temperature will obviously change and the water cannot be regarded as being in an equilibrium state. After an hour, the temperature of the water will be equal to the room temperature. If the room temperature is constant for several hours, then the temperature of the water will accordingly remain constant. Therefore, if the observation time is within this range of several hours, the cup of water can be regarded as being in equilibrium. However, water molecules will continually evaporate from the water surface. In several hours the volume of the water will not change significantly, but if the time is long enough, such as two or three days, all the water will evaporate. So if the observation period is over two or three days, this cup of water cannot be regarded as being in an equilibrium state. After a few days, when all the water has evaporated, then again this empty cup can be regarded as being in an equilibrium state. However, strictly speaking, even this cup is not in an absolute unchanging state, because the molecules of the cup can evaporate, although in a few years' time there will be no change observed.

These arguments show that a system can be shown to be in equilibrium if the observation time is fairly short, while it is no longer possible to consider it in equilibrium for longer observation times. A more curious situation, but one that occurs often, is that the same system can be considered in equilibrium, but with different properties, for different observation times. For example, a block of tar in the course of a period of observation lasting several minutes behaves like an elastic solid. If, however, it is left alone in a container, after several hours have elapsed, it will take the shape of the container like any liquid. It is therefore possible to consider it as if it were a solid for brief periods of time and as a liquid over longer periods of time. Similarly, basalt is commonly considered a solid, but in the context of continental drift (in other words, for observation times lasting several million years), it is possible to consider it a liquid on which the continents float. We shall see that analogous phenomena also occur for much less exotic systems, like gaseous hydrogen. There are some very "stubborn" systems where one cannot forget this fact because the set of thermodynamic properties changes too rapidly with variations in observation times—this is the case for all types of glass, which in a certain sense represent the frontier of thermodynamic description.

As thermodynamics is usually formulated, the observation time τ is not explicitly mentioned; we would, however, do well to remain aware of the fact that the set of thermodynamic variables (and therefore the relations that govern them) is implicitly determined by τ. It is the physicist's task (guided each time by the experiment) to determine on each occasion the most appropriate thermodynamic description of the specific system under examination.

2.2 Extensive Variables

Let us consider two thermodynamic systems, 1 and 2, that can be made to interact with one another. Variables like the volume V, the number of particles N, and the internal energy E, whose value (relative to the total system) is equal to the sum of the values they assume in the single systems, are called **additive** or **extensive**.

Let us note that, strictly speaking, internal energy is *not* extensive. In reality if we put two systems in contact with each other, the total energy of the system $1 \cup 2$ will also receive the contribution of the interaction between the two systems 1 and 2. More generally, we will therefore have

$$E^{(1\cup 2)} = E^{(1)} + E^{(2)} + E_{\text{int}}, \tag{2.1}$$

where the indexes (1) and (2) refer to the two subsystems, and E_{int} is the interaction energy. If the interactions between the components of the two systems are only short-range, however, the interaction between the two systems affects only a surface region, where they are in contact. Therefore, while one might expect $E^{(1)}$ and $E^{(2)}$ to be proportional to the volume of their respective systems, E_{int} will be proportional to their surface. For large systems, as long as they occupy well-defined portions of space, surface grows in size much more slowly than volume, so that ultimately the contribution of E_{int} is shown to be negligible (see figure 2.1). So if a container holds water and oil, the interaction energy between the water and oil is proportional to the container's cross section, while the energy of the

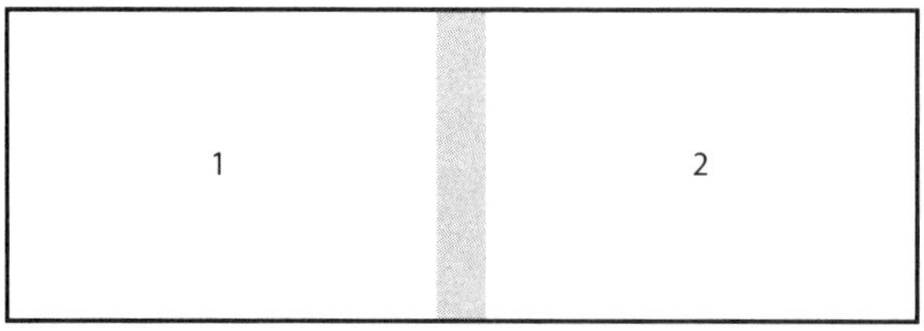

FIGURE 2.1. Systems in contact. The gray area represents the region in which the interaction between the two systems takes place.

water (or oil) is proportional to their respective volumes. In some situations, however, it is possible to obtain structures in which, for instance, oil forms droplets of ~ 50 nm in diameter, which are dispersed in the water and in equilibrium with it. These structures are called **microemulsions**. In this case, the interaction energy between water and oil becomes proportional to the container's volume, because within each element of the system's volume (which is fairly large) a certain amount of interface is present. In this case, it is *not* possible to consider the system as made of water and oil set side by side, but it is more practical to consider it a thermodynamically homogeneous system.

It is not possible to neglect interaction energy in the presence of long-range interactions, as for instance in the case of electrical, magnetic, or gravitational fields. Usually, in these cases, it is possible to consider the effect of these interactions explicitly; one therefore considers the gravitational interaction energy between two bodies (which, as is well known, grows as the *product* of the bodies' masses, and not as their sum) as *mechanical* (and not *internal*) energy.

The **fundamental hypothesis of thermodynamics** is that it should be possible to characterize the state of a thermodynamic system by specifying the values of a certain set (X_0, $X_1, \ldots, X_r$) of *extensive* variables. For example, X_0 could be the internal energy E, X_1 the number of particles N, X_2 the volume V of the system, and so on.

2.3 The Central Problem of Thermodynamics

Classical thermodynamics has *states of equilibrium* as its object. The **central problem of thermodynamics** can be formulated as follows:

> *Given the initial state of equilibrium of several thermodynamic systems that are allowed to interact, determine the final thermodynamic state of equilibrium.*

The interaction between thermodynamic systems is usually represented by idealized walls that allow the passage of one (or more) extensive quantities from one system to the other. Among the various possibilities, the following are usually considered:

Thermally conductive walls: These allow the passage of energy, but not of volume or particles.

Semipermeable walls: These allow the passage of particles belonging to a given chemical species (and consequently also of energy).

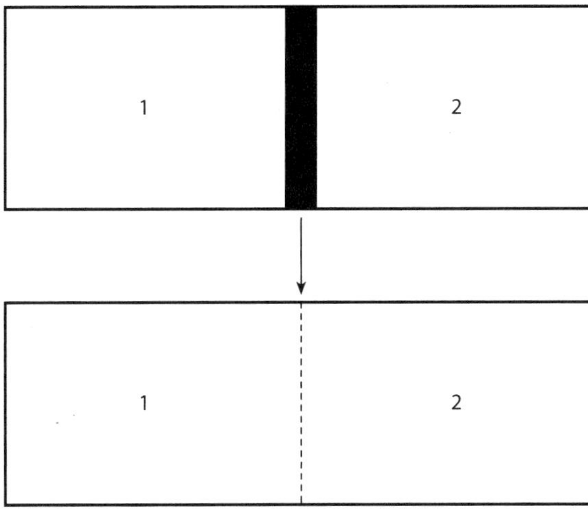

FIGURE 2.2. Generic diagram of a thermodynamic experiment.

Adiabatic walls: These prevent the passage of energy.

Figure 2.2 shows a generic diagram of a thermodynamic experiment. Systems 1 and 2 are each initially in a state of thermodynamic equilibrium characterized by their respective values of the extensive thermodynamic variables $(X_0, X_1, \ldots, X_r)$, or in other words, $(X_{0\text{in}}{}^{(1)}, X_{1\text{in}}{}^{(1)}, \ldots, X_{r\text{in}}{}^{(1)})$ and $(X_{0\text{in}}{}^{(2)}, X_{1\text{in}}{}^{(2)}, \ldots, X_{r\text{in}}{}^{(2)})$. They are put in contact—that is, one allows the exchange of some of these extensive quantities between one system and the other to occur.

After a certain amount of time has elapsed (and its duration is not of concern to thermodynamics), a new state of equilibrium is reached, characterized by the values $(X_{0\text{fin}}{}^{(1)}, X_{1\text{fin}}{}^{(1)}, \ldots, X_{r\text{fin}}{}^{(1)})$ and $(X_{0\text{fin}}{}^{(2)}, X_{1\text{fin}}{}^{(2)}, \ldots, X_{r\text{fin}}{}^{(2)})$, for system 1 and system 2, respectively. The problem is to characterize this state of equilibrium.

The space of *possible* states of equilibrium (compatible with constraints and initial conditions) is called the space of **virtual states**. The initial state is obviously a (specific) virtual state. The central problem of thermodynamics can obviously be restated as follows:

Characterize the actual state of equilibrium among all virtual states.

2.4 Entropy

The solution of the central problem of thermodynamics is provided by the following **entropy postulate**:

*There exists a function S of the extensive variables $(X_0, X_1, \ldots, X_r)$, called the **entropy**, that assumes the maximum value for a state of equilibrium among all virtual states and that possesses the following properties:*

Extensivity: If 1 and 2 are thermodynamic systems, then

$$S^{(1\cup 2)} = S^{(1)} + S^{(2)}. \tag{2.2}$$

Convexity: If $X^1 = (X_0^1, X_1^1, \ldots, X_r^1)$ and $X^2 = (X_0^2, X_1^2, \ldots, X_r^2)$ are two thermodynamic states of the same system, then for any α between 0 and 1, one obtains

$$S((1-\alpha)X^1 + \alpha X^2) \geq (1-\alpha)S(X^1) + \alpha S(X^2). \tag{2.3}$$

From this expression, if we take the derivative with respect to α at $\alpha = 0$, we obtain

$$\sum_{i=0}^{r} \left.\frac{\partial S}{\partial X_i}\right|_{X^1} (X_i^2 - X_i^1) \geq S(X^2) - S(X^1), \tag{2.4}$$

which expresses the fact that the surface $S(X_0, X_1, \ldots, X_r)$ is always *below* the plane that is tangent to each of its points.

Monotonicity[1]: $S(E, X_1, \ldots, X_r)$ is a monotonically increasing function of the internal energy E:

$$\left.\frac{\partial S}{\partial E}\right)_{X_1,\ldots,X_r} = \frac{1}{T} > 0. \tag{2.5}$$

The entropy postulate allows one to solve the central problem of thermodynamics, by referring it back to the solution of a **constrained extremum** problem:

The equilibrium state corresponds to the maximum entropy compatible with the constraints.

In the next section, we will examine some simple applications of this postulate.

2.5 Simple Problems

2.5.1 *Thermal Contact*

Let us consider two systems, 1 and 2, that are in contact by means of a thermally conductive wall—they can therefore exchange internal energy, but not other extensive quantities. The virtual state space is therefore defined by the relations:

$$E^{(1)} + E^{(2)} = X_0^{(1)} + X_0^{(2)} = E = \text{const.}, \tag{2.6}$$

$$X_i^{(1)} = \text{const.}, \quad i = 1, 2, \ldots, r, \tag{2.7}$$

$$X_i^{(2)} = \text{const.}, \quad i = 1, 2, \ldots, r. \tag{2.8}$$

Let us look for the maximum of S as a function, for example, of $E^{(1)}$:

$$\begin{aligned} \frac{\partial S}{\partial E^{(1)}} &= \frac{\partial}{\partial E^{(1)}}\left[S^{(1)}(E^{(1)},\ldots,X_r^{(1)}) + S^{(2)}(E - E^{(1)},\ldots,X_r^{(2)})\right] \\ &= \left.\frac{\partial S^{(1)}}{\partial E^{(1)}}\right|_{E^{(1)}} - \left.\frac{\partial S^{(2)}}{\partial E^{(2)}}\right|_{E^{(2)} = E - E^{(1)}}. \end{aligned} \tag{2.9}$$

[1] There can be some cases, as we will see, in which this condition is violated—these cases are fairly special, because they correspond to the description of certain relatively isolated *subsystems*, in which the energy spectrum has a finite upper bound. In ordinary cases, however, in which energy does not admit a finite upper bound, this condition is always satisfied.

The derivatives are to be understood with $X_i^{(1),(2)}$ (for $i = 1, \dots, r$) kept fixed. We will denote the value of $E^{(1)}$ at equilibrium by $E^{(1)}_{\mathrm{eq}}$. One therefore has

$$\left.\frac{\partial S}{\partial E^{(1)}}\right|_{E^{(1)}_{\mathrm{eq}}} = \left.\frac{\partial S^{(1)}}{\partial E^{(1)}}\right|_{E^{(1)}_{\mathrm{eq}}} - \left.\frac{\partial S^{(2)}}{\partial E^{(2)}}\right|_{E - E^{(1)}_{\mathrm{eq}}} = 0, \tag{2.10}$$

and therefore

$$\left.\frac{\partial S^{(1)}}{\partial E^{(1)}}\right|_{E^{(1)}_{\mathrm{eq}}} = \left.\frac{\partial S^{(2)}}{\partial E^{(2)}}\right|_{E^{(2)}_{\mathrm{eq}}}, \tag{2.11}$$

where $E^{(2)}_{\mathrm{eq}} = E - E^{(1)}_{\mathrm{eq}}$ is the equilibrium value of $E^{(2)}$.

It is useful to ask ourselves which of the two subsystems has released internal energy. Since the initial state is a virtual state, by denoting the relevant values of E with E_{in}, one has

$$S^{(1)}(E^{(1)}_{\mathrm{eq}}) + S^{(2)}(E^{(2)}_{\mathrm{eq}}) \geq S^{(1)}(E^{(1)}_{\mathrm{in}}) + S^{(2)}(E^{(2)}_{\mathrm{in}}), \tag{2.12}$$

and therefore

$$S^{(1)}(E^{(1)}_{\mathrm{eq}}) - S^{(1)}(E^{(1)}_{\mathrm{in}}) + S^{(2)}(E - E^{(1)}_{\mathrm{eq}}) - S^{(2)}(E^{(2)}_{\mathrm{in}}) \geq 0. \tag{2.13}$$

Due to entropy's convexity, for both systems, one has

$$S(E_{\mathrm{eq}}) - S(E_{\mathrm{in}}) \leq \left.\frac{\partial S}{\partial E}\right|_{E_{\mathrm{in}}} (E_{\mathrm{eq}} - E_{\mathrm{in}}). \tag{2.14}$$

We thus obtain

$$\left[\left.\frac{\partial S^{(1)}}{\partial E^{(1)}}\right|_{E^{(1)}_{\mathrm{in}}} - \left.\frac{\partial S^{(2)}}{\partial E^{(2)}}\right|_{E^{(2)}_{\mathrm{in}}}\right]\left(E^{(1)}_{\mathrm{eq}} - E^{(1)}_{\mathrm{in}}\right) \geq 0. \tag{2.15}$$

Energy therefore shifts toward the system with the higher values of $\partial S/\partial E$.

Let us introduce the quantity

$$T = \left(\frac{\partial S}{\partial E}\right)^{-1}. \tag{2.16}$$

According to our hypotheses, this quantity is *positive*. We obtained the following results:

- At equilibrium, T is the same in all subsystems that are in reciprocal contact by means of thermally conductive walls.
- In order to reach equilibrium, energy shifts from systems with higher values of T toward systems with lower values of T.

These properties tie T to empirical temperatures. And in fact, it is possible to deduce that T is a monotonically increasing function of temperature. We will see later that it is possible to identify T with *absolute temperature*.

2.5.2 A Thermally Conductive and Mobile Wall

In this case, the two systems can also exchange volume V, in addition to internal energy E. The two equilibrium conditions are therefore:

1. The condition of maximum entropy relative to exchange of internal energy, as already seen:

$$\frac{\partial S^{(1)}}{\partial E^{(1)}} = \frac{\partial S^{(2)}}{\partial E^{(2)}}. \tag{2.17}$$

2. The condition of maximum entropy relative to exchange of volume, which can be obtained in an analogous fashion:

$$\frac{\partial S^{(1)}}{\partial V^{(1)}} = \frac{\partial S^{(2)}}{\partial V^{(2)}}. \tag{2.18}$$

By introducing the quantity p, defined by the equation

$$\frac{p}{T} = \frac{\partial S}{\partial V}, \tag{2.19}$$

where the partial derivative is taken by keeping the other extensive quantities constant, these conditions can be expressed as follows:

$$T^{(1)} = T^{(2)}, \tag{2.20}$$

$$p^{(1)} = p^{(2)}. \tag{2.21}$$

As we will see, the quantity p can be identified with **pressure**.

One can easily prove that between two systems, both initially at the same temperature, volume is initially released by the system in which pressure is lower to the system in which pressure is higher. In fact, because of the principle of maximum entropy, one has

$$S^{(1)}(E_{\text{eq}}^{(1)}, V_{\text{eq}}^{(1)}) - S^{(1)}(E_{\text{in}}^{(1)}, V_{\text{in}}^{(1)}) + S^{(2)}(E_{\text{eq}}^{(2)}, V_{\text{eq}}^{(2)}) - S^{(2)}(E_{\text{in}}^{(2)}, V_{\text{in}}^{(2)}) \geq 0, \tag{2.22}$$

while because of convexity, for each system one has

$$S(E_{\text{eq}}, V_{\text{eq}}) - S(E_{\text{in}}, V_{\text{in}}) \leq \left.\frac{\partial S}{\partial E}\right|_{E_{\text{in}}} (E_{\text{eq}} - E_{\text{in}}) + \left.\frac{\partial S}{\partial V}\right|_{V_i} (V_{\text{eq}} - V_{\text{in}}). \tag{2.23}$$

When we apply this relation to our case, we obtain

$$\begin{aligned} 0 \leq &\left[\left.\frac{\partial S^{(1)}}{\partial E^{(1)}}\right|_{E_{\text{in}}^{(1)}} - \left.\frac{\partial S^{(2)}}{\partial E^{(2)}}\right|_{E_{\text{in}}^{(2)}} \right] \left(E_{\text{eq}}^{(1)} - E_{\text{in}}^{(1)} \right) \\ &+ \left[\left.\frac{\partial S^{(1)}}{\partial V^{(1)}}\right|_{V_{\text{in}}^{(1)}} - \left.\frac{\partial S^{(2)}}{\partial V^{(2)}}\right|_{V_{\text{in}}^{(2)}} \right] \left(V_{\text{eq}}^{(1)} - V_{\text{in}}^{(1)} \right). \end{aligned} \tag{2.24}$$

The first term vanishes because the initial temperatures in the two systems are equal. We thus obtain:

$$\left[\left.\frac{\partial S^{(1)}}{\partial V^{(1)}}\right|_{V_{\text{in}}^{(1)}} - \left.\frac{\partial S^{(2)}}{\partial V^{(2)}}\right|_{V_{\text{in}}^{(2)}} \right] \left(V_{\text{eq}}^{(1)} - V_{\text{in}}^{(1)} \right) \geq 0. \tag{2.25}$$

Since the temperatures T in the two systems are the same, this relation implies that

$$\left(p^{(1)} - p^{(2)}\right)\left(V_{\text{eq}}^{(1)} - V_{\text{in}}^{(1)}\right) \geq 0. \tag{2.26}$$

2.5.3 *An Adiabatic and Mobile Wall*

Let us now consider the apparently similar case of two systems separated by a wall that is both mobile (allowing the exchange of volume) and adiabatic (preventing the free exchange of internal energy). It turns out that this problem cannot be solved within thermodynamics alone. Let us assume that the resulting complex system cannot exchange either volume or energy with the environment. Then the internal energy of both systems can change only via the mechanical work associated with the change in their respective volumes:

$$\mathrm{d}E^{(1)} = -p^{(1)}\mathrm{d}V^{(1)} = -\mathrm{d}E^{(2)} = p^{(2)}\mathrm{d}V^{(2)}. \tag{2.27}$$

On the other hand, total volume is conserved:

$$\mathrm{d}V^{(1)} = -\mathrm{d}V^{(2)}. \tag{2.28}$$

These two equations are generally incompatible, unless $p^{(1)} = p^{(2)}$. Moreover, for an infinitesimal shift of an adiabatic wall, one has

$$\mathrm{d}S = \left.\frac{\partial S}{\partial E}\right)_V \mathrm{d}E + \left.\frac{\partial S}{\partial V}\right)_E \mathrm{d}V = \left(-\frac{p}{T} + \frac{p}{T}\right)\mathrm{d}V = 0, \tag{2.29}$$

for every value of $\mathrm{d}V$. Consequently, given the hypotheses we made, the system cannot reach equilibrium, unless it is already there, and vice versa, equilibrium conditions do not determine $V^{(1)}$ (or $V^{(2)}$).

Should we attempt to undertake the experiment, the mobile wall would begin to oscillate. Clearly, sooner or later the oscillation would cease, because of friction (which renders the first equation invalid). The final condition of equilibrium will however be determined by the details of energy dissipation due to friction.

2.5.4 A Semipermeable Wall

Let us consider a system composed of several chemical species, and let us introduce the number of molecules $N_1, N_2, \ldots, N_s$ belonging to the chemical species that constitute it as part of the thermodynamic variables. Let us suppose that two systems of this type are separated by a wall that only allows the k-th chemical species to pass. Clearly, it is impossible for the exchange of molecules to occur without an exchange of energy. The equilibrium conditions will therefore be

$$T^{(1)} = T^{(2)}, \tag{2.30}$$

$$\frac{\partial S^{(1)}}{\partial N_k^{(1)}} = \frac{\partial S^{(2)}}{\partial N_k^{(2)}}. \tag{2.31}$$

When we introduce the **chemical potential** μ_k relative to the k-th chemical species by means of the definition

$$\frac{\mu_k}{T} = -\frac{\partial S}{\partial N_k}, \tag{2.32}$$

these equations are seen as equivalent to

$$T^{(1)} = T^{(2)}, \tag{2.33}$$

$$\mu_k^{(1)} = \mu_k^{(2)}. \tag{2.34}$$

Exercise 2.1 Consider two systems, initially at the same temperature, that are in contact by means of a semipermeable wall that allows for the passage of particles of the chemical species k. Show that, in this case, the particles flow from the system with a larger value of μ_k to that with a smaller value.

2.6 Heat and Work

The consistency of thermodynamics requires that internal energy be a function of thermodynamic variables. Some of these variables are *mechanical* in nature, in other words, they are not thermodynamic—for example, the volume V, the number of particles N, but also the total magnetic moment $\boldsymbol{M}$, and so on.

From mechanics (and from electromagnetism), we can derive an expression for the infinitesimal **mechanical work** performed on the system by varying the extensive quantities. One usually adopts a sign convention according to which work W is considered *positive* if the system performs work on the outside, and *negative* if the opposite is true. Following this convention, the expression of infinitesimal work δW is given by

$$\delta W = -\sum_{i=1}^{r} f_i \mathrm{d}X_i. \tag{2.35}$$

The symbol δ reminds us that δW is not an "exact" differential—it is not a function's differential, because the integral $\int \delta W$ depends on the particular path being considered, and not only on the initial and final states.

The quantities f_i that appear in this expression are called **generalized forces**. If the system is thermically isolated, it can exchange energy with the outside only by means of mechanical work. In this case, δW is equal to the opposite of the system's infinitesimal variation of internal energy:

$$\delta W = -\mathrm{d}E = -\sum_{i=1}^{r} f_i \mathrm{d}X_i. \tag{2.36}$$

It is well known that although *perfectly adiabatic walls* that completely prevent the exchange of energy in the form of heat do not exist, using a series of devices, it is possible to approximate this ideal condition.

If the walls are not adiabatic, in general, we will have

$$\mathrm{d}E = \delta Q - \delta W, \tag{2.37}$$

which represents the definition of infinitesimal **heat** δQ exchanged in the process. As highlighted by this notation, not even δQ is an exact differential.

2.6.1 *Temperature*

Let us now show that the quantity T, defined by equation (2.16), which we rewrite as

$$\left.\frac{\partial S}{\partial E}\right)_X = \frac{1}{T},$$

where $X = (X_1, \dots, X_r)$ denotes the extensive thermodynamic variables other than the internal energy, is indeed the *absolute* temperature that one encounters in elementary thermodynamics.

Let us review the *intuitive* properties of the temperature:

Thermal equilibrium: Two bodies in thermal equilibrium have the same temperature.

Heat flux: Heat flows *spontaneously* from bodies at higher temperature to bodies at lower temperatures.

As we have seen in the discussion of equations (2.11) and (2.15), these properties are also satisfied by our definition of T. Thus T is a monotonically increasing function of the thermodynamical temperature.

In order to distinguish momentarily between two definitions of temperature, let us denote by Θ the absolute temperature defined in elementary thermodynamics. This quantity is identified by the efficiency of thermal engines:

The maximal efficiency of a thermal engine working between absolute temperatures Θ_1 and Θ_2 (with $\Theta_1 > \Theta_2$) is given by $\eta = 1 - \Theta_2/\Theta_1$.

We will now show that $\Theta \propto T$, where T is "our" temperature. It is then clear that it is possible to choose the entropy scale in such a way as to put the proportionality constant equal to 1.

Let us consider a system made up of a thermal engine and two heat reservoirs, the first at temperature T_1 and the second at temperature T_2 (with $T_1 > T_2$). A heat reservoir is a system that is assumed to be so large that it can exchange an arbitrary amount of energy without changing its temperature: i.e., for which the temperature $\partial S/\partial E)_X$ is independent of E. The whole compound system is enclosed in a container that allows it to exchange energy with the environment only in a purely mechanical way, e.g., by raising or lowering a macroscopic weight. Let the system evolve from an initial equilibrium condition, in which the first heat reservoir has internal energy E_1, the second has internal energy E_2, and the thermal engine is in some equilibrium state, to a final equilibrium state in which the first heat reservoir has internal energy E_1' (and the same temperature T_1, since it is a heat reservoir), the second has E_2', the weight has its potential energy changed by the amount $W = (E_1 + E_2) - (E_1' + E_2')$, and the thermal engine is back to its initial state. By definition, the efficiency of the engine is given by $\eta = W/(E_1 - E_1')$, i.e., the ratio of the work done on the environment to the energy extracted as heat from the high-temperature reservoir.

In a transformation of this kind, the total entropy of the compound system cannot become smaller. Thus, we have

$$S^{(1)}(E_1) + S^{(2)}(E_2) \le S^{(1)}(E_1') + S^{(2)}(E_2'), \tag{2.38}$$

where $S^{(i)}(E)$ is the fundamental equation for the reservoir i ($i = 1, 2$), because by hypothesis the thermal engine is back to its initial state. On the other hand, since we are dealing with heat reservoirs, we have

$$S^{(i)}(E_i') = S^{(i)}(E_i) + \frac{E_i' - E_i}{T_i}, \qquad i = 1,2. \tag{2.39}$$

Thus, we have

$$\frac{E_1 - E_1'}{T_1} \leq \frac{E_2' - E_2}{T_2}. \tag{2.40}$$

Therefore,

$$W = (E_1 - E_1') + (E_2 - E_2') \leq (E_1 - E_1')\left(1 - \frac{T_2}{T_1}\right). \tag{2.41}$$

This expression should be compared with the maximum efficiency evaluated in elementary thermodynamics that we have seen earlier. We have, therefore,

$$\frac{\Theta_2}{\Theta_1} = \frac{T_2}{T_1}, \tag{2.42}$$

which implies that $T \propto \Theta$. The proportionality constant can be fixed by assigning a definite temperature to a well-identified thermodynamical state. As is well known, one chooses the triple point of pure water, to which one assigns the temperature of 273.16 K. We can thus conclude that

$$T = \left[\left.\frac{\partial S}{\partial E}\right)_{X_1,\dots,X_r}\right]^{-1} \tag{2.43}$$

is the absolute temperature.

2.6.2 *Adiabatic Transformations and the Axiomatic Foundation of the Entropy Principle*

It has been recently realized that the considerations that we described earlier provide a way to establish the existence of the entropy function and of its basic properties on an intellectually and intuitively appealing axiomatic basis. We have considered a transformation in which a thermodynamic system is allowed to exchange energy with the environment in a purely mechanical way, and we have exploited the fact that, in this situation, the only allowed transformations are the ones in which the total entropy is not reduced. One can reverse this logic. Following Lieb and Yngvason [Lieb98], let us define **adiabatic transformations** on a thermodynamic system (which may itself be made up of several subsystems, like the heat reservoirs we have considered) as allowed to exchange energy with the environment in a purely mechanical way. It is possible for the system to interact with an arbitrarily complex auxiliary sytem (like the thermal engine we have introduced) provided that this system is found, at the end of the transformation, in its initial state. On the other hand, no hypothesis is posited about the pace of the transformation itself, which can be arbitrarily fast or violent. It is assumed only that the initial and the final states are both thermodynamical equilibrium states.

Let us consider such a system, and let $X = (X_0, \dots, X_r)$ identify the equilibrium states of such a system. It is then possible to introduce a relation $\prec$ among these states such that $X \prec X'$ if it is possible to reach X' via an adiabatic transformation, starting from X. In this case, one writes

$$X \prec X',$$

and one says X precedes X', or X' follows X. It is quite clear that $\prec$ is a reflexive and transitive relation, but is in general not symmetric. Indeed, as emphasized, e.g., by Carathéodory [Cara09], the second principle of thermodynamics may be traced to the fact that for any equilibrium state X of a thermodynamical system, there are states that cannot be reached from it via an adiabatic transformation, while it can be reached from them.

Let us call two states, X and X' **adiabatically equivalent**, and write $X \overset{A}{\sim} X'$, if $X \prec X'$ and $X' \prec X$. Then, $\overset{A}{\sim}$ is an equivalence relation. If $X \prec X'$, but $X' \not\prec X$, we will write $X \prec\prec X'$. Given two states, X and X', if either $X \prec X'$ or $X' \prec X$, we say that the two states are **comparable**.

Then, it is possible to show that the existence, extensivity, and convexity of the entropy function $S(X)$ follows from a few simple and physically appealing properties of the relation $\prec$:

Reflexivity: $X \overset{A}{\sim} X', \forall X$.

Transitivity: If $X \prec X'$ and $X' \prec X''$, then $X \prec X''$.

Consistency: If $X \prec X'$ and $Y \prec Y'$, then $(X, Y) \prec (X', Y')$.

Scale invariance: If $\lambda > 0$ and $X \prec X'$, then $\lambda X \prec \lambda X'$. This identifies the variables appearing in X as extensive variables, so that the system described by λX is λ times larger than the system described by X.

Splitting and recombination: If $0 < \lambda < 1$, then $X \overset{A}{\sim} (\lambda X, (1 - \lambda)X)$. Thus the system described by X can be adiabatically split into two smaller but similar systems, described by λX and $(1 - \lambda)X$, respectively, and vice versa.

Stability: If $(X, \epsilon Z_0) \prec (X', \epsilon Z_1)$ for some Z_0, Z_1 and for a sequence of ϵ that tends to zero, then $X \prec X'$. This means that the set of allowed adiabatic processes cannot be changed by coupling the system to an arbitrarily small external system.

Then, it can be shown that there is a function $S(X)$ such that if $X \overset{A}{\sim} X'$, then $S(X) = S(X')$, and if $X \prec\prec X'$, then $S(X) < S(X')$. Moreover, if X and Y are the states of two systems and (X, Y) denotes the corresponding state of the compound system, one has

$$S(X, Y) = S(X) + S(Y),$$

i.e., the function $S(X)$ is additive, and for any $\lambda > 0$ and each state X, one has

$$S(\lambda X) = \lambda S(X),$$

i.e., $S(X)$ is extensive.

Let us consider a set Γ of states such that all states in Γ are mutually comparable, i.e., if both $X \in \Gamma$ and $X' \in \Gamma$, then either $X \prec X'$ or $X' \prec X$ (or both). By the axioms, one can then show that the states $((1 - \lambda)X^0, \lambda X^1)$, where $0 \leq \lambda \leq 1$ and X^0 and X^1 belong to Γ, also belong to Γ. It is then possible to give an explicit expression of the entropy function in Γ. One considers two reference states, X^0 and X^1, such that $X^0 \prec\prec X^1$. (If it is not possible to find two such states, then $S(X)$ is the constant function.) Then, for any state X comparable with X^0 and X^1, one sets

$$S(X) = \sup\{\lambda : ((1-\lambda)X^0, \lambda X^1) \prec X\}.$$

Here, we allow $\lambda \leq 0$ or $\lambda \geq 1$ by the conventions that $(X, -Y) \prec X'$ means $X \prec (X', Y)$ and that $(X, 0Y)$ means X.

This construction recalls an old definition by Laplace and Lavoisier in which heat was measured by the amount of ice a body could melt. In our case, the entropy of a state X is measured by the largest amount of X^1 that can be adiabatically transformed into X with the help of X^0.

The entropy function defined in this way is unique up to affine equivalence, i.e., $S(X) \longrightarrow aS(X) + b$, with $a > 0$. In order to define a consistent entropy scale for a set of compound systems, it appears that one has to postulate that all the corresponding states are comparable. If this is assumed, a consistent entropy function for all comparable states of compound systems can be defined. Indeed, Lieb and Yngvason [Lieb98] have shown that this **comparability hypothesis** follows from a few additional axioms concerning the notion of thermal equilibrium and thermal contact, out of which the notion of temperature can be abstracted.

Following this approach, it is possible to establish the basic notions of thermodynamics, and in particular the entropy principle, on a sounder and intuitively appealing basis. A detailed explanation of this approach can be found in [Lieb98] .

2.6.3 Pressure

Let us consider an infinitesimal variation of V while keeping E constant. In this case, mechanics tells us that the work δW performed by the system is given by the product of the pressure P with the infinitesimal variation in volume $\mathrm{d}V$:

$$\delta W = P\mathrm{d}V. \tag{2.44}$$

On the other hand, since $\mathrm{d}E = 0$, one has

$$\delta Q = T\mathrm{d}S = \delta W = P\mathrm{d}V, \tag{2.45}$$

and therefore

$$\left.\frac{\partial S}{\partial V}\right)_{E,\ldots,X_r} = \frac{P}{T}. \tag{2.46}$$

This allows us to identify the pressure P with the quantity p we defined previously.

2.6.4 Chemical Potential

The case of chemical potential μ_k is rather special. If we attempt to vary the number of particles of certain chemical species that belong to a specific system, we must necessarily allow a certain unchecked exchange of energy between the system and the surrounding environment to occur, and therefore a certain exchange of heat. Consequently, the work necessary to vary the number of moles of a chemical species within a system is not strictly speaking of a mechanical nature, and as a result, the fundamental relation that defines chemical potential is the one that describes the variation of **entropy** as the number of particles varies:

$$\left.\frac{\partial S}{\partial N_k}\right)_{E,X_2,\ldots} = -\frac{\mu_k}{T}. \tag{2.47}$$

In order to interpret this expression, let us imagine that we have available some membranes that are impermeable to the chemical species k but allow the flow of the other species. Then we can build up a thought experiment in which the number of moles of species k present within a certain region is changed by mechanical means. Let us imagine, for example, that the system is made up of a region of volume V, contained inside a cylinder of larger dimensions, which contains a solution of two chemical species—we will designate the greater component (the solvent) by 1 and the smaller component (the solute) by 2. Initially, we will have N_1 moles of solvent and N_2 moles of solute within the system's volume V.

Now, let us suppose that we introduce into the cylinder a piston fitted with a membrane that is permeable to the solvent, but not to the solute, and that this piston delimits a region (containing the system), within which there are $N_2 + \mathrm{d}N_2$ moles of solute, and whose volume is $V + \mathrm{d}V$. We can now move the piston slowly (reversibly!), until it delimits the region of volume V that defines our system. In this manner, we will perform some work on the system to change its concentration of solute. In order for this work not to vanish, it is necessary for a pressure p_2 to be exercised on the piston fitted with the semipermeable membrane. This pressure is called **osmotic pressure**.

We will see later how it is possible to obtain an explicit relation between the mechanical work of compression exercised by the piston and the chemical work used to change the concentration.

2.7 The Fundamental Equation

From the postulates of thermodynamics, it follows that if a specific homogeneous system's entropy S is given, as a function of the extensive quantities ($X_0 = E, X_1, \ldots, X_r$) that define the thermodynamic state of the system, we are able to predict the system's thermodynamic behavior. The equation

$$S = S(X_0 = E, X_1, \ldots, X_r), \tag{2.48}$$

is called the **fundamental equation**, and it represents a *complete* description of the thermodynamics of the system being considered.

From the fundamental equation, one can obtain the expression of the temperature T and the generalized forces f by taking derivatives with respect to the extensive quantities. These expressions are called, generically, **equations of state.**

For example, if we take the derivative of the fundamental equation with respect to the volume V, we obtain the dependence of pressure p on the internal energy E, the number of particles N, and the volume. Analogously, if we take the derivative with respect to E, we obtain the temperature's dependence on the same variables. One can resolve this equation with respect to E and substitute it in the first, in order to obtain the habitual equation of state:

$$p = p(T, V, N). \tag{2.49}$$

However, while the fundamental equation is a compendium of *all* the dynamic information available on a system, the equation of state contains only incomplete information. More specifically, if we know only the habitual equation of state, it is not possible to obtain the dependence of internal energy E on p, T, and N. It is necessary to have further data—for example, the specific heat at constant volume for at least one value of volume V.

Exercise 2.2 The following expressions are proposed as possible fundamental equations for various thermodynamic systems. However, some of them violate some of the postulates of entropy and are therefore not acceptable from a physics standpoint. Find the five that are not acceptable from a physics standpoint, and indicate which postulates are violated by each one. The quantities v_0, θ, and R are positive constants. When fractional exponents appear, only the root's positive values should be considered.

1. $S = \left(R^2/v_0\theta\right)^{1/3}[NVE]^{1/3}$.
2. $S = \left(R^2/\theta^2\right)^{1/3}[NE/V]^{2/3}$.
3. $S = (R/\theta)^{1/2}\left[NE - R\theta V^2/v_0^2\right]^{1/2}$.
4. $S = \left(R^2\theta/v_0^3\right)[V^3/NE]$.
5. $S = \left(R^3/v_0\theta^2\right)^{1/5}\left[N^2 VE^2\right]^{1/5}$.
6. $S = NR\ln\left(VE/N^2R\theta v_0\right)$.
7. $S = (R/\theta)^{1/2}[NE]^{1/2}\exp\left(-V^2/2N^2v_0^2\right)$.
8. $S = (R/\theta)^{1/2}[NE]^{1/2}\exp\left(-EV/NR\theta v_0\right)$.

2.8 Energy Scheme

The fundamental equation we discussed is the most natural, also from the point of view of statistical mechanics. However, for essentially historical reasons, we prefer to use a different (but equivalent) formulation of the fundamental principle of thermodynamics, in which entropy assumes the role of an independent variable, while energy becomes a dependent variable that satisfies a variational principle. This formalism is known as the **energy scheme**.

In this formalism, the maximum entropy principle:

Among all states with a specific value of internal energy, *the state of equilibrium is that in which* entropy is maximum.

is replaced by the principle of **minimum internal energy**:

Among all states with a specific entropy *value, the state of equilibrium is that in which* internal energy is minimal.

For the proof, it can be useful to refer to figure 2.3. Let ΔX be a *virtual* variation of the extensive variables (excluding internal energy E) with respect to the equilibrium value X_{eq}. Then,

$$\Delta S = S(E, X_{\text{eq}} + \Delta X) - S(E, X_{\text{eq}}) \leq 0, \tag{2.50}$$

because of the maximum entropy principle. Since S is a monotonically increasing function of E, there exists a value $E' \geq E$ such that $S(E', X_{\text{eq}} + \Delta X) = S(E, X_{\text{eq}})$. Therefore, if S is kept constant, as the system moves out of equilibrium, E cannot but increase. This is what we wanted to prove.

Therefore, the fundamental equation in the energy scheme is defined when the dependence of internal energy E on entropy S and the other extensive variables $(X_1, \ldots, X_r)$ is known:

$$E = E(S, X_1, \ldots, X_r). \tag{2.51}$$

E's differential assumes the form

$$\mathrm{d}E = T\mathrm{d}S + \sum_{i=1}^{r} f_i \mathrm{d}X_i. \tag{2.52}$$

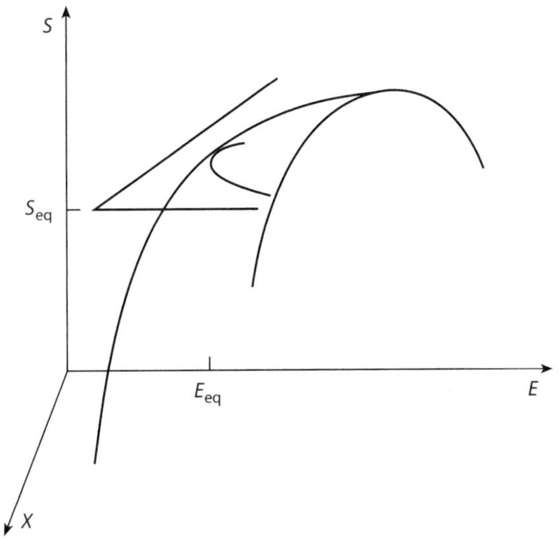

FIGURE 2.3. A schematic representation of the fundamental surface. The equilibrium state can be characterized by the *maximum* of entropy on the plane E = const. or, equivalently, by the *minimum* of internal energy on the plane S = const.

Since TdS is the infinitesimal heat δQ given to the system in a small reversible transformation, and $-\Sigma_i f_i dX_i$ is equal to the corresponding infinitesimal work δW that the system performs during this transformation, we can recognize in this expression the differential version of the **first principle of thermodynamics**:

$$dE = \delta Q - \delta W. \tag{2.53}$$

In this expression, delta reminds us that neither δQ nor δW are exact differentials, whereas dE is.

We know that entropy S is a convex function of the extensive variables $\{X_i\}$—we now see that this implies that energy E is a **concave** function of S and the other extensive variables $\{X_i\}$, $i = 1, \dots, r$. We in fact know that if $X^1 = (X_0^1 = E^1, X_1^1, \dots, X_r^1)$ and $X^2 = (X_2^1 = E^2, X_1^2, \dots, X_r^2)$ are two states of equilibrium, and $0 \le \alpha \le 1$, one has

$$S\big((1-\alpha)X^1 + \alpha X^2\big) \ge (1-\alpha)\,S(X^1) + \alpha S(X^2). \tag{2.54}$$

which expresses the convexity of entropy. Let us denote by S^1 the value of $S(X^1)$ and by S^2 the value of $S(X^2)$. We can then define the states of equilibrium in the energy scheme as follows:

$$Y^1 = \left(S^1, X_1^1, \dots, X_r^1\right), \qquad Y^2 = \left(S^2, X_1^2, \dots, X_r^2\right). \tag{2.55}$$

We therefore get

$$S\big((1-\alpha)E^1 + \alpha E^2, \dots, (1-\alpha)X_r^1 + \alpha X_r^2\big) \ge (1-\alpha)S^1 + \alpha S^2. \tag{2.56}$$

We define E' so that

$$S\big(E', (1-\alpha)X_1^1 + \alpha X_1^2, \dots, (1-\alpha)X_r^1 + \alpha X_r^2\big) = (1-\alpha)S^1 + \alpha S^2. \tag{2.57}$$

Then, since S is an increasing function of E, we will have

$$E' \le (1-\alpha)E^1 + \alpha E^2. \tag{2.58}$$

Therefore,

$$E' = E\big((1-\alpha)Y^1 + \alpha Y^2\big) \le (1-\alpha)\,E(Y^1) + \alpha E(Y^2). \tag{2.59}$$

which implies that the internal energy function is **concave**.

Exercise 2.3 For each of the physically acceptable fundamental equations in exercise 2.2, express its internal energy E as a function of S, V, and N.

2.9 Intensive Variables and Thermodynamic Potentials

The derivatives $f_i = \partial E/\partial X_i$ of the internal energy E with respect to extensive quantities S, $\{X_i\}$ are called **intensive quantities** (or **fields**). For uniformity's sake, we define $f_0 = T = \partial E/\partial S$. A given quantity f_i is called the **conjugate** of the corresponding variable X_i, and vice versa. The temperature T and entropy S are therefore conjugates, as are the pressure (with the opposite sign) $(-p)$ and the volume V.

Since both E and X_i are extensive, in a homogeneous system, intensive variables are not dependent on system size. Moreover, if a system is composed of several subsystems that can exchange the extensive quantity X_i, the corresponding intensive variable f_i assumes the same value in those subsystems that are in contact at equilibrium.

One needs to distinguish between intensive variables, which possess this property, and **densities**, which are quantities independent of system size and which are obtained by taking the ratio of two extensive variables in a homogenous system. For instance, if N is the number of particles, the quantities $s = S/N$, $v = V/N$, $\varepsilon = E/N$ are respectively the entropy, volume, and internal energy densities per particle, and do not depend on the system size. Analogously, one can define some densities with respect to volume, such as $s_V = S/V$. The densities can, however, assume different values in subsystems that are mutually in equilibrium, unlike intensive variables.

We now want to identify the state of equilibrium among all states that exhibit a given value of an intensive variable f_i. Specifically, for $i = 0$, we are confronted with the case of system with a fixed temperature—in practice this can be realized by putting the system in contact with a much larger system (the reservoir). In these conditions, the reservoir's temperature does not, practically speaking, depend on the quantity of energy it exchanges with the system, and so the latter, in equilibrium, assumes the temperature value established by the reservoir.

Let us now define the **Helmholtz free energy** $F(T, X)$ by the relation

$$F(T, X) = E(S(T, X), X) - TS(T, X), \tag{2.60}$$

where $X = (X_1, \ldots, X_r)$ is the list of the extensive thermodynamic variables, apart from the energy, that identify the equilibrium state, and $S(T, X)$ is the equilibrium value of the entropy in the equilibrium state identified by T and X. We can then show that the thermodynamical equilibrium in these conditions is characterized by the following variational principle:

> *The value of the Helmholtz free energy is minimal for the equilibrium state among all virtual states at the given temperature* T.

In order to prove this variational principle, let us denote by X the extensive variables that are free to vary among different virtual states, and let us evaluate the free energy difference between the equilibrium state identified by (T, X^*) (and by the other extensive variables, which are kept fixed and which are understood) and a generic virtual state at the same temperature, identified by (T, X). We obtain

$$\begin{aligned}\Delta F = {} & E(S(T, X), X) - E(S(T, X^*), X^*) \\ & - T\,[S(T, X) - S(T, X^*)]\,.\end{aligned}$$

Let us add and subtract $E(S(T, X^*), X)$ on the right-hand side. We obtain

$$\begin{aligned}\Delta F = {} & [E(S(T, X), X) - E(S(T, X^*), X)] \\ & + [E(S(T, X^*), X) - E(S(T, X^*), X^*)] \\ & - T\,[S(T, X) - S(T, X^*)]\,.\end{aligned}$$

The second term within brackets is nonnegative, because of the variational principle satisfied by the internal energy. Let us consider the first term. Since

$$T = \left.\frac{\partial E}{\partial S}\right)_X,$$

taking into account that E is concave, we obtain

$$E(S(T, X), X) - E(S(T, X^*), X) \geq T[S(T, X) - S(T, X^*)].$$

We have, therefore,

$$\Delta F \geq 0,$$

as we intended to prove.

Let us now consider more generally the Legendre transform of the internal energy E with respect to the intensive variable f_i (see appendix A):

$$\begin{aligned}\Phi(S, f_1, X_2, \dots, X_r) = \\ E(S, X_1(S, f_1, X_2, \dots, X_r), X_2, \dots, X_r) - f_1 X_1(S, f_1, X_2, \dots, X_r),\end{aligned} \tag{2.61}$$

where $X_1(S, f_1, X_2, \dots, X_r)$ is determined by the condition

$$f_1 = \left.\frac{\partial E}{\partial X_1}\right|_{X_1 = X_1(S, f_1, X_2, \dots, X_r)}. \tag{2.62}$$

Then, the state of equilibrium is specified by the following criterion:

Among all the states that have the same value as f_1, *the state of equilibrium is that which corresponds to the minimum value of* Φ.

Because of this criterion, the function Φ is known as the **thermodynamic potential**, since it reaches the minimum value in thermodynamic equilibrium, analogously to mechanical potential, which reaches its minimum in mechanical equilibrium.

In order to prove this criterion, let us suppose that we set the value of f_1, and we allow the extensive variable X_2 to vary. We then get

$$\Phi(S, f_1, X_2) = E(S, X_1(f_1, X_2), X_2) - f_1 X_1(S, f_1, X_2). \tag{2.63}$$

We will designate the equilibrium values of X_1 and X_2 as X_1^* and X_2^*. To a virtual variation ΔX_2 of X_2, there corresponds an analogous variation $\Delta X_1 = X_1(S, f_1, X_2) - X_1^*$. We then have

$$\begin{aligned}\Phi(S, f_1, X_2^* + \Delta X_2) &= E(S, X_1^* + \Delta X_1, X_2^* + \Delta X_2) - f_1(X_1^* + \Delta X_1) \\ &\geq E(S, X_1^*, X_2^* + \Delta X_2) + \left.\frac{\partial E}{\partial X_1}\right|_{X_1^*, X_2^* + \Delta X_2} \Delta X_1 - f_1(X_1^* + \Delta X_1) \\ &= E(S, X_1^*, X_2^* + \Delta X_2) - f_1 X_1^* \geq E(S, X_1^*, X_2^*) - f_1 X_1^* = \Phi(f_1, X_2^*).\end{aligned} \tag{2.64}$$

In this derivation, we first applied the concavity of function E and then the equilibrium condition in the energy scheme.

Let us observe that the partial derivative of Φ, performed with respect to f_1, with the other extensive variables kept fixed, yields the value of the extensive variable X_1 (see appendix A):

$$\left.\frac{\partial \Phi}{\partial f_1}\right)_{S,X_2,\dots,X_r} = \left.\frac{\partial E}{\partial X_1}\right)_{S,X_2,\dots,X_r} \left.\frac{\partial X_1}{\partial f_1}\right)_{S,X_2,\dots,X_r} - X_1(f_1, X_2, \dots, X_r) - f_1 \left.\frac{\partial X_1}{\partial f_1}\right)_{S,X_2,\dots,X_r} = -X_1(S, f_1, X_2, \dots, X_r). \tag{2.65}$$

Nothing prevents us from considering two or more intensive variables as fixed—for example, $f_0 = T$ and f_1. Similar considerations will then lead us to introduce the thermodynamic potential $\Phi(T, f_1, X_2, \dots, X_r)$, obtained as a Legendre transform of E with respect to S and X_1:

$$\Phi(T, f_1, X_2, \dots) = E - TS - f_1 X_1. \tag{2.66}$$

This thermodynamic potential assumes at equilibrium the minimum value among all the states with the same values of T and f_1. We can therefore obtain a whole series of thermodynamic potentials, by using a Legendre transform with respect to the extensive variables X_i. We cannot however eliminate *all* extensive variables in this manner. We will see later that if we did this, the resulting thermodynamic potential would identically vanish. For the time being, it is sufficient to observe that the Φ potentials are *extensive* quantities, and one cannot see how they could be a function only of intensive quantities like the f_i's.

For each thermodynamic potential, a set of "natural" variables has therefore been defined—some are *intensive*, which are conjugates of the variables with respect to which the Legendre transformation has been performed, and the others are *extensive*.

It is interesting to ask oneself what the convexity properties of thermodynamic potentials are. One can see that the thermodynamic potential

$$\Phi(S, f_1, f_2, \dots, f_k, X_{k+1}, \dots, X_r) = E - \sum_{i=1}^{k} f_i X_i, \tag{2.67}$$

obtained as the Legendre transform of E with respect to $X_1, \dots, X_k$, is *concave* as a function of the remaining extensive variables, for fixed values of the intensive variables $f_1, \dots, f_k$. More particularly, for any small virtual variation $\delta X = (\delta X_{k+1}, \dots, \delta X_r)$, one gets

$$\delta^2 \Phi = \sum_{i,j=k+1}^{r} \left.\frac{\partial^2 \Phi}{\partial X_i \partial X_j}\right)_{f_1,\dots,f_k} \delta X_i \delta X_j \geq 0. \tag{2.68}$$

Φ on the other hand is *convex* as a function of the intensive variables f_i's, when the extensive variables are fixed. This follows from the properties of the Legendre transformation (see appendix A), and its proof can be the subject of an exercise.

Exercise 2.4 Consider the thermodynamic potential

$$\Phi(f_0, f_1, \dots, f_k, X_{k+1}, \dots, X_r) = E - \sum_{i=0}^{k} f_i X_i.$$

Prove that the matrix of the second derivatives $(\partial^2 \Phi / \partial f_i \partial f_j)$ is negative definite at each point.

2.10 Free Energy and Maxwell Relations

We can now be slightly more detailed. If we take the temperature T as an independent variable in place of the entropy S, we obtain, as the relevant thermodynamic potential, the Helmholtz free energy which we have defined in equation (2.61). In this equation, the equilibrium value $S(T, X)$ of the entropy in the equilibrium state identified by (T, X) is implicitly defined by the relation

$$T = \left.\frac{\partial E}{\partial S}\right|_{S = S(T,X)}. \tag{2.69}$$

As we have seen, the Helmholtz free energy takes on its minimal value among all virtual states corresponding to the same value of the temperature T.

The natural variables of F are the temperature T and the extensive variables $X_1, \ldots, X_r$, entropy excluded.

Entropy can be obtained by taking the derivative of F with respect to T:

$$\left.\frac{\partial F}{\partial T}\right)_{X_1,\ldots,X_r} = -S. \tag{2.70}$$

We thus obtain the expression for the differential of F:

$$\mathrm{d}F = -S\mathrm{d}T + \sum_{i=1}^{r} f_i \mathrm{d}X_i. \tag{2.71}$$

More specifically, by setting $X_1 = V$, one has

$$-p = \left.\frac{\partial F}{\partial V}\right)_{T,X_2,\ldots,X_r}, \tag{2.72}$$

where $X_2, \ldots, X_r$ are the other extensive variables (excluding both entropy and volume).

If we now take this equation's derivative with respect to T and we use the theorem of the equality of mixed derivatives, we obtain

$$\left.\frac{\partial p}{\partial T}\right)_{V,X_2,\ldots,X_r} = -\frac{\partial^2 F}{\partial T \partial V} = -\frac{\partial^2 F}{\partial V \partial T} = \left.\frac{\partial S}{\partial V}\right)_{T,X_2,\ldots,X_r}. \tag{2.73}$$

This is a thermodynamic identity that relates the heat absorbed by a system in isothermal expansion with the variation in pressure and temperature at constant volume. In effect, if in an isothermal expansion volume varies by $\mathrm{d}V$, the heat absorbed is given by $\delta Q = T(\partial S/\partial V)_T \mathrm{d}V$. We thus obtain

$$\left.\frac{1}{T}\frac{\delta Q}{\delta V}\right)_T = \left.\frac{\partial p}{\partial T}\right)_V. \tag{2.74}$$

These relations between thermodynamic derivatives that derive from the equality of mixed derivatives of thermodynamic potentials are called **Maxwell relations.**

The *free energy* designation is derived from the following property. If a system is put in contact with a reservoir at temperature T, the maximum quantity of work $W_{\max}$ that it can perform on its environment is equal to the variation in free energy between the initial and final states. In other words, one has

$$W \leq W_{max} = F_{in} - F_{fin}. \tag{2.75}$$

More specifically, if the system relaxes toward equilibrium, $F_{fin} = F_{eq} \leq F_{in}$, and therefore, $W_{max} \geq 0$—in other words, the system is able to perform work on its environment.

In order to prove this assertion, let us suppose that the global system (the actual system + the reservoir) is adiabatically isolated. In this situation, we have

$$W = E_{in} - E_{fin}, \tag{2.76}$$

where $E_{fin}(E_{in})$ is the global system's final (initial) internal energy. One therefore also arrives at

$$S_{fin} \geq S_{in}, \tag{2.77}$$

where $S_{fin}(S_{in})$ is the global system's final (initial) entropy. Expressed in greater detail, we have

$$S_{fin} = S_{fin}^{(s)} + S_{fin}^{(r)} \geq S_{in}^{(s)} + S_{in}^{(r)}, \tag{2.78}$$

where (*s*) and (*r*) refer to system and reservoir, respectively. However, since (*r*) is a heat reservoir (with a fixed temperature T), one arrives at

$$\begin{aligned} S_{fin}^{(r)} &= S_{in}^{(r)} + \frac{\partial S^{(r)}}{\partial E^{(r)}}\left(E_{fin}^{(r)} - E_{in}^{(r)}\right) \\ &= S_{in}^{(r)} + \frac{1}{T}\left(E_{fin}^{(r)} - E_{in}^{(r)}\right). \end{aligned} \tag{2.79}$$

Following equation (2.76), we get

$$E_{in}^{(r)} - E_{fin}^{(r)} = W - E_{in}^{(s)} + E_{fin}^{(s)}, \tag{2.80}$$

and therefore

$$S_{fin}^{(r)} = S_{in}^{(r)} - \frac{1}{T}\left(W - E_{in}^{(s)} + E_{fin}^{(s)}\right). \tag{2.81}$$

By substituting this expression in (2.78) and rearranging the terms, we obtain

$$S_{fin}^{(r)} - S_{in}^{(r)} = -\frac{1}{T}\left(W - E_{in}^{(s)} + E_{fin}^{(s)}\right) \geq S_{in}^{(s)} - S_{fin}^{(s)}, \tag{2.82}$$

and, since $T \geq 0$, multiplying by $-T$, we obtain

$$W \leq \left(E_{in}^{(s)} - TS_{in}^{(s)}\right) - \left(E_{fin}^{(s)} - TS_{fin}^{(s)}\right). \tag{2.83}$$

2.11 Gibbs Free Energy and Enthalpy

Transforming F according to Legendre with respect to V, we obtain a new thermodynamic potential, called the **Gibbs free energy**:

$$G(T, p, X_2, \ldots, X_r) = F + pV = E - TS + pV, \tag{2.84}$$

where $X_2, \ldots, X_r$ are the other extensive variables (excluding entropy and volume). The sign of the product pV is due to the fact that the intensive variable conjugated to the volume is the opposite of pressure. The variational principle satisfied by the Gibbs free energy is the following:

> *Among all states that have the same* temperature *and* pressure *values, the state of equilibrium is that in which the Gibbs free energy assumes the minimum value.*

This principle can be proven following the model of F's variational principle.

It is easy to see that

$$\left.\frac{\partial G}{\partial T}\right)_{p, X_2, \ldots, X_r} = -S, \tag{2.85}$$

$$\left.\frac{\partial G}{\partial p}\right)_{T, X_2, \ldots, X_r} = V. \tag{2.86}$$

G's differential is therefore expressed as follows:

$$\mathrm{d}G = -S\mathrm{d}T + V\mathrm{d}p + \sum_{i=2}^{r} f_i \mathrm{d}X_i. \tag{2.87}$$

It is also easy to prove that if a system is brought toward equilibrium while temperature and pressure are kept constant, the maximum work that can be performed on its environment is given precisely by the difference between the initial and final values of G. This is what justifies its designation as *free energy*.

Exercise 2.5 Prove the preceding assertion.

If on the other hand, we Legendre transform the internal energy E with respect to V, we obtain a new thermodynamic potential, usually denoted by H and called **enthalpy**:

$$H(S, p, X_2, \ldots, X_r) = E + pV. \tag{2.88}$$

Enthalpy governs the equilibrium of adiabatic processes that occur while pressure is constant:

> *Among all states that have the same* entropy *and* pressure *values, the state of equilibrium is the one that corresponds to the minimum value of enthalpy.*

Enthalpy is particularly important for chemical reactions in open containers (which are therefore subject to atmospheric pressure).

If a system relaxes toward equilibrium while the pressure is kept constant, the maximum heat that can be produced by the system is equal to its variation in enthalpy. For this reason, enthalpy it is also called *free heat*. In fact, let us suppose that a system relaxes toward equilibrium while the pressure is constant. We will then have

$$H_{\text{fin}} - H_{\text{in}} = E_{\text{fin}} - E_{\text{in}} + p(V_{\text{fin}} - V_{\text{in}}). \tag{2.89}$$

If the system can perform mechanical work on the environment only by means of expansion, one arrives at

$$W = p(V_{\text{fin}} - V_{\text{in}}), \tag{2.90}$$

where W is the mechanical work performed on the system (with our sign conventions). If we assume that the transformation is reversible, we can apply the first principle of thermodynamics, and we will therefore have

$$E_{\text{fin}} - E_{\text{in}} = Q_{\text{rev}} - W, \tag{2.91}$$

where Q_{rev} is the heat given off to the system in the course of the reversible transformation at constant pressure. In our case, Q_{rev} is *negative,* because $H_{\text{fin}} \leq H_{\text{in}}$, since according to our hypothesis, the system is relaxing toward equilibrium. Let me remind the reader that the heat released to the environment is the opposite of Q. We therefore arrive at

$$H_{\text{fin}} - H_{\text{in}} = Q_{\text{rev}}. \tag{2.92}$$

Exercise 2.6 Prove that the heat released by a system that is relaxing at equilibrium at constant pressure is less than (or equal to) its change in enthalpy:

$$Q \leq H_{\text{in}} - H_{\text{fin}} = -Q_{\text{rev}}.$$

The differential of H is given by

$$\mathrm{d}H = T\mathrm{d}S + V\mathrm{d}p + \sum_{i=2}^{r} f_i \mathrm{d}X_i. \tag{2.93}$$

The equality of the mixed derivatives of G and H yield two more Maxwell relations:

$$\left.\frac{\partial S}{\partial p}\right)_T = -\left.\frac{\partial V}{\partial T}\right)_p, \quad \left.\frac{\partial T}{\partial p}\right)_S = \left.\frac{\partial V}{\partial S}\right)_p. \tag{2.94}$$

2.12 The Measure of Chemical Potential

We now show how, by using Helmholtz free energy, one can relate differences in chemical potential to the osmotic pressure, thus allowing for their mechanical measurement.

Let us consider a cylinder of section Δ, inside of which there is a piston fitted with a semipermeable membrane that blocks the passage of chemical species A (see figure 2.4). The system is adiabatically isolated from the external environment. The piston exerts an osmotic pressure p_A, which we will regard as positive when exerted toward the left.

Let us now suppose that the piston is shifted reversibly by a small interval $\mathrm{d}\ell$, toward the right, for example. Let us mentally subdivide the system into three subsystems, from left to right:

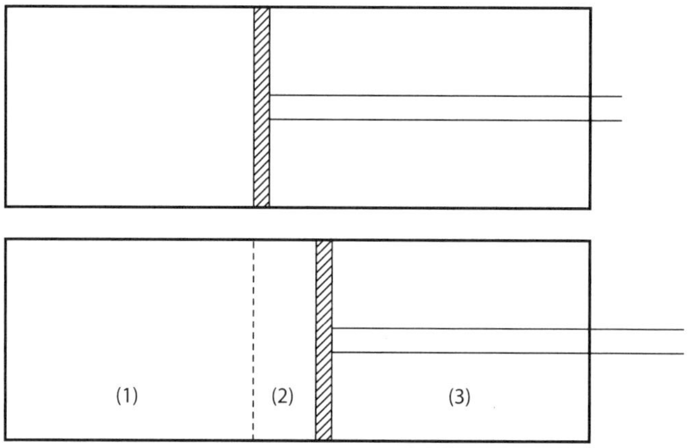

FIGURE 2.4. The measure of chemical potential.

1. The system that occupies the area to the left of the piston's initial position.
2. The system that occupies the area between the piston's two positions.
3. The system that occupies the area to the right of the piston's final position.

Since the transformation is reversible and the system is adiabatically isolated, the total variation of entropy ΔS vanishes, and the variation in internal energy ΔE is given by

$$\Delta E = -p_{\mathrm{A}} \Delta \, d\ell = -p_{\mathrm{A}} V^{(2)}. \tag{2.95}$$

Let us now consider the thermodynamic potential (which we shall denote by $\mathcal{F}$) whose independent variables are the volume V, the temperature T, the chemical potential of the solvent μ, and the number of solute particles N_{A}. It is defined by

$$\mathcal{F} = F - \mu_s N_s. \tag{2.96}$$

The chemical potential of the solute is given by

$$\mu = \left.\frac{\partial \mathcal{F}}{\partial N_{\mathrm{A}}}\right)_{V,T,\mu_s} = \mu(T, \mu_s, c_{\mathrm{A}}), \tag{2.97}$$

where $c_A = N_A/V$ is the solute's concentration. Since the system is at equilibrium at the beginning and the end of the transformation, and the transformation is reversible,

$$\Delta\mathcal{F} = \Delta\mathcal{F}^{(1)} + \Delta\mathcal{F}^{(2)} + \Delta\mathcal{F}^{(3)} = \Delta E = -p_{\mathrm{A}} V^{(2)}. \tag{2.98}$$

In addition, one has $T^{(1)} = T^{(2)} = T^{(3)} = \text{const.}$, while the chemical potential of the solvent is the same in the three systems we are considering.

It is obvious that $\Delta\mathcal{F}^{(3)} = 0$. Therefore, the variation in total free energy $\mathcal{F}$ is due to the shift in $\Delta N = (c_{\mathrm{A}}^{(1)} - c_{\mathrm{A}}^{(3)})\, V^{(2)}$ particles from subsystem (1) to subsystem (2), where $c_{\mathrm{A}}^{(1)}$ is the solute concentration in system (1). System (1) undergoes a variation

$$\Delta\mathcal{F}^{(1)} = -\mu^{(1)}\Delta N = -\mu^{(1)}\left(c_{\rm A}^{(1)} - c_{\rm A}^{(3)}\right)V^{(2)} = -\mu^{(1)}\Delta c_{\rm A}V^{(2)}. \tag{2.99}$$

On the other hand, we arrive at

$$\Delta\mathcal{F}^{(2)} = V^{(2)}\int_{c_{\rm A}^{(3)}}^{c_{\rm A}^{(1)}} \mathrm{d}c\, \mu(T, \mu_s, c). \tag{2.100}$$

If $\Delta c_{\rm A}$ is small (with respect to both $c_A^{(1)}$ and $c_A^{(3)}$), one can set

$$\mu(T, \mu_s, c) \simeq \mu^{(3)} + \left.\frac{\mathrm{d}\mu}{\mathrm{d}c}\right|_{c_{\rm A}^{(3)}}\left(c - c_{\rm A}^{(3)}\right). \tag{2.101}$$

By integrating, we obtain

$$\Delta\mathcal{F}^{(2)} \simeq V^{(2)}\left[\mu^{(3)}\Delta c_{\rm A} + \frac{1}{2}\left.\frac{\mathrm{d}\mu}{\mathrm{d}c}\right|_{c_{\rm A}^{(3)}}\Delta c_{\rm A}^2\right], \tag{2.102}$$

and therefore,

$$\begin{aligned}\Delta\mathcal{F} &\simeq V^{(2)}\left[\left(\mu^{(3)} - \mu^{(1)}\right)\Delta c_{\rm A} + \frac{1}{2}\left.\frac{\mathrm{d}\mu}{\mathrm{d}c}\right|_{c_{\rm A}^{(3)}}\Delta c_{\rm A}^2\right]\\ &\simeq \frac{1}{2}V^{(2)}\left(\mu^{(3)} - \mu^{(1)}\right)\Delta c_{\rm A}.\end{aligned} \tag{2.103}$$

One therefore arrives at

$$p_{\rm A} \simeq \frac{1}{2}\left(\mu^{(1)} - \mu^{(3)}\right)\Delta c_{\rm A}. \tag{2.104}$$

In this slightly convoluted manner, the difference in the chemical potential of the solute in the two subsystems can be measured by means of the osmotic pressure.

2.13 The Koenig Born Diagram

Relations between thermodynamic potentials, extensive variables and the corresponding intensive variables are summed up in the **Koenig Born diagram**, which is shown in figure 2.5. In this diagram, the thermodynamic potentials are written on the sides of a square, while the relevant thermodynamic variables are written on the corners, in such a fashion that each thermodynamic potential is accompanied by its natural variables. The diagonals associate conjugated pairs of variables [S with T, V with $(-p)$], and the arrows determine the sign of the corresponding differential element—if the arrow points away from the potential being considered, the sign is (+); otherwise, it is (–). So E's differential is $T\mathrm{d}S + (-p)\mathrm{d}V$, while H's differential is $T\mathrm{d}S - V\mathrm{d}(-p)$.

It is easy to read Maxwell's relations from this diagram; they correspond to *triangles* that share one side—for example, the relation

$$\left.\frac{\partial S}{\partial(-p)}\right)_T = \left.\frac{\partial V}{\partial T}\right)_p, \tag{2.105}$$

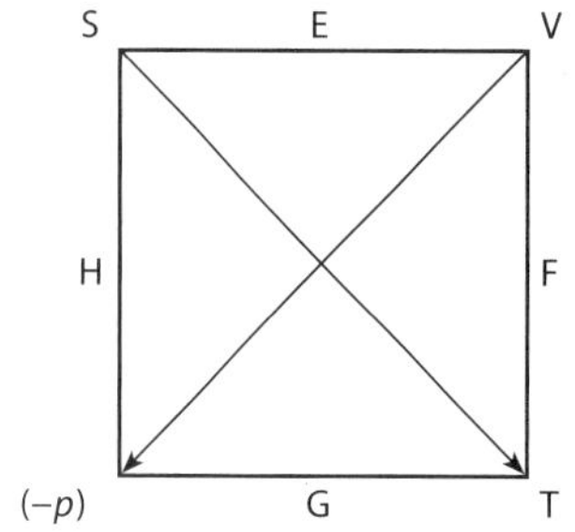

FIGURE 2.5. The Koenig Born diagram for the variables entropy S and volume V. The thermodynamic potentials on the sides are accompanied by their respective "natural" variables at the vertices. Note that the potentials follow one another in alphabetical order, in a clockwise pattern, and that the conjugate variable of volume V is the opposite of pressure, following our conventions.

corresponds to the bottom side, which can be obtained from the equality of the second-order derivatives of Gibbs free energy G. There are no extra signs to account for, since the arrows' structure is the same for both derivatives.

If instead we consider the F side, for example, we obtain

$$\left.\frac{\partial(-p)}{\partial T}\right)_V = -\left.\frac{\partial S}{\partial V}\right)_T. \tag{2.106}$$

Here there is an explicit "minus" sign, because on the one hand, we have an arrow that points *toward* the side being considered, and on the other, one that points *away* from it.

2.14 Other Thermodynamic Potentials

The Legendre transforms with respect to the number of particles N (for each chemical species) produce other thermodynamic potentials, which are extremely important in chemistry and statistical mechanics but which do not have either a name or a standard notation.

The Legendre transform of F with respect to N produces a thermodynamic potential (often written as Ω) that depends on T, on volume V, on chemical potential μ, and on the other extensive variables:

$$\Omega(T, V, \mu) = F(T, V, N) - \mu N. \tag{2.107}$$

Its differential is expressed as follows:

$$\mathrm{d}\Omega = -\,S\mathrm{d}T - p\mathrm{d}V - N\mathrm{d}\mu. \tag{2.108}$$

If one transforms E instead, one obtains a rarely used potential that depends on S, V, and μ, which we will designate as Φ (see figure 2.6):

$$\Phi(S, V, \mu) = E(S, V, N) - \mu N. \tag{2.109}$$

Its differential is given by

$$\mathrm{d}\Phi = T\mathrm{d}S - p\mathrm{d}V - N\mathrm{d}\mu. \tag{2.110}$$

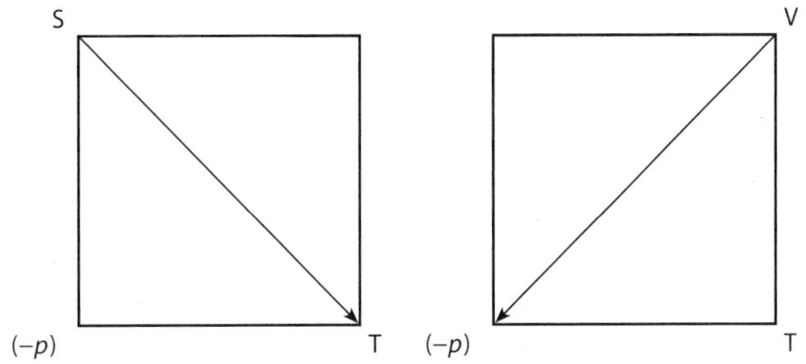

FIGURE 2.6. Koenig Born diagram and Maxwell relations.

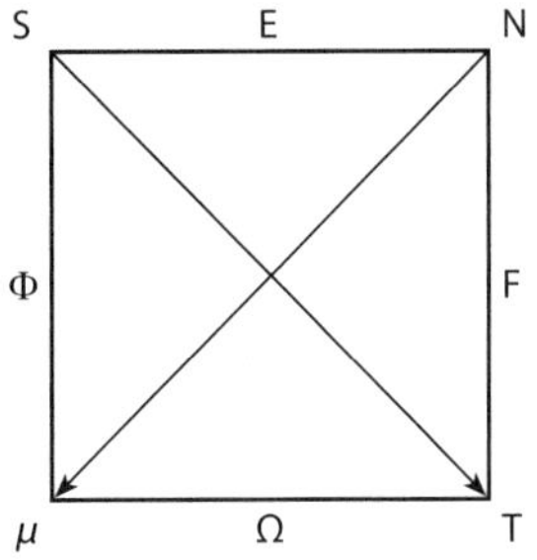

FIGURE 2.7. Koenig Born diagram relative to the variables S and N. The positions of E and F and of the arrows is the same as in figure 2.6.

The E, F, Ω, and Φ potentials allow one to write a Koenig Born diagram analogous to the one we saw in the preceding section, as shown in figure 2.7. We therefore arrive, for example, at the following Maxwell relations:

$$\left.\frac{\partial S}{\partial \mu}\right)_T = \left.\frac{\partial N}{\partial T}\right)_\mu, \tag{2.111}$$

$$\left.\frac{\partial S}{\partial N}\right)_T = -\left.\frac{\partial \mu}{\partial T}\right)_N. \tag{2.112}$$

2.15 The Euler and Gibbs-Duhem Equations

Let us imagine that we are putting a homogeneous system, characterized by the extensive variables $X_0 = S, X_1, \ldots, X_r$ and the corresponding internal energy value E, in contact with other $(m-1)$ identical systems. Due to the extensivity of the variables X and the function E, we obtain a system characterized by the values $X_i' = mX_i$ of the extensive variables and by the internal energy value $E' = mE$. Following the same line of reasoning, we can mentally subdivide the system into n identical systems, each with the values $(1/n)X_i$ for the X_i variables and with the value $(1/n)E$ for the internal energy. Because of continuity, we can conclude that internal energy E, since it is an extensive function of the extensive variables X_i, must, for each positive number λ, satisfy a homogeneity relation:

$$E(\lambda S, \lambda X_1, \ldots, \lambda X_r) = \lambda E(S, X_1, \ldots, X_r). \tag{2.113}$$

By taking the derivative of this equation with respect to λ and setting $\lambda = 1$, we obtain the **Euler equation**:

$$S\frac{\partial E}{\partial S} + \sum_{i=1}^{r} X_i \frac{\partial E}{\partial X_i} = E. \tag{2.114}$$

Making it explicit, we obtain

$$TS + \sum_{i=1}^{r} f_i X_i = E. \tag{2.115}$$

More particularly, for simple fluids, one obtains

$$TS - pV + \mu N = E, \tag{2.116}$$

which among other things implies that

$$\mu = (E - TS + pV)/N = G/N, \tag{2.117}$$

$$\Omega = E - TS - \mu N = -pV. \tag{2.118}$$

More particularly, from the Euler equation, it follows that the Legendre transform of E with respect to *all* extensive variables vanishes identically. Let us note that the interpretation of the chemical potential as a per particle density of Gibbs free energy is valid only in the case of *simple fluids*—in the case of a mixture of several chemical species, it is no longer valid.

If we take the Euler equation's differential, we obtain

$$\mathrm{d}E = T\mathrm{d}S + S\mathrm{d}T + \sum_{i=1}^{r} (f_i \mathrm{d}X_i + X_i \mathrm{d}f_i). \tag{2.119}$$

By subtracting both sides of this equation from the usual expression of $\mathrm{d}E$, we obtain the **Gibbs-Duhem equation**:

$$S\mathrm{d}T + \sum_{i=1}^{r} X_i \mathrm{d}f_i = 0. \tag{2.120}$$

By dividing this equation by any (nonvanishing) extensive variable, we realize that it relates intensive variables to densities. It therefore plays a key role in the interpretation of thermodynamic data.

In the case of simple fluids, for example, one arrives at

$$S\mathrm{d}T - V\mathrm{d}p + N\mathrm{d}\mu = 0, \tag{2.121}$$

which allows one to calculate the chemical potential (up to an additive constant) if the equation of state $V = V(N, p, T)$ and the value of entropy $S(N, p^*, T)$ for at least one value of the pressure p^* at each temperature are known. By dividing with respect to the number of particles N, one obtains the Gibbs-Duhem equation in the form

$$\mathrm{d}\mu = v\mathrm{d}p - s\mathrm{d}T, \tag{2.122}$$

where v represents volume per particle and s entropy per particle.

2.16 Magnetic Systems

In statistical mechanics, one is often dealing with systems whose thermodynamic variables include the total magnetic moment (usually denoted by M). It is an extensive quantity that can therefore be added to the other extensive variables S, V, N, and so on.

Let us suppose that we have a sample in the form of a narrow cylinder, placed inside a solenoid, identified by a number of coils per unit of length equal to ν, inside which a current I is circulating. In this case, the sample is subjected to a uniform magnetic field $\boldsymbol{H}$, oriented so that it is parallel to the cylinder's axis, whose modulus is νI. Within the geometry we have been examining, if the cylinder is homogeneous, it develops a uniform magnetization, with a certain value $\boldsymbol{m}$ of magnetic moment density per unit of volume. As a consequence, the magnetic induction vector $\boldsymbol{B}$ is uniform, parallel to $\boldsymbol{H}$, and takes the form $\boldsymbol{B} = \mu_0(\boldsymbol{H} + \boldsymbol{m})$, where $\mu_0 = 4\pi 10^{-7}\mathrm{Wbm^{-1}A^{-1}}$ is the magnetic permeability of the vacuum.[2]

Now, let us suppose that we vary the current circulating in the solenoid by ΔI. The work performed by the system and associated with this variation is given by

$$\delta W = -\int_{t_{\mathrm{in}}}^{t_{\mathrm{fin}}} \mathrm{d}t'\, I(t')\, \mathcal{V}_{\mathrm{ind}}(t'), \tag{2.123}$$

where $\mathcal{V}_{\mathrm{ind}}$ is the induced electromotive force, which can be expressed in terms of the flux Φ of the magnetic induction vector $\boldsymbol{B}$ concatenated with the solenoid as follows:

$$\mathcal{V}_{\mathrm{ind}} = -\frac{\mathrm{d}\Phi}{\mathrm{d}t}. \tag{2.124}$$

We thus obtain

$$\delta W = \int_{t_{\mathrm{in}}}^{t_{\mathrm{fin}}} \mathrm{d}t'\, I(t') \frac{\mathrm{d}\Phi(t')}{\mathrm{d}t'} \simeq I\Delta\Phi. \tag{2.125}$$

In this expression, we have neglected higher-order contributions in ΔI. The value of the concatenated flux is $\Phi = B\nu\ell A = \nu BV$ for a solenoid of length ℓ and cross section A, and therefore of volume V. We arrive at

$$\Delta\Phi = \nu V \Delta B = \nu V \mu_0(\Delta H + \delta m) = \nu V \mu_0 \Delta H + \nu\mu_0 \Delta M. \tag{2.126}$$

As a consequence, the work δW can be expressed as

$$\delta W = \nu V \mu_0 I \Delta H + \nu \mu_0 I \Delta M = \mu_0 V H \Delta H + \mu_0 H \Delta M. \tag{2.127}$$

The first term represents the variation in the energy of the magnetic field due to variations in the magnetic field itself. In fact, this energy is equal to

$$E_{\mathrm{magn}} = \int \mathrm{d}\boldsymbol{r} \frac{1}{2}\mu_0 H^2, \tag{2.128}$$

and the value of its variation is therefore $\Delta E_{\mathrm{magn}} \simeq \mu_0 \int \mathrm{d}\boldsymbol{r}\, \boldsymbol{H}\cdot\Delta\boldsymbol{H}$. The second term instead represents the work performed on the sample in order to change its magnetization value.

[2] We are following the standards of the International System of Units.

The contribution of the extensive variable $\boldsymbol{M}$ to "mechanical" work is therefore given by

$$\delta W_{\text{magn}} = \mu_0 \boldsymbol{H} \cdot \mathrm{d}\boldsymbol{M}. \tag{2.129}$$

We can see that the variable conjugated to magnetization (in other words, to the total magnetic dipole) is the *magnetic field* (up to the factor μ_0).

Since dipolar interactions (which occur between regions that are magnetically polarized) are long range, a certain caution is necessary when dealing with the magnetization $\boldsymbol{M}$ as an ordinary thermodynamic variable. More specifically, given equal amounts of material and volume, samples with different shapes possess different energies even if immersed in the same magnetic field. One usually accounts for this effect by introducing a **demagnetization factor**, which is dependent on shape.

It is also useful to introduce the Legendre transform of E with respect to M:

$$E'(S, \boldsymbol{H}, X_1, \ldots, X_r) = E(S, \boldsymbol{M}, X_1, \ldots, X_r) - \mu_0 \boldsymbol{H} \cdot \boldsymbol{M}. \tag{2.130}$$

This differential of this thermodynamic potential has the expression

$$\mathrm{d}E' = T\mathrm{d}S - \mu_0 \boldsymbol{M} \cdot \mathrm{d}\boldsymbol{H} + \sum_{i=1}^{r} f_i \mathrm{d}X_i. \tag{2.131}$$

One should note that given these conventions, there are no extra signs to be applied to the differentials.

2.17 Equations of State

Relations between the densities and the intensive variables obtained by deriving the fundamental equation (in one of its various representations) are called **equations of state.**

If, for example, we consider the Gibbs free energy for a simple fluid, we arrive at

$$V = \left.\frac{\partial G}{\partial p}\right)_{T,N}, \tag{2.132}$$

from which we obtain

$$v = \frac{V}{N} = v(p, T), \tag{2.133}$$

where we have made use of the fact that G is extensive, and therefore proportional to N.

In the case of the simple fluid, we have another independent equation of state:

$$s = \frac{S}{N} = -\frac{1}{N}\left.\frac{\partial G}{\partial T}\right)_p, \tag{2.134}$$

which expresses the entropy per particle s as a function of p and T. In reality, the two equations of state are not completely independent, because of the Maxwell relations:

$$\left.\frac{\partial v}{\partial T}\right)_p = -\left.\frac{\partial s}{\partial p}\right)_T. \tag{2.135}$$

And actually, by integrating the first equation of state, we can obtain the entropy as a function of p and T, up to an additive term independent of p:

$$\begin{aligned} s(p,T) &= \frac{1}{N}\int_{(p_0,T_0)}^{(p,T)}\left[\left.\frac{\partial S}{\partial p}\right)_{T,N} \mathrm{d}p + \left.\frac{\partial S}{\partial T}\right)_{p,N} \mathrm{d}T\right] + s(p_0,T_0) \\ &= \frac{1}{N}\left\{-\int_{(p_0,T)}^{(p,T)} \left.\frac{\partial v}{\partial T}\right)_p \mathrm{d}p + \int_{(p_0,T_0)}^{(p_0,T)} \left.\frac{C_p \mathrm{d}T}{T}\right)_{T,N}\right\} + s(p_0,T_0). \end{aligned} \tag{2.136}$$

We have used a Maxwell relation and the relation

$$\left.\frac{\partial S}{\partial T}\right)_{p,N} = \frac{C_p}{T}, \tag{2.137}$$

where C_p is the heat capacity at constant pressure. The first term of this expression can therefore be obtained from the equation of state; in the case of the second term, however, an independent measure is necessary. And vice versa, Maxwell relations allow one to check the validity of thermodynamic data.

The **ideal gas** is a simple fluid that satisfies the equation of state

$$p = \frac{Nk_B T}{V}, \tag{2.138}$$

where $k_B = 1.384\ 10^{-23}\ \mathrm{JK}^{-1}$ is **Boltzmann's constant.**

Maxwell relations allow us to prove that in an ideal gas, the internal energy only depends on N and T (and not on the volume V). In fact, one arrives at

$$\left.\frac{\partial E}{\partial V}\right)_T = \left.\frac{\partial E}{\partial V}\right)_S + \left.\frac{\partial E}{\partial S}\right)_V \left.\frac{\partial S}{\partial V}\right)_T = -p + T\left.\frac{\partial p}{\partial T}\right)_V = 0. \tag{2.139}$$

Moreover, entropy is the sum of a term that depends only on T with one that depends only on V. In fact, one gets

$$\left.\frac{\partial S}{\partial V}\right)_T = \left.\frac{\partial p}{\partial T}\right)_V = \frac{Nk_B}{V}, \tag{2.140}$$

and therefore,

$$\frac{\partial^2 S}{\partial T \partial V} = 0, \tag{2.141}$$

which is what we wanted to prove.

2.18 Stability

Let us now consider two systems, 1 and 2, in contact with each other, and which can exchange energy, particles and volume. Following the line of reasoning pursued in section 2.5, we must have at equilibrium

$$T^{(1)} = T^{(2)}, \tag{2.142}$$

$$p^{(1)} = p^{(2)}, \tag{2.143}$$

$$\mu^{(1)} = \mu^{(2)}. \tag{2.144}$$

Continuing to follow the same line of reasoning from section 2.5, we can conclude that if the other intensive qualities are the same in the two systems, then

- If initially $T^{(1)} > T^{(2)}$, system 1 releases energy to system 2.
- If initially $p^{(1)} > p^{(2)}$, system 1 expands at the expense of system 2—in other words, system 2 releases volume to system 1.
- If initially $\mu^{(1)} > \mu^{(2)}$, system 1 releases particles to system 2.

These results are due to the concavity of surface $E(S, V, \ldots)$.

The concavity is connected to the *stability* of thermodynamic equilibrium. Stability actually implies that the thermodynamic potential cannot decrease in the course of any virtual displacement. Let us in fact consider an infinitesimal virtual variation δX_i of the extensive variable X_i, and let us suppose that the intensive variables $f_1, \ldots, f_{i-1}$ and the extensive variables $X_{i+1}, \ldots, X_r$ are kept fixed.

Let us denote the corresponding thermodynamic potential as follows:

$$\Psi(f_1, \ldots, f_{i-1}, X_i, X_{i+1}, \ldots, X_r) = E - \sum_{k=1}^{i-1} f_k X_k. \tag{2.145}$$

One then must obtain

$$\delta^2 \Psi = \frac{1}{2} \left.\frac{\partial^2 \Psi}{\partial X_i^2}\right)_{f_1, \ldots, f_{i-1}, X_{i+1}, \ldots, X_r} \delta X_i^2 \geq 0. \tag{2.146}$$

More particularly, the specific heat is always positive (or vanishing), whatever constraint is placed on the system:

$$C = T \frac{\partial S}{\partial T} \geq 0. \tag{2.147}$$

If this were not the case, a small fluctuation in temperature that, for example, might make one of the system's regions colder would lead to this system claiming heat from surrounding regions. With negative specific heat, this energy would make the temperature of the region in question diminish further, and in this manner, the energy fluctuation would be amplified.

Moreover, the specific heat with a fixed intensive variable f_i is always greater than that of the corresponding fixed *extensive* variable. In fact, one has

$$C_{f_i} = T \left.\frac{\partial S}{\partial T}\right)_{f_i} = T \left.\frac{\partial S}{\partial T}\right)_{X_i} + T \left.\frac{\partial S}{\partial X_i}\right)_T \left.\frac{\partial X_i}{\partial T}\right)_{f_i}. \tag{2.148}$$

By applying the Maxwell relation

$$\left.\frac{\partial S}{\partial X_i}\right)_T = -\left.\frac{\partial f_i}{\partial T}\right)_{X_i}, \tag{2.149}$$

and the identity

$$\left.\frac{\partial x}{\partial y}\right)_z = -\left.\frac{\partial x}{\partial z}\right)_y \left.\frac{\partial z}{\partial y}\right)_x, \tag{2.150}$$

valid for each triplet of variables (x, y, z) connected by a relation of the type $F(x, y, z) = 0$, we obtain

$$C_{f_i} = T\left.\frac{\partial S}{\partial T}\right)_{X_i} + T\left.\frac{\partial f_i}{\partial T}\right)_{X_i}^2 \left.\frac{\partial X_i}{\partial f_i}\right)_T. \tag{2.151}$$

Exercise 2.7 Prove relation (2.150).

The last term of this expression is undoubtedly positive. We have therefore arrived at

$$C_{f_i} \geq C_{X_i}, \tag{2.152}$$

and more particularly, the specific heat at constant pressure is always greater than that at constant volume.

Exercise 2.8 (Law of Adiabatic Processes) Prove that for an ideal gas, one gets

$$C_p - C_V = Nk_B.$$

By defining $\gamma = C_p/C_V$ (which is only a function of temperature), deduce the **law of reversible adiabatic processes**:

$$\left.\frac{\partial p}{\partial V}\right)_S = -\gamma\frac{p}{V}.$$

If we assume that $C_V = \text{const.}$, one can integrate this relation, and obtain

$$pV^{\gamma} = \text{const.}$$

Deduce that in a reversible adiabatic process one has

$$pT^{\gamma/(1-\gamma)} = \text{const.}$$

Exercise 2.9 (Adiabatic Atmosphere) It is a well-known fact that atmospheric temperature decreases with an increase in elevation. As a first approximation, given air's low thermal conductivity, we can assume that the atmosphere is at equilibrium with respect to vertical *adiabatic* shifts of small quantities of air. By using the equation of state for ideal gases, the law of adiabatic processes, and Stevin's law:

$$\frac{\partial p}{\partial z} = -m_{\text{mol}}g\frac{N}{V},$$

where m_{mol} is air's average molecular mass $(0.67\, m_{N_2} + 0.33\, m_{O_2})$, prove that the atmosphere's thermal gradient is given by

$$\left.\frac{\partial T}{\partial z}\right)_S = -\frac{m_{\text{mol}}g}{k_B}(1 - \gamma^{-1}).$$

The value of this quantity is about –9.9 Kkm^{-1}.

Exercise 2.10 Let us consider a container whose walls are adiabatic, in which there is an opening closed by a faucet. A void is created inside the container, then it is placed in contact with the atmosphere at pressure p_0 and temperature T_0. At this point, the faucet is opened and air rapidly enters the container. After a brief interval, internal pressure has equalized external pressure. Calculate the temperature of the air inside the container under these conditions. We remind readers that for air, $\gamma = C_p/C_V = 7/5$.

2.19 Chemical Reactions

Let us now consider a mixture of r chemical species, $A_1, \dots, A_r$, which can be transformed into one other by a reaction of the following type:

$$\nu_1 A_1 + \cdots + \nu_k A_k \rightleftharpoons \nu_{k+1} A_{k+1} + \cdots + \nu_r A_r. \tag{2.153}$$

The interpretation of this formula, as is well known, is that one takes ν_1 molecules of the species A_1, ν_2 of the species $A_2, \dots, \nu_k$ of the species A_k (the **reagents**), and one obtains ν_{k+1} of the species $A_{k+1}, \dots, \nu_r$ of the species A_r (the **products**). The integers ν_1 to ν_r are called **stoichiometric coefficients**.

There is no fundamental distinction between reagents and products, because each reaction can go both in one direction and in the reverse, as is indicated by the two arrows in the formula. We can conventionally assign negative stoichiometric coefficients to the products, so as to write this formula as a formal equation:

$$\sum_{i=1}^{r} \nu_i A_i = 0. \tag{2.154}$$

The idea of this representation is that, if the reaction takes place, at each moment of the variation δN, the number of particles of the species i is proportional to the corresponding stoichiometric coefficient, taken with its own sign:

$$\frac{\delta N_1}{\nu_1} = \frac{\delta N_2}{\nu_2} = \cdots = \frac{\delta N_r}{\nu_r}. \tag{2.155}$$

Obviously, the variation will be negative for the reagents and positive for the products, if the reaction goes from left to right, and vice versa.

If temperature and pressure are kept fixed, we can calculate the variation of Gibbs free energy for a certain variation in the number of particles due to the reaction

$$\delta G = \sum_i \left(\frac{\partial G}{\partial N_i}\right)_{p,T} \delta N_i \propto \sum_i \left(\frac{\partial G}{\partial N_i}\right)_{p,T} \nu_i = \sum_i \mu_i \nu_i. \tag{2.156}$$

We have thus obtained the equilibrium criterion with respect to the reaction described by equation (2.153). Since at equilibrium one must have $\delta G = 0$ for any virtual variation of the N_i, one will have

$$\sum_i \nu_i \mu_i = 0, \tag{2.157}$$

where the stoichiometric coefficients ν_i are taken with their corresponding signs.

2.20 Phase Coexistence

It frequently happens that two systems characterized by different thermodynamic density values can maintain thermodynamic equilibrium even in the absence of constraints on the mutual exchange of extensive quantities. This situation is called **phase coexistence**. In the case of a simple fluid, it is realized, for example, when a liquid coexists with its vapor inside a container. In this case, the intensive variables assume the same value in both systems, while densities (for example, the volume v per particle given by $v = V/N$) assume different values. In these cases, we refer to each of the coexisting *homogeneous* systems as a **phase**.

One can describe phase coexistence by saying that the equation of state

$$v = v(p, T) \tag{2.158}$$

does not admit of a unique solution, but instead allows for at least the two solutions $v = v_{\text{liq}}$ and $v = v_{\text{vap}}$, which correspond to the liquid and vapor, respectively.

The postulates of thermodynamics then imply that *all* intermediate values of v between v_{liq} and v_{vap} satisfy the equation of state. In effect, since the liquid and vapor coexist and can exchange particles, the chemical potential of the liquid has to be equal to that of the vapor:

$$\mu_{\text{liq}}(p, T) = \mu_{\text{vap}}(p, T). \tag{2.159}$$

On the other hand, we know that for a simple fluid, the chemical potential is equal to the Gibbs free energy per particle:

$$\mu = \frac{G}{N}. \tag{2.160}$$

The Gibbs free energy is the thermodynamic potential that is suited for describing equilibrium when pressure p and temperature T are fixed, as a function of the number of particles N. Therefore, in the situation we are considering, the Gibbs free energy in the total system does not depend on the number of particles N_{liq} and N_{vap} that make up the liquid and the vapor system, respectively:

$$G = G_{\text{liq}} + G_{\text{vap}} = N_{\text{liq}}\mu_{\text{liq}} + N_{\text{vap}}\mu_{\text{vap}} = (N_{\text{liq}} + N_{\text{vap}})\mu = N\mu. \tag{2.161}$$

The volume per particle of a system in which N_{liq} particles are part of the liquid and N_{vap} particles are part of the vapor is given by

$$v = \frac{V}{N} = \frac{V_{\text{liq}} + V_{\text{vap}}}{N} = \frac{N_{\text{liq}}v_{\text{liq}} + N_{\text{vap}}v_{\text{vap}}}{N}. \tag{2.162}$$

We thus obtain

$$v = x_{\text{liq}} v_{\text{liq}} + x_{\text{vap}} v_{\text{vap}}, \tag{2.163}$$

where $x_{\text{liq}} = N_{\text{liq}}/N$ is the fraction of particles in the liquid and $x_{\text{vap}} = 1 - x_{\text{vap}}$ that of the particles in the vapor. As a consequence, the value of v lies somewhere between $v_{\text{liq}} = v_{\text{vap}}$. Obviously, an analogous result obtains for the density $\rho = 1/v$ and for the other densities one could define.

We will now discuss different aspects of phase coexistence in a simple fluid.

In the equation of state $p = p(v, T)$, the discussion we just completed shows that phase coexistence appears as a horizontal segment, for a given value of T, and for values of v included between v_{liq} and v_{vap} (see figure 2.8). Let us now consider the Helmholtz free energy F. The pressure p is obtained as the derivative of F with respect to V:

$$p = -\left.\frac{\partial F}{\partial V}\right)_{T,X}. \tag{2.164}$$

Phase coexistence therefore manifests itself because, for a given value of T, and other extensive variables X remaining set, the curve $F = F(T)$ exhibits a straight segment with slope $(-p_t)$, lying between $V = Nv_{\text{liq}}$ and $V = Nv_{\text{vap}}$ (see figure 2.9a).

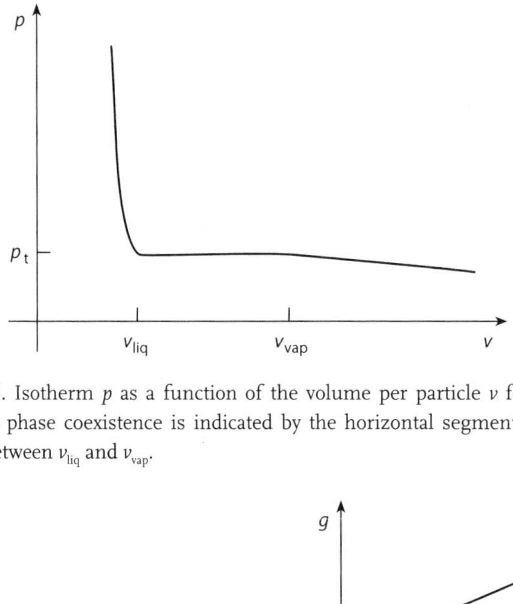

FIGURE 2.8. Isotherm p as a function of the volume per particle v for a simple liquid. The phase coexistence is indicated by the horizontal segment for $p = p_t$ included between v_{liq} and v_{vap}.

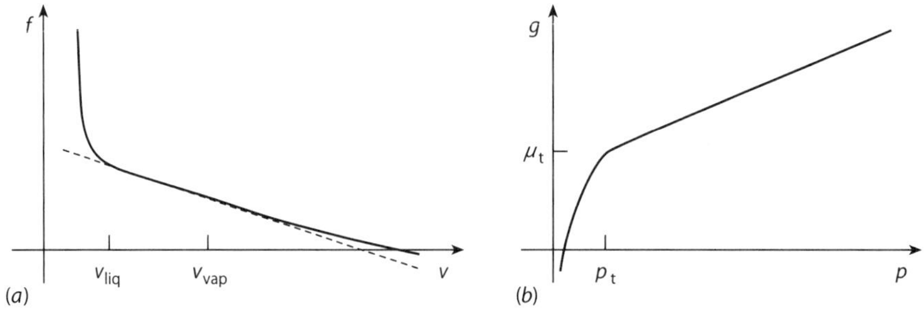

FIGURE 2.9. (a) Isotherm of the Helmholtz free energy per particle; $f = F/N$, as a function of volume per particle $v = V/N$. The straight line corresponds to phase coexistence. (b) Isotherm of the Gibbs free energy per particle; $g = G/N$ as a function of pressure p. The angular point with respect to p_t corresponds to the phase coexistence.

The properties of the Legendre transform (see appendix A) imply that the Gibbs free energy has an angular point at $p = p_1$, as a function of p at a fixed value of T (see figure 2.9b). The two slopes that coexist at the angular point correspond to v_{liq} and v_{vap}, respectively. Since $\mu = G/N$, the aspect of the isotherm $\mu = \mu(p)$ is identical. Convexity leads to the smaller specific volume (and therefore the smaller derivative) always being in the stable phase at higher values of p.[3]

2.21 The Clausius-Clapeyron Equation

If we consider a thermodynamic system as a function of its intensive variables, we can identify some regions in which the thermodynamic properties vary regularly with variations of their values. These regions represent thermodynamically stable phases and are limited by curves that represent **phase transitions**. The phase transitions can be **discontinuous**, like the phase coexistence we just discussed, or **continuous**. In the first case, the densities present a discontinuity at the transition, while in the second, they vary with continuity, even though their derivatives can exhibit some singularities. One often also employs the following terminology: discontinuous transitions are called **first-order transitions**, while continuous ones are called **second-order transitions**.

In the case of a simple fluid, it is possible to identify the transition curve within the plane of the intensive variables (p, T)—in other words, the curve $p = p_t(T)$—from the condition of equality of the chemical potential μ between the two coexisting phases:

$$\mu_{\text{liq}}(p_t(T), T) = \mu_{\text{vap}}(p_t(T), T). \tag{2.165}$$

The pressure $p_t(T)$ at which the transition occurs is also called **vapor pressure**. It is possible to relate this curve locally with the discontinuity of densities at transition. To obtain this relation, let us take the total derivative of this equation with respect to T, along the transition line $p = p_t(T)$. We obtain

$$\left.\frac{\partial \mu_{\text{liq}}}{\partial p}\right)_T \frac{\mathrm{d}p_t}{\mathrm{d}T} + \left.\frac{\partial \mu_{\text{liq}}}{\partial T}\right)_p = \left.\frac{\partial \mu_{\text{vap}}}{\partial p}\right)_T \frac{\mathrm{d}p_t}{\mathrm{d}T} + \left.\frac{\partial \mu_{\text{vap}}}{\partial T}\right)_p. \tag{2.166}$$

Therefore,

$$v_{\text{liq}} \frac{\mathrm{d}p_t}{\mathrm{d}T} - s_{\text{liq}} = v_{\text{vap}} \frac{\mathrm{d}p_t}{\mathrm{d}T} - s_{\text{vap}}, \tag{2.167}$$

where v and s are the volume and entropy per particle in the two coexisting phases, respectively. We thus arrive at

$$\frac{\mathrm{d}p_t}{\mathrm{d}T} = \frac{s_{\text{vap}} - s_{\text{liq}}}{v_{\text{vap}} - v_{\text{liq}}}. \tag{2.168}$$

[3] This observation is not as trivial as it seems, since many textbooks reproduce grossly erroneous diagrams.

We can see that under usual conditions, since $v_{vap} > v_{liq}$ and $s_{vap} = s_{liq} + Q_{evap}/T > s_{liq}$ (where $Q_{evap} > 0$ is the latent evaporation heat), the vapor pressure $p_t(T)$ increases with the temperature:

$$\frac{dp_t}{dT} > 0. \tag{2.169}$$

This equation, which can be applied to each case of phase coexistence, is called the **Clausius-Clapeyron equation.**

In the case of the melting of water, in which, as is well known, the phase at low temperature (ice) has a volume per particle that is larger than the phase at higher temperature (liquid), we have

$$s_{liq} > s_{gh}, \tag{2.170}$$

$$v_{liq} > v_{gh}. \tag{2.171}$$

The Clausius-Clapeyron equation therefore predicts that

$$\frac{dp_t}{dT} < 0. \tag{2.172}$$

The curve of the melting pressure versus temperature for water at $T = 0$°C has therefore a negative slope. In other words, at higher than atmospheric pressure, the melting temperature is lower than 0°C. This effect is fairly reasonable, since it is a manifestation of stability—given an increase in pressure, the system reacts by assuming the phase that occupies less volume.

2.22 The Coexistence Curve

We can represent the phase diagram in the plane (v, T), in which the intensive variable T is accompanied by the density v, the volume per particle, rather than in the intensive variables plane (p, T). In this manner, phase coexistence is represented by the existence of a *forbidden region* $v_{liq}(T) < v < v_{vap}(T)$ in the plane. Outside this region, it is possible to obtain any given value of v in a *homogeneous* system. Within this region, instead, the system separates into a liquid and a vapor phase (see figure 2.10). The x_{liq} fraction of particles in the liquid phase (and the analogous fraction in the vapor phase) are determined by the condition that the entire system's volume per particle be equal to v. One thus obtains

$$x_{liq} = \frac{v_{vap} - v}{v_{vap} - v_{liq}}, \qquad x_{vap} = \frac{v - v_{liq}}{v_{vap} - v_{liq}} = 1 - x_{liq}. \tag{2.173}$$

This result is known as the **lever rule.** In effect, if one considers the segment included between v_{liq} and v_{vap} as the arm of a lever, with its fulcrum corresponding to v, the fractions x_{liq} and x_{vap} are proportional to the weights one would need to place at each extremity so that the lever would be in equilibrium.

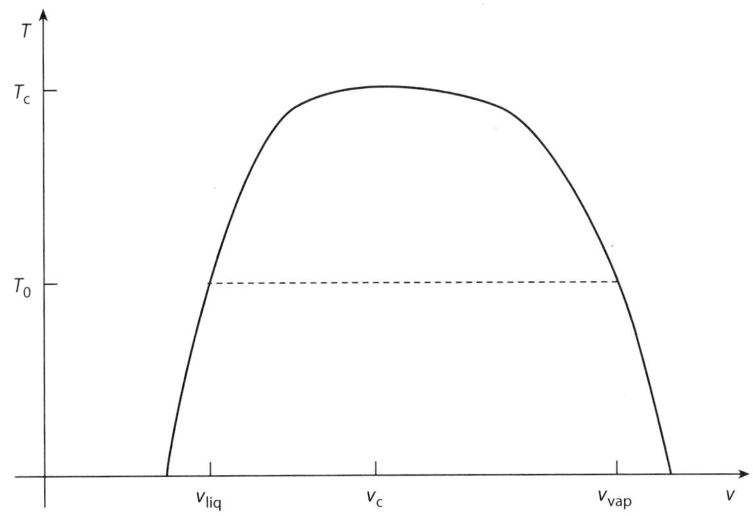

FIGURE 2.10. Coexistence curve for a simple fluid. The temperature T is on the y axis, and the volume per particle v is on the x axis. The critical point corresponds to (v_c, T_c).

Let us note that if the coexistence curve is defined without ambiguities in the plane (v, T), because the points that represent the phases in coexistence must correspond to the same value of T (an intensive variable), this situation is no longer valid if we place ourselves in a plane characterized by *two* densities—for instance, (v, s) (where s is the entropy per particle). In this case, the coexistence of two phases occurs along straight segments that connect the points $(v_{\text{liq}}, s_{\text{liq}})$ and $(v_{\text{vap}}, s_{\text{vap}})$, which correspond to equal temperature values. These segments must be represented explicitly, and are called **tie lines**.

2.23 Coexistence of Several Phases

It is also possible for several phases to coexist for certain values of the intensive variables. For example, in the case of a simple fluid, one can observe the coexistence of a solid, a liquid, and a vapor for a specific pressure value p_{tr} and temperature value T_{tr}. This point is called the **triple point**. At the triple point, the chemical potentials corresponding to the solid, liquid, and vapor phases must be equal:

$$\mu_{\text{sol}}(p_t, T_t) = \mu_{\text{liq}}(p_t, T_t) = \mu_{\text{vap}}(p_t, T_t). \tag{2.174}$$

Here we have *two* equations in the two intensive variables p and T, instead of the single equation that we had obtained for the coexistence of the liquid and the vapor. For a simple fluid, therefore, the coexistence of three phases occurs only for isolated points of the plane (p, T). There can be several triple points, corresponding to the different possible solid phases. The most familiar example is water's triple point—$T_{\text{tr}} = 273.16$ K; $p_{\text{tr}} = 0.6113$ kPa—which as we stated in section 2.6 was chosen to define the temperature unit (one degree Kelvin).

Let us now consider a more general case—in other words, a *mixture* of particles belonging to r different chemical species. Let us suppose that we are looking for the coexistence of q phases ($q \geq 2$). At equilibrium, all the intensive variables must assume the same value in the coexisting phases. We will therefore have a specific value for the pressure and the temperature, and in addition the chemical potential of each species will have to assume the same value in all the different phases. If we denote the chemical potential of species i in phase α as $\mu_i^{(\alpha)}$, we will then have

$$\mu_i^{(\alpha)} = \mu_i, \qquad i = 1, \ldots, r, \qquad \alpha = 1, \ldots, q. \tag{2.175}$$

In this equation, μ_i is the shared value taken by the chemical potential of species i. We thus obtain $r(q - 1)$ equations for $q(r - 1) + 2$ unknown values. These unknown values are p, T, and the $q(r - 1)$ independent densities $x_i^{(\alpha)}$ of species i in phase α. Generically speaking, therefore, $f = 2 - q + r$ free parameters remain. For $f = 0$, coexistence will occur in isolated points of the phase diagram, for $f = 1$, along a line, and so on. The quantity f is called **variance.** The reasoning we have followed therefore allows us to conclude that

$$f = 2 - q + r, \tag{2.176}$$

which is called the **Gibbs phase rule.**

The Gibbs phase rule must be used with caution, because cases in which it appears to be violated are not uncommon. A classic example is the coexistence of two solid phases of carbon (graphite and diamond) at ambient temperature. Logically, they should coexist along a coexistence line, and therefore, once the temperature has been fixed, for a specific value of pressure. They instead appear to coexist at any pressure (as long as it is low) and even in a "vacuum"—in other words, when their vapor is present. The solution to this paradox lies in the fact that the transformation of one phase into the other at ambient temperature is extremely slow. Thermodynamically speaking, the phase that is stable at ambient temperature and pressure is graphite; however, as is well known, the time necessary for a diamond (even a very small one) to turn into graphite is on the order of millions of years.

2.24 The Critical Point

Let us once again consider the simple fluid and observe that the coexistence of liquid and vapor cannot be obtained for temperatures higher than a certain temperature T_c, called the **critical temperature.** To be more precise, the transition curve ends at a point (p_c, T_c), where p_c is the critical pressure. For $T < T_c$, the difference $v_{\text{vap}} - v_{\text{liq}}$ tends continuously toward zero when T gets close to T_c—the discontinuity of thermodynamic densities tends to vanish, or in other words, the transition goes from being discontinuous to being continuous (and finally to disappear at higher temperatures).

The critical point is a thermodynamic state with exceptional properties. For example, since for $T < T_c$ and $p = p_t(T)$ one gets $\partial p/\partial V)_T = 0$ (within the coexistence curve), this relation must ultimately be valid also at the critical point—in other words, the system's compressibility diverges:

$$\chi = -\frac{1}{V}\frac{\partial V}{\partial P}\bigg)_T \to \infty \qquad \text{for } T \to T_c. \tag{2.177}$$

Various thermodynamic properties exhibit analogous singularities.

The study of the thermodynamic properties of the critical point cannot proceed very far without appealing to genuine statistical mechanics. For this reason, we will discuss these properties at the appropriate moment.

In addition to the critical point in the liquid–gas transition, there are many examples of analogous situations (the Curie point in ferromagnets, the λ point in liquid helium, and so on). They form the simplest class of continuous transitions.

2.25 Planar Interfaces

An **interface** forms between two different phases in equilibrium. The contributions of the interface to thermodynamic properties are negligible for large systems (because they are proportional to the system's surface, not its volume), but they are interesting by themselves.

Let us consider a container in which the liquid and vapor phases of a simple fluid coexist. If the container is large enough, the interface is horizontal for the greater part of the container's cross section. We can measure the local density of the fluid as a function of height z and thus obtain a curve like the one shown in figure 2.11. At great distances from the interface, the density is constant and equal to the density of the liquid or the vapor. However, if the distance from the interface is small (but still very large when compared to atomic scales), the density varies with continuity. We can therefore introduce the notion of an *interface width*.

From the thermodynamic point of view, however, it is useful to be able to consider the interface as a geometric surface, placed at a certain height z_0. Let us consider the system composed of phases 1 and 2, with the addition of an interface, as shown in figure 2.12.

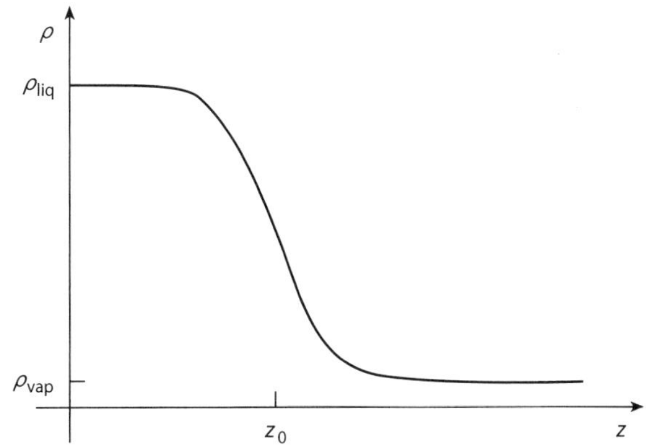

FIGURE 2.11. Local density profile in a liquid–vapor interface; ρ as a function of height z.

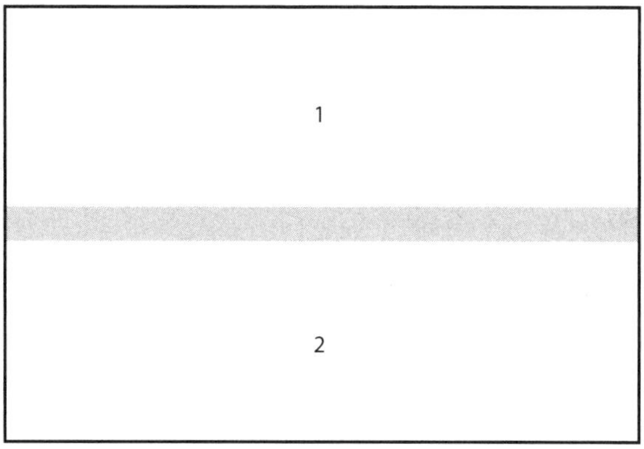

FIGURE 2.12. Coexistence of two phases, with their interface.

The internal energy is given by the sum of the contributions from each phase, to which the interface's is then added:

$$E = E^{(1)} + E^{(2)} + E^{\text{int}}. \tag{2.178}$$

The interface's contribution must depend on its area σ. Let us then write E's differential in the form

$$\mathrm{d}E = T\mathrm{d}S - p\mathrm{d}V + \mu\mathrm{d}N + \gamma\mathrm{d}\sigma. \tag{2.179}$$

The quantity

$$\gamma = \left.\frac{\partial E}{\partial \sigma}\right)_{S,V,N} \tag{2.180}$$

is called **surface tension**. It must remain positive—otherwise, the interface would tend to increase its area, filling all space available, and we could no longer describe the system as the coexistence of two distinct phases. Situations of this type occur in microemulsions.

As was the case with internal energy, the other extensive variables can also be obtained by adding the contributions of the two phases to that of the interface. By using Euler's equation, we arrive at

$$\begin{aligned} E = {} & T\left(S^{(1)} + S^{(2)} + S^{\text{int}}\right) \\ & - p\left(V^{(1)} + V^{(2)} + V^{\text{int}}\right) + \mu\left(N^{(1)} + N^{(2)} + N^{\text{int}}\right) + \gamma\sigma. \end{aligned} \tag{2.181}$$

If we want to treat the interface as a geometric surface, we must set $V^{\text{int}} = 0$. Moreover, if we choose the height z_0 of the interface in a suitable fashion, we can arrange things so that

$$\int_{-\infty}^{z_0} \mathrm{d}z\left[\rho_{\text{liq}} - \rho(z)\right] = \int_{z_0}^{\infty} \mathrm{d}z\left[\rho(z) - \rho_{\text{vap}}\right]. \tag{2.182}$$

This condition is equivalent to imposing $N^{\text{int}} = 0$ and is known as **Gibbs condition**. We thus obtain

$$E^{\text{int}} = TS^{\text{int}} + \gamma\sigma. \tag{2.183}$$

Due to extensivity, S^{int} must also be proportional to the area σ of the interface. From this equation, we can obtain the physical interpretation of γ:

$$\gamma = \frac{E^{\text{int}} - TS^{\text{int}}}{\sigma}. \tag{2.184}$$

Therefore, γ is equal to the Helmholtz free energy per unit area of the interface.

Let us now assume that we are dealing with a system with two coexisting phases, but one in which several chemical species are present. We can still define the position of the interface so as to have $N_1^{\text{int}} = 0$ for species 1 (which we will assume is prevalent—the solvent). In general, however, we will have to assign a certain quantity of particles of the other species to the interface. We will then have

$$\mathrm{d}E^{\text{int}} = T\mathrm{d}S^{\text{int}} + \sum_{i=1}^{r} \mu_i \mathrm{d}N_i^{\text{int}} + \gamma \mathrm{d}\sigma. \tag{2.185}$$

In the case of only two species, we will have

$$\mathrm{d}E^{\text{int}} = T\mathrm{d}S^{\text{int}} + \mu_2 \mathrm{d}N_2^{\text{int}} + \gamma \mathrm{d}\sigma, \tag{2.186}$$

and as a result of Euler's equation:

$$E^{\text{int}} = TS^{\text{int}} + \mu_2 N_2^{\text{int}} + \gamma\sigma. \tag{2.187}$$

We thus obtain the analog of the Gibbs-Duhem equation for surface tension:

$$\mathrm{d}\gamma = -s^{\text{int}}\mathrm{d}T - \rho_2^{\text{int}}\mathrm{d}\mu_2, \tag{2.188}$$

where s^{int} is the entropy per unit area, and ρ_2^{int} is the surface density of particles of species 2 on the interface.

At constant temperature T, we have the equation

$$\left.\frac{\partial \gamma}{\partial \mu_2}\right)_T = -\rho_2^{\text{int}}, \tag{2.189}$$

whose solution is called the **Gibbs isotherm**. Let us note that one has

$$0 < \rho_2^{\text{int}} = -\left.\frac{\partial \gamma}{\partial \mu_2}\right)_T = -\left.\frac{\partial \gamma}{\partial \rho_2^{\text{int}}}\right)_T \left.\frac{\partial \rho_2}{\partial \mu_2}\right)_T. \tag{2.190}$$

The last factor is positive due to stability. We thus obtain

$$\left.\frac{\partial \gamma}{\partial \rho_2}\right)_T < 0, \tag{2.191}$$

which implies that surface tension diminishes when a solute accumulates on the interface. This effect is utilized to diminish the surface tension and favor mixing between liquids, by introducing particles that prefer to accumulate at the interface (the **tensioactives**).

Exercise 2.11 (Theory of Elasticity) Let us consider an elastic band with a fixed cross section as a thermodynamic system. The thermodynamic variables that characterize it within the energy scheme are entropy S, length L, and the mass (or number of moles) n. The expression of $\mathrm{d}E$ is given by

$$\mathrm{d}E = T\mathrm{d}S + f\mathrm{d}L + \mu \mathrm{d}n,$$

where f is the tension acting on the band, and μ is the chemical potential.

1. Derive the analogue of the Gibbs-Duhem equation.

   One can observe experimentally that the tension f, when the length L and the mass m are constant, increases proportionally to temperature. This suggests that the equation of state is

   $$E = cnT,$$

   where c is a constant. Hooke's law, on the other hand, implies that the relative lengthening is proportional to tension (at least in the case of increases in length that are not too large), and therefore,

   $$f = \phi \frac{L - L_0}{n},$$

   where $L_0 = n\ell_0$ is the length at rest.

2. Show that ϕ is proportional to temperature T:

   $$\phi = bT.$$

3. Derive the expression for the entropy differential as a function of E, L, and N.

4. Derive the fundamental equation in both the entropy and energy schemes.

5. Calculate the relative variation of $L - L_0$ for a small increment in temperature $\mathrm{d}T$.

Recommended Reading

The preceding discussion was inspired by H. B. Callen's *Thermodynamics and an Introduction to Thermostatistics*, 2nd ed., New York: J. Wiley & Sons, 1985. Callen's ideas are in their turn rooted in those of L. Tisza; see L. Tisza, *Generalized Thermodynamics*, Cambridge, MA: The MIT Press, 1966. More "traditional" discussions follow a different path; they tend to *construct* the axiomatic structure starting from the phenomenology. An excellent presentation can be found in A. B. Pippard, *The Elements of Classical Thermodynamics*, Cambridge, UK: Cambridge University Press, 1957.

3 The Fundamental Postulate

When he woke up, the dinosaur was still there.
—Augusto Monterroso

The aim of statistical mechanics is to describe the thermodynamic properties of complex systems, composed of a large number of particles. The characteristic evolution times of these systems are microscopic, and for this reason, measuring mechanical quantities in an experiment of reasonable length is equivalent to the measurement of these quantities' average during the corresponding time interval. In this manner, it is possible to define the thermodynamic observables as time averages of mechanical quantities. The pressure applied by a fluid on a wall, for example, can be expressed (by analogy with what we saw in chapter 1) as the time average of the force applied on the wall, divided by the surface area of the wall itself.

We have seen, however, that entropy is not a mechanical quantity, in the sense we just expressed. The key to the relation between mechanical and thermodynamic properties is the microscopic expression of entropy. This expression was found by Ludwig Boltzmann, and it has such an important role that we have decided to call it the **fundamental postulate** of statistical mechanics. Since we are dealing with a postulate, it cannot be proven—we can, however, make it plausible by showing that the properties of entropy that derive from it agree with the properties that we postulated in thermodynamics.

In order to make the fundamental postulate more explicit, we must resort to an abstract representation of the dynamics of a system with many particles—this representation uses the concepts of phase space and of a system's representative point.

3.1 Phase Space

Let us consider a simple fluid represented as a classical system composed of N identical particles, of mass m, interacting with one another and with the walls of the container. Let us suppose that the system is energetically isolated so that there are no exchanges of energy between it and the external environment. At each instant, the state of the system can be described by giving each particle's position $\boldsymbol{r}$ and momentum $\boldsymbol{p}$: $6N$ scalar variables in total. These identify a point, the system's **representative point**, within a $6N$-dimensional space, called the **phase space**. The evolution of the system is represented by the motion of the representative point within this space and is described by Hamilton's system of $6N$ differential equations of motion. If we denote the system's Hamiltonian by H:

$$H = \sum_{i=1}^{N} \left[\frac{p_i^2}{2m} + U(\mathbf{r}_1, \dots, \mathbf{r}_N) \right], \tag{3.1}$$

these equations have the expression

$$\frac{d\mathbf{r}_i}{dt} = \frac{\partial H}{\partial \mathbf{p}_i}, \tag{3.2}$$

$$\frac{d\mathbf{p}_i}{dt} = -\frac{\partial H}{\partial \mathbf{r}_i}. \tag{3.3}$$

Let us denote the generic point of the space phase by $x = (\boldsymbol{p}_1, \boldsymbol{r}_1, \dots, \boldsymbol{p}_N, \boldsymbol{r}_N)$. With the passage of time, the system's representative point moves along a trajectory $x = x(t; x^0)$, identified by the initial condition $x(0; x^0) = x^0 = (\boldsymbol{p}_1^0, \boldsymbol{r}_1^0, \dots, \boldsymbol{p}_N^0, \boldsymbol{r}_N^0)$.

Generally speaking, this trajectory is extremely complicated. The only thing we can state a priori is that, if the system is isolated, it lies entirely on the surface defined by the equation

$$H = E = \text{const.} \tag{3.4}$$

The Hamiltonian's constant value E can be identified with the system's internal energy.

Strictly speaking, for a system of $6N$ equations of motion, there exist $6N - 1$ integrals of motion, which define a *line* in phase space along which the system must move. To a very large extent, however, these integrals cannot be measured or theoretically predicted, because a small disturbance to the system causes their values to vary greatly. In a system composed of many particles, a small (experimentally imperceptible) shift of each particle corresponds to a large shift of the system's representative point in phase space. It is obvious that quantities that change a lot due to a shift of this kind cannot be relevant to a thermodynamic description of the system. Therefore, the properties that are dealt with by thermodynamics and statistical mechanics must vary only slightly when the system's representative point is changed slightly. We will make this condition more explicit further on.

In order for our argument to be consistent with thermodynamics, it is necessary that given a certain (not very large) number of these quantities, we can predict their value (and the value of other analogous quantities) after thermodynamic transformations have occurred, at least for sufficiently large systems. How these quantities can be characterized is a well-defined problem, but one that has not yet been satisfactorily resolved in the general case. In order to obtain a microscopic equivalent of the postulates of thermodynamics, however, we can assume that these quantities are the thermodynamic observables, only this time considered as functions of the *microscopic* state of the system, rather than its thermodynamic state. We shall therefore formulate the following hypothesis:

The region in phase space in which the system is evolving can be completely identified, for thermodynamic purposes, by the values of certain observable quantities.

In the case of the simple fluid we considered, for example, this region can be identified by the condition that the positions of all particles be situated inside a region of volume V, which represents the inside of the container, and by the value E of internal energy. If the thermodynamics of simple fluids that we discussed in chapter 2 is valid, it is impossible to identify other physically important quantities that vary slowly in the phase space and that would restrict this region in a significant manner.

3.2 Observables

Let us call **observables** in statistical mechanics those functions, defined on the phase space, that vary smoothly enough when the representative point varies. The total kinetic energy of a mechanical system, for example, defined by

$$E_c = \sum_{i=1}^{N} \frac{p_i^2}{2m}, \tag{3.5}$$

is an observable—a particle's variation in momentum causes a relative increase of E_c on the order of N^{-1} (which is therefore negligible when N is large).

An example of a *nonobservable* function is the following:

$$\Theta = \prod_{i=1}^{N} \theta\left(r_0^2 - r_i^2\right). \tag{3.6}$$

This quantity is equal to 1 if *all* the particles are within a sphere of radius r_0 around the origin and is otherwise equal to 0. A shift applied to just one particle is sufficient to change the value of this function by a finite quantity.

As we saw previously, the internal energy corresponds to the observable H (the Hamiltonian). The number of particles can be expressed as an observable as follows:

$$N = \int \mathrm{d}\mathbf{r} \sum_{i=1}^{N} \delta(\mathbf{r} - \mathbf{r}_i). \tag{3.7}$$

We are thus led to formulate the following hypothesis:

All the extensive thermodynamic variables as a function of which the fundamental equation is expressed are observables.

Since the fundamental equation is the expression of entropy as a function of extensive variables, what we are missing is a mechanical interpretation of entropy. It will be provided by the fundamental postulate.

3.3 The Fundamental Postulate: Entropy as Phase-Space Volume

The fundamental postulate of statistical mechanics expresses entropy as a function of the accessible volume in phase space—in other words, of the volume of the phase space in which the thermodynamic observables have values compatible with a specific thermodynamic state.

Let us assume that the thermodynamic state is determined by the extensive variables $(X_0, \ldots, X_r)$. Each of these variables—X_i, for example—is represented (in our hypotheses) by a function $X_i(x)$ defined over the space of microscopic states. The region of phase space that is accessible is defined by the values that these variables assume and that we will denote by $(X_0, \ldots, X_r)$. In effect, this region is defined by the values $(X_0, \ldots, X_r)$, up to errors ΔX_i, $i = 1, \ldots, r$, which can be assumed to be small but macroscopic. The region of phase space in which these observables have those specific values (within the errors) will be denoted by Γ, and its volume by $|\Gamma|$. Then the fundamental postulate states that

$$S = k_B \ln |\Gamma|. \tag{3.8}$$

In this formula,

- S is thermodynamic entropy (expressed as a function of extensive variables).
- k_B is Boltzmann's constant ($1.384\ 10^{-23}\ \mathrm{JK}^{-1}$).
- The equals sign must be understood as implying "except for nonextensive terms, which become negligible."

Since it is a postulate, this relation cannot be "proven"—however, one can attempt to justify it by showing that $\ln |\Gamma|$ has the properties of thermodynamic entropy.

The preceding discussion therefore allows one to give the following expression to thermodynamic entropy, as a function of the extensive variables $(X_0, \ldots, X_r)$:

$$S(X_0, \ldots, X_r) = k_B \ln \int_\Gamma \mathrm{d}x = k_B \ln \int \mathrm{d}x \prod_{i=0}^{r} \left[\theta(X_i(x) - (X_i - \Delta X_i))\, \theta(X_i - X_i(x)) \right]. \tag{3.9}$$

The product of the theta functions is such that the integrand assumes a value of 1 if the microscopic state x is such that the values of the observables belong to the interval $X_i - \Delta X_i \le X_i(x) \le X_i$, $i = 0, \ldots, r$, and is otherwise equal to zero.

3.3.1 *Other Expressions of Entropy*

If we assume that the fundamental postulate is valid, as a corollary we can conclude that the quantity that appears on the left-hand side of equation (3.9) is extensive. As a consequence, it is possible to obtain other expressions of the same quantity, ones that differ from it for terms that grow less rapidly than S with the system's dimensions. Since in what follows we will make use of these expressions in an interchangeable manner, it is useful to discuss them explicitly here. For simplicity's sake, we will start with a situation in which all the extensive thermodynamic variables $(X_1, \ldots, X_r)$ have a well-defined value, and we explicitly consider entropy solely as a function of internal energy E.

In this situation, equation (3.9) takes the form

$$S(E) = k_B \ln \int \mathrm{d}x\, \theta(H(x) - E + \Delta E)\, \theta(E - H(x)). \tag{3.10}$$

A fairly surprising first result is that if the energy has a lower bound (we can assume, for example, that it cannot be negative), one obtains the same value for entropy by extending the integral to the entire region $H(x) \leq E$:

$$S(E) = k_B \ln \int \mathrm{d}x\, \theta(E - H(x)) + \text{nonextensive terms.} \tag{3.11}$$

In effect, as we know, $S(E)$ is a monotonically increasing function of E, and it is extensive. The accessible phase space volume $|\Gamma(E)| = e^{S(E)/k_B}$ increases with E, and it increases all the more rapidly the greater the size of the system being considered. As a consequence, the greater part of the accessible phase space volume corresponds to the largest values of E. In effect, given two values of internal energy E and E', the relation between the corresponding phase space volumes is given by

$$\frac{|\Gamma(E)|}{|\Gamma(E')|} = e^{[S(E) - S(E')]/k_B}, \tag{3.12}$$

and if $E < E'$, this relation is exponentially small in the system's size.

We have thus obtained expression (3.11). Given the same hypotheses, we can conclude that entropy can also be expressed as a function of the surface area defined by $H(x) = E$:

$$S(E) = k_B \ln \int \mathrm{d}x\, \delta(H(x) - E) + \text{nonextensive terms,} \tag{3.13}$$

where $\delta(x)$ is the Dirac delta "function." In effect, let us temporarily denote the quantity defined on the right-hand side of equation (3.13) by $\mathcal{S}(E)$. By substituting this expression in (3.9), we obtain

$$S(E) = k_B \ln \int_{E-\Delta E}^{E'} \mathrm{d}E' \mathrm{e}^{\mathcal{S}(E')/k_B}. \tag{3.14}$$

If ΔE is sufficiently small but finite, the integrand varies slowly in the interval $E - \Delta E < E' < E$, and one therefore arrives at

$$S(E) \simeq k_B \ln\left(\mathrm{e}^{\mathcal{S}(E)/k_B} \Delta E\right) = \mathcal{S}(E) + \ln \Delta E. \tag{3.15}$$

The last term is obviously negligible in the thermodynamic limit, and therefore $\mathcal{S}(E) = S(E)$. By using this result, we can obtain a more explicit derivation of (3.11).

3.4 Liouville's Theorem

In the fundamental postulate, entropy is expressed in terms of the accessible volume *in phase space*. One may ask why phase space plays this privileged role, rather than other possible spaces that also describe the microscopic state of a mechanical system. Let us consider, e.g., a system composed of identical particles of mass m. In this case, the momentum $\boldsymbol{p}$ of a particle is simply proportional to its velocity $\boldsymbol{v}$, and it is a matter of taste to describe the system in the phase space, i.e., by the collection $(\boldsymbol{r},\boldsymbol{p})$ of the position and momentum vectors of the particles or by the space of position and velocity vectors. But let us now consider a mixture of particles of two different masses—say, m and m'. When particles collide, they exchange momentum in the center-of-mass frame. If they have the same mass, this is tantamount to a velocity exchange. But this is not true when the particles have a different mass. Indeed, we can show that the phase space of positions and momenta possesses a very special property with respect to time evolution. This property is expressed by **Liouville's theorem**.

Let us consider a mechanical system described in the phase space $x = (\boldsymbol{r}, \boldsymbol{p})$, where $\boldsymbol{r} = (\boldsymbol{r}_1, \ldots, \boldsymbol{r}_N)$ and $\boldsymbol{p} = (\boldsymbol{p}_1, \ldots, \boldsymbol{p}_N)$. Let the system's Hamiltonian be $H(\boldsymbol{r}, \boldsymbol{p}, t)$, which may depend on the time t in an arbitrary way. Then the representative point of the system evolves according to the canonical equations of motion:

$$\dot{\boldsymbol{r}} = \frac{\partial H}{\partial \boldsymbol{p}}, \qquad \dot{\boldsymbol{p}} = -\frac{\partial H}{\partial \boldsymbol{r}}. \tag{3.16}$$

In this way, to each point x in phase space is assigned a velocity vector $b(x, t)$ in the same space such that if the representative point of the system lies at x at a given time t, it will be at $x + b(x, t)\,\mathrm{d}t$ at time $t + \mathrm{d}t$, where $\mathrm{d}t$ is infinitesimal. The vector $b(x, t)$ has $6N$ coordinates. To simplify the following discussion, let us identify all pairs $(r_{i\alpha}, p_{i\alpha})$ of conjugate coordinates and momenta, where i denotes the particle and $\alpha = x, y, z$ denotes the coordinate, by a single index i running from 1 to $3N$. Thus the canonical equations of motion assign a local velocity $b(x) = (b_{ri}(x), b_{pi}(x))$ $(i = 1, \ldots, 3N)$ to each phase-space point $x = (r_i, p_i)$ $(i = 1, \ldots, 3N)$ where

$$b_{ri}(x) = \frac{\partial H}{\partial p_i}, \qquad b_{pi}(x) = -\frac{\partial H}{\partial r_i}, \qquad i = 1, \ldots, 3N. \tag{3.17}$$

Let us consider a small region Γ^0 of phase space, identified by the conditions

$$r_i^0 < r_i < r_i^0 + \mathrm{d}r_i, \qquad p_i^0 < p_i < p_i^0 + \mathrm{d}p_i, \qquad i = 1, \ldots, 3N.$$

We define the **evolute** Γ^t at time t as the set of all points that satisfy the following condition: Let $x(t; x_0)$ be the solution of the canonical equations of motion that satisfies the initial condition $x(t = 0; x_0) = x_0$. Then, Γ^t contains all the points x such that $x = x(t; x_0)$ for some $x_0 \in \Gamma^0$. In practice, we can say that all the points that made up Γ^0 at time 0 have moved to Γ^t at time t.

Liouville's theorem states that the volume in phase space of Γ^t is equal to that of Γ^0:

$$|\Gamma^t| = |\Gamma^0|. \tag{3.18}$$

It will be sufficient to prove the result for an infinitesimal time change $\mathrm{d}t$, since it is clear that if the volume of the evolute remains constant for a small time change, iterating the statement will prove that it remains constant forever.

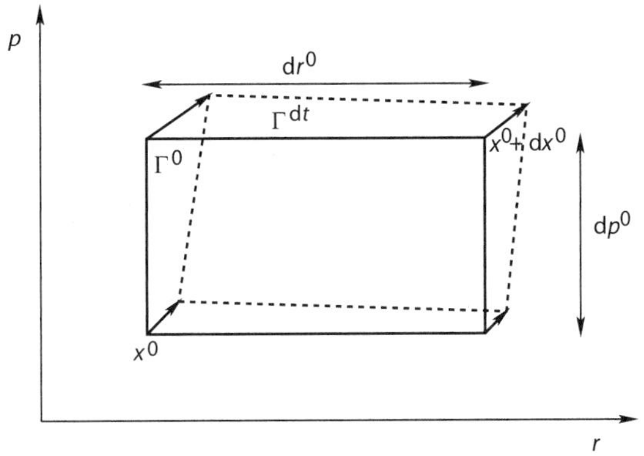

FIGURE 3.1. Evolution of a small phase–space region during an infinitesimal time interval.

To fix one's ideas, it is useful to consider the case of a single particle in one dimension, i.e., of a two-dimensional phase space (see figure 3.1). Then, let the infinitesimal region $d\Gamma^0$ be defined by the inequalities

$$r^0 < r < r^0 + dr^0, \qquad p^0 < p < p^0 + dp.$$

Let us denote by x^0 the point (r^0, p^0) and by dx^0 the vector $(dr^0, dp^0) = dr^0\boldsymbol{i} + dp^0\boldsymbol{j}$, where $\boldsymbol{i} = (1,0)$ and $\boldsymbol{j} = (0,1)$ are the base vectors of the space. The local velocity is given by

$$b_r(r,p) = \frac{\partial H}{\partial p}, \qquad b_p(r,p) = -\frac{\partial H}{\partial r}. \tag{3.19}$$

The evolute region $d\Gamma^{dt}$ is the parallelogram whose vertices are, respectively, the points $x^0 + b(x^0)dt$; $x^0 + dp^0\boldsymbol{j} + b(x^0 + dp^0\boldsymbol{j})dt$; $x^0 + dx^0 + b(x^0 + dx^0)dt$; and $x^0 + dr^0\boldsymbol{i} + b(x^0 + dr^0\boldsymbol{i})$. Its area is therefore approximately given by

$$\begin{aligned} |d\Gamma^{dt}| = dr\,dp &= \{dr^0 + [b_r(r^0 + dr^0, p') - b_r(r^0, p')]\,dt\} \\ &\quad \times \{dp^0 + [b_p(r', p^0 + dp^0) - b_p(r', p^0)]\,dt\} \\ &\simeq dr^0\,dp^0 + dt\left(\frac{\partial b_r}{\partial r} + \frac{\partial b_p}{\partial p}\right)dr^0\,dp^0 + \text{higher orders}, \end{aligned}$$

where r' and p', respectively, differ from r^0 and p^0 by less than dr^0 or dp^0. One has, however, from equations (3.19),

$$\frac{\partial b_r}{\partial r} + \frac{\partial b_p}{\partial p} = \frac{\partial^2 H}{\partial r \partial p} - \frac{\partial^2 H}{\partial p \partial r} = 0. \tag{3.20}$$

Thus $|d\Gamma^{dt}| = dr^0\,dp^0 = |d\Gamma^0|$, up to terms that are higher order in dt or in the region's size.

We can now consider the full $6N$-dimensional phase space flow. By the same considerations, we obtain, up to higher-order terms,

$$|\mathrm{d}\Gamma^{\mathrm{d}t}| = |\mathrm{d}\Gamma^0| \left\{ 1 + \mathrm{d}t \sum_{i=1}^{3N} \left[\frac{\partial b_{r_i}(x,t)}{\partial r_i} + \frac{\partial b_{p_i}(x,t)}{\partial p_i} \right] \right\}. \tag{3.21}$$

By evaluating the derivatives, we obtain

$$\sum_{i=1}^{3N} \left[\frac{\partial b_{r_i}(x,t)}{\partial r_i} + \frac{\partial b_{p_i}(x,t)}{\partial p_i} \right] = \sum_{i=1}^{3N} \left[\frac{\partial^2 H(x,t)}{\partial r_i \partial p_i} + \frac{\partial^2 H(x,t)}{\partial p_i \partial r_i} \right] = 0, \tag{3.22}$$

and thus $|\mathrm{d}\Gamma^{\mathrm{d}t}| = |\mathrm{d}\Gamma^0|$. Therefore, for a finite t, we have

$$|\mathrm{d}\Gamma^t| = |\mathrm{d}\Gamma^0| = \text{const.} \tag{3.23}$$

which corresponds to Liouville's theorem for an infinitesimal phase space region. The theorem can be extended to arbitrary regions by decomposing them into the union of infinitesimal ones.

The fact that the canonical equations conserve the volume of regions in phase space is the special property of phase space we were looking for.

Liouville's theorem has an important corollary. Let $\rho(x, t)$ be the local probability density of finding a system in the neighborhood of phase space point x. Given a small region $\mathrm{d}\Gamma$ of phase space around point x, the probability of finding the given system in $\mathrm{d}\Gamma$ at time t is given by

$$\mathrm{d}P = \rho(x,t)\,|\mathrm{d}\Gamma|. \tag{3.24}$$

We want to evaluate the way $\rho(x, t)$ changes if the system evolves according to the canonical equations. Let us consider an arbitrary small region $\mathrm{d}\Gamma^0$ around x_0, and let $\rho(x, 0)$ be the initial distribution at $t = 0$. Then, the probability $\mathrm{d}P(t)$ that the system is found at time t within the evolute $\mathrm{d}\Gamma^t$ of Γ^0 does not vary by definition, since if its initial condition was lying within $\mathrm{d}\Gamma^0$ at $t = 0$, its evolved point must lie within $\mathrm{d}\Gamma^t$ at time t. We have, therefore,

$$\mathrm{d}P = \rho(x(t;x_0),t)\,|\mathrm{d}\Gamma^t| = \rho(x_0,0)\,|\mathrm{d}\Gamma^0|.$$

Thus,

$$\rho(x(t;x_0),t) = \rho(x_0,0)\frac{|\mathrm{d}\Gamma^0|}{|\mathrm{d}\Gamma^t|} = \rho(x_0,0), \tag{3.25}$$

where in the last equality we have taken into account Liouville's theorem. Therefore, the local probability density does not change if one moves along with the local velocity dictated by the canonical equations.

Let us evaluate the expression of the local probability density change. We have

$$\begin{aligned}\frac{\mathrm{d}\rho}{\mathrm{d}t} &= \frac{\partial\rho}{\partial t} + \sum_{i=1}^{3N}\left[\frac{\partial\rho}{\partial r_i}b_{ri}(x,t) + \frac{\partial\rho}{\partial p_i}b_{pi}(x,t)\right]\\ &= \frac{\partial\rho}{\partial t} + \sum_{i=1}^{3N}\left(\frac{\partial\rho}{\partial r_i}\frac{\partial H}{\partial p_i} - \frac{\partial\rho}{\partial p_i}\frac{\partial H}{\partial r_i}\right).\end{aligned}$$

Since Liouville's theorem implies that $\mathrm{d}\rho/\mathrm{d}t = 0$, we obtain

$$\frac{\partial \rho}{\partial t} = -\sum_{i=1}^{3N} \left(\frac{\partial H}{\partial p_i} \frac{\partial \rho}{\partial r_i} - \frac{\partial H}{\partial r_i} \frac{\partial \rho}{\partial p_i} \right) = -[H, \rho]_{\text{PB}}. \tag{3.26}$$

The expression on the right-hand side is called the **Poisson bracket** of the Hamiltonian H with ρ, and is usually denoted by $[\,,\,]_{\text{PB}}$. This equation is known as **Liouville's equation.**

Let $A(x)$ be a function defined on the phase space of the system, e.g., an observable. Let us evaluate the change of its instantaneous value as the system evolves according to the canonical equations of motion. We have

$$A(t) = A(x(t; x_0)),$$

and

$$\frac{dA}{dt} = \sum_{i=1}^{3N} \left[b_{ri}(x) \frac{\partial A}{\partial r_i} + b_{pi}(x) \frac{\partial A}{\partial p_i} \right] = \sum_{i=1}^{3N} \left(\frac{\partial H}{\partial p_i} \frac{\partial A}{\partial r_i} - \frac{\partial H}{\partial r_i} \frac{\partial A}{\partial p_i} \right) = [H, A]_{\text{PB}}. \tag{3.27}$$

Thus, the derivative of A with respect to time is expressed in terms of its Poisson bracket with the Hamiltonian. Notice, however, the change in sign with respect to Liouville's equation (3.26). In particular, the constants of the motion are characterized by the vanishing of their Poisson bracket with the Hamiltonian.

Let us now assume that $\rho(x, t)$ is such that, at $t = 0$, it depends only on the values of a set $(X_1, \ldots, X_r)$ of constants of the motion, i.e.,

$$\rho(x, 0) = \Phi(X_1(x), \ldots, X_r(x)),$$

where, for $k = 1, \ldots, r$, one has

$$\frac{dX_k}{dt} = [H, X_k]_{\text{PB}} = 0.$$

We then have

$$\frac{\partial \rho}{\partial t} = -\sum_{i=1}^{3N} \left(\frac{\partial H}{\partial p_i} \frac{\partial \rho}{\partial r_i} - \frac{\partial H}{\partial r_i} \frac{\partial \rho}{\partial p_i} \right) = -\sum_{k=1}^{r} \sum_{i=1}^{3N} \left(\frac{\partial H}{\partial p_i} \frac{\partial X_k}{\partial r_i} - \frac{\partial H}{\partial r_i} \frac{\partial X_k}{\partial p_i} \right) \frac{\partial \Phi}{\partial X_k} = 0.$$

Therefore, $\rho(x)$ does not change with time. This provides us with a recipe to obtain probability distribution functions that do not change with time—it is sufficient that they depend on x only via some constants of the motion X_k.

3.5 Quantum States

We know that in classic thermodynamics, entropy would be defined up to an additive constant if one did not add a third postulate to the principles discussed in the preceding chapter, a postulate called either **Nernst's postulate** or the **third law of thermodynamics**, which stipulates the following property:

The entropy of any system tends to zero for $T \to 0$.

According to this postulate, each transformation that is isotherm to $T = 0$ is also an adiabatic transformation. (From this, it follows that one cannot reach absolute zero.)

On the other hand, from the expression $S = k_B \ln|\Gamma|$ of the fundamental postulate, it appears obvious that changing the phase space's units of measurement results in the addition of a constant to the entropy. As a consequence, if Nernst's postulate univocally sets the value of entropy (except in cases involving subextensive terms), it must also implicitly fix the unit of measurement for phase space's volume.

The value of this unit is obtained from considerations that involve quantum mechanics and that I will try to describe briefly in this section. We have to keep in mind that we are still far from disposing of a satisfying set of postulates for the foundation of quantum statistical mechanics. The informal remarks that follow have only a heuristic value.

A *microscopic* state in quantum mechanics is described by a vector (normalized and defined up to a phase) $|\Psi\rangle$ that belongs to a Hilbert space $\mathcal{H}$. It evolves according to the time-dependent Schrödinger equation:

$$i\hbar\frac{\partial}{\partial t}|\Psi\rangle = \widehat{H}|\Psi\rangle, \tag{3.28}$$

where $\widehat{H}$ is the Hamiltonian operator. (We will always denote quantum operators with a $\widehat{\ }$.) The **quantum observables** A are represented by Hermitian operators $\widehat{A}$ defined on $\mathcal{H}$, whose eigenvalues correspond to the possible results of the measurement of A.

More particularly, if a quantum observable $\widehat{A}$ commutes with the Hamiltonian $\widehat{H}$, it is a constant of motion, and its average value does not change over time—in addition, it is possible to identify some common eigenstates $|E, A\rangle$ of both the Hamiltonian and $\widehat{A}$, and in which both the energy E and the observable A have well-defined values. If a system is prepared in one of these eigenstates, the system's evolution (dictated by the Schrödinger equation [3.28]) lets it remain in the same state.

Let us then suppose that it is possible to identify a certain number of quantum observables $(\widehat{X}_0, \dots, \widehat{X}_r)$ that commute among themselves as well as with the Hamiltonian and that represent the extensive thermodynamic properties $(X_0, \dots, X_r)$. Obviously, following our conventions, one has $\widehat{X}_0 = \widehat{H}$.

Let us consider the set of all microscopic states in which the thermodynamic observables $(\widehat{X}_0, \dots, \widehat{X}_r)$ all have a well-defined value, which is equal to $(X_0, \dots, X_r)$ up to the small uncertainties ΔX_i, $i = 0, \dots, r$. These states are the common eigenstates of the Hamiltonian and of the macroscopic observables $(\widehat{X}_0, \dots, \widehat{X}_r)$, whose eigenvalues Ξ_i satisfy the inequalities $X_i \le \Xi_i \le X_i + \Delta X_i$, $i = 0, \dots, r$. Their number $\mathcal{N}$ is the quantum analog of the accessible volume of phase space.

The fundamental postulate, in its quantum version, then stipulates that

$$S = k_B \ln \mathcal{N}. \tag{3.29}$$

In this formulation, the arbitrariness associated with the unit of measurement of the phase space's volume has disappeared. More specifically, the Nernst postulate has a clear interpretation: when the temperature tends to zero, only those states that correspond to the lowest energy values become accessible. The number of these states is finite, or grows more slowly than an exponential when the system's site increases, so that the extensive part of entropy vanishes.[1]

We can now see how in the semiclassical limit, the number $\mathcal{N}$ of quantum states is proportional to the volume $|\Gamma|$ of the phase space, and we can calculate the relevant constant of proportionality. In effect, Heisenberg's indeterminacy principle stipulates that for each pair (q, p) formed by the generalized coordinate q and the relevant conjugate momentum p, the quantum uncertainties must satisfy

$$\Delta p \Delta q \geq O(h). \tag{3.30}$$

Each quantum state therefore "occupies" a region of the phase space whose projection on the plane (q, p) has an area proportional to h. For a system with N degrees of freedom, the volume therefore becomes proportional to h^N, where N is the number of coordinate pairs and relevant conjugate momentums.

In order to determine the constant of proportionality (which must be universal), one can consider some simple systems. Let us take the harmonic oscillator, for example. The eigenstates of the Hamiltonian of a harmonic oscillator with angular frequency ω correspond to the values $E_n = \hbar\omega(n + 1/2)$ of energy, where $n = 0, 1, \ldots$. We can evaluate the volume of phase space "occupied" by each eigenstate by calculating the volume included between the curves $H = n\hbar\omega$ and $E = (n+1)\hbar\omega$. (Since we are in two dimensions, the "volume" is, obviously, an area.) We have

$$H(p, q) = \frac{p^2}{2m} + \frac{1}{2} m\omega^2 q^2, \tag{3.31}$$

and therefore

$$\Delta|\Gamma| = \int_{n\hbar\omega}^{(n+1)\hbar w} \mathrm{d}E \int \mathrm{d}q\mathrm{d}p \; \delta(H(p, q) - E). \tag{3.32}$$

We can now introduce the variables

$$x^2 = \frac{p^2}{m}, \qquad y^2 = m\omega^2 q^2, \tag{3.33}$$

and therefore the integral becomes

$$\begin{aligned} \Delta|\Gamma| &= \int_{n\hbar\omega}^{(n+1)\hbar\omega} \mathrm{d}E \int \frac{\mathrm{d}x\mathrm{d}y}{\omega} \, \delta\left(\frac{1}{2}(x^2 + y^2) - E\right) \\ &= \int_{r^2 = 2n\hbar\omega}^{r^2 = 2(n+1)\hbar\omega} \frac{2\pi r}{\omega} \mathrm{d}r = 2\pi\hbar = h. \end{aligned} \tag{3.34}$$

The "volume" of each quantum state is therefore equal to h for a system with only one degree of freedom, as we had anticipated.

[1] One often comes across the statement that Nernst's postulate is a consequence of the fact that the ground state of a quantum system is unique. This is not correct, because the postulate concerns the behavior of the extensive part of $\ln \mathcal{N}$ for a system with a large N, in the limit $T \to 0$. (In other words, one *first* passes to the "thermodynamic limit" $N \to \infty$, and one then puts T to 0.) The uniqueness of the ground state is not sufficient to guarantee that entropy vanishes within this order of limits.

Exercise 3.1 (Particle in a Potential Well) Let us consider a particle of mass m, constrained to moving in one dimension along a segment of length n. In this case, the possible energy values are $E_n = \hbar^2\pi^2 n^2/L^2$, where $n = 1, 2, \ldots$. Prove that in this case also, the phase space volume "occupied" by each quantum state is equal to h. (Be careful to calculate the classic volume correctly!)

More generally, one will be able to obtain the correct factor for calculating the volume of quantum states by comparing the entropy calculated classically with the quantum entropy in the semiclassical limit. (Usually, at high temperatures both system behave "classically.") Since we do not intend to deal with quantum statistical mechanics in this text, we will not delve more deeply into this topic.

3.6 Systems in Contact

We will now discuss the extensivity of the entropy. Let us consider two systems, 1 and 2, such that the thermodynamic state of 1 corresponds to the region $\Gamma^{(1)}$ of phase space, and that of 2 to $\Gamma^{(2)}$. Then the Cartesian product $\Gamma^{(1\cup 2)} = \Gamma^{(1)} \times \Gamma^{(2)}$ will correspond to the region of phase space describing the state of the compound system $1\cup 2$ (see figure 3.2). The volume of this region is equal to $|\Gamma^{(1\cup 2)}| = |\Gamma^{(1)}||\Gamma^{(2)}|$.

In this manner, the fundamental postulate gives us:

$$\begin{aligned} S &= k_B \ln|\Gamma^{(1\cup 2)}| = k_B \ln\left(|\Gamma^{(1)}||\Gamma^{(2)}|\right) \\ &= k_B \ln|\Gamma^{(1)}| + k_B \ln|\Gamma^{(2)}| = S^{(1)} + S^{(2)}, \end{aligned} \tag{3.35}$$

thus conforming to entropy's additivity.

One should note that this result implicitly presupposes that no relation between the dynamic states of each of the two subsystems exists, so that the variables that define the overall state of the system are given simply by those that define the state of each subsystem taken separately. This condition of *dynamic independence* of the subsystems is in principle an experimentally based hypothesis, essential for statistical mechanics' validity, and can be considered as a supplement to the fundamental postulate.

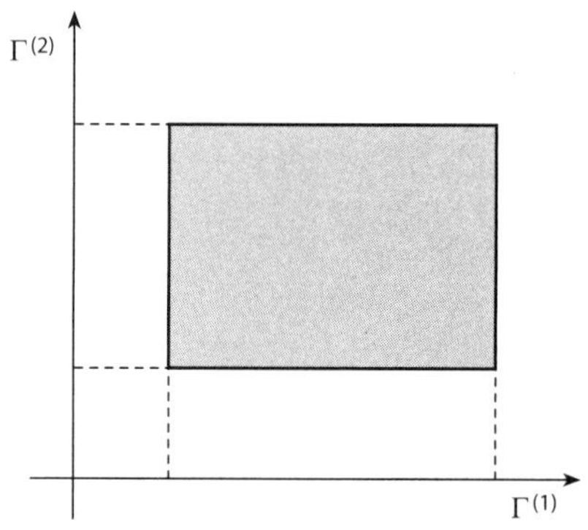

FIGURE 3.2. Schematic representation of the phase–space distribution $\Gamma^{(1)} \times \Gamma^{(2)}$ of two systems in contact.

In the last analysis, this hypothesis corresponds to the independence of the observables that pertain to systems (in the sense given it by probability theory) in contact. A little later, we will see to what extent it may be possible to justify this hypothesis on the basis of the chaos theory of dynamic systems.

3.7 Variational Principle

The other fundamental property of entropy is the **variational principle**, according to which it reaches a maximum at thermodynamic equilibrium. Let us see how this principle is compatible with the interpretation of entropy in terms of accessible volume in phase space.

Let us consider a system whose thermodynamic state is identified by the values of a certain collection (X_i) of extensive variables. At thermodynamic equilibrium, certain observables, which we shall collectively denote by ξ, will have the equilibrium values ξ^*:

The variational principle for the entropy states that the equilibrium values ξ^* correspond to the maximum value of the entropy.

For example, let us assume that our system is composed of two systems in contact, 1 and 2, that can exchange energy. The total internal energy, E_{tot}, has a fixed value: at equilibrium, the value $E^{(1)}_{\text{eq}}$ of the internal energy of system 1 will also be fixed. This value is determined by the condition that the entropy reaches a maximum. In order to simplify, we will assume that each system's internal energy is determined up to a (macroscopic) uncertainty ΔE, and that the thermodynamic states will correspond to discrete values of energy that differ by ΔE.

If we consider the two systems as isolated from each other, the accessible volume of the phase space is given by

$$|\Gamma_0| = \left|\Gamma^{(1)}\left(E^{(1)}_{\text{in}}\right)\right|\left|\Gamma^{(2)}\left(E^{(2)}_{\text{in}}\right)\right|. \tag{3.36}$$

If we allow for the energy exchange between the two systems, the accessible volume becomes

$$|\Gamma| = \sum_{E^{(1)}} |\Gamma^{(1)}(E^{(1)})||\Gamma^{(2)}(E_{\text{tot}} - E^{(1)})|. \tag{3.37}$$

The sum runs over all possible values of $E^{(1)}$. It is obvious that $|\Gamma| \geq |\Gamma_0|$, since the term that appears in equation (3.36) is one of the terms that appears in (3.37), and therefore the removal of the constraints leads to an increase of the accessible volume of the phase space—in other words, of the entropy. We thus obtain the property that entropy cannot decrease spontaneously.

At equilibrium, however, the value of the internal energy of system 1 is well determined, while in this equation *all* possible values appear. If we were to take this equation at face value, the result would be disastrous, because the value of the system's entropy would come to depend on its history, and not only on the values of the observables. In effect, this equation implies

$$|\Gamma| > |\Gamma^{(1)}(E^{(1)}_{\text{eq}})||\Gamma^{(2)}(E^{(2)}_{\text{eq}})|, \tag{3.38}$$

instead of the equality we would expect from thermodynamics.

This problem is solved by considering that the accessible volume of the phase space corresponds *almost entirely* to the states in which $E^{(1)} = E^{(1)}_{\text{eq}}$, $E^{(2)} = E^{(2)}_{\text{eq}}$, so that the difference between the two expressions of $\ln|\Gamma|$ is negligible for large values of N. This remark is based in its turn on the extensivity of entropy.

Let us in fact make the sum explicit, reading the fundamental postulate backwards. We thus obtain

$$\begin{aligned} |\Gamma| &= \sum_{E^{(1)}} |\Gamma^{(1)}(E^{(1)})||\Gamma^{(2)}(E_{\text{tot}} - E^{(1)})| \\ &= \sum_{E^{(1)}} \exp\left\{\frac{1}{k_B}\left[S^{(1)}(E^{(1)}) + S^{(2)}(E_{\text{tot}} - E^{(1)})\right]\right\}. \end{aligned} \tag{3.39}$$

We know that the total entropy—and therefore the summand—reaches its maximum for the equilibrium values of $E^{(1)}$ and $E^{(2)}$. Let us approximate the sum with an integral and expand the exponential's argument into a Taylor series around these values. The first-order term vanishes, and we thus obtain

$$\begin{aligned} |\Gamma| \approx \int \frac{dE^{(1)}}{\Delta E} \exp\left\{\frac{1}{k_B}\left[S^{(1)}(E^{(1)}_{\text{eq}}) + S^{(2)}(E^{(2)}_{\text{eq}})\right.\right. \\ \left.\left. + \frac{1}{2}\left(\left.\frac{\partial^2 S^{(1)}}{\partial E^{(1)}}\right|_{E^{(1)}_{\text{eq}}} + \left.\frac{\partial^2 S^{(2)}}{\partial E^{(2)}}\right|_{E^{(2)}_{\text{eq}}}\right)\left(E^{(1)} - E^{(1)}_{\text{eq}}\right)^2\right]\right\}. \end{aligned} \tag{3.40}$$

Due to the extensivity of entropy and internal energy, the derivatives $\partial^2 S/\partial E^2$ are of order $1/N$, where N is a measure of the two subsystems' sizes, and therefore the contribution of the integral in $E^{(1)}$ is a factor of order $N^{1/2}$. This produces an additive contribution to entropy on the order of $\ln N$, and is therefore negligible.

This reasoning gives us another important piece of information: that each system's internal energy values can differ from the equilibrium value of quantities of $N^{1/2}$ order and are therefore negligible with respect to the value of internal energy, which is of order N. This is a first step toward a theory of the fluctuations of thermodynamic quantities.

We therefore see that the a priori uncertainty ΔE over the value of internal energy is not relevant, at least so long as it is not much smaller than the thermodynamic uncertainty, which is of order $N^{1/2}$—in accordance with thermodynamics, we therefore arrive at the conclusion that the value of entropy does not depend on the uncertainty by which we know the values of the thermodynamic variables.

3.8 The Ideal Gas

Let us consider the ideal gas. In this case, the extensive variables to be considered are the number of particles N, the volume V, and internal energy E. The system is described by the Hamiltonian:

$$H = \sum_{i=1}^{N} \left[\frac{p_i^2}{2m} + u(\mathbf{r}_i) \right], \tag{3.41}$$

where the single particle potential $u(\mathbf{r})$ is given by

$$u(\mathbf{r}) = \begin{cases} 0, & \text{if } \boldsymbol{r} \text{ is inside the box of volume } V, \\ \infty, & \text{otherwise.} \end{cases} \tag{3.42}$$

The accessible volume of the phase space is given by

$$|\Gamma| = \frac{1}{N!} \frac{1}{h^{3N}} \int \prod_i \mathrm{d}r_i \mathrm{d}p_i \, \delta\left(\sum_i \frac{p_i^2}{2m} - E \right). \tag{3.43}$$

In this expression, the integral over the spatial coordinates is extended to the box of volume V, the delta function in the momenta imposes the internal energy value, and the h^{-3N} factor is used to normalize the number of states, as previously mentioned. The factor $1/N!$ take into account the fact that, because of the indistinguishability of identical particles, states that differ only by a permutation of particle indices correspond to the same quantum state and do not need to be counted separately. It is worthy of note that Gibbs was led to introduce this factor on the basis of considerations on entropy's extensivity, well before the introduction of quantum mechanics.

Easily performing the integral over the coordinates, we are led to

$$|\Gamma| = \frac{V^N}{N! h^{3N}} \int \prod_i \mathrm{d}p_i \, 2m\delta\left(\sum_i p_i^2 - 2mE \right). \tag{3.44}$$

We can interpret the right-hand side integral as the surface area of a sphere of radius $R = (2ME)^{1/2}$ in a space with $3N$ dimensions. It is easy to see (see appendix B) that this area is equal to $1/2\pi^{3N/2} R^{3N-1} / \Gamma_E(3N/2)$, where Γ_E is Euler's gamma function. We finally obtain

$$|\Gamma| = \frac{V^N \sqrt{\pi} (2mE)^{(3N-1)/2}}{N! h^{3N} \Gamma_E(3N/2)}, \tag{3.45}$$

up to factors that vary more slowly with N. Shifting to logarithms and taking the asymptotic expression of Euler's function (Stirling's formula; see appendix B) into account, we obtain

$$S = Nk_B \ln\left\{ \left(\frac{eV}{N} \right) \left[\frac{2\pi me}{h^2} \left(\frac{2E}{3N} \right) \right]^{3/2} \right\}, \tag{3.46}$$

up to terms that are negligible for large values of N. We can see from this expression that the entropy thus obtained is actually extensive (while it would not have been so had the factor $1/N!$ in the definition of $|\Gamma|$ been absent).

It is interesting to remark that the expression of the entropy of the ideal gas, including the quantum factors h, has been obtained in the early twentieth century by O. Sackur and H. M. Tetrode [Sack11, Tetr14], before the formulation of quantum mechanics. This expression is therefore known as the **Sackur-Tetrode formula**.

Having thus arrived at the fundamental equation, we can deduce the equations of state by taking derivatives:

$$\frac{1}{T} = \left.\frac{\partial S}{\partial E}\right|_{V,N} = \frac{3}{2}\frac{Nk_B}{E}; \tag{3.47}$$

$$\frac{p}{T} = \left.\frac{\partial S}{\partial V}\right|_{E,N} = \frac{Nk_B}{V}. \tag{3.48}$$

These equations correspond to the usual equations for the ideal monoatomic gas—let us note that they imply that the entropy has the expression $S = S^{(1)}(E) + S^{(2)}(V)$, and, moreover, that the internal energy of the ideal gas depends only on the temperature $E = E(T)$.

3.9 The Probability Distribution

The power of statistical mechanics is highlighted by the possibility of providing predictions for observables that are not explicitly considered in thermodynamics.

Let us assume that the state of our system is identified by a collection $X = (X_i)$ of extensive variables, which is associated with a region Γ of volume $|\Gamma|$. We want to calculate the value taken by another observable, A, which we will assume is extensive. This observable is defined by a function, $A(x)$, over the phase space. According to our hypothesis, in the thermodynamic state specified by the (X_i), the observable A assumes a specific equilibrium value, which we will denote by a^*. If we impose the constraint that the value of A be a, not necessarily identical to the equilibrium value a^*, on our system, the fundamental hypotheses of thermodynamics tell us that entropy can only decrease:

$$S(X; a) \leq S(X; a^*). \tag{3.49}$$

On the other hand, the fundamental postulate of statistical mechanics relates entropy $S(X; a)$ to the accessible volume $|\Gamma(a)|$ of the phase space once the further constraint that the value of A be equal to a has also been imposed:

$$S(X; a) = k_B \ln \int_\Gamma dx\, \delta(A(x) - a). \tag{3.50}$$

Since the equilibrium value of A is equal to a^*, however, one must get

$$S(X) = S(X; a^*) \geq S(X; a). \tag{3.51}$$

These relations are compatible only if essentially the entire volume of the accessible region Γ of phase space corresponds to the equilibrium value a^* of the observable A.

One can become convinced that this is how things stand by calculating the ratio of the volumes of the regions being considered:

$$\begin{aligned}\frac{|\Gamma(a)|}{|\Gamma|} &= \frac{1}{|\Gamma|}\int_\Gamma dx\, \delta(A(x) - a) \\ &= \exp\left\{\frac{1}{k_B}[S(X; a) - S(X; a^*)]\right\} \\ &\simeq \exp\left\{\frac{1}{2k_B}\left[\left.\frac{\partial^2 S}{\partial A^2}\right|_{a^*}(a - a^*)^2\right]\right\}.\end{aligned} \tag{3.52}$$

The first-order contribution vanishes because of the maximum entropy principle. The second-order derivative is negative and of order $1/N$; the characteristic values of the additive observable A are also of order N. The exponential function therefore rapidly becomes negligible as soon as the value of A differs from the equilibrium value.

But if almost all the accessible volume corresponds to A's equilibrium value, we can calculate this value simply by taking A's average over the entire accessible region of the phase space:

$$a^* = \langle A(x) \rangle = \frac{1}{|\Gamma|} \int_\Gamma \mathrm{d}x \, A(x). \tag{3.53}$$

This gives us a very convenient "recipe" for calculating thermodynamic quantities. Let us observe that the average appearing in equation (3.53) corresponds to the average over a probability distribution that is uniform within Γ and vanishes outside—this distribution is called **microcanonical**. In statistical mechanics, one often uses the word *ensemble* as a synonym for probability distribution in the phase space. The ensemble we have described, in which the internal energy has a fixed value, is called the **microcanonical ensemble** and was introduced by Gibbs.

Let us remark that conceptually, the ensemble is nothing more than a rule to calculate the thermodynamic value of the observables, but that any other distribution, so long as it is fairly smooth within the phase space and only parametrized by the observables, would lead to equivalent results. We will see some additional examples further on.

3.10 Maxwell Distribution

Let us consider for example the distribution of velocity in the ideal gas. We define the observable *density of particles having a certain value of velocity*:

$$\mathcal{F}(\mathbf{v}) = \sum_{i=1}^{N} \delta\left(\frac{\mathbf{p}_i}{m} - \mathbf{v}\right). \tag{3.54}$$

Let us calculate the average value of $\mathcal{F}$ over the accessible region of phase space:

$$\langle \mathcal{F}(\mathbf{v}) \rangle = \frac{1}{|\Gamma|} \int_\Gamma \prod_i \mathrm{d}\mathbf{p}_i \, \delta\left(\frac{\mathbf{p}_i}{m} - \mathbf{v}\right). \tag{3.55}$$

The right-hand-side integrals belong to the type that includes projections onto the axes of a $3N$-dimensional sphere and are discussed in appendix B. Using those results, we obtain

$$\langle \mathcal{F}(\mathbf{v}) \rangle = N \left(\frac{2\pi E}{3Nm}\right)^{-3/2} \exp\left[-\frac{v^2}{2E/(3Nm)}\right], \tag{3.56}$$

which is the Maxwell distribution.

3.11 The Ising Paramagnet

Let us now consider a simple model of paramagnet. We can disregard the molecules' translational degrees of freedom as irrelevant to our purposes, and assume that the molecules are arranged on a lattice. Let us also assume that each molecule possesses an elementary magnetic moment of magnitude μ that can be arranged (for quantum reasons) parallel or antiparallel to the direction of the z axis. Let us additionally assume that we can disregard all interactions between these magnetic moments and, as a matter of convenience, that we have rescaled the magnetic field h by the magnetic permeability of the vacuum μ_0. These assumptions result in the following definition of the model called the **Ising paramagnet**:

- Each molecule at site i of the lattice is associated with a variable σ_i, that can assume the values $+1$ and -1, depending on whether the corresponding magnetic moment is parallel or antiparallel to the z axis.

- The energy is due only to the interaction with the magnetic field h (which we assume is parallel to the z axis) and is equal to $-h\sum_{i=1}^{N}\mu\sigma_i$

In this case, the internal energy vanishes, and the only extensive variable is the total magnetization $M=\sum_{i=1}^{N}\mu\sigma_i$. The relation between entropy and the magnetic field h can be obtained from

$$\mathrm{d}E = T\mathrm{d}S + h\mathrm{d}M. \tag{3.57}$$

Taking the fact that $E=0=\text{const.}$ into account, we obtain

$$\frac{h}{T} = -\frac{\mathrm{d}S}{\mathrm{d}M}. \tag{3.58}$$

We must therefore evaluate the entropy as a function of M. Given M's value, the number of molecules L whose magnetic moment is oriented parallel to the z axis is determined:

$$L = \frac{1}{2}\left(N + \frac{M}{\mu}\right). \tag{3.59}$$

Strictly speaking, since L is an integer, M cannot assume an arbitrary value. Since the quantities we are examining vary slowly with L, however, it is allowable to consider that M is known with a small uncertainty, so as to permit the existence of a certain number of possible values of L. This uncertainty produces a (nonextensive) additive term within entropy, which we will not take into account.

The number of ways in which these L molecules can be chosen is given by

$$|\Gamma| = \frac{N!}{L!(N-L)!}. \tag{3.60}$$

By using Stirling's formula, therefore, we obtain

$$\begin{aligned} S &= k_B \ln|\Gamma| \\ &= -Nk_B\left[\frac{1}{2}\left(1-\frac{M}{N\mu}\right)\ln\frac{1}{2}\left(1-\frac{M}{N\mu}\right)+\frac{1}{2}\left(1+\frac{M}{N\mu}\right)\ln\frac{1}{2}\left(1+\frac{M}{N\mu}\right)\right]. \end{aligned} \tag{3.61}$$

By taking the derivative of S with respect to M, we obtain

$$\frac{h}{T} = Nk_B \cdot \frac{1}{\mu N}\tanh^{-1}\left(\frac{M}{\mu N}\right), \tag{3.62}$$

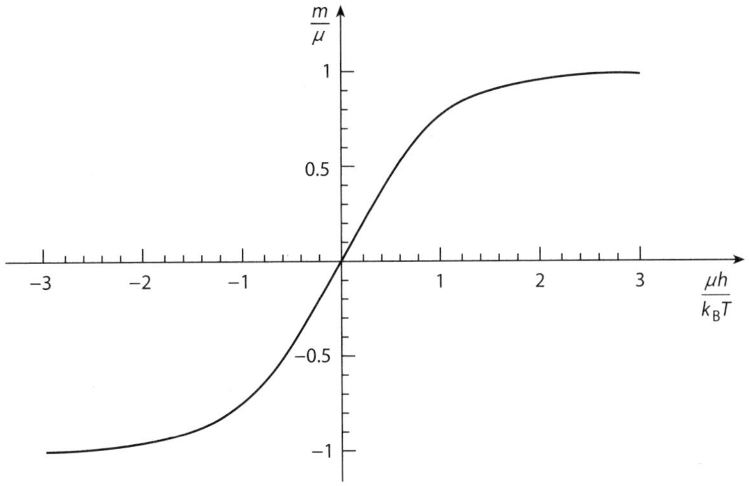

FIGURE 3.3. Equation of state of the Ising paramagnet: m/μ as a function of $\mu h/k_B T$.

from which the equation of state follows:

$$m = \frac{M}{N} = \mu \tanh\left(\frac{\mu h}{k_B T}\right). \tag{3.63}$$

Figure 3.3 shows the equation of state (3.63) m/μ as a function of $\mu h/k_B T$. One should note that for $|\mu h/k_B T| \ll 1$ one has $m \propto h$, while for $|\mu h/k_B T| \gg 1$, m tends to saturate at values of $m = \pm\mu$.

The specific heat with magnetic field held constant can be obtained by taking the derivative of the entropy with respect to T:

$$\frac{C_h}{T} = \left.\frac{\partial S}{\partial T}\right|_h = \frac{dS}{dM}\cdot\left.\frac{\partial M}{\partial T}\right|_h = \frac{1}{T} N k_B \left(\frac{\mu h}{k_B T}\right)^2 \left[1 - \tanh^2\left(\frac{\mu h}{k_B T}\right)\right]. \tag{3.64}$$

Figure 3.4 shows the behavior of C_h/Nk_B as a function of $k_B T/\mu h$.

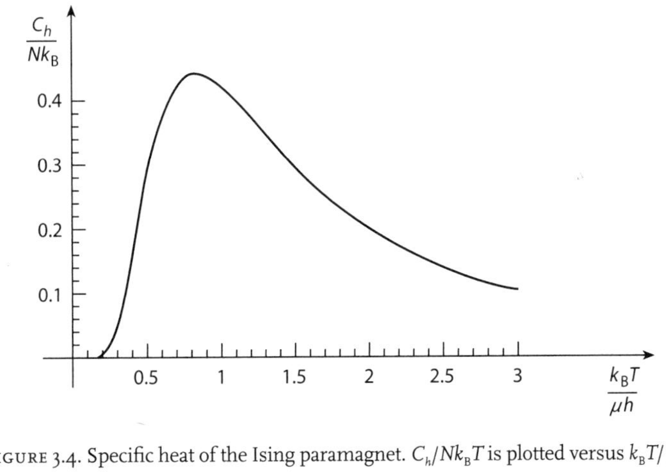

FIGURE 3.4. Specific heat of the Ising paramagnet. $C_h/Nk_B T$ is plotted versus $k_B T/\mu h$.

3.12 The Canonical Ensemble

The microcanonical ensemble that we have described in the preceding sections corresponds to the most direct description of thermodynamic systems, in which the thermodynamic equilibrium states are identified by the values of a collection of extensive thermodynamic variables. We have seen that it is possible, in thermodynamics, to provide a different but equivalent description of the same thermodynamic state by considering the values of some intensive variables, rather than those of their conjugate extensive ones, as fixed. In an analogous way, it is possible to define different ensembles in statistical mechanics that correspond to these representations of thermodynamic equilibrium states.

The most important case is that in which the intensive variable we want to introduce is temperature and the corresponding intensive one is internal energy. In this, case, we will suppose that we put the system we are interested in, and which we will denote by an S and call *system*, in contact with a much larger system, which we will denote by an R and which we will call *reservoir*. System and reservoir can freely exchange energy. At equilibrium, they will have the same temperature—since the reservoir is much larger than the system, however, we can assume that its temperature does not vary independently from the amount of energy it exchanges with the system. In this fashion, we have imposed a fixed temperature value on the system. We want to be able to calculate the average a^* of any observable A relative to the system S only. In order to do this, we observe that the global system's state (S∪R) is defined by the pair (x_S, x_R), where x_S is the system's representative point, and x_R is the reservoir's. Therefore, by applying the rule we derived in the preceding chapter, we obtain:

$$a^* = \frac{1}{|\Gamma|}\int_\Gamma \mathrm{d}x_S\,\mathrm{d}x_R\, A(x_S). \tag{3.65}$$

The region Γ is defined by the value E of total internal energy and by the values of the other extensive variables that we will not specify. Let us rewrite the integral as follows:

$$\langle A\rangle = \frac{1}{|\Gamma|}\int \mathrm{d}x_S\,\mathrm{d}x_R\, A(x_S)\,\delta\big(H^{(S)}(x_S) + H^{(R)}(x_R) - H^{(S)}(x_S)\big). \tag{3.66}$$

This integral can be rearranged in the following manner:

$$\langle A\rangle = \frac{1}{|\Gamma|}\int \mathrm{d}x_S\, A(x_S) \times \int \mathrm{d}x_R\,\delta\big(H^{(R)}(x_R) - (E - H^{(S)}(x_S))\big). \tag{3.67}$$

The integral in $\mathrm{d}x_R$ that appears in this equation is the volume of phase space accessible to the reservoir when its internal energy assumes the value $E - H^{(S)}$. This quantity can be expressed by the entropy:

$$\int \mathrm{d}x_R\,\delta\big(H^{(R)}(x_R) - (E - H^{(S)}(x_S))\big) \simeq \exp\left\{\frac{1}{k_B} S^{(R)}\big(E - H^{(S)}\big)\right\}. \tag{3.68}$$

When we remember that the system is much smaller than the reservoir, we can expand the exponent into a Taylor series:

$$\exp\left\{\frac{1}{k_B} S^{(R)}(E - H^{(S)}(x_S))\right\} \simeq \exp\left[\frac{1}{k_B} S^{(R)}(E)\right]\exp\left[-\frac{1}{k_B}\left.\frac{\partial S^{(R)}}{\partial E}\right|_E H^{(S)}(x_S)\right]. \tag{3.69}$$

At this point, we should note that $\partial S^{(R)}/\partial E^{(R)} = 1/T$, and that in our hypotheses, this quantity is fixed. We can then write:

$$\langle A \rangle = \frac{1}{Z}\int \mathrm{d}x_S \, A(x_S) \exp\left[-\frac{H^{(S)}(x_S)}{k_B T}\right]. \tag{3.70}$$

The normalization constant Z is called the **partition function** and is equal to:

$$Z = \int \mathrm{d}x_S \exp\left[-\frac{H^{(S)}(x_S)}{k_B T}\right]. \tag{3.71}$$

We see that we can now freely integrate over the system's entire phase space but that the different regions of the phase space are weighed proportionally to the factor

$$\exp\left[-\frac{H^{(S)}(x_S)}{k_B T}\right], \tag{3.72}$$

which is called the **Boltzmann factor**. We can interpret this factor as the probability density of a certain point in the phase space. In this way, we have defined a new probability distribution—in other words, a new ensemble, which is called the **canonical ensemble**. In order for these results to be compatible with those we obtained from the microcanonical ensemble, it is necessary that the contribution to A's average are dominated by the region of phase space in which the system's internal energy assumes the equilibrium value. In order to prove that this is how things stand, let us rearrange the integral according to the values of internal energy. From here on, we will omit the S indices, because we will be examining only the system itself explicitly. We therefore obtain

$$\langle A \rangle = \frac{1}{Z}\int \mathrm{d}E \int \mathrm{d}x \, \delta(H(x) - E)\, A(x) \exp\left(-\frac{E}{k_B T}\right). \tag{3.73}$$

The integral in x is proportional to the average microcanonical value of A, and we will denote it with $a^*(E)$, in the state in which the internal energy is equal to E. The proportionality factor is the corresponding volume of phase space, which we can express by means of the entropy. This integral can therefore be written

$$\langle A \rangle = \frac{1}{Z}\int \mathrm{d}E' a^*(E') \exp\left[-\frac{E' - TS(E')}{k_B T}\right]. \tag{3.74}$$

Since $a^*(E)$ varies slowly, the integral is determined by the behavior of the exponential, in which the extensive functions E' and S appear. If we evaluate it by means of the saddle point method (see appendix B), we notice that it is dominated by a region in which $E' \simeq E^*$, where E^* is the extremal of $E - TS(E)$ and therefore satisfies

$$\left.\frac{\partial S}{\partial E}\right|_{E^*} = \frac{1}{T}. \tag{3.75}$$

We can therefore identify E^* with the thermodynamic average value of the energy—in other words, with the internal energy. The contribution of E's other values rapidly goes to zero as soon as $E' - E^*$ is on the order of $N^{1/2}$. Considering the integral that defines the partition function, we have

$$Z \simeq \exp\left[-\frac{E^* - TS(E^*)}{k_B T}\right] = \exp\left(-\frac{F}{k_B T}\right). \tag{3.76}$$

It is therefore connected to the Helmholtz free energy expressed in the relevant natural variables. Knowledge of the partition function therefore allows us to obtain thermodynamics.

3.12.1 *Energy Fluctuations*

More particularly, the value of internal energy can be obtained by differentiating $\ln Z$ with respect to $\beta = 1/k_B T$:

$$\left.\frac{\partial \ln Z(\beta)}{\partial \beta}\right|_{\beta = 1/k_B T} = -\frac{1}{Z}\int \mathrm{d}x\, H(x) \exp\left[-\frac{H(x)}{k_B T}\right] = -\langle H(x) \rangle = -E. \tag{3.77}$$

The relation one obtains by taking this equation's derivative is even more interesting:

$$\left.\frac{\partial^2 \ln Z}{\partial \beta^2}\right|_{\beta = 1/k_B T} = \langle H^2(x) \rangle - \langle H(x) \rangle^2. \tag{3.78}$$

In this manner, we have obtained the variance of internal energy in the canonical ensemble. It is proportional to N. Moreover, since it cannot be negative, it follows that $\ln Z$ must be a concave function of its argument $1/k_B T$.

It is easy to relate this result to the stability properties of Helmholtz free energy. In fact, by interpreting the derivative differently, we obtain

$$\langle H^2 \rangle - \langle H \rangle^2 = -\frac{\partial E}{\partial (1/k_B T)} = k_B T^2 \frac{\partial E}{\partial T} = k_B T^2 C, \tag{3.79}$$

where C is the specific heat. We have thus related a *statistical* quantity (variance), connected with the fluctuations of internal energy, to a *thermodynamic* quantity (specific heat). The variance's positiveness implies the specific heat's positiveness. This is the first of a series of important relations that come under the general designation of **fluctuation-dissipation theorems**.

3.12.2 *The Ideal Gas in the Canonical Ensemble*

The advantage of using the canonical ensemble becomes obvious when we are dealing with systems whose Hamiltonian has the form

$$H = \sum_\alpha h(x_\alpha). \tag{3.80}$$

In other words, it decomposes into terms, each of which is a function of only one or a few degrees of freedom. In fact, while in the microcanonical ensemble, we must satisfy a *global* constraint in which all the degrees of freedom appear, in the canonical ensemble, we can split our integrals into factors in which only a few degrees of freedom appear at each time, and which can be evaluated explicitly.

The ideal gas obviously falls within this category. In fact, we have

$$H = \sum_{\alpha=1}^{N} \frac{p_\alpha^2}{2m} = \sum_{\alpha=1}^{N}\sum_{i=1}^{3} \frac{p_{i\alpha}^2}{2m}. \tag{3.81}$$

Let us calculate the partition function:

$$Z = \frac{1}{N!}\int \prod_{\alpha=1}^{N}\prod_{i=1}^{3}\left(\frac{\mathrm{d}r_{i\alpha}\,\mathrm{d}p_{i\alpha}}{h}\right)\exp\left(-\frac{1}{k_B T}\sum_{\alpha=1}^{N}\sum_{i=1}^{3}\frac{p_{i\alpha}^2}{2m}\right). \tag{3.82}$$

The integral over the coordinates yields V^N as in the microcanonical case, while the integral over momenta can be factorized into $3N$ identical integrals, which give

$$Z = \frac{V^N}{N!}\left[\int \frac{\mathrm{d}p}{h}\exp\left(-\frac{p^2}{2mk_B T}\right)\right]^{3N} = \frac{V^N}{N!}\left(\frac{2\pi m k_B T}{h^2}\right)^{3N/2}. \tag{3.83}$$

By taking the logarithm and using Stirling's formula, we obtain

$$\ln Z = \frac{F}{k_B T} = N\ln\left[\frac{eV}{N}\left(\frac{2\pi m k_B T}{h^2}\right)^{3/2}\right] = N\ln\left(\frac{ev}{\lambda^3}\right), \tag{3.84}$$

where $e = 2.178\ldots$, i.e., the basis of natural logarithms. In this equation, we introduced the volume per particle $v = V/N$ and the **thermal de Broglie wavelength**:

$$\lambda = \left(\frac{h^2}{2\pi m k_B T}\right)^{1/2}. \tag{3.85}$$

From this equation, it is easy to get free energy F and, by taking derivatives, obtain the equations of state:

$$F = -Nk_B T \ln\left(\frac{ev}{\lambda^3}\right), \tag{3.86}$$

$$S = -\left.\frac{\partial F}{\partial T}\right|_V = Nk_B \ln\left(\frac{ev}{\lambda^3}\right) - \frac{3}{2}Nk_B, \tag{3.87}$$

$$p = -\left.\frac{\partial F}{\partial V}\right|_T = \frac{Nk_B T}{V}. \tag{3.88}$$

From the definition $F = E - TS$ or, equivalently, by taking derivatives of $\ln Z$ with respect to $1/k_B T$, we obtain the value of internal energy:

$$E = \frac{3}{2}Nk_B T. \tag{3.89}$$

By taking further derivatives of this result, we obtain the specific heat at constant volume:

$$C_V = \left.\frac{\partial E}{\partial T}\right|_V = \frac{3}{2}Nk_B. \tag{3.90}$$

3.13 Generalized Ensembles

Let us now suppose that we want to keep fixed the value of the variable f_i, which is conjugated with the extensive variable X_i. In order to physically realize this constraint, we can follow the procedure in the preceding section. We imagine that we are putting our system S (the system) in contact with a system R (the reserve) that is much larger, with which it can exchange the quantity X_i, and that is characterized by the value f_i of the corresponding conjugate variable. Keeping in mind that f_i is determined by the relation

$$\left.\frac{\partial S^{(R)}}{\partial X_i^{(R)}}\right)_E = -\frac{f_i}{T}, \tag{3.91}$$

we can write

$$\begin{aligned}\langle A\rangle &= \frac{1}{|\Gamma|}\int_\Gamma \mathrm{d}x_R\,\mathrm{d}x_S\, A(x_S)\\ &= \frac{1}{|\Gamma|}\int \mathrm{d}x_S\, A(x_S)\int \mathrm{d}x_R\,\delta\left(X_i^{(R)}(x_R) + X_i^{(S)}(x_S) - X_i\right)\\ &= \frac{1}{|\Gamma|}\int \mathrm{d}x_S\, A(x_S)\exp\left[-\frac{1}{k_B}S^{(R)}\left(X_i - X_i^{(S)}(x_S)\right)\right]\\ &\simeq \frac{1}{Z}\int \mathrm{d}x_S\, A(x_S)\exp\left[-\frac{1}{k_B}\frac{\partial S^{(R)}}{\partial X_i}X_i^{(S)}(x_S)\right]\\ &= \frac{1}{Z}\int \mathrm{d}x_S\, A(x_S)\exp\left[\frac{f_i X_i^{(S)}(x_S)}{k_B T}\right].\end{aligned} \tag{3.92}$$

where Z is a normalization constant. We can now see that the different points in phase space are weighted proportionally to the factor $\exp(f_i X_i / k_B T)$. We have thus introduced a new ensemble, parametized by the intensive variable f_i.

It is a consequence of the principles of thermodynamics that practically the entire weight is concentrated in the region of phase space in which X_i assumes the equilibrium value. One can become convinced of this fact by again following the line of reasoning we just laid out.

The normalization constant Z can be easily interpreted in thermodynamic terms. We have (omitting the S index):

$$Z = \int \mathrm{d}x\,\exp\left[\frac{f_i X_i(x)}{k_B T}\right] = \int \mathrm{d}\xi \int \mathrm{d}x\,\delta\left(X_i(x) - \xi\right)\exp\left(\frac{f_i \xi}{k_B T}\right). \tag{3.93}$$

The integral over phase space in this expression is equal to the entropy expressed as a function of the extensive variable X_i (as well as of the other extensive thermodynamic variables). Therefore,

$$Z = \int \mathrm{d}\xi\,\exp\left[\frac{TS(\xi) + f_i\xi}{k_B T}\right]. \tag{3.94}$$

Since both S and X_i are extensive variables (and are therefore proportional to N), we can calculate this integral with the saddle point method, and we obtain

$$Z \simeq \exp\left[\frac{TS(X_i^*) + f_i X_i^*}{k_B T}\right]. \tag{3.95}$$

where X_i^* is the value of X_i for which the exponent becomes a maximum—in other words,

$$\left.\frac{\partial S}{\partial X_i}\right|_{X_i^*} = -\frac{f_i}{T}. \tag{3.96}$$

We therefore see that the exponential's argument is a Legendre transform of entropy. These thermodynamic functions are known as **Massieu functions**—we will see, however, that in the most useful cases we will end up once again with the usual Legendre internal energy transforms that we saw in chapter 2.

3.13.1 *Average Values and Fluctuations*

The normalization constant allows us to obtain the values for the quantities of interest by simple differentiation. One therefore has

$$\langle X_i \rangle = \frac{1}{Z}\int \mathrm{d}x\; X_i(x)\exp\left[\frac{f_i X_i(x)}{k_B T}\right] = \frac{\partial \ln Z}{\partial (f_i/k_B T)}. \tag{3.97}$$

By taking the derivative of this expression, we obtain

$$\frac{\partial^2 \ln Z}{\partial (f_i/k_B T)^2} = \langle X_i^2 \rangle - \langle X_i \rangle^2 = k_B T \frac{\partial \langle X_i \rangle}{\partial f_i}, \tag{3.98}$$

which is another fluctuation–dissipation relation. Let us note how the positivity of the variance is reflected in the positivity of X_i's derivative with respect to the conjugate variable f_i, which is in its turn a consequence of stability. In this latest relation, the ratio between the thermodynamic derivative $\partial X_i/\partial f_i$ and X_i's variance contains the Boltzmann constant—therefore, in principle, it is possible to measure Boltzmann's constant (and thus Avogadro's constant) if one finds a way to measure the fluctuations of the thermodynamic variables. This is in effect the core of the first methods used to measure Avogadro's constant.

3.13.2 *The Paramagnet*

In the case of the paramagnet, we want to proceed from the ensemble with a fixed magnetization M to that in which the magnetic field h is fixed. According to our rules, we obtain

$$\begin{aligned} Z &= \sum_{\{\sigma_i\}} \exp\left[\frac{h}{k_B T} M(\{\sigma_i\})\right] = \sum_{\{\sigma_i\}} \exp\left(\frac{1}{k_B T}\sum_{i=1}^{N} \mu h \sigma_i\right) \\ &= \prod_{i=1}^{N}\left[\sum_{\sigma_i} \exp\left(\frac{1}{k_B T}\sum_{i=1}^{N} \mu h \sigma_i\right)\right] = \left[\mathrm{e}^{\mu h/k_B T} + \mathrm{e}^{-\mu h/k_B T}\right]^N \\ &= \left[2\cosh\left(\frac{\mu h}{k_B T}\right)\right]^N. \end{aligned} \tag{3.99}$$

The corresponding free energy is given by

$$-\frac{F}{k_B T} = \ln Z = N \ln 2\cosh\left(\frac{\mu h}{k_B T}\right). \tag{3.100}$$

The average magnetization is given by

$$\langle M \rangle = \frac{\partial \ln Z}{\partial (h/k_{\mathrm{B}}T)} = N\mu \tanh\left(\frac{\mu h}{k_{\mathrm{B}}T}\right), \tag{3.101}$$

in agreement with what has been obtained from the microcanonical ensemble, while M's variance is given by

$$\langle M^2 \rangle - \langle M \rangle^2 = \frac{\partial^2 \ln Z}{\partial (h/k_{\mathrm{B}}T)^2} = N\mu^2 \left[1 - \tanh^2\left(\frac{\mu h}{k_{\mathrm{B}}T}\right)\right] = k_{\mathrm{B}}T\chi, \tag{3.102}$$

where χ is the isothermal susceptibility:

$$\chi = \left.\frac{\partial \langle M \rangle}{\partial h}\right)_T. \tag{3.103}$$

We have thus obtained another fluctuation-dissipation relation. From this relation, we can see that isothermal susceptibility at $h = 0$ is inversely proportional to absolute temperature—this result is known as **Curie's law** and is the basis of the methods used to measure low temperatures.

3.14 The *p-T* Ensemble

Let us now suppose that we fix the value of the temperature T and of the pressure p (and therefore let the internal energy and the volume fluctuate). The corresponding ensemble is called the ***p-T* ensemble** and is described by the following formula:

$$\langle A \rangle = \frac{1}{Z}\int \mathrm{d}x\, A(x) \exp\left[-\frac{E(x) + pV(x)}{k_{\mathrm{B}}T}\right]. \tag{3.104}$$

The partition function is connected to the Gibbs free energy:

$$\ln Z = -\frac{E - TS + pV}{k_{\mathrm{B}}T} = -\frac{G}{k_{\mathrm{B}}T}. \tag{3.105}$$

We can easily obtain the following relations:

$$E = \langle H \rangle = -\left.\frac{\partial \ln Z}{\partial (1/k_{\mathrm{B}}T)}\right)_{p/k_{\mathrm{B}}T}, \tag{3.106}$$

$$V = -\left.\frac{\partial \ln Z}{\partial (p/k_{\mathrm{B}}T)}\right)_{1/k_{\mathrm{B}}T}. \tag{3.107}$$

By taking further derivatives, we obtain:

$$\begin{aligned} \langle H^2 \rangle - \langle H \rangle^2 &= -\left.\frac{\partial E}{\partial \beta}\right)_{p/k_{\mathrm{B}}T} = k_{\mathrm{B}}T^2\left[\left.\frac{\partial E}{\partial T}\right)_p + \left.\frac{\partial E}{\partial p}\right)_T \left.\frac{\partial p}{\partial T}\right)_{p/k_{\mathrm{B}}T}\right] \\ &= k_{\mathrm{B}}T^2\left[\left.\frac{\partial E}{\partial T}\right)_V - \frac{1}{T}\left.\frac{\partial E}{\partial V}\right)_T^2 \left.\frac{\partial V}{\partial p}\right)_T\right] \geq k_{\mathrm{B}}T^2 C_V, \end{aligned} \tag{3.108}$$

$$\langle V^2\rangle - \langle V\rangle^2 = \frac{\partial^2 \ln Z}{\partial (p/k_BT)^2} = -k_BT\left.\frac{\partial V}{\partial p}\right)_T = k_BTV\kappa_{T,} \quad (3.109)$$

where

$$\kappa_T = -\frac{1}{V}\left.\frac{\partial V}{\partial p}\right|_T, \quad (3.110)$$

is the isothermal compressibility.

In order to obtain the expression in the second line, we can make use of the fact that, in the p-T ensemble, the fluctuations of internal energy are also induced by fluctuations in volume. A small fluctuation δV in volume therefore causes a corresponding fluctuation $\partial E/\partial V)_T \delta V$ in energy. As a consequence, the first term of the expression (3.108) corresponds to fluctuations in H at constant volume, while second term corresponds to the contribution of the volume fluctuations—in other words to $\partial E/\partial V)_T^2 (\langle V^2\rangle - \langle V\rangle^2)$, as can be seen by comparing with equation (3.109). Let us remark that these equations imply that fluctuations in energy are larger than in the canonical ensemble—we removed some constraints, and therefore the fluctuations can only increase.

Exercise 3.2 Making use of the Maxwell relations and the expression $p = -\partial F/\partial V)_T$ of pressure, prove the validity of the third equality in equation (3.108).

Last, we have the mixed derivatives:

$$\begin{aligned}\langle HV\rangle - \langle H\rangle\langle V\rangle &= \frac{\partial^2 \ln Z}{\partial (1/k_BT)\,\partial (p/k_BT)} \\ &= k_BT^2\left.\frac{\partial V}{\partial T}\right)_p = k_BT\left.\frac{\partial E}{\partial p}\right)_T.\end{aligned} \quad (3.111)$$

We see that the Maxwell relations are connected to the symmetry of correlations. Moreover, the fact that the correlation matrix is positive definite is related to the concavity of the partition function.

3.14.1 *One-Dimensional Hard Sphere Gas*

As an application of the p-T ensemble, we can solve our first problem with interacting particles. Let us suppose that we have some particles arranged on a straight line. Each particle is a hard sphere of diameter ℓ. Two particles that do not touch do not interact.

The particles' potential energy is therefore given by

$$U(\{r_i\}) = \frac{1}{2}\sum_{(i\neq j)} \phi(r_i - r_j), \quad (3.112)$$

where

$$\phi(r) = \begin{cases}\infty, & \text{if } |r| \le \ell,\\ 0, & \text{otherwise.}\end{cases} \quad (3.113)$$

We will see that calculating the partition function becomes very simple in the p-T ensemble.

Since the particles cannot exchange position, we can imagine numbering them consecutively starting with the one situated most to the left. If we additionally assume that we put the origin in correspondence of the particle with the number zero, then the system's volume can be expressed as the coordinate of the last particle. We thus obtain the following expression of the partition function:

$$Z = \int \prod_{i=1}^{N} \frac{dp_i}{h} \int_{\ell}^{\infty} dr_1 \int_{r_1+\ell}^{\infty} dr_2 \cdots \int_{r_{N-1}+\ell}^{\infty} dr_N \times \exp\left(-\frac{1}{k_B T} \sum_{i=1}^{N} \frac{p_i^2}{2m} - \frac{p r_N}{k_B T} \right). \tag{3.114}$$

One can evaluate first the integral over momentum, and this gives us as a contribution λ^{-N}, where λ is the thermal de Broglie wavelength. We still have to calculate the integral over the r's, which is also called the configuration integral. Let us change integration variables:

$$\xi_1 = r_1, \quad \xi_i = r_i - r_{i-1}, \quad i > 1. \tag{3.115}$$

The partition function can then be written

$$Z = \lambda^{-N} \int_{\ell}^{\infty} d\xi_1 \int_{\ell}^{\infty} d\xi_2 \cdots \int_{\ell}^{\infty} d\xi_N \exp\left(-\frac{p}{k_B T} \sum_{i=1}^{N} \xi_i \right). \tag{3.116}$$

The integral then factorizes into N identical integrals:

$$Z = \lambda^{-N} \left[\int_{\ell}^{\infty} d\xi \exp\left(-\frac{p\xi}{k_B T} \right) \right]^N = \left[\frac{1}{\lambda} \frac{k_B T}{p} \exp\left(-\frac{p\ell}{k_B T} \right) \right]^N. \tag{3.117}$$

By taking the derivative of $\ln Z$ with respect to $p/k_B T$, we obtain the equation of state:

$$V = -\frac{\partial \ln Z}{\partial (p/k_B T)} = N\left(\frac{k_B T}{p} + \ell \right). \tag{3.118}$$

In other words,

$$p(V - N\ell) = k_B T. \tag{3.119}$$

The internal energy is obtained by taking the derivative of $\ln Z$ with respect to $1/k_B T$ (at $p/k_B T$ constant) and is equal to $3k_B T/2$.

Exercise 3.3 Extend the preceding result to the case in which, in addition to the hard core, there is an interaction $u(r)$, between the particles, with a range between ℓ and 2ℓ (such that only close nearest-neighbor particles can interact).

3.15 The Grand Canonical Ensemble

The generalized ensemble in which the number of particles fluctuates and the chemical potential is fixed is called the **grand canonical ensemble**. We can directly write

$$\langle A \rangle = \frac{1}{Z} \sum_{N=1}^{\infty} \int \mathrm{d}x \, A(x) \exp\left[-\frac{H_N(x) - \mu N}{k_B T} \right]. \tag{3.120}$$

The grand canonical partition function is connected to the potential Ω:

$$\ln Z = -\frac{E - TS - \mu N}{k_B T} = -\frac{\Omega}{k_B T}. \tag{3.121}$$

As a reminder, let me add that it follows from Euler's equation that $\Omega = -pV$ for simple fluids (see equation [2.118]). The average values of H and N are obtained, as usual, by taking derivatives:

$$E = -\left.\frac{\partial \ln Z}{\partial\left(\frac{1}{k_B T}\right)}\right)_{\frac{\mu}{k_B T}}, \tag{3.122}$$

$$N = \left.\frac{\partial \ln Z}{\partial (\mu / k_B T)}\right)_{1/k_B T}. \tag{3.123}$$

By taking further derivatives, we obtain the expressions for the variances and the corresponding correlations. This derivation is left as an exercise.

Exercise 3.4 Consider a simple fluid, described by a grand canonical ensemble. Express the following quantities in terms of the grand canonical partition function:

$$\begin{aligned} \langle \Delta E^2 \rangle &= \langle E^2 \rangle - \langle E \rangle^2; \\ \langle \Delta N^2 \rangle &= \langle N^2 \rangle - \langle N \rangle^2; \\ \langle \Delta E \Delta N \rangle &= \langle EN \rangle - \langle \Delta E \rangle \langle \Delta N \rangle. \end{aligned}$$

Interpret the quantities thus obtained in terms of thermodynamic derivatives.

Exercise 3.5 Show that the energy fluctuations in the grand canonical ensemble are always greater than the corresponding fluctuations in the canonical ensemble:

$$\langle \Delta E^2 \rangle_{GC}(\mu) \geq \langle \Delta E^2 \rangle_{C}(N),$$

where N is the average value of the number of particles corresponding to μ.

Since one usually tries to obtain the thermodynamic properties of gases as a function of the number of particles N rather than of the thermodynamic potential μ, it is worthwhile to adopt the following strategy. One calculates the grand canonical partition function, thus obtaining $\ln Z\ (T, u, V)$. One then calculates the average number of particles N as a function of μ, T and V using equation (3.123). One inverts this relation, obtaining μ as a function of N, T, and V. This is always possible, since stability implies that N is a monotonically increasing function of μ when the other quantities are kept fixed. One performs the substitutions in equations (3.121) and (3.122) so as to obtain the equation of state for the pressure p and the internal energy E.

Let us calculate the properties of the ideal gas in the grand canonical ensemble:

$$\begin{aligned} Z &= \sum_{N=1}^{\infty} e^{\mu N/k_B T} \frac{1}{N!} \int \prod_{i=1}^{N} \left(\frac{d\mathbf{r}_i d\mathbf{p}_i}{h^3} \right) \exp\left(-\frac{1}{k_B T} \sum_{i=1}^{N} \frac{p_i^2}{2m} \right) \\ &= \sum_{N=1}^{\infty} z^N Z_N, \end{aligned} \tag{3.124}$$

where Z_N is a function of the canonical partition of a gas of N particles (calculated in section 3.12) and where we have also introduced the **fugacity** z:

$$z = \exp\left(\frac{\mu}{k_B T} \right). \tag{3.125}$$

From section 3.12.2 we obtain:

$$Z_N = \frac{V^N}{N! \lambda^{3N}}, \tag{3.126}$$

and therefore,

$$Z = \sum_{N=1}^{\infty} \frac{1}{N!} \left(\frac{zV}{\lambda^3} \right)^N = \exp\left(\frac{zV}{\lambda^3} \right). \tag{3.127}$$

By taking the logarithm, we obtain the equation of state expressed as a function of z:

$$\frac{pV}{k_B T} = \ln Z = \frac{zV}{\lambda^3}. \tag{3.128}$$

In order to express it in terms of the usual variables, let us calculate the average value of the number of particles as a function of fugacity:

$$N = \left. \frac{\partial \ln Z}{\partial (\mu / k_B T)} \right)_{1/k_B T} = z \frac{\partial \ln Z}{\partial z} = \frac{zV}{\lambda^3}. \tag{3.129}$$

From this equation, we can obtain z as a function of N and substitute it in the preceding equation, so as to obtain the usual equation of state. Obviously, in this case it is trivial, and we immediately obtain the equation of state for ideal gases.

3.16 The Gibbs Formula for the Entropy

In this section, we want to show how in any generalized ensemble, entropy can be expressed in terms of the probability density that defines the ensemble. In order to simplify the notations, let us assume that we are dealing with a discrete phase space.

Let us assume that our generalized ensemble is obtained by setting the value of the intensive variable f, which is conjugated with the extensive variable X. The expression of any observable's average is

$$\langle A \rangle = \frac{1}{Z} \sum_{\nu} A_{\nu} \exp\left(\frac{f X_{\nu}}{k_B T} \right), \tag{3.130}$$

where the partition function Z is given by

$$Z = \sum_{\nu} \exp\left(\frac{fX_{\nu}}{k_{\mathrm{B}}T}\right), \tag{3.131}$$

and the sum runs over all of the system's ν microstates. As we saw, the relation between the partition function and the thermodynamic potentials is

$$\ln Z = \frac{TS + f\langle X\rangle}{k_{\mathrm{B}}T}, \tag{3.132}$$

where we make explicit the fact that the thermodynamic value of the extensive variable X is given by its average.

A microstate's ν probability is therefore given by:

$$p_{\nu} = \frac{1}{Z}\exp\left(\frac{fX_{\nu}}{k_{\mathrm{B}}T}\right). \tag{3.133}$$

Let us evaluate the logarithm of this expression:

$$\ln p_{\nu} = \frac{fX_{\nu}}{k_{\mathrm{B}}T} - \ln Z = \frac{1}{k_{\mathrm{B}}T}(fX_{\nu} - TS - f\langle X\rangle). \tag{3.134}$$

If we calculate the average of both sides of this equation, we obtain

$$\langle \ln p_{\nu}\rangle = \sum_{\nu} p_{\nu}\ln p_{\nu} = \frac{1}{k_{\mathrm{B}}T}(f\langle X\rangle - TS - f\langle X\rangle) = -\frac{S}{k_{\mathrm{B}}}. \tag{3.135}$$

We have thus obtained the **Gibbs formula for the entropy**:

$$S = -k_{\mathrm{B}}\sum_{\nu} p_{\nu}\ln p_{\nu}. \tag{3.136}$$

This elegant equation is at the origin of modern information theory. One of the most obvious reasons of its interest is the fact that, although it was derived only in the context of generalized ensembles, its right-hand side is defined for any probability distribution. In effect, the quantity defined as *entropy* in information theory is defined precisely by this expression. The temptation to use it to define the thermodynamic entropy of any probability distribution over the phase space is great, even outside thermodynamic equilibrium. One has to resist this temptation as much as possible. In fact, the probability distribution in phase space is only a conceptual instrument that we introduced to calculate the values of the thermodynamic observables, and not a physically observable quantity.

In information theory, it is in principle possible to sample the probability distribution p_{ν} and therefore provide a reliable estimate of S. In our case, the phase space is simply too vast to allow us to perform this program even for the simplest system. We must therefore estimate entropy by resorting to some hypotheses about the independence of degrees of freedom and so on that are valid only if the state we want to describe is close to thermodynamic equilibrium.

Another objection against the generalized use of this relation is that the quantity S obtained in this way is a constant of the motion for an isolated system, with an arbitrary initial probability distribution. This is a consequence of Liouville's theorem. This is of course at variance with the behavior of thermodynamic entropy, which, as we know, increases as the system approaches equilibrium.

3.17 Variational Derivation of the Ensembles

One can use the Gibbs formula with the necessary care, being careful that we consider only thermodynamic equilibrium and states close to it. In this section, we show how the ensembles' probability distributions can be derived from a variational principle over the S defined by the Gibbs formula.

Let us prove the variational principle for the microcanonical ensemble. Let us define the functional $\mathcal{S}$ of the probability distribution p_ν:

$$\mathcal{S} = -k_{\mathrm{B}} \sum_{\nu} p_\nu \ln p_\nu. \tag{3.137}$$

This functional is maximum for the (uniform) microcanonical distribution among all the distributions that are supported (in other words, that do not vanish) only within the accessible region Γ of the phase space. In this fashion, we can relate the choice of distribution associated with the ensembles to a principle of maximum entropy.

This is a typical problem of constrained extrema. The constraint is due to probability's normalization condition:

$$\sum_{\nu \in \Gamma} p_\nu = 1. \tag{3.138}$$

By introducing the Lagrange multiplier α, we look for the functional's unconstrained extremum:

$$\Phi = \mathcal{S} - \alpha\left(\sum_{\nu} p_\nu - 1\right). \tag{3.139}$$

We obtain

$$0 = \frac{\partial \Phi}{\partial p_\nu} = -k_{\mathrm{B}}(\ln p_\nu + 1) - \alpha. \tag{3.140}$$

The solution is obviously

$$p_\nu = \text{cost} = \frac{1}{|\Gamma|} \tag{3.141}$$

(the constant's value follows from the normalization condition), which corresponds to

$$\mathcal{S} = k_{\mathrm{B}} \ln |\Gamma|. \tag{3.142}$$

If we consider a generalized ensemble defined by the extensive variable X, we must take the following constraint into account

$$\langle X_\nu \rangle = \sum_{\nu} p_\nu X_\nu = X. \tag{3.143}$$

Let us introduce the Lagrange multiplier λ and look for the extremal of

$$\Phi = \mathcal{S} - \alpha\left(\sum_\nu p_\nu\right) - \lambda\left(\sum_\nu p_\nu X_\nu\right). \tag{3.144}$$

(We omitted some inessential additive constants.) We obtain

$$\ln p_\nu = \text{const.} - \lambda X_\nu, \tag{3.145}$$

which corresponds to the distributions of the generalized ensembles. It is easy to see that, in the case in which X is the internal energy, the corresponding Lagrange multiplier (usually denoted by β) is equal to $1/k_B T$.

3.18 Fluctuations of Uncorrelated Particles

In this section, we attempt to explain the universality of the equation of state of ideal gases by showing that it is a consequence of some very simple hypotheses about the particle distribution.

Let us suppose in fact that the phase space can be considered a product of single-particle states and that the global system's state can be defined by giving the **occupation number** for each single-particle's state. For example, in the ideal quantum gas, the global system's state is determined when the number of particles that possess a specific (quantized) value of momentum has been determined. Let us additionally assume that the occupation numbers of the single-particle states are independent random variables (not necessarily identically distributed!) and that its average is small (so that for every state i, it is possible to disregard $\langle n_i \rangle^2$ with respect to $\langle n_i \rangle$). From these hypotheses, it follows that the system obeys the equation of state of ideal gases.

Let us note that nothing is said in these hypotheses about the nature of single-particle states—we can assume that they correspond to larger or smaller cells in configuration space, or in phase space, or in momentum space. One cannot even say with absolute certainty that the occupation numbers define the system's microstate **univocally**—there is nothing to prevent different microstates from corresponding to the same occupation numbers, and so on.

Let us consider the grand canonical ensemble and express the total number of particles by means of the n_i:

$$N = \sum_i \langle n_i \rangle. \tag{3.146}$$

We can now calculate N's variance:

$$\langle N^2 \rangle - \langle N \rangle^2 = \sum_{ij} \left(\langle n_i n_j \rangle - \langle n_i \rangle \langle n_j \rangle\right). \tag{3.147}$$

In this sum, all the terms with $i \neq j$ vanish because of the hypothesis about the n_i's independence. On the other hand, if $\langle n_i \rangle$ is small, the greatest contributions are those for which n_i's value is 0 or 1, and therefore $n_i^2 = n_i$. We thus obtain

$$\langle N^2 \rangle - \langle N \rangle^2 = \sum_i \left(\langle n_i \rangle - \langle n_i \rangle^2\right). \tag{3.148}$$

If we can neglect $\langle n_i \rangle^2$ with respect to $\langle n_i \rangle$, this relation gives us

$$\langle N^2 \rangle - \langle N \rangle^2 = \sum_i \langle n_i \rangle = N. \tag{3.149}$$

The variance N is therefore equal to its average. (Here, we recognize a property of the Poisson distribution.) This relation implies the state equation for ideal gases. In fact, one has

$$\begin{aligned} \langle N^2 \rangle - \langle N \rangle^2 &= \left.\frac{\partial^2 \ln Z}{\partial (\mu/k_B T)^2}\right)_{1/k_B T} \\ &= \left.\frac{\partial N}{\partial (\mu/k_B T)}\right)_{1/k_B T} = V k_B T \left.\frac{\partial (N/V)}{\partial \mu}\right)_{T,V}. \end{aligned} \tag{3.150}$$

Then, by introducing the volume per particle $v = V/N$, one gets

$$\left.\frac{\partial \mu}{\partial v}\right)_{T,V} = -v^2 \left.\frac{\partial \mu}{\partial (N/V)}\right)_{T,V} = \frac{V k_B T}{N} = v k_B T. \tag{3.151}$$

As a consequence of the Gibbs-Duhem equation, on the other hand, we have

$$\left.\frac{\partial \mu}{\partial v}\right)_{T,V} = -p. \tag{3.152}$$

Therefore,

$$p = \frac{k_B T}{v} = \frac{N k_B T}{V}. \tag{3.153}$$

Recommended Reading

I have followed the point of view laid forth in S.-K. Ma, *Statistical Mechanics,* Singapore: World Scientific, 1985, which is close to that in L. D. Landau and E. M. Lifshitz, *Statistical Physics,* 3rd ed., Part I, Oxford, UK: Pergamon Press, 1980. Another point of view can be found in R. Balian, *From Microphysics to Macrophysics: Methods and Applications of Statistical Physics* (Berlin/New York: Springer, 1991).

In this volume, I have chosen not to discuss the many subtle problems posed by the foundations of statistical mechanics. An interesting discussion can be found in chapter 26 of Ma's text, *Statistical Mechanics.* Another point of view is expressed in G. Gallavotti, *Statistical Mechanics: A Short Treatise,* Berlin: Springer, 1999.

4 Interaction-Free Systems

> The simplest possible variable is one that can have two values. (If there is only one value, no variation is possible.)
>
> —Shang-Keng Ma

In this chapter, we will study some simple systems in which the partition function can be expressed in terms of integrals over a small number of variables. This is possible when, in some representation, the Hamiltonian can be written as a sum of independent functions, where each one depends on a small number of arguments. In this case, by means of a suitable choice of ensemble, the partition function factorizes into many independent integrals that can be calculated explicitly. We have already found this situation in the case of the ideal gas.

4.1 Harmonic Oscillators

In this section, we consider a system composed of N one-dimensional independent harmonic oscillators. The system's microstate is defined by the collection of all r_i coordinates (which are defined so as to vanish at the equilibrium point) and the conjugate momenta p_i. The Hamiltonian has the following expression:

$$H = \sum_{i=1}^{N}\left(\frac{p_i^2}{2m} + \frac{1}{2}\kappa r_i^2\right), \tag{4.1}$$

where m is the particles' mass, and κ is Hooke's constant, connected to the angular frequency of oscillations ω_0 by

$$\omega_0^2 = \frac{\kappa}{m}. \tag{4.2}$$

The partition function is given by:

$$Z = \int \prod_{i=1}^{N}\left(\frac{\mathrm{d}r_i \mathrm{d}p_i}{h}\right) \exp\left[-(k_B T)^{-1} \sum_{i=1}^{N}\left(\frac{p_i^2}{2m} + \frac{1}{2} m\omega_0^2 r_i^2\right)\right]$$

$$= \prod_{i=1}^{N} \int \left(\frac{\mathrm{d}r_i \mathrm{d}p_i}{h} \right) \exp\left[-(k_B T)^{-1} \left(\frac{p_i^2}{2m} + \frac{1}{2} m \omega_0^2 r_i^2 \right) \right]$$

$$= \left[\left(\frac{2\pi m k_B T}{h^2} \right)^{1/2} \left(\frac{2\pi k_B T}{m \omega_0^2} \right)^{1/2} \right]^N$$

$$= \left[\frac{k_B T}{\hbar \omega_0} \right]^N. \tag{4.3}$$

We can therefore easily obtain the internal energy and specific heat:

$$E = -\frac{\partial \ln Z}{\partial (1/k_B T)} = N k_B T, \tag{4.4}$$

$$C = \frac{\partial E}{\partial T} = N k_B. \tag{4.5}$$

4.1.1 *The Equipartition Theorem*

It is interesting to note the "universal" aspect of these results relative to E and C—each specific quantity in the system, like the angular frequency ω_0, has disappeared. This is a special case of a more general result, called the **equipartition theorem**: when the systems' Hamiltonian depends on a degree of freedom (coordinate or momentum) only via a quadratic term, the average of this term in the canonical ensemble is given by $k_B T/2$.

Let us assume that we can divide the N degrees of freedom of a classic Hamiltonian H into two classes: $x_1, \ldots, x_r$ and $x_{r+1}, \ldots, x_N$, and write H as the sum of two terms:

$$H = H_0(x_1, \ldots, x_r, x_{r+1}, \ldots, x_N) + H_1(x_{r+1}, \ldots, x_N). \tag{4.6}$$

Let us additionally assume that the H_0 is quadratic in the $x_1, \ldots, x_r$, and is positive definitive, while the H_1 does not depend on them. In these hypotheses, H_0 satisfies Euler's equation:

$$\sum_{i=1}^{r} x_i \frac{\partial H_0}{\partial x_i} = 2H_0. \tag{4.7}$$

Then, the average of H_0 is equal to

$$\langle H_0 \rangle = \frac{r}{2} k_B T. \tag{4.8}$$

In order to prove this result, let us express $\langle H_0 \rangle$ in the canonical ensemble. By introducing the partition function $Z = \int \prod_{i=1}^{N} \mathrm{d}x_i \exp(-H/k_B T)$, we obtain

$$\begin{aligned} \langle H_0 \rangle &= \frac{1}{Z} \int \prod_{i=1}^{r} \mathrm{d}x_i \int \prod_{j=r+1}^{N} \mathrm{d}x_j\, H_0 \exp\left(-\frac{H_0 + H_1}{k_B T} \right) \\ &= \frac{1}{Z} \int \prod_{i=1}^{r} \mathrm{d}x_i \int \prod_{j=r+1}^{N} \mathrm{d}x_j \left(\frac{1}{2} \sum_{\ell=1}^{r} x_\ell \frac{\partial H_0}{\partial x_\ell} \right) \exp\left(-\frac{H_0 + H_1}{k_B T} \right) \\ &= -\frac{k_B T}{2} \sum_{\ell=1}^{r} \frac{1}{Z} \int \prod_{i=1}^{r} \mathrm{d}x_i \int \prod_{j=r+1}^{N} \mathrm{d}x_j\, x_\ell \frac{\partial}{\partial x_\ell} \exp\left(-\frac{H_0 + H_1}{k_B T} \right). \end{aligned} \tag{4.9}$$

In this fashion, we obtain r terms, each of which can be integrated by parts. Each of them has the form

$$-\frac{k_B T}{2}\int\prod_{i=1}^{r}\mathrm{d}x_i\int\prod_{j=r+1}^{N}\mathrm{d}x_j x_\ell\frac{\partial}{\partial x_\ell}\exp\left(-\frac{H_0+H_1}{k_B T}\right).$$

The integrated term contains a factor $\exp(-H_0/k_B T)$ evaluated for $x_\ell \to \pm\infty$, which vanishes following the hypothesis that the quadratic form H_0 is positive definite. Therefore, we are left with the differentiated term, which yields $k_B T/2$ for each degree of freedom.

This theorem's importance lies in the fact that any classic stable Hamiltonian is quadratic in the momenta, and "almost quadratic" in the coordinates' deviations from the point of equilibrium (in other words, of the potential energy's minimum). Therefore, if the internal energy is small enough (in other words, for small temperature values) we can expect that it be equal to $k_B T/2$ per degree of freedom, up to an additive constant, and that the specific heat therefore be equal to $k/2_B$ per degree of freedom. The results obtained for the ideal gas and the harmonic oscillators are only special cases of this result.

4.1.2 *Quantum Harmonic Oscillators*

The situation changes if we consider quantum systems. We know from quantum mechanics that it is possible to specify a base $|n\rangle$ in the Hilbert space of the harmonic oscillator, in which each state $|n\rangle$ is an eigenstate of the Hamiltonian, belonging to the eigenvalue $(n+1/2)\hbar\omega_0$, $n=0,1,2,\ldots$. Since we are dealing solely with the Hamiltonian operator, our problem reduces to a classical one, in which these states are weighted with the Boltzmann factor. The situation would be different if we had to deal with several operators that do not mutually commute.

Let us consider a system of N identical and independent quantum harmonic oscillators. In this case, the partition function can be expressed as follows:

$$\begin{aligned} Z &= \sum_{n_1=0}^{\infty}\sum_{n_2=0}^{\infty}\cdots\sum_{n_N=0}^{\infty}\exp\left[-\sum_{i=1}^{N}\frac{(n_i+\frac{1}{2})\hbar\omega_0}{k_B T}\right] \\ &= \left[\sum_{n=0}^{\infty}\exp\left(-\frac{(n+\frac{1}{2})\hbar\omega_0}{k_B T}\right)\right]^N = \zeta^N. \end{aligned} \tag{4.10}$$

Apart from a factor, ζ is a geometric series:

$$\begin{aligned} \zeta &= \exp\left(-\frac{\hbar\omega_0}{2k_B T}\right)\sum_{n=0}^{\infty}\left[\exp\left(-\frac{\hbar\omega_0}{k_B T}\right)\right]^n \\ &= \exp\left(-\frac{\hbar\omega_0}{2k_B T}\right)\left[1-\exp\left(-\frac{\hbar\omega_0}{k_B T}\right)\right]^{-1}. \end{aligned} \tag{4.11}$$

Therefore,

$$\begin{aligned} E &= -\left.\frac{\partial \ln Z}{\partial \beta}\right|_{\beta=1/k_BT} = -N\left.\frac{\partial \ln \zeta}{\partial \beta}\right|_{\beta=1/k_BT} \\ &= N\frac{\hbar\omega_0}{2} + N\hbar\omega_0\left[\exp\left(\frac{\hbar\omega_0}{k_BT}\right) - 1\right]^{-1}. \end{aligned} \tag{4.12}$$

The first term represents the **zero-point energy**—in other words, the energy the harmonic oscillators possess even in their ground states as a result of the uncertainty principle. The second term represents the dependence on temperature. Taking the derivative with respect to T, we obtain the specific heat:

$$C = \frac{\partial E}{\partial T} = Nk_B\left(\frac{\hbar\omega_0}{k_BT}\right)^2 \exp\left(\frac{\hbar\omega_0}{k_BT}\right)\left[\exp\left(\frac{\hbar\omega_0}{k_BT}\right) - 1\right]^{-2}. \tag{4.13}$$

In figure 4.1, the specific heat per oscillator is shown (expressed in k_B units) as a function of the ratio $k_BT/\hbar\omega_0$. We see that while for $T \to \infty$ one recovers the classic value $C = Nk_B$, the specific heat rapidly tends to zero as soon as the temperature falls below a characteristic temperature $T_E = \hbar\omega_0/k_B$.

The preceding derivation is due to Einstein, whose intent was to explain the fact that in certain solids (including diamond), the specific heat is much lower than the expected value, equal to $3Nk_B$—in other words, about 6 calories per mole (**Dulong-Petit law**). He therefore supposed that in a solid, atoms would oscillate around their equilibrium point with a characteristic angular frequency ω_0. Below a temperature T_E (subsequently known as **Einstein temperature**), the specific heat is much smaller than the one we could expect on the basis of classical statistical mechanics. Since the characteristic frequency is presumably greater in the harder and lighter solids ($\omega_0 = \sqrt{\kappa/m}$), this effect is more evident in them at room temperature.

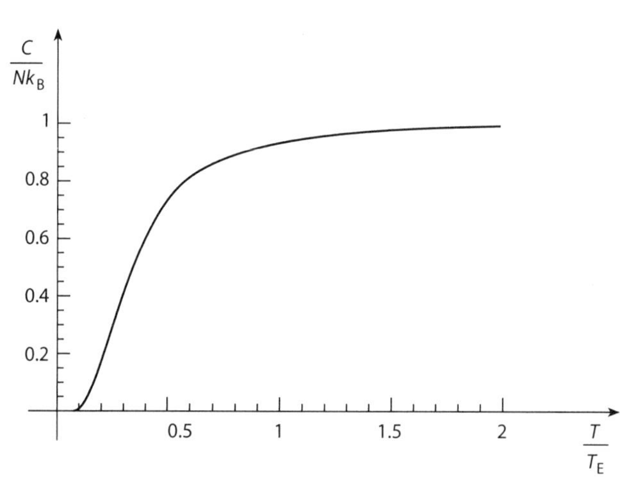

FIGURE 4.1. Specific heat in the Einstein model, in k_B units, as a function of T/T_E, where $T_E = \hbar\omega_0/k_B$.

4.2 Photons and Phonons

4.2.1 *Black Body Radiation: Photons*

A physical realization of a perfectly black body is a cavity in a metal body filled with electromagnetic radiation, in equilibrium with the body at a given temperature T. In the present section, we will derive Planck's law of radiation for such a system, by building on the analogy with a system of quantum oscillators. Let us consider a cubic cavity of side length L: its volume V is therefore equal to L^3. To simplify the discussion, we impose periodic boundary conditions. Therefore, the oscillations of the electromagnetic field in the cavity must be spatially periodic, with period equal to L in each direction. As a consequence, these oscillations can be represented by a Fourier series, as a combination of elementary oscillations, each characterized by a wave vector of the form $\boldsymbol{k} = (2\pi/L)\,(\ell_x, \ell_y, \ell_z)$, where the ℓ are integers. In addition, each oscillation is characterized by a polarization state: we choose as base states those of circular polarization, which are identified by a scalar quantity $\alpha = \pm 1$, according to whether the field rotates counterclockwise (+) or clockwise (−) as seen from $\boldsymbol{k}$. Thus $(\boldsymbol{k}, \alpha)$ identifies an oscillation mode, with a characteristic (angular) frequency $\omega(\boldsymbol{k}, \alpha) = c\,|\boldsymbol{k}|$, where c is the speed of light.

We can represent this system's microstate by assigning each harmonic oscillator, identified by its wave vector $\boldsymbol{k}$ and by its polarization state α, the corresponding quantum occupation number $n(\boldsymbol{k}, \alpha) = 0, 1, 2, \ldots$.

The energy of this microstate is given by:

$$H(\{n(\boldsymbol{k}, \alpha)\}) = \sum_{\boldsymbol{k},\alpha} \left[n(\boldsymbol{k}, \alpha) + \frac{1}{2} \right] \hbar\omega(\boldsymbol{k}, \alpha). \tag{4.14}$$

We can now calculate the partition function in the canonical ensemble:

$$Z = \prod_{\{\boldsymbol{k},\alpha\}} \sum_{n(\boldsymbol{k},\alpha)} \exp\left\{ -(k_{\mathrm{B}}T)^{-1} \sum_{\boldsymbol{k},\alpha} \left[n(\boldsymbol{k}, \alpha) + \frac{1}{2} \right] \hbar\omega(\boldsymbol{k}, \alpha) \right\}. \tag{4.15}$$

As in the preceding section, we can break up this sum into factors, in each of which only a single occupation number n appears—the resulting sums are geometric series that can be easily calculated. We thus obtain the following expression of $\ln Z$:

$$\ln Z = \sum_{\boldsymbol{k},\alpha} \left\{ -\frac{\hbar\omega}{2k_{\mathrm{B}}T} - \ln\left[1 - \exp\left(-\frac{\hbar\omega}{k_{\mathrm{B}}T} \right) \right] \right\}. \tag{4.16}$$

In order to calculate this sum, we regroup those terms that correspond to values close to the frequency ω

$$\ln Z = \int_0^\infty \mathrm{d}\omega \sum_{\boldsymbol{k},\alpha} \delta(\omega - \omega(\boldsymbol{k}, \alpha)) \left\{ -\frac{\hbar\omega}{2k_{\mathrm{B}}T} - \ln\left[1 - \exp\left(-\frac{\hbar\omega}{k_{\mathrm{B}}T} \right) \right] \right\}. \tag{4.17}$$

In wave-vector space, there are two states with different polarizations for each small cube of volume $(2\pi/L)^3 = (2\pi)^3/V$. Making use of the relation between ω and $\boldsymbol{k}$, we obtain

$$\sum_{\boldsymbol{k},\alpha} \delta(\omega - \omega(\boldsymbol{k}, \alpha)) = 2\frac{V}{(2\pi)^3} \int \mathrm{d}k\, 4\pi k^2 \delta(\omega - ck) = 2\frac{V}{(2\pi c)^3} 4\pi\omega^2. \tag{4.18}$$

By substituting in the preceding equation, we obtain

$$\ln Z = \frac{V}{(\pi c)^3} \int_0^\infty d\omega\, \omega^2 \left\{ -\frac{\hbar\omega}{2k_B T} - \ln\left[1 - \exp\left(-\frac{\hbar\omega}{k_B T}\right)\right]\right\}. \tag{4.19}$$

The first term (which is infinite) corresponds to the zero-point energy and does not interest us for the moment. (It is, however, at the origin of the so-called Casimir effect—in other words, of the attractive forces, that arise among the conductive walls delimiting a cavity.)

If we consider only the effect of second term, we obtain the internal energy:

$$\begin{aligned} E &= \frac{V}{(\pi c)^3} \int_0^\infty d\omega\, \omega^3 \left[\exp\left(\frac{\hbar\omega}{k_B T}\right) - 1\right]^{-1} = \frac{V}{(\hbar\pi c)^3} (k_B T)^4 \int_0^\infty dx \frac{x^3}{e^x - 1} \\ &= \frac{V}{(\hbar\pi c)^3} (k_B T)^4 6\zeta_R(4), \end{aligned} \tag{4.20}$$

where $\zeta_R(s)$ is the Riemann zeta function, defined by

$$\zeta_R(s) = \sum_{n=1}^\infty \frac{1}{n^s}. \tag{4.21}$$

More specifically, one has

$$\zeta_R(4) = \frac{\pi^4}{90}. \tag{4.22}$$

This expression can be rewritten in a more transparent manner, by remembering that the characteristic energy ε of the frequency mode ω is equal to $\hbar\omega$:

$$E = \frac{V}{(2\pi)^3} \sum_\alpha \int d^3k\, \hbar\omega(\mathbf{k}, \alpha) \left\{\exp\left[\frac{\hbar\omega(\mathbf{k}, \alpha)}{k_B T}\right] - 1\right\}^{-1}. \tag{4.23}$$

We can see from this expression that for each wave vector's "mode" and fixed polarization status, there are on average $f_-(\varepsilon)$ "quanta," where

$$f_-(\epsilon) = \left[\exp\left(\frac{\epsilon}{k_B T}\right) - 1\right]^{-1}, \tag{4.24}$$

is the **Bose factor.** Its behavior is plotted in figure 4.2. This expression admits two classic limits: When $\varepsilon \ll k_B T$, each mode's contribution to internal energy is equal to $k_B T$ in accordance with the equipartition principle (each mode behaves like a classical harmonic oscillator). When $\varepsilon \gg k_B T$, the probability of having an energy quantum ε is proportional to the Boltzmann factor $\exp(-\epsilon/k_B T)$. The quanta then behave like particles. This last observation was made by Einstein on the basis of the black body radiation law and led him to introduce the concept of the **photon.**

It is useful to calculate the power $W(\omega)d\omega$ of the radiation emitted within a certain frequency interval. One must keep in mind that this radiation escapes the cavity, through a small hole for example, at the speed of light c.

To evaluate the power emitted by the cavity through a hole of area S, in a frequency interval of size $d\omega$, we must calculate the energy density $\mathcal{E}(\omega)$ per unit volume of all the oscillation modes whose frequency lies in this interval, and whose wave vectors are directed

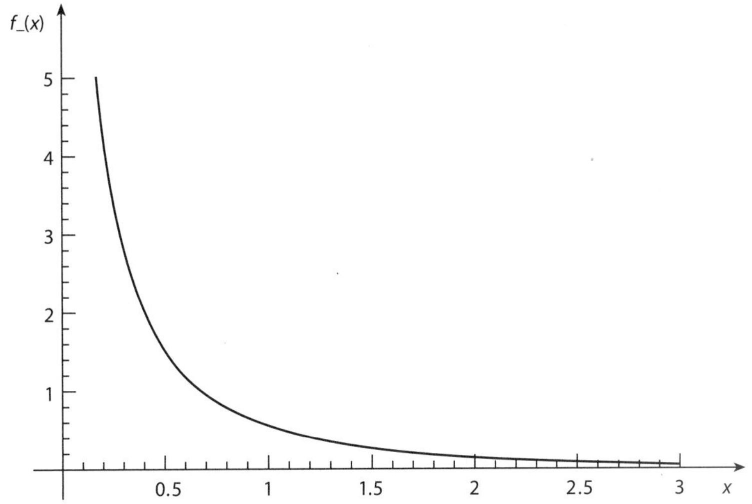

FIGURE 4.2. Bose factor: $f_{-}(x) = (e^{x} - 1)^{-1}$.

toward the exterior of the cavity. The wave vector $\boldsymbol{k}$ of these modes satisfies the relation $\omega \leq c|\boldsymbol{k}| \leq \omega + d\omega$. Summing over the polarization modes a and over the directions of $\boldsymbol{k}$ toward the exterior, we obtain that the power emitted per unit frequency and per unit area of the hole is given by

$$W(\omega) = \frac{\hbar}{4\pi^2 c^2} \frac{\omega^3}{\exp(\hbar\omega/k_B T) - 1}, \tag{4.25}$$

which is Planck's formula for the black body radiation.

In figure 4.3, the intensity W of the black body radiation is plotted (in units $(k_B T)^3/(hc)^2$) as a function of the frequency ω (in units" $k_B T/\hbar$). One should remark the classical regime

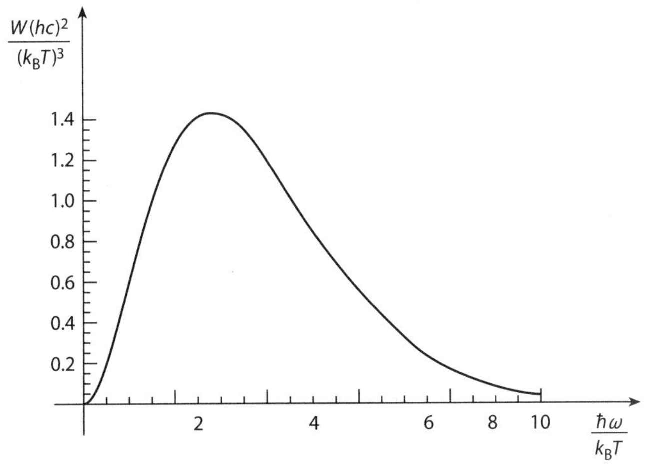

FIGURE 4.3. Planck's law for the black body radiation.

at low frequencies ($\hbar\omega \ll k_B T$), where $W \sim \omega^2$, and the "photonic" regime at high frequencies, where $W \sim e^{-\hbar\omega/k_B T}$.

Exercise 4.1 (Stefan-Boltzmann Law) By integrating equation (4.25), one obtains the total power emitted by a body at temperature T per unit surface:

$$W_{\text{tot}} = \int_0^\infty d\omega\, W(\omega) = \kappa T^4,$$

where the constant κ is equal to

$$\kappa = \frac{2\pi^5 k_B^4}{15c^2 h^3} = 5.67\, 10^{-8}\,\text{Jm}^{-2}\text{s}^{-1}\text{K}^4.$$

Show that this result (except for the value of the constant κ) follows from thermodynamics and the following hypotheses:

1. The internal energy of the black body radiation is proportional to the system's volume: the energy's density depends only on the temperature:

   $$E = \epsilon(T)\, V.$$

2. The pressure p is equal to $1/3$ of the energy's density:

   $$p = \frac{1}{3}\epsilon(T).$$

This hypothesis can be justified by taking account of the fact that if the distribution of light is isotropic, only one-third of the light present contributes to the pressure on any given element of the surface.

First obtain the expression of $\varepsilon(T)$, and from it, derive the value of W_{tot}.

4.2.2 *Phonons and the Specific Heat of Solids*

Einstein's model is not very efficient in describing in a quantitative way the behavior of specific heat in solids. The fundamental reason for this state of affairs was identified at the same time by Born and von Kármán [Born12] on the one hand and by Debye [Deby12] on the other. Debye's model, although less rigorous, is more intuitive. The idea is that the oscillations of atoms in a crystal lattice are not independent from one another, but are collective oscillations, and can be characterized by a wave vector $\boldsymbol{k}$ and a polarization index α. Once these quantities are assigned, the oscillations are harmonic, with a frequency $\omega = \omega(\boldsymbol{k}, \alpha)$. The oscillations of the lattice can be assimilated to sound waves. The relation between the wave vector and the frequency is given, at least for small wave vectors, by

$$\omega = c_\alpha |\boldsymbol{k}|, \tag{4.26}$$

where c_α is the speed of sound for oscillations with polarization α.

The model thus obtained for a solid at temperature T is fairly similar to that of a radiating cavity, which is filled by photons. By analogy, the quanta of the oscillation of the crystal lattice are called **phonons**.

The fundamental difference with respect to the case of the radiant cavity (apart from the fact that the states of polarization are three in solids and only one in fluids, and that the speed of sound can be different in the three different states) is that the number of degrees of freedom of oscillation of a solid composed of N atoms is finite, and equal to $3N$. It is therefore necessary to cut off the sums over the oscillatory degrees of freedom so as to remain with no more than $3N$ total degrees of freedom. From a geometrical point of view, this is equivalent to excluding the oscillations for wavelengths smaller than the lattice step.

We have thus defined the **Debye model**:

$$\ln Z = -\frac{V}{(2\pi h)^3}\sum_{\alpha}\int_0^{k_D} 4\pi k^2 \mathrm{d}k \ln\left[1 - \exp\left(-\frac{\hbar c_\alpha k}{k_B T}\right)\right], \tag{4.27}$$

where the cutoff or Debye wave vector k_D is implicitly defined by

$$\frac{V}{(2\pi h)^3}\int_0^{k_D} 4\pi k^2 \mathrm{d}k = N. \tag{4.28}$$

If we assume for simplicity's sake that the speed of sound does not depend on polarization, we can introduce the **Debye frequency** $\omega_D = ck_D$ and the **Debye temperature** $k_B T_D = \hbar\omega_D$. Making use of these quantities, one obtains the following expression for the solid's internal energy:

$$E = 3\frac{3N(k_B T)^4}{(\hbar\omega_D)^3}\int_0^{\hbar\omega/k_B T} \mathrm{d}x \frac{x^3}{e^x - 1}. \tag{4.29}$$

It is convenient to introduce the **Debye function**:

$$D(x) = \frac{3}{x^3}\int_0^x \mathrm{d}t \frac{t^3}{e^t - 1}. \tag{4.30}$$

The expression (4.29) can then be rewritten as follows:

$$E = 3Nk_B T D\left(\frac{T_D}{T}\right), \tag{4.31}$$

and the specific heat assumes the value

$$\begin{aligned} C &= 3Nk_B\left[D\left(\frac{T_D}{T}\right) - x\frac{\mathrm{d}D(x)}{\mathrm{d}x}\bigg|_{x = T_D/T}\right] \\ &= 3Nk_B\left[4D\left(\frac{T_D}{T}\right) - \frac{3(T_D/T)}{e^{T_D/T} - 1}\right]. \end{aligned} \tag{4.32}$$

In figure 4.4, C/Nk_B is plotted as a function of T/T_D. For high temperatures, we once again obtain the classical value $C = 3Nk_B$, while for small temperatures, we obtain

$$C \simeq Nk_B\frac{12\pi^4}{5}\left(\frac{T}{T_D}\right)^3. \tag{4.33}$$

The specific heat is therefore proportional to T^3 at low temperatures—this behavior is much slower than the exponential obtained from the Einstein model, and it is in agreement with the experimental data for insulating materials.

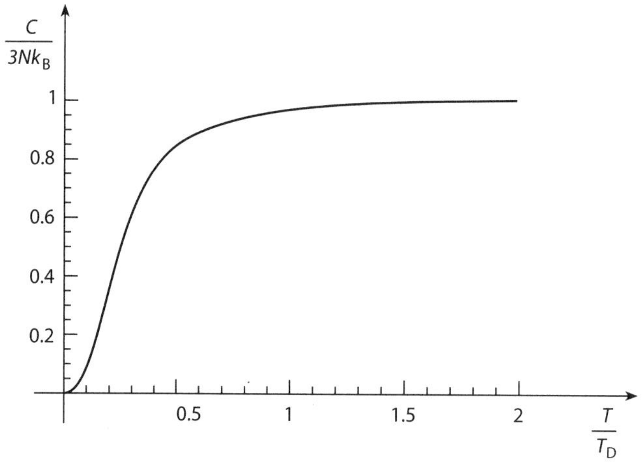

FIGURE 4.4. Specific heat per particle, $C/3Nk_B$, in the Debye model, as a function of T/T_D, where $T_D = \hbar\omega_D/k_B$ is the Debye temperature.

In metals, the behavior is instead linear in T, due the electrons' contribution, as we will see later.

4.3.2 Phonons and Second Sound

The phonons' *particulate* nature manifests itself in the possibility that the phonon gas will itself exhibit oscillation waves just as a regular particle gas would. This phenomenon is called *second sound.*

In ordinary gases, because of mass conservation, the local mass density $\rho(\boldsymbol{r}, t)$ satisfies the continuity equation:

$$\frac{\partial \rho}{\partial t} + \nabla \cdot (\rho \boldsymbol{v}) = 0, \tag{4.34}$$

where $\boldsymbol{v}(\boldsymbol{r}, t)$ is the fluid's local velocity. On the other hand, because of Newton's law applied to a small amount of fluid, we have

$$\frac{\partial}{\partial t}(\rho \boldsymbol{v}) + \rho\,(\boldsymbol{v}\cdot\nabla)\,\boldsymbol{v} = -\nabla p, \tag{4.35}$$

where p is the local pressure. In this equation, the left-hand side represents the derivative of the local momentum density, and the right-hand side the total forces applied to the small fluid element.

By linearizing these equations, one obtains

$$\rho = \rho_0 + \rho', \tag{4.36}$$

$$\frac{\partial \rho'}{\partial t} = -\rho_0(\nabla \cdot \boldsymbol{v}), \tag{4.37}$$

$$\rho_0 \frac{\partial \boldsymbol{v}}{\partial t} = -\nabla p = -\left.\frac{\partial p}{\partial \rho}\right)_S \nabla \rho'. \tag{4.38}$$

In the last equation, the quantity

$$-\left.\frac{\partial p}{\partial \rho}\right)_S = \rho_0^{-1}\kappa_S^{-1}, \tag{4.39}$$

was introduced, where κ_S is the adiabatic compressibility: $\kappa_S = -\partial \ln V/\partial p)_S$. The adiabatic compressibility κ_S rather than the isothermal one κ_T appears in this expression: this is due to the fact that when the lattice oscillates, the local compressions and dilatations happen too fast, and thermal diffusion is not able to reestablish thermal equilibrium.

From these equations, one can immediately derive the wave equation:

$$\frac{\partial^2 \rho'}{\partial t^2} = c^2 \nabla^2 \rho', \tag{4.40}$$

where the speed of sound c is given by

$$c^2 = \left.\frac{\partial p}{\partial \rho}\right)_S. \tag{4.41}$$

For an ideal gas, we obtain

$$c = \sqrt{\frac{5p}{3\rho}} = \sqrt{\frac{5k_B T}{3m}}. \tag{4.42}$$

Let us now consider the possibility that the phonon gas itself can oscillate. Since the number of phonons is not conserved, it will rather be the energy density ε that will oscillate. Since energy is conserved, its density satisfies a continuity equation of the form

$$\frac{\partial \varepsilon}{\partial t} + \nabla \cdot \boldsymbol{j}_\varepsilon = 0, \tag{4.43}$$

where $\boldsymbol{j}_\varepsilon$ is the energy current density. On the other hand, the local momentum density, which we shall denote by $\boldsymbol{q}$, satisfies Newton's law:

$$\frac{\partial \boldsymbol{q}}{\partial t} = -\nabla p. \tag{4.44}$$

In order to close these equations, we must relate the three quantities ε, $\boldsymbol{j}_\varepsilon$, and $\boldsymbol{q}$ to each other (and to the local velocity $\boldsymbol{v}$). Since the effect of the pressure gradients is to put the medium locally in motion, in order to reach this objective, we will briefly discuss the statistical mechanics of a system in motion.

We can consider the total momentum $\boldsymbol{P}$ as a new extensive variable. Entropy will generally be a function both of total energy and of total momentum. In effect, it is a function only of *internal* energy, calculated in reference to the center of mass frame—once we have denoted the system's total mass by M, we have

$$S(E_{\text{tot}}, \boldsymbol{P}) = S\left(E_{\text{tot}} - \frac{P^2}{2M}, 0\right). \tag{4.45}$$

This is due to the fact that entropy is a statistical quantity (the volume of phase space), and it must be independent of the system of reference.[1]

[1] We are considering **nonrelativistic** systems. The discussion of relativistic systems is more difficult, and, in certain respects, still open.

The intensive variable conjugate to the total momentum is given by

$$\left.\frac{\partial S}{\partial \boldsymbol{P}}\right)_{E_{\mathrm{tot}}} = \frac{\partial S}{\partial E}\left(-\frac{\boldsymbol{P}}{M}\right) = -\frac{\boldsymbol{v}}{T}. \tag{4.46}$$

Therefore, this conjugate variable is nothing other than the body's velocity—using the same reasoning employed in chapter 2, one can prove that at equilibrium, it must be the same in all the parts of the body being considered. (Analogously, one can prove that the conjugate variable of the angular momentum is angular velocity, which must be a uniform quantity at equilibrium.)

Having established these premises, it becomes obvious that the average number of phonons characterized by the polarization state α and the wave vector $\boldsymbol{k}$ in a system endowed with an average velocity $\boldsymbol{v}$ is given by the Bose factor $f_{-}(\epsilon) = (e^{\epsilon/k_B T} - 1)^{-1}$ expressed as a function of the phonons' energy at rest, which is given by

$$\epsilon(\boldsymbol{k}, \boldsymbol{v}) = \hbar\omega_{\alpha k} - \hbar\boldsymbol{k}\cdot\boldsymbol{v}, \tag{4.47}$$

where $\omega_{\alpha k} = c_\alpha |\boldsymbol{k}|$. In what follows, we will assume that the speed of sound does not depend on the polarization.

In order to obtain this result, it is sufficient to retrace the derivation we performed previously on the system at rest and then transform in the reference frame in which the system moves with a translation velocity $\boldsymbol{v}$.

We can now formulate the following expressions for energy density, momentum, and energy flow:

$$\epsilon = \frac{3}{(2\pi)^3}\int \mathrm{d}\boldsymbol{k}\, f_{-}(\hbar\omega_k - \hbar\boldsymbol{k}\cdot\boldsymbol{v})\,\hbar\omega_k, \tag{4.48}$$

$$\boldsymbol{q} = \frac{3}{(2\pi)^3}\int \mathrm{d}\boldsymbol{k}\, f_{-}(\hbar\omega_k - \hbar\boldsymbol{k}\cdot\boldsymbol{v})\,\hbar\boldsymbol{k}, \tag{4.49}$$

$$\boldsymbol{j}_\epsilon = \frac{3}{(2\pi)^3}\int \mathrm{d}\boldsymbol{k}\, f_{-}(\hbar\omega_k - \hbar\boldsymbol{k}\cdot\boldsymbol{v})\,\hbar\omega_k c\frac{\boldsymbol{k}}{|\boldsymbol{k}|}. \tag{4.50}$$

The factor 3 in these equations comes from the sum over the three polarization states. (Let us recall that the phonon *gas* behaves like a fluid, but the underlying solid does not!) The last equation is due to the fact that the energy contained in the phonon moves at the speed of sound and parallel to the wave vector $\boldsymbol{k}$. Expanding to the first order in $\boldsymbol{v}$, one obtains

$$\epsilon(\boldsymbol{v}) \simeq \epsilon. \tag{4.51}$$

Let us now evaluate $\boldsymbol{q}$'s expansion to first order:

$$\begin{aligned} q_i &= \frac{3}{(2\pi)^3}\int \mathrm{d}\boldsymbol{k}\, f_{-}(\hbar - \omega_k\hbar\boldsymbol{k}\cdot\boldsymbol{v})\,\hbar k_i, \\ &\simeq -\frac{3}{(2\pi)^3}\sum_{j=1}^{3}\int \mathrm{d}\boldsymbol{k}\, k_j v_j \left.\frac{\partial f_{-}}{\partial \epsilon}\right|_{\epsilon=\hbar\omega_k} \hbar k_i. \end{aligned} \tag{4.52}$$

The integral over $\boldsymbol{k}$ of an integrand proportional to $k_i k_j$ vanishes, except for $i = j$. In order to evaluate this integral, let us calculate its trace—one has $\sum_i k_i^2 = k^2$, and therefore each element is equal to $1/3$ of the integral in which k^2 replaces $k_i k_j$. We thus obtain

$$q \simeq \rho v, \tag{4.53}$$

where

$$\rho = -\frac{1}{(2\pi)^3}\int \mathrm{d}k\, k^2 \frac{\partial f_-}{\partial \omega}, \tag{4.54}$$

and more or less plays the role of "mass density" in the phonon gas. In the same fashion, we obtain

$$j_\varepsilon \simeq \frac{3}{(2\pi)^3}\int \mathrm{d}k \left(-\, k \cdot v \frac{\partial f_-}{\partial E}\right) \hbar\omega_k c\hat{k} \simeq c^2 \rho v. \tag{4.55}$$

On the other hand, we have

$$\frac{\partial \varepsilon}{\partial t} = \frac{\partial \varepsilon}{\partial T}\frac{\partial T}{\partial t}, \tag{4.56}$$

$$\nabla p = \frac{1}{3}\nabla \varepsilon = \frac{1}{3}\frac{\partial \varepsilon}{\partial T}\nabla T, \tag{4.57}$$

(this relation comes from the $pV = E/3$ law, which is valid for phonons as a result of the same isotropy reasons that make it valid for photons), and last

$$\frac{\partial \varepsilon}{\partial T} = \frac{3}{(2\pi)^3}\int \mathrm{d}k \frac{\partial f_-}{\partial T}\hbar\omega_k = -\frac{3}{(2\pi)^3}\int \mathrm{d}k \frac{\partial f_-}{\partial E}\frac{(\hbar\omega_k)^2}{k_\mathrm{B}T} = \frac{3c^2}{k_\mathrm{B}T}\rho. \tag{4.58}$$

By substituting, we obtain

$$\frac{\partial T}{\partial t} = \frac{1}{3}T\nabla \cdot v, \tag{4.59}$$

$$\frac{\partial v}{\partial t} = -c^2 \nabla T, \tag{4.60}$$

and therefore

$$\frac{\partial^2 T}{\partial t^2} = -c_{II}^2 \nabla^2 T, \tag{4.61}$$

where $c_{\mathrm{II}} = c/\sqrt{3}$ is the second sound velocity. These temperature vibrations will be observable if the hypotheses we made are satisfied. These hypotheses state that

- *The variation in momentum due to phonons changes only because of pressure gradients.*
- *The interaction between phonons and between phonons and the lattice is fairly weak, to the extent that one can consider phonon gas to be an ideal gas.*

These hypotheses are satisfied if the sample is fairly pure (the interaction with impurities does not conserve momentum) and the temperature is fairly low (low enough that one can disregard those interactions among phonons that do not conserve momentum). If the temperature is too low, however, the phonon gas becomes too dilute and will not be able to reach equilibrium.

4.3 Boson and Fermion Gases

Let us consider a set of noninteracting quantum particles, enclosed in a cubic box of side L (and whose volume is given by $V = L^3$). We want to study the thermodynamic properties of this system. The microstate of the system is identified by the collection of the occupation numbers of each single-particle state, as a consequence of the particles' indistinguishability. Moreover, since there is no interaction, such a state is an eigenstate of the Hamiltonian, and therefore possesses a well-defined value for energy.

The first problem is therefore to classify the single-particle states. If we assume that the particles do not have internal degrees of freedom, only the translational degrees of freedom and spin remain.

In order to fix the translational degrees of freedom, let us assume that we impose periodic boundary conditions. In this case, the single-particle wave function must satisfy the condition

$$\psi(x + L, y, z) = \psi(x, y + L, z) = \psi(x, y, z + L) = \psi(x, y, z). \tag{4.62}$$

We can therefore decompose our function into eigenfunctions of momentum, belonging to the eigenvalues

$$\boldsymbol{p} = \hbar \boldsymbol{k} = \left(\frac{\hbar}{L} n_x, \frac{\hbar}{L} n_y, \frac{\hbar}{L} n_z\right), \tag{4.63}$$

where n_x, n_y, and n_z are integers. The system's microstate is therefore defined by giving the occupation number $\nu(\boldsymbol{p}, \alpha)$ of the single-particle state characterized by the momentum $\boldsymbol{p}$ and by the value $\hbar\alpha$ of the spin's projection onto the z axis.

At this point, we can sort the particles into two types:

- Particles that have an integer spin can have any occupation number. These particles are called **bosons**; photons and phonons are of this type. (More generally, elementary particles with boson-like characteristics are *interaction quanta*—in addition to the photon, there are particles like the W or the Z_0 that are carriers of the electroweak interaction, the *gluons* that are part of the strong interaction, and so on.)

- For particles with a semi-integer spin, the wave function's antisymmetry condition dictates that the occupation number can be equal only to either 0 or 1. These particles satisfy the Pauli **exclusion principle** and are called **fermions**. (The elementary particles with fermion-like characteristics are the basic building-blocks of matter [protons, electrons, neutrons, and so on]). The "exceptions" are the neutrinos, because of the weakness of their interaction.

Given the complete collection of occupation numbers $\nu(\boldsymbol{p}, \alpha)$ of the various single-particle states, the total energy is given by

$$H = \sum_{\boldsymbol{p},\alpha} \frac{p^2}{2m} \nu(\boldsymbol{p}, \alpha), \tag{4.64}$$

while the total number of particles is obviously given by

$$N = \sum_{p,\alpha} \nu(\boldsymbol{p}, \alpha). \tag{4.65}$$

Let us provide the expression for the canonical partition function:

$$Z_N = \sum_{\{\nu(\boldsymbol{p},\alpha)\}} \delta\left(\sum_{(\boldsymbol{p},\alpha)} \nu(\boldsymbol{p}, \alpha) - N\right) \exp\left[-\frac{1}{k_B T}\sum_{(\boldsymbol{p},\alpha)} \frac{p^2}{2m}\nu(\boldsymbol{p}, \alpha)\right]. \tag{4.66}$$

The difficulty in evaluating this sum lies in the constraint that the total number of particles is fixed (expressed by the delta function, which in reality is a Kronecker delta). This makes it impossible to calculate the sum as it stands. We are therefore forced to soften the constraint, shifting to the grand canonical ensemble and calculating the corresponding partition function:

$$\begin{aligned} Z_{GC} &= \sum_{N=0}^{\infty} Z_N \exp\left(\frac{\mu N}{k_B T}\right) \\ &= \sum_{\{\nu(\boldsymbol{p},\alpha)\}} \exp\left[-\frac{1}{k_B T}\sum_{(\boldsymbol{p},\alpha)}\left(\frac{p^2}{2m} - \mu\right)\nu(\boldsymbol{p}, \alpha)\right]. \end{aligned} \tag{4.67}$$

The partition function now factorizes into sums involving the occupation numbers of the individual single-particle states:

$$Z_{GC} = \prod_{(\boldsymbol{p},\alpha)} \sum_{\nu(\boldsymbol{p},\alpha)} \exp\left[-\frac{\nu}{k_B T}\left(\frac{p^2}{2m} - \mu\right)\right] = \prod_{(\boldsymbol{p},\alpha)} \zeta(\epsilon(\boldsymbol{p}, \alpha) - \mu). \tag{4.68}$$

We have introduced the *single-particle partition function* defined by

$$\zeta(\epsilon - \mu) = \sum_{\nu} \exp\left[-\frac{\nu(\epsilon - \mu)}{k_B T}\right], \tag{4.69}$$

where $\epsilon(\boldsymbol{p}, \alpha)$ is the energy of the single-particle state identified by $(\boldsymbol{p}, \alpha)$. In the case of fermions, we have only two terms, and the sum is immediate:

$$\sum_{\nu=0,1} \exp\left[-\frac{\nu(\epsilon - \mu)}{k_B T}\right] = 1 + \exp\left(-\frac{\epsilon - \mu}{k_B T}\right) = \zeta_+(\epsilon - \mu). \tag{4.70}$$

In the case of bosons, we are dealing with a geometric series with a common ratio of $q = e^{-(\epsilon - \mu)/k_B T}$. In order for the series to converge it is necessary that $q < 1$, and therefore $\mu < \epsilon(\boldsymbol{p}, \alpha), \forall(\boldsymbol{p}, \alpha)$. In practice, since the minimum energy vanishes, this means that $\mu < 0$.

We therefore have a constraint on the chemical potential. If this constraint is satisfied, we can sum the series, and we obtain

$$\zeta_-(\epsilon - \mu) = \sum_{\nu=0}^{\infty} \exp\left[-\frac{\nu}{k_B T}(\epsilon - \mu)\right] = \left[1 - \exp\left(-\frac{\epsilon - \mu}{k_B T}\right)\right]^{-1}. \tag{4.71}$$

We thus obtain the general expression of the grand canonical partition function:

$$\begin{aligned} \ln Z_{GC} &= \sum_{(\boldsymbol{p},\alpha)} \ln \zeta_{\pm}(\epsilon(\boldsymbol{p}, \alpha) - \mu) \\ &= (\pm)\sum_{(\boldsymbol{p},\alpha)} \ln\left\{1 \pm \exp\left[-\frac{\epsilon(\boldsymbol{p}, \alpha) - \mu}{k_B T}\right]\right\}. \end{aligned} \tag{4.72}$$

The upper signs apply for fermions; the lower ones for bosons.

Let us note that the photon (or phonon) gas that we examined in the preceding section corresponds to the case in which the chemical potential vanishes. This is because the total number of particles is not fixed. In effect, the chemical potential is given (except for the sign) by the derivative of free energy (either Helmholtz's or Gibbs's) with respect to the number of particles. If this number is not fixed, thermodynamic equilibrium is reached for that value of N where these potentials reach their minimum. At equilibrium, we therefore have

$$\mu = \left.\frac{\partial F}{\partial N}\right)_V = 0, \tag{4.73}$$

or the analogous expression that contains G.

The sum over s different spin states ($\alpha = 1, \ldots, s$) can be performed immediately, and we therefore are left with

$$\ln Z_{\mathrm{GC}} = (2s+1)\sum_{p} \ln \zeta_{\pm}\left(\frac{p^2}{2m} - \mu\right). \tag{4.74}$$

In order for us to calculate this quantity, let us assume that no term gives a very large contribution. This is always the case with fermions, while it is true only if μ is strictly less than zero in the case of bosons—if at the end of our calculations this condition is violated, we will have to reconsider this point. Having established this hypothesis, let us transform the sum into an integral, keeping in mind that in the volume element $\mathrm{d}\boldsymbol{p}$, there are $(L/2\pi\hbar)^3\mathrm{d}\boldsymbol{p} = (V/h^3)\mathrm{d}\boldsymbol{p}$ terms in the sum:

$$\begin{aligned} \ln Z_{\mathrm{GC}} &= \frac{V}{h^3}(2s+1)\int \mathrm{d}\boldsymbol{p}\, \ln \zeta_{\pm}\left(\frac{p^2}{2m} - \mu\right) \\ &= (\pm)\frac{V}{h^3}(2s+1)\int_0^\infty 4\pi p^2 \mathrm{d}p\, \ln\left[1 \pm z\exp\left(-\frac{p^2}{2mk_{\mathrm{B}}T}\right)\right], \end{aligned} \tag{4.75}$$

where in the last step, we introduced the **fugacity** $z = \exp(\mu/k_{\mathrm{B}}T)$.

By performing the change of variables $p \to x = (p^2/2mk_{\mathrm{B}}T)$, we obtain

$$\begin{aligned} \ln Z_{\mathrm{GC}} &= (\pm)\frac{V}{h^3}(2s+1)\,4\pi\,(2mk_{\mathrm{B}}T)^{3/2}\int_0^\infty x^2\mathrm{d}x\, \ln\left(1 \pm z\mathrm{e}^{-x^2}\right) \\ &= (\pm)\frac{V}{h^3}(2s+1)\,4\pi\,(2mk_{\mathrm{B}}T)^{3/2}\sum_{n=1}^{\infty}\int_0^\infty x^2\mathrm{d}x\,(-)\,\frac{(\mp z)^n \mathrm{e}^{-nx^2}}{n} \\ &= (\mp)\frac{V}{h^3}(2s+1)(2mk_{\mathrm{B}}T)^{3/2}\sum_{n=1}^{\infty}\frac{(\mp z)^n}{n^{5/2}} \\ &= \frac{V}{\lambda^3}(2s+1)\,(\mp)\,\phi_{5/2}(\mp z), \end{aligned} \tag{4.76}$$

where the upper signs hold for the fermions and the lower for the bosons. We have denoted by ϕ_α the **polylog function**, defined by

$$\phi_\alpha(z) = \sum_{n=1}^{\infty}\frac{z^n}{n^\alpha}. \tag{4.77}$$

The average number of particles can be obtained by taking the derivative of the preceding expression with respect to z:

$$N = z\frac{\partial \ln Z_{GC}}{\partial z} = \frac{V}{h^3}(2s+1)\int_0^\infty 4\pi p^2 \mathrm{d}p \frac{1}{z^{-1}\mathrm{e}^{p^2/2mk_BT} \pm 1} = \frac{V}{\lambda^3}(2s+1)(\mp)\,\phi_{3/2}(\mp z). \tag{4.78}$$

This equation can be interpreted by saying that the average occupation number of the single-particle state defined by $(\mathbf{p}, \alpha)$ is given for the fermions by the **Fermi factor**:

$$f_+\left(\frac{\varepsilon(\mathbf{p},\alpha)-\mu}{k_BT}\right) = \left\{\exp\left[\frac{\varepsilon(\mathbf{p},\alpha)-\mu}{k_BT}\right]+1\right\}^{-1}, \tag{4.79}$$

and for the bosons by the Bose factor, which we already encountered:

$$f_-\left(\frac{\varepsilon(\mathbf{p},\alpha)-\mu}{k_BT}\right) = \left\{\exp\left[\frac{\varepsilon(\mathbf{p},\alpha)-\mu}{k_BT}\right]-1\right\}^{-1}, \tag{4.80}$$

The Bose factor's behavior is shown in figure 4.2, and that of the Fermi factor in figure 4.5.

In order to obtain the equation of state, it is necessary to solve equation (4.78) with respect to Z (considering N as known) and substitute in equation (4.76), since, as we know, we have

$$\ln Z_{GC} = \frac{pV}{k_BT}. \tag{4.81}$$

We see that for bosons, pressure is *lower* than that of a classical gas in equivalent conditions.

In figure 4.6, we show the equation of state for a fermion gas. Let us note that the pressure is now always *higher* than that of a classical gas in equivalent conditions. The classical

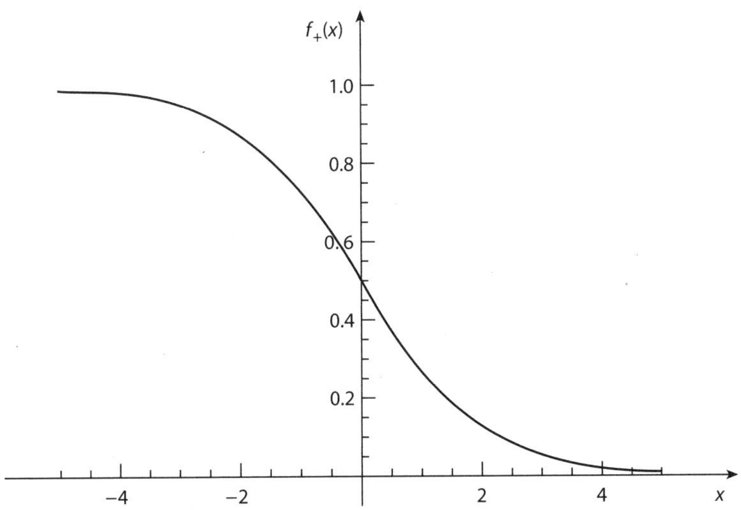

FIGURE 4.5. Fermi factor: $f_+(x) = (\mathrm{e}^x + 1)^{-1}$.

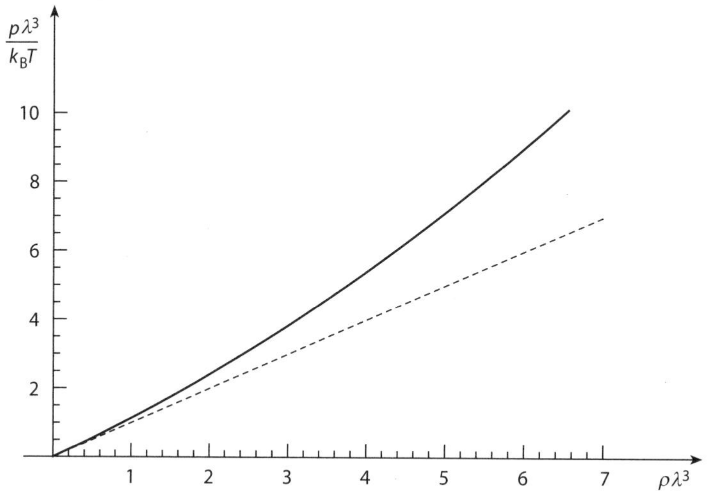

FIGURE 4.6. Equation of state for a fermion gas, $p\lambda^3/k_BT$ expressed as a function of $\rho\lambda^3$, where $\lambda = (\hbar^2/2\pi m k_B T)^{1/2}$ is the de Broglie thermal wavelength. The dashed line represents the equation of state of classical gases, $p\lambda^3/k_BT = \rho\lambda^3$.

limit of these relations can be obtained by considering small values of z. In this case, if we stop at the first term of the expansion of $\phi_{3/2}(\mp z) \simeq \mp z$, we obtain, for both bosons and fermions,

$$z \simeq \frac{N\lambda^3}{(2s+1)V}, \tag{4.82}$$

and by substituting this result in the expression for pressure, we once again obtain the classical equation for ideal gases. We can therefore introduce a parameter ξ that tells us when the classic theory of ideal gases is reliable—this parameter is the number of particles contained in a volume equal to the cube of de Broglie's thermal wavelength:

$$\xi = \frac{N\lambda^3}{V}. \tag{4.83}$$

The classical results will be reliable only if $\xi \ll 1$; otherwise, the quantum properties become evident, and the gas is said to be **degenerate**.

It is also easy (and we leave it as an exercise) to calculate the first quantum corrections to the equation of ideal gases—one should keep in mind that usually the corrections due to interactions are much more important than those of quantum origin.

4.3.1 *Electrons in Metals*

Electrons in metals form an approximately ideal and very degenerate fermion gas—in other words, one that is very far from a classical behavior. The first statement is surprising, since electrons are subject to strong Coulomb interactions with the lattice's atomic nuclei, as well as with other electrons. It is beyond the scope of this book to explain why, as a first approximation, one can avoid making explicit consideration of the interactions among electrons. Insofar as interactions with the atoms of the lattice are concerned, since they are represented by a spatially periodic potential, their effect can be summarized by changing the expression of the electron's energy as a function of the wave vector. We here refer the reader to any good book on solid state—for example, Ashcroft and Mermin [Ashc76], to explain how the periodic potential's effect is to break up the intervals of the allowable values for single-particle energy into bands, and how each band is run over when the particle's wave vector varies in the region known as the **first Brillouin zone**.

It is easy, on the other hand, to calculate the characteristic values of the degeneration parameter $\xi = (N\lambda^3/V)$, which defines the classical limit. In the case of electrons, we have $m = 9.11 \cdot 10^{-31}$ kg, and therefore, at room temperature (300 K), the thermal de Broglie wavelength is given by

$$\begin{aligned}\lambda &= \left(\frac{h^2}{2\pi m k_B T}\right)^{1/2} = \left[\frac{(6.63 \cdot 10^{-34})^2}{2 \cdot 3.14 \cdot 9.11 \cdot 10^{-31} \cdot 1.38 \cdot 10^{-23} \cdot 300}\right]^{1/2} \\ &= 4.31 \cdot 10^{-9}\ \text{m} = 43.1\ \text{Å}.\end{aligned} \tag{4.84}$$

As a result, since characteristic distances between electrons are on the order of the angstrom, the value of parameter ξ will be on the order of $40^3 \sim 6.4 \cdot 10^4$, and the gas will therefore be very far from the classical limit.

Let us therefore attempt to investigate the behavior of a strongly degenerate fermion gas. In order to clarify our ideas, let us assume that the particles are free so that the relation between momentum $\boldsymbol{p}$ and energy is the usual one:

$$\epsilon(\boldsymbol{p}, \alpha) = \frac{p^2}{2m}. \tag{4.85}$$

In the limit $T \to 0$, the Fermi factor tends to 1, if $\epsilon - \mu < 0$, and otherwise, it tends to 0—this means that all available states with energy below the chemical potential are completely occupied, while those at higher energies are empty. The maximum value of the energy of the occupied states is called **Fermi energy** and is denoted by ϵ_F. The total number of particles, expressed as a function of the chemical potential μ, is therefore given by

$$N = V(2s + 1)\int_0^{\mu} \omega(\epsilon)\, \mathrm{d}\epsilon, \tag{4.86}$$

where $\omega(\epsilon)$ is the **density of states**—in other words, the number of single-particle states with a given spin, with an energy situated between ϵ and $\epsilon + \mathrm{d}\epsilon$, per unit of volume and energy. In the case of free electrons, one gets

$$\omega(\epsilon) = \frac{4\pi m (2m\epsilon)^{1/2}}{h^3}. \tag{4.87}$$

If the density of states at the Fermi energy does not vanish, as in this case, the Fermi energy is equal to the chemical potential at zero temperature. We can therefore calculate the total number of particles as a function of the chemical potential μ, or (which is the same thing) of the Fermi energy ϵ_F:

$$N = \frac{4\pi V}{3h^3}(2s+1)(2m\varepsilon_F)^{3/2}. \tag{4.88}$$

By inverting this relation, we obtain the expression for the Fermi energy:

$$\varepsilon_F = \frac{h^2}{2m}\left[\frac{3N}{4\pi V(2s+1)}\right]^{2/3}. \tag{4.89}$$

It is interesting to note that the Fermi energy is inversely proportional to the fermions' mass—and this is the reason that phenomena related to the Fermi distribution are more obvious in electrons. We can associate a temperature T_F, which is called the **Fermi temperature**, to the Fermi energy by means of the relation

$$\varepsilon_F = k_B T_F. \tag{4.90}$$

Another way to define the degree of quantum degeneration of electrons in metals is to consider the ratio T_F/F. The values characteristic of T_F for metals go from the $1.84 \cdot 10^4$ K of cesium to the $16.0 \cdot 10^4$ K of beryllium.

In order to calculate the properties of fermion gas at low temperatures, one uses a technique developed by Sommerfeld. From now on, we will make use of the fact that, for electrons, one has $s = 1/2$, and therefore $2s + 1 = 2$. Let us expand the Fermi factor around the origin, and then integrate both sides—in other words, both the expression for the thermodynamic potential (the logarithm of the partition function) and that for the number of particles. By defining the integral of the density of states $\Phi(\varepsilon)$ via the expression:

$$\Phi(\varepsilon) = \int_0^{\varepsilon} \omega(\varepsilon')\,d\varepsilon', \tag{4.91}$$

we obtain

$$\begin{aligned} N &= 2V\int_0^{\infty} d\varepsilon\,\frac{d\Phi}{d\varepsilon} f_+(\varepsilon-\mu) \\ &= 2V\left[\frac{d\Phi}{d\varepsilon} f_+(\varepsilon-\mu)\right]_0^{\infty} - 2V\int_0^{\infty} d\varepsilon\,\Phi(\varepsilon) f'_+(\varepsilon-\mu). \end{aligned} \tag{4.92}$$

The first term vanishes, while in the second term, the $\Phi(\varepsilon)$ can be expanded into a series of $\varepsilon - \mu$, since the derivative of the Fermi factor is not equal to zero only within a small interval on the order of $k_B T$ around the origin. We thus obtain

$$N = -2V\int_0^{\infty} d\varepsilon \sum_{n=0}^{\infty} \frac{1}{n!}\left.\frac{d^n\Phi}{d\varepsilon^n}\right|_{\varepsilon=\mu} (\varepsilon-\mu)^n f'_+(\varepsilon-\mu). \tag{4.93}$$

We can therefore extend the integration over ε to the entire real axis. Since $f'_+(x)$ is an even function of x, only terms with even n survive in this expression and yield

$$\begin{aligned} N &= -2V\Phi(\mu) - 2V\sum_{n=1}^{\infty} \frac{1}{(2n)!}\left.\frac{d^{2n}\Phi}{d\varepsilon^{2n}}\right|_{\varepsilon=\mu} \int_{-\infty}^{\infty} d\varepsilon\,(\varepsilon-\mu)^{2n} f'_+(\varepsilon-\mu) \\ &= 2V\Phi(\mu) + 2V\sum_{n=1}^{\infty} \frac{(k_B T)^{2n}}{(2n)!}\left.\frac{d^{2n}\Phi}{d\varepsilon^{2n}}\right|_{\varepsilon=\mu} \int_{-\infty}^{\infty} dx\,\frac{x^{2n}e^x}{(e^x+1)^2}. \end{aligned} \tag{4.94}$$

This last integral can be evaluated explicitly:

$$a_n = \int_{-\infty}^{\infty} dx \frac{x^{2n}}{(2n)!} \frac{e^x}{(e^x + 1)^2} = \left(2 - \frac{1}{2^{2(n-1)}}\right) \zeta_R(2n). \tag{4.95}$$

We therefore see that the chemical potential does not change at the first order in T/T_F. By substituting this result in the partition function and taking the derivative, we obtain that the specific heat is linear in T and is given by

$$C_V = \frac{\pi^2 V}{3} k_B T \omega(\varepsilon_F), \tag{4.96}$$

where we made explicit the dependence of density from the Fermi energy states. This expression is valid in general. In the case of free electrons, one has

$$C_V = \frac{\pi^2}{2} \frac{T}{T_F} N k_B. \tag{4.97}$$

Since the ratio T/T_F, as we saw, is very small, the contribution of the electronic degrees of freedom to the specific heat of metals is usually negligible. Since, however, it decreases linearly with temperature, it tends to dominate the phononic contribution (which, as we saw, is proportional to T^3) at low temperatures.

4.3.2 *Relation between Pressure and Internal Energy*

In all the cases in which the relation between wave vector (or momentum) and energy in single-particle states takes the form of a power law, it is possible to prove the existence of a proportionality relation between the pressure and the density of internal energy in ideal gases (classical or quantum). For generality's sake, we will consider some particles in d-dimensional space that possess a relation between energy and momentum of the form

$$\varepsilon(\mathbf{p}, \alpha) = \kappa |\mathbf{p}|^{\gamma}, \tag{4.98}$$

where κ is a constant. In the case of particles with mass, such as electrons, one has $\gamma = 2$ and $\kappa = (2m)^{-1}$; in the case of photons (or of ultrarelativistic particles), one has $\gamma = 1$ and $\kappa = c$ (where c is the speed of light). The same relation (where c, however, is the speed of sound) is valid for phonons.

We know that the pressure is given by the logarithm of the partition function:

$$\ln Z = \frac{pV}{k_B T} = \frac{V}{h^d}(2s+1) \int_0^{\infty} S_d p^{d-1} dp \, (\pm) \ln\left\{1 \pm \exp\left[-\frac{\varepsilon(p) - \mu}{k_B T}\right]\right\}, \tag{4.99}$$

where S_d is the surface area of a sphere with a unit radius in d dimensions. By introducing the fugacity z and the change of variables

$$p \to x = \left(\frac{\kappa p^{\gamma}}{k_B T}\right), \tag{4.100}$$

we obtain

$$\ln Z = \frac{V}{h^d}(2s+1)\left(\frac{\kappa}{k_B T}\right)^{d/\gamma} S_d \int_0^{\infty} dx \, x^{(d/\gamma)-1} (\pm) \ln(1 - z e^{-x}). \tag{4.101}$$

As we know, the internal energy is given by

$$E = -\left.\frac{\partial \ln Z}{\partial(1/k_B T)}\right)_z. \tag{4.102}$$

Since $\ln Z$ is a homogenous function of degree d/γ in $k_B T$, we obtain

$$E = -\frac{\partial \ln Z}{\partial(1/k_B T)} = \frac{d}{\gamma} k_B T \ln Z = \frac{d}{\gamma} pV, \tag{4.103}$$

which is the result we wanted to prove. This result is independent of the nature (fermion-like or boson-like) of the particles being considered.

More particularly, in the case of electrons at high densities, the internal energy is of the order of the Fermi energy, which, as we have seen, is proportional to the density to $2/3$. We thus obtain

$$pV \propto V^{2/3}, \tag{4.104}$$

and in a strongly degenerate electron gas, we therefore have

$$pV^{5/3} = \text{const.} \tag{4.105}$$

This equation recalls the law of adiabatics of the classical gas $pV^{(C_p/C_V)}$. Indeed, Nernst's postulate implies that adiabatics and isotherms coincide at $T = 0$, and one can show that for our system one has $C_p/C_V = 1 + d/\gamma$. The derivation of this relation is left to the reader as an exercise.

4.3.3 *Variational Derivation of Fermi and Bose Statistics*

In the present section, we show that it is possible to obtain the Fermi and Bose distributions (4.79) and (4.80) from a variational principle on the quantity $\mathcal{S} = k_B \ln \mathcal{N}$, where $\mathcal{N}$ is the number of microscopic configurations associated with a given distribution of the occupation numbers of single-particle states.

Fermi statistics. Let us suppose that we have a system of N fermions, each of which can be in a single-particle state i, which is associated with energy ϵ_i. We want to evaluate the average value n_i of the occupation number of state i. Although this problem can be resolved more easily in the grand canonical ensemble, it is instructive to obtain it by using the variational principle for the generalized ensembles. In this case, the system's state is univocally defined by the collection of the occupied states $\{i_1, \ldots, i_N\}$. Since the probability of occupation ultimately depends only on the state's energy ϵ_i, we can collect the single-particle states together in groups such that the number of states G_j that are found in the same group j will be large, the number of particles N_j that are found in group j will also be large, but the energy of all the states that belong to group j will be virtually identical. The ensemble is therefore well defined when, for each group j, the number of states N_j, the corresponding energy ϵ_j, and the average occupation number $n_j = N_j/G_j$ are all known. Once N_j is fixed, the number of ways in which we can distribute the particles within group j is clearly given by $G_j!/N_j!/(G_j - N_j)!$, which corresponds to the number of ways in which it is possible to choose the N_j states that are occupied among all the G_j possible ones. The logarithm of this quantity is given by

$$s_j = -G_j[n_j \ln n_j + (1 - n_j) \ln (1 - n_j)].$$

and is similarly obtained by applying Stirling's formula. The entropy associated with the distribution $\{n_j\}$ is given by

$$\mathcal{S} = k_B \sum_j s_j = -k_B \sum_j G_j[n_j \ln n_j + (1 - n_j) \ln (1 - n_j)].$$

We want to look for this expression's maximum with respect to n_j, with the conditions

$$\sum_j N_j = \sum_j G_j n_j = N, \qquad \sum_j G_j \epsilon_j n_j = E.$$

By introducing the Lagrange multipliers α and β, we obtain the equation

$$\ln \frac{n_j}{1 - n_j} = \alpha - \beta\epsilon_j,$$

which admits as a solution

$$n_j = \frac{1}{e^{-\alpha + \beta\epsilon_j} + 1}.$$

This is the Fermi distribution, expressed as a function of the parameters $\alpha = \mu/k_B T$ and $\beta = 1/k_B T$.

Bose statistics. Let us now consider a system of N bosons, and let us proceed once again to grouping the single-particle states together. In this case, the number of ways in which it is possible to arrange the N_j particles in the G_j states is given by the number of ways in which the N_j states can be chosen *with repetition.* This number can be evaluated by considering that it is equal to the number of ways one can arrange N_j identical marbles in G_j boxes. We represent an arrangement of this sort in the following manner:

|• • |•||• • •| etc.,

where the vertical bars sequentially mark the borders of the first, the second, ..., the n-th box, and the dots represent the marbles. We have N_j marbles and $G_j + 1$ bars in total, but the position of the first and last bar are fixed, at the beginning and end of the sequence, respectively. The number of arrangements is therefore equal to the number of ways in which we can place the N_j marbles in $N_j + G_j - 1$ possible positions. This number is equal to $(N_j + G_j - 1)!/N_j!(G_j - 1)!$, and its logarithm is given by

$$s_j = G_j[(1 + n_j) \ln (1 + n_j) - n_j \ln n_j],$$

where $n_j = N_j/G_j$ is the average occupation number, and we have assumed $G_j \gg 1$.

The entropy associated with this distribution is therefore

$$\mathcal{S} = k_B \sum_j s_j = k_B \sum_j G_j[(1 + n_j) \ln (1 + n_j) - n_j \ln n_j].$$

By optimizing this expression with the usual constraints, we obtain

$$n_j = \frac{1}{e^{-\alpha + \beta\epsilon_j} - 1}.$$

By interpreting the parameters α and β as we did earlier, we obtain the Bose distribution

$$n_j = \frac{1}{e^{(\epsilon_j - \mu)/k_B T} - 1}.$$

where we assumed that

$$\mu \leq \min_j \epsilon_j,$$

because the n_j's must not be negative.

4.4 Einstein Condensation

Let us now consider the behavior of a strongly degenerate boson gas. We analyze the expression that we derived for the average number of particles as a function of z, and we recall that since $\mu < 0$, one has $z < 1$. For generality's sake, let us assume that we have a system in a d-dimensional space. We thus have

$$\begin{aligned} N &= \frac{V}{\lambda^d}(2s+1)\, S_d \frac{1}{\pi^{d/2}} \int_0^\infty \mathrm{d}x\, x^{d-1} \left(\frac{e^{-x^2}}{z} - 1 \right)^{-1} \\ &= \frac{V}{\lambda^d}(2s+1)\phi_{d/2}(z), \end{aligned} \tag{4.106}$$

where $\lambda = (h^2/2\pi m k_B T)^{1/2}$ is the thermal de Broglie wavelength. It is easy to see that the integral on right-hand side is an increasing function of z, which therefore assumes its maximum value at $z = 1$. For $d > 2$, this maximum value is finite; for $d = 3$, for example, one has

$$\begin{aligned} N &\leq \frac{V}{\lambda^3}(2s+1)\frac{S_3}{\pi^{3/2}} \int_0^\infty \mathrm{d}x\, x^2 \left(e^{-x^2} - 1 \right)^{-1} \\ &= \frac{V}{\lambda^3}(2s+1)\phi_{3/2}(1) = \frac{V}{\lambda^3}(2s+1)\zeta_R(3/2) = \frac{V}{\lambda^3}(2s+1)\, 2.612\ldots \end{aligned} \tag{4.107}$$

It therefore seems that the quantity $N\lambda^d/V$ cannot grow above a certain value. On the other hand, when T decreases, the thermal de Broglie wavelength increases, and therefore, for a system with N particles contained in a fixed volume, this quantity will eventually become large at low temperatures. Einstein recognized this problem while reading an article sent to him by an obscure Bengali physicist, Satyendranath Bose, asking for his approval for publication, and Einstein suggested the solution (see [Bose24] and [Eins25]).

Let us recall that approximating the sum with the integral when expressing N is allowed only if each term in the sum gives a negligible contribution to the total. So if $\mu \to 0$ (and in other words, if $z \to 1$), the Bose factor, for the lowest single-particle energy values—in particular, for $\varepsilon(\mathbf{p} = 0)$—becomes very large (as can be seen from the plot of the Bose factor that diverges for $x \to 0$). It is therefore useful to separate the contribution of the term $\mathbf{p} = 0$ and calculate the corresponding value of z:

$$N = (2s+1)\left[\frac{z}{1-z} + \frac{V}{\lambda^d}\phi_{d/2}(z)\right]. \tag{4.108}$$

Normally, the first term is negligible when compared to the rest, which is proportional to V, but since the second term cannot be larger than $(V/\lambda^d)\phi_{d/2}(1)$, one needs to account for it if N becomes too large (or T too small). In this situation, we see that one has to choose a value of z that is very close to 1, and it is therefore possible to obtain the approximation

$$N \simeq (2s+1)\left[\frac{1}{1-z} + \frac{V}{\lambda^d}\phi_{d/2}(1)\right] = N_0 + (N - N_0). \tag{4.109}$$

We thus obtain the occupation number N_0 of the single-particle state at $\boldsymbol{p} = 0$:

$$N_0 = N - \frac{V(2s+1)}{\lambda^d}\phi_{d/2}(1) = N\left[1 - \frac{(2s+1)}{\rho}\left(\frac{2\pi m k_B T}{h^2}\right)^{d/2}\phi_{d/2}(1)\right], \tag{4.110}$$

where $\rho = N/V$. Clearly, this expression makes sense only if $N_0 > 0$—in other words, if

$$T < T_0 = \frac{2\pi m k_B T}{h^2}\left[\frac{\rho}{(2s+1)\phi_{d/2}(1)}\right]^{2/d}, \tag{4.111}$$

where we have introduced the **transition temperature** T_0. For $T > T_0$, the number of particles that occupy the minimum energy state is macroscopic: one says that the particles have **condensed.** This phenomenon is called **Einstein condensation** (sometimes, illegitimately, Bose-Einstein condensation).

Below the transition temperature, the number of particles in a state of minimum energy ("in the condensate") is given by

$$N_0 = N\left[1 - \left(\frac{T}{T_0}\right)^{d/2}\right], \tag{4.112}$$

and the pressure depends only on temperature:

$$p = \frac{k_B T}{\lambda^d}\phi_{(d+2)/2}(1) = \frac{k_B T}{\lambda^d}\zeta_R\left(\frac{d+2}{2}\right). \tag{4.113}$$

In particular, in three dimensions one has $p = k_B T \zeta_R(5/2)/\lambda^3 \simeq 1.342\, k_B T/\lambda^3$.

The equation of state of an ideal Bose gas in three dimensions is shown in figure 4.7. The phenomenon of Einstein condensation is the basis for the macroscopic quantum properties of helium (superfluidity) and of superconductivity. In these cases, however, the interactions among particles are very important, to the extent that they qualitatively change the phenomenon. We cannot therefore discuss them here.

We shall conclude by observing that, if $d \le 2$, the Einstein condensation does not take place—in this case, in fact $\lim_{z\to 1}\phi_{d/2}(z) = \infty$. Therefore, for any value of N, it is possible to find a value of $z < 1$, and the contribution of the minimum energy state is negligible.

Exercise 4.2 Evaluate the behavior of the internal energy and specific heat of a boson gas in the vicinity of the Einstein condensation.

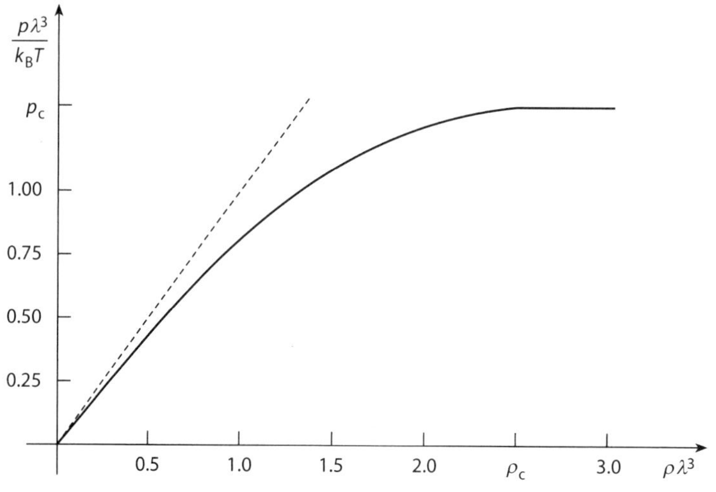

FIGURE 4.7. Equation of state of a Bose gas. Rescaled pressure $p\lambda^3/k_BT$ expressed as a function of the rescaled density $p\lambda^3$. The dashed line shows the equation of state of classical ideal gases. The Einstein condensation takes place at the point (ρ_c, p_c).

4.5 Adsorption

Let us consider a wall that can adsorb molecules, placed in contact with a container that holds those same molecules in a gaseous state. To say that the wall can adsorb molecules is the equivalent of saying that their energy when they are on the wall is equal to $-\varepsilon_0$, and is therefore lower than when they are in the gas. On the other hand, the wall's capacity is finite. Let us assume that there can, at most, be only N particles deposited on it. When a molecule has settled on the wall, another molecule cannot adhere to it at the same point.

These observations lead to the introduction of the following Hamiltonian to describe the wall's state:

- One introduces a variable $\tau_i = 0{,}1$ for each adsorption site: 0 if the site is empty, 1 if the site is full.
- The energy of the molecules adsorbed is equal to $H = -\epsilon_0 \sum_{i=1}^{N} \tau_i$.
- Since the gas can freely exchange particles with the surrounding gas, the number of adsorbed particles is not fixed—we must in other words use the grand canonical ensemble.

The partition function is therefore given by

$$Z = \sum_{\{\tau_i\}} \exp\left[k_B T \sum_{i=1}^{N} (\epsilon_0 + \mu)\, \tau_i \right], \tag{4.114}$$

and recalls that of a fermion system. By evaluating the sum, we obtain

$$\ln Z = N \ln\left[1 + \exp\left(\frac{\epsilon_0 + \mu}{k_B T}\right)\right]. \tag{4.115}$$

The average number of adsorbed particles is given by

$$\langle M \rangle = \frac{\partial \ln Z}{\partial (\mu/k_B T)} = N\left[\exp\left(-\frac{\epsilon_0 + \mu}{k_B T}\right) + 1\right]^{-1}. \tag{4.116}$$

This expression can be related to the pressure of the gas surrounding the wall by exploiting the fact that, for a classical ideal gas, one has

$$\exp\left(\frac{\mu}{k_B T}\right) = z = \frac{p\lambda^3}{k_B T}, \tag{4.117}$$

where λ is the thermal de Broglie wavelength. We thus obtain

$$\langle M \rangle = N \frac{p}{p + p_0}, \tag{4.118}$$

where we have introduced the characteristic pressure p_0 by means of the expression

$$\exp\left(-\frac{\epsilon_0}{k_B T}\right) = \frac{p_0 \lambda^3}{k_B T}. \tag{4.119}$$

We observe that the characteristic pressure p_0 is that in which 50% of sites are occupied. This law is known as the **Langmuir isotherm**. In figure 4.8, we show its characteristic behavior. On the x axis, we have p/p_0, and on the y axis, we have $\langle M \rangle / N$.

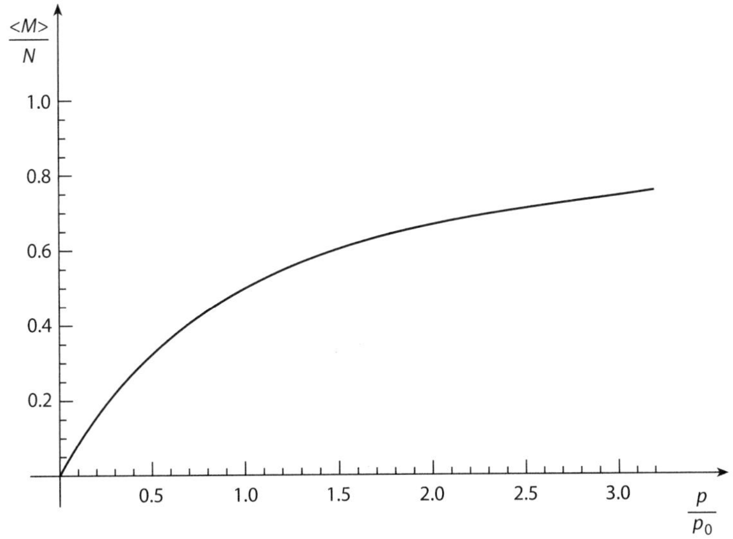

FIGURE 4.8. The Langmuir isotherm. The average occupation number of each site $\langle M \rangle / N$ as a function of rescaled pressure p/p_0 is shown.

4.5.1 *Myoglobin and Hemoglobin*

The Langmuir isotherm has important applications in biophysics. Some proteins specialize in transporting oxygen—the simplest is myoglobin, which is also the first protein whose three-dimensional structure was solved. Myoglobin can be found in muscles, and it is responsible for their characteristic red color. In one of the protein's "pockets" there is a complex organic group, called **heme**, which also includes an iron ion, which is responsible for the protein's activity. A set of N proteins behaves like the adsorbing wall that we just discussed. The plot of the average number of oxygen molecules adsorbed by myoglobin as a function of the partial pressure of oxygen has a typical Langmuir trend like that of figure 4.8.

Hemoglobin is another protein, which is present in the blood, and whose task is to carry oxygen from the lungs to the muscles. It is composed of four units, each with a structure similar to myoglobin. It is important that hemoglobin can adsorb oxygen where the partial pressure of oxygen is highest and release it where it is lowest. One can observe that the corresponding isotherm has a fairly different behavior from that of myoglobin at low oxygen pressures—this feature, due to the cooperative effort of the four units, increases the protein's efficiency. Hemoglobin's characteristic pressure moreover decreases as pH levels become lower, especially when muscles are tired—this is the **Bohr effect**, named for Niels Bohr's father, who was a famous physiologist.

A fetus's hemoglobin has a smaller characteristic pressure than that of an adult, so it can take oxygen from its mother. In certain wild geese who migrate above the Himalayas, a single mutation in an amino acid significantly decreases p_0 so as to increase the oxygen adsorbed at great altitudes.

Exercise 4.3 (A Simplified Model of Hemoglobin) Let us associate each of the four units i ($i = 1, \ldots, 4$) that make up a molecule of hemoglobin with a variable τ_i, which is equal to 1 if an O_2 molecule is adsorbed, and is otherwise 0. The energy of the hemoglobin molecule is given by

- $-\varepsilon_0$ times the number of O_2 molecules adsorbed.
- $-J$ times the number of pairs of nearest-neighbor units, each containing a molecule of adsorbed O_2. (The i and $i+1$ sites are nearest neighbors, and so are sites 4 and 1).

Consider a system of $N/4$ molecules of hemoglobin. Evaluate the average number $\langle M \rangle = \sum_{\alpha=1}^{N/4} \sum_{i=1}^{4} \tau_{\alpha i}$ of adsorbed molecules as a function of $x = \exp[(\epsilon_0 + \mu)/k_B T]$ and of $y = \exp(J/k_B T)$.

4.6 Internal Degrees of Freedom

4.6.1 *Atomic Gases*

Up to now, we assumed that the particles being considered had no internal structure. The case in which each particle has a certain number of internal degrees of freedom can be treated in a simple fashion, as long as we can disregard the interactions between the different particles (and therefore between the relevant internal degrees of freedom).

The first case of interest is that of a diluted atomic gas. In this case, in addition to the translational degrees of freedom, one also needs to consider the effects of the electronic and nuclear degrees of freedom. These degrees of freedom are independent of the translational ones (which are those we have treated so far), with excellent approximation. We can therefore identify the atom's state by means of the quantity

$$\nu = (\boldsymbol{p}, n, \sigma), \tag{4.120}$$

where $\boldsymbol{p}$ is momentum and defines the translational state of the center of mass, n denotes the atom's electronic state collectively, and σ denotes the nucleus' state. The atom's energy is given by

$$\varepsilon(\nu) = \frac{p^2}{2m} + \epsilon_{n,\sigma}, \tag{4.121}$$

which expresses the independence of the internal degrees of freedom.

Let us now assume that the gas is diluted enough that we can make use of the canonical partition function for independent particles:

$$\begin{aligned} Z &= \frac{1}{N!}\left\{\sum_{\nu} \exp\left[-\frac{\varepsilon(\nu)}{k_B T}\right]\right\}^N \\ &= \frac{1}{N!}\left[\sum_{p} \exp\left(-\frac{p^2}{2mk_B T}\right)\right]^N \left[\sum_{n,\sigma} \exp\left(-\frac{\epsilon_{n,\sigma}}{k_B T}\right)\right]^N \\ &= \frac{1}{N!}\zeta_{\text{trasl}}^N(T, V)\,\zeta_{\text{int}}^N(T), \end{aligned} \tag{4.122}$$

where we have introduced the single-particle *translational* and *internal* partition functions:

$$\begin{aligned} \zeta_{\text{transl}}(T, V) &= \sum_{p} \exp\left(-\frac{p^2}{2mk_B T}\right), \\ \zeta_{\text{int}}(T) &= \sum_{n,\sigma} \exp\left(-\frac{\epsilon_{n,\sigma}}{k_B T}\right) \end{aligned} \tag{4.123}$$

$$= \exp\left(-\frac{\epsilon_{0,0}}{k_B T}\right) \sum_{n,\sigma} \exp\left(-\frac{\epsilon_{n,\sigma} - \epsilon_{0,0}}{k_B T}\right). \tag{4.124}$$

The translational partition function is the one we examined in the preceding sections. In calculating the internal partition function, one has to keep in mind that the differences in energy that appear in its expression are usually much larger than $k_B T$ (in order of magnitude, 1 eV corresponds to about 10^4 K). It is therefore necessary to consider only those states that have the same energy as the ground state, and we then obtain

$$\begin{aligned} \zeta_{\text{int}}(T) &\simeq \exp\left(-\frac{\epsilon_{0,0}}{k_B T}\right) \times \text{(ground state multiplicity)} \\ &= \exp\left(-\frac{\epsilon_{0,0}}{k_B T}\right) g_{\text{el}}^0 g_{\text{nucl}}^0, \end{aligned} \tag{4.125}$$

where g^0_{el} and g^0_{nucl} are the multiplicities of the electronic and nuclear ground states, respectively. If we assume that the only important nuclear degree of freedom of the nucleus is its spin, we obtain

$$g^0_{nucl} = 2I + 1, \tag{4.126}$$

where I is the value of the nuclear spin. We thus obtain the expression for the canonical partition function and, consequently, that for the Helmholtz free energy:

$$\ln Z = -\frac{F}{k_B T} = \ln\left[\frac{1}{N!}\zeta^N_{trasl}(T, V)\right] + N \ln\left[(2I+1)g^0_{elettr}\right] - \frac{N_{\epsilon_{0,0}}}{k_B T}. \tag{4.127}$$

Since the dependence on volume is entirely contained in the translational partition function, we obtain that the pressure (and therefore the equation of the isotherms) is independent of the internal degrees of freedom, which do however influence the value of the entropy and the internal energy.

4.6.2 Molecular Gases

Let us now consider a gas composed of molecules of a specific chemical species. The situation becomes more complicated for two reasons:

1. One has to take into account the molecules' rotational degrees of freedom, which have characteristic energies that are comparable to $k_B T$.

2. The degrees of freedom for both the electronic and nuclear spin are not independent from the rotational ones in the case of molecules composed of identical atoms, because of the Pauli principle.

First, let us attempt an accurate description of a molecule's quantum behavior. Strictly speaking, we ought to consider the whole problem of the interaction between nuclei and electrons; however, since the ratio of the nuclei's masses to those of the electrons is large, we can introduce the **Born-Oppenheimer approximation**. This consists of supposing that the positions of the nuclei are fixed, which allows us to solve the time-independent Schrödinger equation for the electrons' coordinates. We thus obtain a fixed value for energy once the positions of the nuclei and the quantum numbers of the electrons have been fixed. This energy is then considered as an effective potential in the nuclei's Schrödinger equation.

More explicitly, the system's total wave function (nuclei + electrons) is written in the following manner:

$$\Psi(r, R) = \Phi_n(r, R)\,\psi_{n,\nu}(R), \tag{4.128}$$

where r denotes the electronic coordinates and R those of the nuclei. The electronic wave function $\Phi_n\,(r, R)$ satisfies the equation

$$[\hat{K}(r) + V(r) + V(R) + V(r, R)]\Phi(r, R) = \varepsilon_n(R)\Phi(r, R), \tag{4.129}$$

where $\hat{K}(r)$ is the kinetic energy operator for electrons only. The nuclei's wave function satisfies the equation

$$[\hat{K}(R) + \epsilon_n(R)]\psi_{n,\nu}(R) = \epsilon_{n,\nu}\psi(R), \tag{4.130}$$

where $\hat{K}(R)$ is the nuclei's kinetic energy operator. In the context of this approximation, we will have, more generally

$$\zeta_{\text{int}}(T) = \sum_{n,\nu} \exp\left(-\frac{\epsilon_{n,\nu}}{k_B T}\right). \tag{4.131}$$

As we saw previously, it is usually possible to disregard all the excited electronic states, since the energies that come into play are much larger than $k_B T$. We therefore obtain

$$\begin{aligned} \zeta_{\text{int}}(T) &\simeq \sum_{\nu} \exp\left(-\frac{\epsilon_{0,\nu}}{k_B T}\right) \\ &\simeq g^0 \exp\left(-\frac{\epsilon_{0,0}}{k_B T}\right) \sum_{\nu} \exp\left(-\frac{\epsilon_{0,\nu} - \epsilon_{0,0}}{k_B T}\right). \end{aligned} \tag{4.132}$$

It is usually possible to consider the vibrational and rotational degrees of freedom as independent, because the vibrations take place much more rapidly than the rotations. We thus obtain

$$\Delta\epsilon_\nu = \epsilon_{0,\nu} - \epsilon_{0,0} \approx \left(\lambda + \frac{1}{2}\right)\hbar\omega + \ell(\ell+1)\hbar^2/2I. \tag{4.133}$$

Here, $\lambda = 0, 1, \ldots$ is the quantum vibration number, and ω is the vibration frequency around the Born-Oppenheimer potential's minimum, which corresponds to a distance R_0 between nuclei:

$$\omega^2 = \left(\frac{1}{m_A} + \frac{1}{m_B}\right) \frac{\partial^2 \epsilon_0(R)}{\partial R^2}\bigg|_{R_0}. \tag{4.134}$$

The first factor is the inverse reduced mass of the two nuclei. The second term of the preceding equation is the rotational energy, where $\ell = 0, 1, \ldots$ is the angular quantum number, and I is the moment of inertia. The level identified by ℓ is $(2\ell + 1)$ times degenerate.

If the molecule is composed of different atoms, we can sum independently over the vibrational and rotational degrees of freedom. If we are dealing with identical atoms, we must be careful to sum only over those states that are either even or odd with respect to nuclear exchange, respectively, if the nuclei are bosons or fermions. In the classical limit, it is sufficient to divide by a symmetry factor that takes into account the double counting. If we assume that we can perform this approximation, we obtain

$$\zeta_{\text{int}}(T) = \frac{g^0}{\sigma_{AB}} \exp\left(-\frac{E_{0,0}}{k_B T}\right)(2I_A + 1)(2I_B + 1)\zeta_{\text{vib}}(T)\zeta_{\text{rot}}(T), \tag{4.135}$$

where the symmetry factor σ_{AB} is defined by

$$\sigma_{AB} = \begin{cases} 1, & \text{if } A \neq B, \\ 2, & \text{if } A = B. \end{cases}$$

Calculating the vibrational partition function is simple:

$$\begin{aligned}\zeta_{\text{vib}} &= \sum_{\lambda=0}^{\infty} \exp\left[-\frac{(\lambda+\frac{1}{2})\hbar\omega}{k_B T}\right] \\ &= \exp\left(-\frac{\hbar\omega}{2k_B T}\right)\left[1-\exp\left(-\frac{\hbar\omega}{k_B T}\right)\right]^{-1}.\end{aligned} \tag{4.136}$$

We can now calculate the contribution of the rotational degrees of freedom. As we saw, the quantum states of a rotator are specified by two integers: the **angular** quantum number ℓ, which can be equal to $0, 1, 2, \ldots$, and the **magnetic** quantum number m, which can be equal to $-\ell, \ell+1, \ldots, \ell$. The kinetic energy of a quantum rotator in the (ℓ, m) state is given by $\epsilon(\ell, m) = \hbar^2 \ell(\ell+1)/2I$, where I is the rotator's moment of inertia. This energy does not depend on the magnetic quantum number, and for this reason, the energy levels are $2\ell+1$ times degenerate.

In this fashion, we can introduce the rotational partition function ζ_{rot}, defined by

$$\begin{aligned}\zeta_{\text{rot}} &= \sum_{\ell=0}^{\infty}\sum_{m=\ell}^{\ell} \exp\left[-\frac{\hbar^2\ell(\ell+1)}{2Ik_B T}\right] \\ &= \sum_{\ell=0}^{\infty}(2\ell+1)\exp\left[-\frac{\hbar^2\ell(\ell+1)}{2Ik_B T}\right].\end{aligned} \tag{4.137}$$

This expression cannot be represented in terms of elementary functions, but it is easy to calculate it numerically and also obtain its behavior for large and small temperature values. We see that the characteristic temperature θ_{rot} for rotational degrees of freedom is given by

$$\theta_{\text{rot}} = \frac{\hbar^2}{2Ik_B}. \tag{4.138}$$

In figure 4.9, we show the behavior of the contribution of rotational degrees of freedom to specific heat—in other words, C_{rot}/k_B as a function of T/θ_{rot}. We see that asymptotically, one has $C_{\text{rot}} \to k_B$, while $C_{\text{rot}} \to 0$ when the temperature falls below θ_{rot}. Since θ_{rot} is usually quite a bit smaller than T, it is worthwhile to expand $\zeta_{\text{rot}}(T)$ in a power series of θ_{rot}/T. This can be obtained by exploiting the Euler-MacLaurin series to evaluate the expression of ζ_{rot}. The result is

$$\zeta_{\text{rot}} = \frac{T}{\theta_{\text{rot}}}\left[1+\frac{1}{3}\left(\frac{\theta_{\text{rot}}}{T}\right)+\frac{1}{15}\left(\frac{\theta_{\text{rot}}}{T}\right)^2+\cdots\right]. \tag{4.139}$$

4.6.3 Ortho-Hydrogen and Para-Hydrogen

The effects of the quantum exclusion principle that we alluded to are particularly obvious in the case of hydrogen. In this case, moreover, it is necessary to discuss some of the subtleties of thermodynamic equilibrium fairly carefully.

Let us consider the H_2 molecule. The two nuclei are protons (with a spin 1/2) and therefore fermions. In the Born-Oppenheimer approximation, we can consider only the wave function of the two nuclei (in an appropriate electronic state) and factorize the dependence

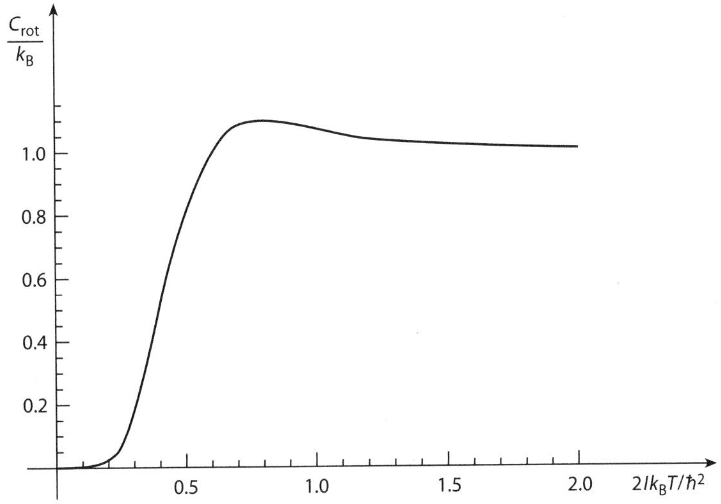

FIGURE 4.9. Specific heat per particle (in k_B units) versus $2Ik_BT/\hbar^2$ for rotational degrees of freedom. Note that at high temperatures, the classic value $C_{rot} = k_B$ is recovered. Also note the typical "hump" at intermediate temperatures.

on the center of mass's motion. The two nuclei's state will therefore be described by a wave function that is a function of the two spins and the position vector relative to the two nuclei:

$$\Psi_{\sigma_1,\sigma_2}(\boldsymbol{r}). \tag{4.140}$$

The Pauli principle requires that this wave function be antisymmetrical with respect to the exchange of the two nuclei, which implies that

$$\Psi_{\sigma_2,\sigma_1}(-\boldsymbol{r}) = -\Psi_{\sigma_1,\sigma_2}(\boldsymbol{r}). \tag{4.141}$$

Since the interaction between the nuclear spins and the other degrees of freedom is weak, one can choose a base in which the orbital degrees of freedom and those relating to spin are factorized, and one obtains

$$\Psi_{\sigma_1,\sigma_2}(\boldsymbol{r}) = \chi_{\sigma_1,\sigma_2}\phi(\boldsymbol{r}). \tag{4.142}$$

One can moreover choose the orbital wave function $\phi(\boldsymbol{r})$ so that it is an eigenfunction of angular momentum:

$$\varphi(\boldsymbol{r}) = \psi(r)\,Y_{\ell m}(\theta, \phi). \tag{4.143}$$

Let us recall that as a consequence of the transformation $\boldsymbol{r} \to -\boldsymbol{r}$, one has $\theta \to \pi - \theta$ and $\phi \to \phi + \pi$, and therefore:

$$Y_{\ell m} \to (-1)^{\ell}\,Y_{\ell m}. \tag{4.144}$$

Morever, the spin wave function can correspond to a total state of spin $s = 1$, and therefore be symmetrical, or to a spin $s = 0$, and be antisymmetrical. In order for the total wave function to be antisymmetrical, we must have

$$\ell = \begin{cases} 0, 2, 4, \ldots & \text{if } s = 0; \\ 1, 3, 5, \ldots & \text{if } s = 1. \end{cases} \tag{4.145}$$

On the other hand, the multiplicity of states with $s = 0$ is equal to 1, while that of states with $s = 1$ is equal to 3. The contribution of the rotational degrees of freedom to the specific heat of hydrogen can therefore be obtained by considering the rotational partition function:

$$\begin{aligned} \zeta_{\text{rot}} &= \sum_{\ell \text{ pari}} (2l+1) \exp\left[-\frac{\hbar^2}{2Ik_B T}\ell(\ell+1)\right] \\ &\quad +3 \sum_{\ell \text{ dispari}} (2l+1) \exp\left[-\frac{\hbar^2}{2Ik_B T}\ell(\ell+1)\right]. \end{aligned} \tag{4.146}$$

From this expression, we can obtain the rotational specific heat of hydrogen:

$$C_{\text{rot}} = Nk_B T \frac{\partial^2 (T \ln \zeta_{\text{rot}})}{\partial T^2}. \tag{4.147}$$

This expression, however, does not agree with experimental data at low temperatures (see figure 4.10).

The reason for this discrepancy is related to the fact that, since the interactions with the nuclear spins are very weak, the different states of nuclear spin *do not equilibrate* in the characteristic time of these specific heat experiments. The quantity of molecules N_0 in the state $s = 0$, and N_1 in the state $s = 1$ *have to be considered as independent thermodynamic variables.* It is as if molecular hydrogen were obtained as a mixture of two types of particles, **para-hydrogen** with a nuclear spin equal to 0, and **ortho-hydrogen** with a nuclear spin equal to 1.

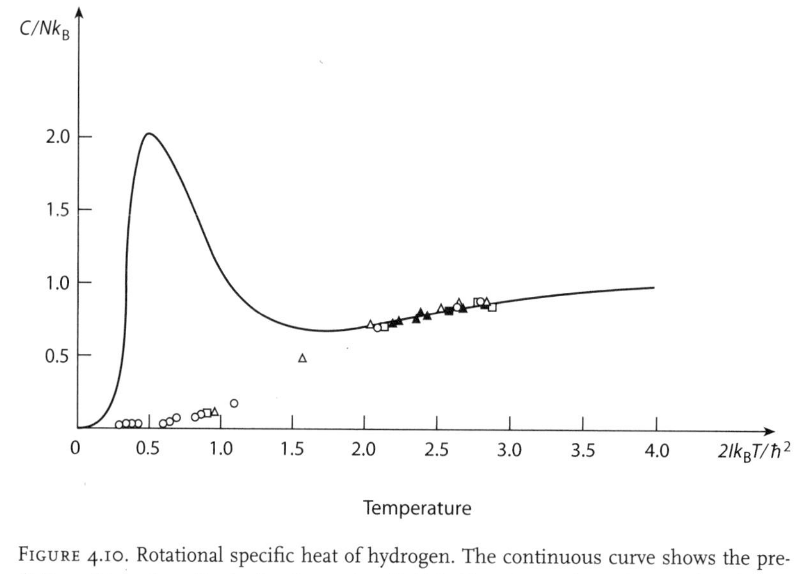

FIGURE 4.10. Rotational specific heat of hydrogen. The continuous curve shows the prediction obtained on the basis of equation (4.146). Experimental results are also shown. Adapted from [Wann66].

In this fashion, one obtains the following expression for rotational specific heat:

$$C_{\rm rot} = Nk_{\rm B}T\left[\frac{N_0}{N}\frac{\partial^2(T\ln\zeta_0)}{\partial T^2}+\frac{N_1}{N}\frac{\partial^2(T\ln\zeta_1)}{\partial T^2}\right], \tag{4.148}$$

where ζ_0 is the first and ζ_1 is the second term in the expression of $\zeta_{\rm rot}$ that we considered previously, and they correspond to the contributions of para-hydrogen and ortho-hydrogen, respectively. This expression can be interpreted by saying that it is necessary to average the *logarithm* of the partition function over the "frozen" variables N_0 and N_1. The experimental data shown in figure 4.11 match fairly well the values

$$N_0 = \frac{1}{4}, \qquad N_1 = \frac{3}{4}. \tag{4.149}$$

These are the equilibrium values at high temperatures. The data were obtained in the early twentieth century by several experimenters whose names are reported in the figure.

4.7 Chemical Equilibria in Gases

Let us consider a chemical reaction defined by the formula

$$\nu_A A + \nu_B B \longleftrightarrow \nu_C C + \nu_D D, \tag{4.150}$$

where the ν_i are the stoichiometric coefficients that we discussed in section 2.19. This expression can be rewritten in a more symmetrical manner:

$$\sum_i \nu_i X_i = 0, \tag{4.151}$$

where the X_i denote the different chemical species, and the stoichiometric coefficients relative to the right-hand-side member of the reaction (the *products*) are negative. The

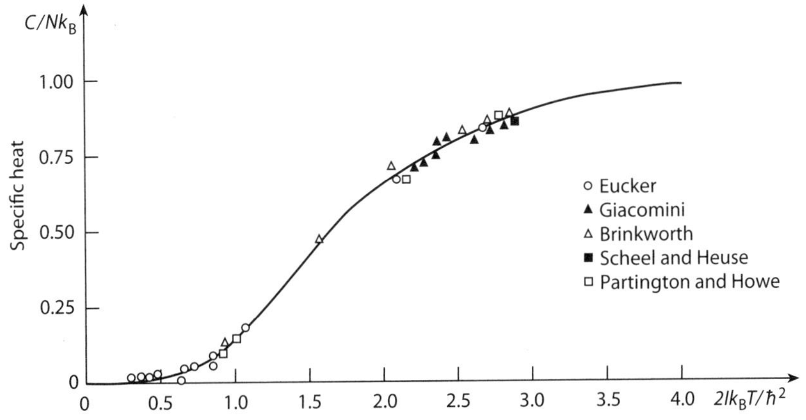

FIGURE 4.11. Rotational specific heat of hydrogen at low temperatures. The continuous curve shows the prediction obtained on the basis of equation (4.148). Adapted from [Wann66].

existence of the reaction allows the system to move along a line in the space identified by the number of particles of the various chemical species. In fact, by denoting the variation in the number of particles of species i with ΔN_i, we will have

$$\frac{\Delta N_i}{\nu_i} = \text{const.}, \tag{4.152}$$

for each allowable variation in the number of particles.

The equilibrium condition (at fixed temperature and pressure) is that the total Gibbs free energy be minimal with respect to all the variations in the number of particles that respect these constraints. By taking into account the fact that the derivative of the Gibbs free energy with respect to the number of particles is the chemical potential, we obtain the condition

$$0 = \sum_{i=1}^{r} \frac{\partial G}{\partial N_i} \Delta N_i = \text{const.} \sum_{i=1}^{r} \mu_i \nu_i. \tag{4.153}$$

The equilibrium condition with respect to the chemical reaction we are examining can therefore be expressed by

$$\sum_i \mu_i \nu_i = 0. \tag{4.154}$$

We can now use a relation that is valid for gases at low density:

$$\mu = \ln\left[\rho \lambda^3(T)\right] + \ln \zeta_{\text{int}}(T). \tag{4.155}$$

We thus obtain the **law of mass-action**:

$$\prod_i \rho_i^{\nu_i} = \prod_i \left[\lambda_i^3 \zeta_{\text{int}}(T)\right]^{\nu_i} = K(T). \tag{4.156}$$

The quantity on the right-hand-side depends only on temperature and is in principle calculable. Historically, this was one of the first triumphs of statistical mechanics.

Recommended Reading

The arguments discussed in the present chapter are fairly standard and can be found in all standard textbooks on statistical mechanics. I therefore recommend just two exercise books: C. Chahine and P. Devaux, *Thermodynamique Statistique*, Paris: Dunod, 1976, and Y.-K. Lim, *Problems and Solutions in Statistical Mechanics*, Singapore: World Scientific, 1990.

5 Phase Transitions

The lobster demands to be cut while alive.

—Brillat-Savarin

5.1 Liquid–Gas Coexistence and Critical Point

I borrow from M. E. Fisher this vivid description of carbon dioxide's critical point [Fish83, p. 6]:

> The first critical point to be discovered was in carbon dioxide. Suppose one examines a sealed tube containing CO_2 at an overall density of about 0.5 gm/cc, and hence a pressure of about 72 atm. At a temperature of about 29° C one sees a sharp meniscus separating liquid (below) from vapor (above). One can follow the behavior of the liquid and vapor densities if one has a few spheres of slightly different densities close to 0.48 gm/cc floating in the system. When the tube is heated up to about 30°C one finds a large change in the two densities since the lighter sphere floats up to the very top of the tube, i.e. up into the vapor, while the heaviest one sinks down to the bottom of the liquid. However, a sphere of about "neutral" density (in fact "critical density") remains floating "on" the meniscus. There is, indeed, still a sharp interface between the two fluids, but they have approached one another closely to density. Further slight heating to about 31°C brings on the striking phenomenon of critical opalescence. If the carbon dioxide, which is quite transparent in the visible region of the spectrum, is illuminated from the side, one observes a strong intensity of scattered light. This has a bluish tinge when viewed normal to the direction of illumination, but has a brownish-orange streaky appearance, like a sunset on a smoggy day, when viewed from the forward direction (i.e., with the opalescent fluid illuminated from behind). Finally, when the temperature is raised a further few tenths of a degree, the opalescence disappears and the fluid becomes completely clear again. Moreover, the meniscus separating "liquid" from "vapor" has completely vanished: no trace of it remains! *All* differences between the two phases have gone: indeed only one, quite homogeneous, "fluid" phase remains above the critical temperature ($T_c = 31.04$°C).

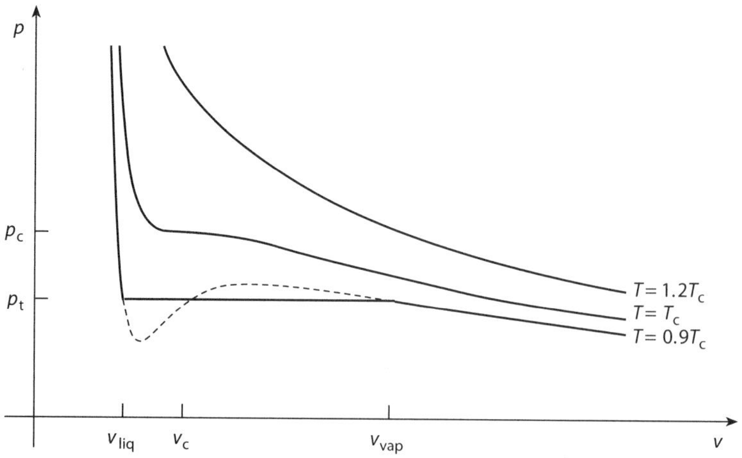

FIGURE 5.1. Pressure-volume isotherm of a fluid according to the van der Waals equation. The dashed line represents the nonphysical part of the isotherm in the coexistence region.

We can describe this behavior in a pressure–volume diagram like the one shown in figure 5.1. This diagram is based on the (approximate) van der Waals equation of state, which we will discuss in the next section. Above a characteristic temperature value T_c, called the **critical temperature**, the pressure p is a monotonically decreasing function of the specific volume per particle $v = V/N$. For large values of v, p is inversely proportional to v and therefore follows the equation of state for perfect gases. When p becomes very large, v cannot decrease beyond a certain limit, which is the molecule's *intrinsic volume*. As temperature diminishes, the isotherm tends to become flatter in the intermediate region, until for $T = T_c$, it shows a horizontal inflection point corresponding to the value $v = v_c = 1/\rho_c$. This value of density is called the **critical density**. The value p_c for pressure corresponds to this value of specific volume. Below the critical temperature, the isotherm displays a flat region, between $v = v_{\text{liq}} = 1/\rho_{\text{liq}}$ and $v = v_{\text{vap}} = 1/\rho_{\text{vap}}$, where $v_{\text{liq,vap}}$ are the densities of the liquid and vapor, respectively. Only one pressure value corresponds to each value of density included between these two values; it is called the **transition pressure** (or **vapor pressure**), and we will denote it by $p_1(T)$. If, as mentioned earlier, we consider a sample of critical density and we place it at a temperature below T_c, we observe the coexistence of two phases at different densities. By increasing the temperature, the difference in density decreases until it disappears at T_c.

By plotting this behavior in a pressure–temperature diagram, we obtain the **phase diagram** shown in figure 5.2. Let us note that in this case, the diagram's axes both correspond to *intensive* quantities. The curve $p = p_t(T)$ in the plane (p, T) defines the line in which there is a coexistence of the two phases: vapor on the left and liquid on the right. This curve terminates in correspondence with the point (p_c, T_c), without exhibiting any singularities.

Other curves in the same plane describe the coexistence of fluid and solid phases—they meet in the **triple point** $(p_{\text{tr}}, T_{\text{tr}})$, where all three phases coexist. One should note that

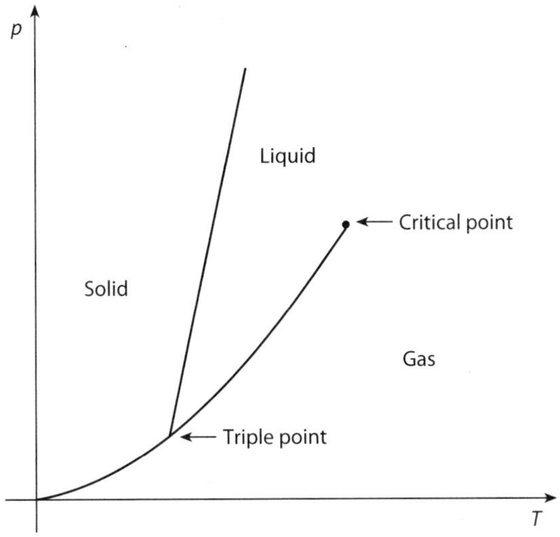

FIGURE 5.2. Phase diagram (schematic, not to scale) of a fluid on the plane (p, T). Note the liquid–gas coexistence curve, which terminates at the critical point, and the fusion (solid–liquid coexistence) and sublimation (solid–gas coexistence) curves, which meet the preceding curve at the triple point.

neither the solid–liquid nor the solid–vapor coexistence exhibit a critical point, unlike the liquid–vapor coexistence.

5.2 Van der Waals Equation

A first interpretation of this curious thermodynamic behavior can be found in van der Waals' doctoral thesis [Waal73]. James C. Maxwell was so enthusiastic about this work (to which he made a substantial contribution [Maxw74] as we will see) that he said it was worthwhile to learn Dutch even just to be able to read it. (It has since been translated.) Van der Waals' idea is to start from the equation of state for ideal gases (we denote by v the volume per particle V/N):

$$pv = k_B T. \tag{5.1}$$

We will take two effects into account:

1. The *intrinsic volume* effect of the gas particles: the volume actually available to each particle is not equal to v, but is smaller, and equal to $(v - b)$, where b (the **covolume**) is a parameter related to the particles' intrinsic volume.

2. The effect of the particles' reciprocal attraction: the pressure exerted by the particles on the walls is lower than the actual pressure acting on them within the sample, because they tend to be attracted to the other particles. In order to evaluate this effect, let us denote by u the potential of a particle within the fluid, due to the attraction by the other particles, and by u_w

the corresponding value of the potential for a particle close to the wall. One then expects $u_w = u/2$, since only the particles within a half-space attract a particle close to the wall:

$$u_w = \frac{u}{2} = -\frac{N}{2V}u_0, \tag{5.2}$$

where u_0 is the integral of the pair potential $u(r)$:

$$u_0 = \int d^3r\, u(r). \tag{5.3}$$

This allows us to evaluate the local density near the wall:

$$\begin{aligned} \left.\frac{N}{V}\right|_w &= \frac{N}{V}\exp\left[-\frac{1}{k_BT}(u_w - u)\right] \\ &= \frac{N}{V}e^{-Nu_0/2Vk_BT}. \end{aligned} \tag{5.4}$$

Supposing that the pressure exerted by the molecules near the wall is given at least approximately by the ideal gas law, we obtain

$$\begin{aligned} p &= k_BT\left.\frac{N}{V}\right|_w = k_BT\frac{N}{V}e^{-Nu_0/2Vk_BT} \simeq k_BT\frac{N}{V}\left(1 - \frac{Nu_0}{2Vk_BT}\right) \\ &= k_BT\frac{N}{V} - \frac{1}{2}\left(\frac{N}{V}\right)^2 u_0. \end{aligned} \tag{5.5}$$

One thus obtains the well-known **van der Waals equation**:

$$\left(p + \frac{a}{v^2}\right)(v - b) = k_BT, \tag{5.6}$$

where $a = u_0/2$ and b are two parameters characteristic of the material.

Maxwell realized that the isotherms' behavior, below a temperature $T_c = 8a/27b$, was no longer monotonically decreasing, and that this contradicted thermodynamic stability, according to which the isothermal compressibility:

$$K_T = -\left.\frac{\partial v}{\partial p}\right)_T \tag{5.7}$$

must be a positive quantity. He therefore suggested that below T_c, a portion of the isotherm should be replaced with a horizontal segment, which represented phase coexistence. One could determine the position of this straight line by making use of the construction (which was later given his name) whereby the total work performed by first going along the horizontal line and then along the isothermal curve would vanish. This construction is equivalent to imposing the equality of Helmholtz free energy per molecule for the two coexisting phases:

$$\mu_{liq} = \mu_{vap}. \tag{5.8}$$

The van der Waals equation of state is just a clever interpolation formula, which does however possess a significant heuristic value. It is, in effect, possible to estimate the two parameters, a and b that appear in the equation, from measures on gases at low densities, and then substitute the results in the expression of the critical parameters:

$$k_B T_c = \frac{8a}{27b}, \qquad v_c = 3b, \qquad p_c = \frac{a}{27b^2}. \tag{5.9}$$

The results one obtains are reasonable. Moreover, one can choose to measure the temperature, pressure, and volume in terms of the corresponding critical values, introducing the ratios:

$$\mathcal{T} \equiv \frac{T}{T_c}, \qquad \mathcal{P} \equiv \frac{p}{p_c}, \qquad \mathcal{V} \equiv \frac{v}{v_c}. \tag{5.10}$$

One thus obtains a "universal" equation, independent of the nature of the fluid being considered:

$$\left(\mathcal{P} + \frac{3}{\mathcal{V}^2}\right)\left(\mathcal{V} - \frac{1}{3}\right) = \frac{8}{3}\mathcal{T}. \tag{5.11}$$

This equation is called the **law of corresponding states**. It yields good results if one is not too close to the critical temperature.

More particularly, it follows from the van der Waals equation that for T close enough to T_c, the difference between the gas's and the liquid's specific volume vanishes as a power of $|T - T_c|$:

$$v_{\text{vap}} - v_{\text{liq}} \propto |T - T_c|^{\beta}. \tag{5.12}$$

It follows from the van der Waals equation (as one can easily see) that $\beta = 1/2$. Experimentally, one can observe that this quantity actually behaves as a power of $|T - T_c|$, but with an exponent β closer to 1/3 than to 1/2. It is also interesting to remark that the value of this exponent is apparently the same (within the errors) for several different liquids, but also that the equation of state has an apparently universal form, as can be seen in figure 5.3.

5.3 Other Singularities

As a consequence of the van der Waals equation, one can show that isothermal compressibility exhibits a singularity at the critical point, which can be verified experimentally:

$$\chi = -\left.\frac{\partial v}{\partial p}\right)_T \sim |T - T_c|^{-\gamma}. \tag{5.13}$$

The van der Waals equation, however, implies that $\gamma = 1$, while experimentally, γ is closer to 1.25.

As the system approaches the critical temperature, also the specific heat C exhibits a singularity that is well described by a power law of the form

$$C \sim |T - T_c|^{-\alpha}, \tag{5.14}$$

where the exponent α has a value included between 1/8 and 1/9 and is the same (within the errors) for all fluids. As we will see, this behavior contrasts with the simpler theories of phase transitions.

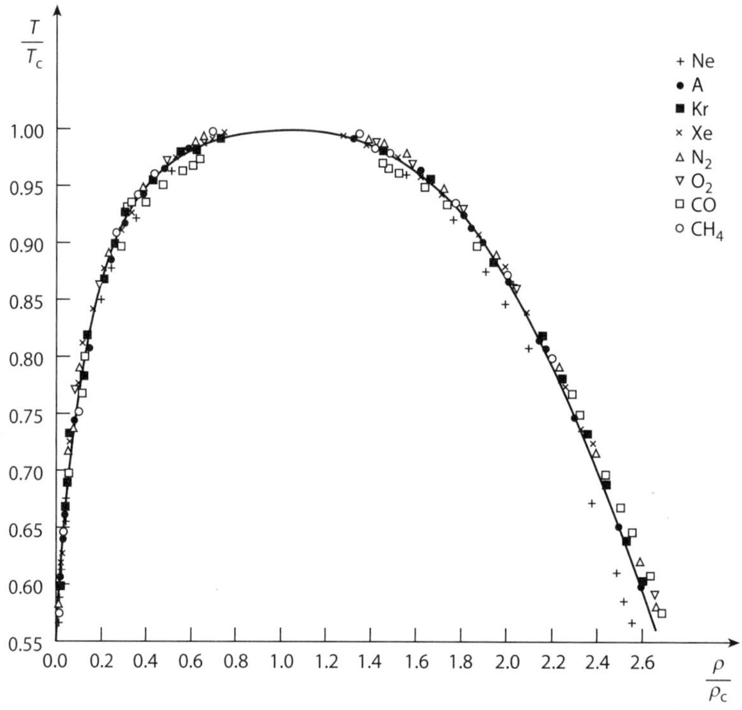

FIGURE 5.3. Coexistence curve ρ/ρ_c as a function of T/T_c for eight fluids in the proximity of the critical temperature. The continuous curve corresponds to $\rho_{\text{liq,vap}} - \rho_c \propto |T - T_c|^{\beta}$, with $\beta = 1/3$. From [Gugg45], with permission.

5.4 Binary Mixtures

Another system that exhibits an analogous behavior is a fluid mixture composed of two chemical species. One usually uses organic compounds whose names are difficult to pronounce—one of the most convenient (also from a phonetic point of view) is aniline–cyclohexane, because its critical point is close to ambient temperature at atmospheric pressure. We will refer to one of the components as A and to the other as B. By fixing the pressure, it is possible to observe the coexistence of *three* phases within a certain temperature interval, and they are arrayed from the bottom up in a decreasing order of density: an α phase rich in A, a β phase rich in B, and a vapor phase. As the temperature varies, the composition (and therefore also the density) of the α and β phases move closer to each other until, for a particular temperature value T_c (as long as we have selected the correct composition of the sample) the two liquid phases are replaced by a homogeneous phase. The point at which this occurs is called the **consolution point**.

The behavior of this system is fairly similar to that of the liquid–gas system. More particularly, it also exhibits critical opalescence in proximity of the consolution point. Let us note some differences:

1. The quantity that distinguishes the two phases is the relative concentration of one of the components—A, for example. We refer to A's molar fraction as x_{A} and its value in phase α as x_{A}^{α}—then, $x_{\mathrm{A}}^{\alpha} - x_{\mathrm{A}}^{\beta} \to 0$ when $T - T_{\mathrm{c}}$ along the coexistence curve.

2. Fairly frequently, the consolution point corresponds to a temperature *lower* than the coexistence temperature. In other words, a homogeneous phase is replaced by a phase coexistence when the sample is heated. This does not occur with simple fluids.

3. One can observe, however, that the exponent β that describes the behavior of this system is also close to $1/3$ and indistinguishable from the value it assumes in simple fluids.

5.5 Lattice Gas

In order to obtain a more microscopic description of phase coexistence, let us build a simple statistical model of a "fluid." We consider a system of N particles, described by the Hamiltonian:

$$H = K + U, \tag{5.15}$$

where K is the kinetic energy

$$K = \sum_{\alpha=1}^{N} \frac{p_{\alpha}^{2}}{2m}, \tag{5.16}$$

and U is the potential energy, which depends only on the particles' positions. When we calculate the partition function:

$$Z = \frac{1}{N!} \int \prod_{\alpha=1}^{N} \frac{\mathrm{d}\boldsymbol{p}_{\alpha}\mathrm{d}\boldsymbol{r}_{\alpha}}{h^{3}} \exp\left(-\frac{H}{k_{\mathrm{B}}T}\right), \tag{5.17}$$

the integral over momenta $\boldsymbol{p}_{\alpha}$ factorizes and results in λ^{-3N}, where λ is the thermal de Broglie wavelength:

$$\lambda^{2} = \frac{h^{2}}{2\pi m k_{\mathrm{B}} T}. \tag{5.18}$$

In the two previous equations, h is Planck's constant, which appears to account for the volume of phase space occupied by the quantum states. In this fashion, we obtain the partition function expressed as a function of an integral over the system's configurations:

$$Z = \frac{\lambda^{-3N}}{N!} \int \prod_{\alpha=1}^{N} \mathrm{d}\boldsymbol{r}_{\alpha} \exp\left(-\frac{U}{k_{\mathrm{B}}T}\right). \tag{5.19}$$

The interaction between particles is usually described by a potential that exhibits a hard core—it becomes large and positive when the particles get closer than a certain distance a_0, where $a_0/2$ is the particle radius—and an attractive tail, i.e., it becomes negative at intermediate distances, eventually vanishing when the particles are far apart.

We can provide a drastically simplified description of this state of affairs with the following hypotheses:

1. The particles can occupy only the sites on a simple cubic lattice whose lattice constant is a_0.
2. At every site, there can be one particle at most.
3. For each pair of particles that occupy nearest-neighbor sites, the system's potential energy decreases by a quantity ϵ.

Let us denote the state of occupation of site i by a variable τ_i, which is equal to 0 if the position is empty and 1 if it is occupied. Then the partition function assumes the form

$$Z_N = \lambda^{-3N} \sum_{\{\tau_i\}} \delta_{N,\sum_i \tau_i} \exp\left(\sum_{\langle ij \rangle} \frac{\epsilon}{k_B T} \tau_i \tau_j \right). \tag{5.20}$$

In this equation, the Kronecker delta expresses the condition that the total number of particles in the system is exactly N:

$$\sum_i \tau_i = N. \tag{5.21}$$

The expression $\sum_{\langle ij \rangle}$ indicates that the sum runs over all pairs of nearest-neighbor sites on the lattice. The combinatorial factor $N!$ has disappeared because this expression is invariant with respect to the permutation of particle labels. The model defined in this manner is called the **lattice gas**.

In order to make this expression more manageable, it is useful to eliminate the Kronecker delta—one can do this by shifting to the grand canonical ensemble. In this fashion, we define the grand partition function Z_{GC}:

$$Z_{GC} = \sum_{N=0}^{\infty} \exp\left(\frac{\mu N}{k_B T} \right) Z_N. \tag{5.22}$$

The grand partition function Z_{GC} is related to the pressure p by the equation

$$\ln Z_{GC} = pV/k_B T, \tag{5.23}$$

where V is the system's volume. It is easy to see that Z_{GC} can be expressed as follows:

$$Z_{GC} = \sum_{\{\tau_i\}} \exp\left(-\frac{H}{k_B T} \right), \tag{5.24}$$

where the Hamiltonian H is defined by

$$-\sum_{\langle ij \rangle} \epsilon \tau_i \tau_j - \sum_i \tilde{\mu} \tau_i. \tag{5.25}$$

In this equation, we have introduced the expression

$$\frac{\tilde{\mu}}{k_B T} = \frac{\mu}{k_B T} - 3 \ln \lambda. \tag{5.26}$$

The particle density per site is given by the quantity

$$x = \langle \tau \rangle = \frac{1}{N}\sum_i \langle \tau_i \rangle. \tag{5.27}$$

Along the coexistence curve $\tilde{\mu} = \tilde{\mu}^*(T)$, for $T < T_c$, this quantity exhibits a discontinuity:

$$x^+ = \lim_{\tilde{\mu} \to \tilde{\mu}^*(T)^+} x(\tilde{\mu}, T) > x^- = x^+ = \lim_{\tilde{\mu} \to \tilde{\mu}^*(T)^-} x(\tilde{\mu}, T). \tag{5.28}$$

The discontinuity tends to zero when the temperature gets close to the critical temperature. If the model correctly describes the experimental data, we can expect the discontinuity to tend to zero as a power of $|T < T_c|$:

$$x^+ - x^- \propto |T - T_c|^{\beta}. \tag{5.29}$$

5.6 Symmetry

The model we just described cannot be solved exactly. It is possible, however, to manipulate it to highlight certain properties. In place of the variable τ_i, which assumes the values 0 and 1, let us consider the variable σ_i, defined by

$$\sigma_i = 2\tau_i - 1. \tag{5.30}$$

This variable assumes the values -1 (when $\tau_i = 0$) and $+1$ (when $\tau_i = 1$). Expressing the τ_i as a function of the σ_i, and substituting in the Hamiltonian (5.25), we obtain the expression (up to an irrelevant additive constant):

$$H = -\sum_{\langle ij \rangle} J\sigma_i\sigma_j - \sum_i h\sigma_i, \tag{5.31}$$

We have defined

$$J = \epsilon/4, \tag{5.32}$$

$$h = (\tilde{\mu} + \epsilon\zeta)/2, \tag{5.33}$$

where ζ is the lattice's **coordination number**—in other words, the number of each site's nearest neighbors. In the case of the simple cubic lattice, $\zeta = 6$ (and it is equal to $2d$ in the case of a cubic lattice in d dimensions). This Hamiltonian can also describe binary mixtures, if one stipulates that $\sigma_i = +1$ corresponds to the presence of a molecule of type A in i, $\sigma_i = -1$ to a molecule of type B, and $2h = \mu_B - \mu_A$.

This expression makes explicit the fact that if $h = 0$, the value that H assumes for a certain configuration $\sigma = \{\sigma_i\}$ remains invariant with respect to the transformation I, where

$$\mathsf{I}\sigma_i = -\sigma_i, \quad \forall i. \tag{5.34}$$

We can express this fact by saying that when $h = 0$, the Hamiltonian H is **symmetrical** with respect to the transformation group **G** composed by I and the identity E. Since $\mathsf{I}^2 = \mathsf{E}$, it is not necessary in this case to consider other transformations.

The model we obtained is the most famous model of statistical mechanics. It was invented by Lenz, who proposed it as a thesis to one of his students, and for this reason, it is given the student's name, *Ising*. It was actually proposed to describe the behavior of ferromagnetic systems. In this case, σ_i is interpreted as the component z of the magnetic moment of an electron (apart from a factor), and h is proportional to the applied magnetic field. The factor J represents the tendency of electrons belonging to contiguous atoms to align themselves parallel because of the exchange interaction—it is therefore known as the exchange integral. In spite of the quantum origin of this interaction the Ising model is a classical model (in the sense that all the quantities commute).

5.7 Symmetry Breaking

We can now look at phase coexistence from another point of view. In the case of the ferromagnet, the two coexisting phases are equivalent from a physical standpoint and differ only because of the prevailing direction of their spin alignment. In other words, although the Hamiltonian is invariant with respect to the transformation group **G**, the thermodynamic state is not. This situation is called **spontaneous symmetry breaking**. It can be made manifest by identifying some thermodynamic quantities that should vanish in a thermodynamic state that is symmetrical with respect to the transformation group **G** and that instead do not vanish.

Let us consider, for instance, the **magnetization** M in the Ising model:

$$\langle M \rangle = \sum_i \langle \sigma_i \rangle = \frac{1}{Z} \sum_\sigma \sum_i \sigma_i \exp\left[-\frac{H(\sigma)}{k_B T} \right], \tag{5.35}$$

where the first sum runs over all of the system's 2^N configurations $\sigma = (\sigma_i)$, and $H(\sigma_i)$ is Ising's Hamiltonian (5.31).

Let us assume that $h = 0$, and let us consider the contributions to $\langle \sigma_i \rangle$. For each configuration σ in which $\sigma_i = +1$, there will be a configuration $I\sigma_i = (-\sigma_i)$ in which $\sigma_i = -1$. Since $H(I\sigma) = H(\sigma)$, this configuration will, in principle, have the same Boltzmann weight as the preceding one. We can therefore expect $\langle \sigma_i \rangle$ to always vanish. But then how can phase coexistence occur? The point is that in certain circumstances, a small nonsymmetrical perturbation is sufficient to cause a macroscopic violation of symmetry. It is clear that states in which the spin is prevalently oriented up or down have a much lower energy than those in which the spins are arranged in a disordered fashion. The latter are much more numerous, and therefore prevail at high temperatures. But at low temperatures, they are not accessible (they cost too much energy). The system therefore prefers that the spins be aligned prevalently in one direction. Which direction is indifferent from an energetic point of view—a small perturbation (due to boundary conditions that slightly favor the up direction, or to a weak magnetic field, for instance) is therefore sufficient to make the system prefer one direction over another.

It is possible in this manner to obtain a thermodynamic state in which magnetization is not equal to zero, even though the external macroscopic field h vanishes:

$$\langle M \rangle = \sum_i \langle \sigma_i \rangle = Nm \neq 0. \tag{5.36}$$

The Hamiltonian's symmetry implies that if $M_0 = \langle M \rangle \neq 0$ defines a state of thermodynamic equilibrium, there exists another state of thermodynamic equilibrium (which corresponds to the same values of the intensive variables, but perhaps to different boundary conditions) in which the magnetization is equal to $-M_0$. Therefore, if $\langle M \rangle \neq 0$ for $h = 0$, several states characterized by different values of the magnetization can coexist.

The magnetization's value can be fixed by means of an artifice called **quasi-average**. One applies an external field h—a positive one, for example—which explicitly breaks symmetry, and one evaluates the magnetization in the limit in which this field goes to zero:

$$Nm_0(T) = \lim_{h \to 0^+} \langle M \rangle. \tag{5.37}$$

Because of the symmetry, one has

$$\lim_{h \to 0^-} \langle M \rangle = -Nm_0(T). \tag{5.38}$$

Therefore, $m_0(T)$ is a measure of the degree of spontaneous symmetry breaking. This quantity is known as the **spontaneous magnetization** per particle and vanishes at the critical point.

Let us note that the formal equivalence between the Ising model and the lattice gas allows us to identify the coexistence curve for the latter in the plane (μ, T):

$$\tilde{\mu}_c(T) = -\epsilon\zeta. \tag{5.39}$$

5.8 The Order Parameter

Let us now consider the same problem from a slightly more abstract point of view. Let us suppose that we have a system described by the Hamiltonian H, which is invariant with respect to a transformation group $\mathbf{G}$ applied to the microscopic states σ:

$$H(\mathrm{g}\sigma) = H(\sigma), \qquad \forall \mathrm{g} \in \mathbf{G}. \tag{5.40}$$

If the thermodynamic state is not invariant with respect to all the group's transformations, one says that spontaneous symmetry breaking is present. In order to define the degree of symmetry breaking, one considers an observable M that has the following characteristics:

1. M is extensive: M's value for a system composed of several subsystems in contact is equal to the sum of each subsystem's values.

2. M is transformed according to a nontrivial representation of the group $\mathbf{G}$:

$$M(\mathrm{g}\sigma) = \mathsf{T}(\mathrm{g})M(\sigma), \qquad \forall \mathrm{g} \in \mathbf{G}, \tag{5.41}$$

where $\mathsf{T}(\mathrm{g})$ is a linear operator belonging to some representation of $\mathbf{G}$.

The symmetry breaking then manifests itself by a means of a nonvanishing value of $\langle M \rangle$. With the quasi-average method, it is possible to obtain various values of $\langle M \rangle$ that correspond to the same value of temperature (and of other possible intensive parameters). Because of symmetry, they can be obtained from one another by applying a T(g) transformation.

A quantity endowed with these characteristics is called an **order parameter**. The introduction of this concept into phase transition theory is due to Lev D. Landau (see [Land80], ch. XIV).

Let us now consider a certain number of systems, the transitions that characterize them, and the corresponding order parameters:

1. **Ising antiferromagnet**: Let us assume that the exchange integral J in the Ising Hamiltonian is negative. In the lowest energy states, the neighboring spins tend to be aligned in *antiparallel* fashion. In this case, reflection symmetry ($\sigma_i \to -\sigma_i$) and translational symmetry $\sigma_i \to \sigma_{i+\delta}$ are broken at the same time.

   One can then use the **staggered magnetization** as the order parameter. In a simple cubic lattice in d dimensions, one can identify two intertwined sublattices, such that each point of each sublattice is surrounded by nearest-neighbor points belonging to the other sublattice. Let us now assign a factor ϵ_i, whose value is $+1$ in one sublattice and -1 in the other, to each point i. The staggered magnetization is then defined by

   $$N \equiv \sum_i \epsilon_i \sigma_i. \tag{5.42}$$

   The symmetry group **G** is still the group with two elements (E, I) that correspond to spin identity and spin inversion.

2. **Vector order parameter**: The Ising model describes a ferromagnet in which the spins are constrained to align in parallel or antiparallel fashion in a fixed direction. One often finds situations in which the spins have more degrees of freedom and can point in arbitrary directions in a plane or in space. We will disregard here the quantum nature of the spins. If we denote the spin's components by $\sigma_i^\alpha (\alpha = 1, 2, 3)$, we can define the order parameter $\boldsymbol{M}$ as follows:

   $$M^\alpha = \sum_i e^\alpha \sigma_i^\alpha, \tag{5.43}$$

   where $\boldsymbol{e}^\alpha$ is the versor of the α axis. More generally, one can consider vectorial order parameters with n components, which are transformed by the rotational symmetry in n dimensions, represented by the group $O(n)$. The generalization of Ising's model for $n = 2$ is often called the **XY model**, while that for $n = 3$ is called the **Heisenberg model**.

3. **Einstein condensation**: Sometimes the order parameter is not directly accessible to the experiment. In the Einstein condensation, for example, the order parameter is the condensate's wave function, and the broken symmetry is the gauge symmetry of the first kind, which expresses the invariance with respect to the multiplication of the wave functions by a phase:

   $$\Psi \to e^{i\alpha} \Psi. \tag{5.44}$$

   Since the order parameter is a complex number, and the gauge transformation is isomorphic to a rotation in the plane, it can be assimilated to a vector order parameter with $n = 2$.

There are more complex situations (for example, in liquid crystals) that require the introduction of less intuitive order parameters. What needs to be emphasized is that the identification of the order parameter can be a fairly difficult problem by itself, because there is no method that allows one to identify it a priori. One needs to be guided by physical intuition.

5.9 Peierls Argument

The existence of spontaneous symmetry breaking in the two-dimensional Ising model can be proven by means of an argument made by Peierls [Peie34]. This argument can easily be generalized to more than two dimensions.

I will try to describe this argument without dwelling excessively on rigor—the argument expounded in this fashion can, however, easily be made rigorous. This section will therefore also serve as an introduction to the problems and methods of a rigorous statistical mechanics.

Let us then consider an Ising model defined over a square lattice with $L \times L$ spin. A spin variable $\sigma_i = \pm 1$ is associated with each lattice site i. The system's microstate σ is defined by the collection of the values (σ_i) of these variables for each site i. The system's Hamiltonian is the usual Ising Hamiltonian:

$$\mathcal{H} = -\sum_{\langle ij \rangle} J\sigma_i\sigma_j, \tag{5.45}$$

where the sum over $\langle ij \rangle$ runs over all the pairs of nearest-neighbor sites. This Hamiltonian is invariant with respect to spin inversions defined by

$$\sigma' = \mathrm{I}\sigma, \qquad \sigma_i' = -\sigma_i, \qquad \forall i. \tag{5.46}$$

Spontaneous symmetry breaking occurs when, in the thermodynamic limit, one can have different states of equilibrium in correspondence with the same temperature value T.

In order to clarify our ideas, let us assume that we impose on the system the boundary conditions $+$, in which the spins external to the system are directed up. In this situation, we expect that there will be a slight prevalence of up spins in the entire system at any temperature. On the other hand, if we impose the boundary conditions $-$, in which the spins external to the system are directed down, we will have a prevalence of down spins. The problem is whether this prevalence due to the boundary conditions will remain in the thermodynamic limit.

Let us consider the spin placed at the origin, and let us denote by p the probability that it is equal to $+1$, with the boundary conditions $+$. Symmetry breaking occurs if p remains strictly larger than $1/2$ in the thermodynamic limit. The symmetry of the Hamiltonian guarantees in fact that p is also equal to the probability that the spin at the origin is equal to -1 with the boundary conditions $-$. Therefore, if $p > 1/2$, there will be two different equilibrium states at the same temperature T, which are selected by the boundary conditions.

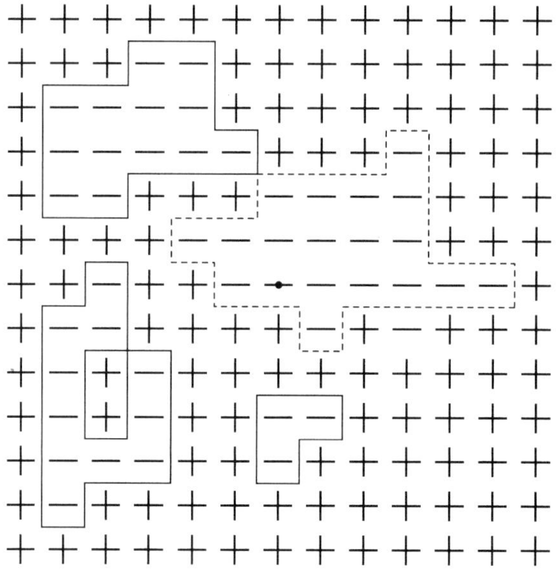

FIGURE 5.4. Configuration of boundaries associated with a spin configuration. The spin at the origin is marked by a black dot.

Let us now evaluate p with the boundary conditions $+$ and show that, if the temperature is low enough, one has $p > 1/2$. Let us consider an arbitrary configuration σ of the spins of the system. We can associate a configuration Γ of boundaries between up and down spins with each such configuration. When the boundary conditions are fixed, this correspondence is such that to each σ corresponds one Γ and vice versa.

Let us focus our attention on the spin at the origin (denoted by a black dot in figure 5.4), and distinguish the configurations into two sets: the set $\sum^+$ contains all σ configurations in which this spin is equal to $+1$, and the set $\sum^-$ contains all the others in which the spin at the origin is equal to -1. It is possible to associate a precise configuration $\sigma^+ \in \sum^+$ with each configuration $\sigma^- \in \sum^-$, thus flipping the direction of all the spins inside the smallest boundary that also contains the origin within it. In the case shown in figure 5.4, it is the discontinuous boundary. The same configuration σ^+ can obviously be obtained starting from several configurations $\sigma^- \in \sum^-$.

One obviously has

$$p = \frac{1}{Z} \sum_{\sigma \in \Sigma^+} e^{-\mathcal{H}(\sigma)/k_B T}, \tag{5.47}$$

where Z is the partition function, and the sum runs over all the configurations belonging to $\sum^+$. Analogously,

$$q = 1 - p = \frac{1}{Z} \sum_{\sigma \in \Sigma^-} e^{-\mathcal{H}(\sigma)/k_B T}. \tag{5.48}$$

It is clear that the energy of a configuration σ is equal to the energy E_0 of the ground state ($\sigma_i = +1, \forall i$) plus the contribution of the boundaries. Since each pair of antiparallel nearest-neighbor spins contributes $+2J$ to energy, one has

$$H(\sigma) = E_0 + 2J\,|\,\Gamma\,|, \tag{5.49}$$

where $|\Gamma|$ is the total length of the boundaries that appear in σ. We will denote the set of all configurations $\sigma' \in \sum^-$ that are applied in σ, by removing the smallest boundary that contains the origin, with $\Psi(\sigma)$, where $\sigma \in \sum^+$, as shown earlier. We will denote this boundary with $\gamma_0(\sigma')$ and its length with $|\gamma_0(\sigma')|$. We will then have

$$p - q = \frac{1}{Z} \sum_{\sigma \in \Sigma^+} \mathrm{e}^{-\mathcal{H}(\sigma)/k_B T} \left[1 - \sum_{\sigma' \in \Psi(\sigma)} \mathrm{e}^{-2J|\gamma_0(\sigma')|/k_B T} \right]. \tag{5.50}$$

We now want to show that the sum $\sum$ over σ' that appears in the right-hand side is certainly smaller than 1 in the thermodynamic limit, if the temperature is low enough. We collect the configurations σ' according to the length $\ell = |\gamma_0(\sigma')|$ of the smallest boundary that contains the origin. Obviously, $\ell \geq 4$. Let us now evaluate the number $g(\ell)$ of configurations σ' that correspond to a certain value of ℓ. Let us assume that the boundary starts from a point i—the first segment can go in four directions, because the boundary cannot retrace its steps. If we neglect the constraint that the path closes, the total number of paths of ℓ length that leave from and return to i is certainly smaller than $4 \times 3^{\ell-1} = 4/3\ 3^\ell$. The starting point i, however, can be arbitrarily chosen among the ℓ boundary points, and the boundary itself can be traveled in two directions. The number of boundaries is therefore certainly smaller than $(4/3)(1/2\ell)3^\ell = 2/(3\ell)3^\ell$. Since the origin could be in any point inside the boundary, one needs to multiply by the area included in the boundary, which is certainly smaller than or equal to the area of a square whose side is $\ell/4$. We thus obtain

$$g(\ell) < \frac{4}{3\ell} 3^\ell \left(\frac{\ell}{4} \right)^2 = \frac{\ell}{24} 3^\ell. \tag{5.51}$$

One should note that in following this line of reasoning, we omitted the constraints imposed by the presence of other boundaries determined by the $\sigma \in \sum^+$ configuration, over which σ' is applied—we are therefore (significantly) overestimating the number of configurations σ'.

To conclude,

$$\sum = \sum_{\sigma' \in \Psi(\sigma)} \mathrm{e}^{-2J|\gamma_0(\sigma')|/k_B T} < \sum_{\ell=4}^{\infty} \frac{\ell}{24} w^\ell, \tag{5.52}$$

where

$$w = 3\mathrm{e}^{-2J/k_B T}. \tag{5.53}$$

Therefore,

$$\sum < \frac{1}{24} \sum_{n=2}^{\infty} (2n)\, w^{2n} = \frac{w^4 (2 - w^2)}{12 (1 - w^2)^2}. \tag{5.54}$$

When $w < w_0 = 0.869756\ldots$, the expression in the right-hand side is certainly smaller than 1. This occurs for temperatures below

$$T_0 = 2J/[k_B \ln(3/w_0)] \simeq 1.61531\ J/k_B. \tag{5.55}$$

We therefore see that $p - q > 0$ for temperatures below T_0, and that therefore spontaneous symmetry breaking occurs, and the corresponding critical temperature is higher than T_0. It is obviously possible to obtain a better bound on T_c with little effort. The exact result, for the two-dimensional Ising model, is $T_c = 2.269\ J/k_B$.

Exercise 5.1 Generalize the preceding result to $d = 3$ dimensions.

5.10 The One-Dimensional Ising Model

The preceding argument cannot be applied in one dimension. It is possible in effect to show that the Ising model in one dimension does not exhibit spontaneous symmetry breaking. The most convincing line of reasoning entails the exact solution of the model, but it is possible to show, by means of an argument due to Landau, that there cannot be discrete symmetry breaking in a one-dimensional system with short-range interactions. This also holds for continuous symmetry. In this section, we arrive at the exact solution of the one-dimensional Ising model, and we describe a (nonrigorous) version of Landau's argument.

5.10.1 *Solution of the One-Dimensional Ising Model*

The model is defined by the Hamiltonian:

$$\mathcal{H} = -J\sum_{i=1}^{N} \sigma_i \sigma_{i+1}, \tag{5.56}$$

where we have imposed periodic boundary conditions:

$$\sigma_{i+N} = \sigma_i, \qquad \forall i. \tag{5.57}$$

The partition function is expressed by

$$Z = \sum_{\sigma} e^{-\mathcal{H}(\sigma)/k_BT} = \sum_{\{\sigma_1,\ldots,\sigma_N} \prod_{i=1}^{N} e^{K\sigma_i,\sigma_{i+1}}, \tag{5.58}$$

where $K = J/k_BT$. By introducing the **transfer matrix** $\mathsf{T} = (T_{\sigma\sigma'})$, defined by

$$T_{\sigma\sigma'} = e^{K\sigma\sigma'}, \tag{5.59}$$

we see that the preceding expression assumes the form of a product of the matrices T, which, taking into account the boundary conditions that impose $\sigma_{n+1} = \sigma_1$, can be written

$$Z = \mathrm{Tr}\,\mathsf{T}^N. \tag{5.60}$$

The transfer matrix T has two eigenvalues, $t_+ = 2\cosh K$ and $t_- = 2\sinh K$, with $t_+ > t_-$. Therefore for $N \gg 1$,

$$\mathrm{Tr}\,\mathrm{T}^N = t_+^N + t_-^N \simeq t_+^N. \tag{5.61}$$

In the thermodynamic limit, therefore,

$$f = \lim_{N\to\infty} \frac{\ln Z}{N} = \ln 2\cosh K. \tag{5.62}$$

This expression does not exhibit singularities for $0 \le K < \infty$, and the model therefore does not exhibit phase transitions at finite temperature.

Exercise 5.2 Consider the spin correlation function

$$C_{ij} = \langle \sigma_i \sigma_j \rangle - \langle \sigma_i \rangle \langle \sigma j \rangle,$$

in the one-dimensional Ising model with a vanishing field. Show that one has

$$\lim_{N\to\infty} C_{ij} = (\tanh K)^{|i-j|}.$$

This expression decays as $e^{-|r_j - r_i|/\xi}$, where ξ is called the coherence length and is equal to

$$\xi = a_0 / |\ln \tanh K|,$$

where a_0 is the lattice step.

Exercise 5.3 Show that, in the presence of a nonvanishing magnetic field h, in the one-dimensional Ising model, one has

$$f = \lim_{N\to\infty} \frac{\ln Z}{N} = \ln\left[\mathrm{e}^K \cosh\lambda + \sqrt{\mathrm{e}^{2K}\sinh^2\lambda + \mathrm{e}^{-2K}} \right],$$

where $\lambda = h/k_B T$.

Exercise 5.4 Obtain the same result with the following method, introduced by Lifson [Lifs64]. Let us consider the system with $\sigma_0 = +1$, and with σ_N free. Then (analogously with what was done with the Peierls argument), the spin configurations σ are univocally defined by the configurations Γ of the boundaries—in other words, of the pairs of nearest-neighbor spins with opposite values. The system is subdivided into n intervals, of length $\ell_i \ge 1$, such that sites belonging to the same interval have the same value as σ_i of the spin.

1. Write the canonical partition function as a function of the ℓ_i's.
2. Move to the corresponding grand canonical partition function by introducing the spin fugacity z.
3. Evaluate the grand canonical partition function as a function of z.
4. Evaluate $\langle N \rangle$ and prove that, in order to obtain the thermodynamic limit $N \to \infty$, it is necessary that $z \to z^*$, where z^* is the value of z closest to the origin for which Z admits a singularity. Evaluate z^*.

5. Show that $f = \lim_{N\to\infty} \ln Z/N = -\ln z^*$, and compare the result with what was already obtained.

5.10.2 *The Landau Argument*

At the temperature $T = 0$, a one-dimensional Ising system will be in one of the two states of minimal energy—for example, in the one in which all spins are aligned in the + direction. The problem is whether this state can persist in the presence of thermal fluctuations at $T > 0$. Let us assume that we switch the direction of a certain number of consecutive spins—we will obtain an excited state, with an energy that is $2J$ above the minimum. The number of states that have this energy is proportional to $\ln N$, because we can arrange each boundary between + and − spins in N different fashions. Therefore, as soon as $k_B T \ln N > 2J$, reversing a macroscopic number of spins leads to an advantage in free energy. But for $N \to \infty$, this will be true for any positive value of T. This argument can immediately be generalized to all cases of short-range interaction.

Exercise 5.5 Let us suppose that the interaction is long range, so that the coupling constant J_{ij} between the i spin and the j spin decays as $|i - j|^{-a}$, where a is a positive constant. Show that Landau's argument is not valid if $a < 2$.

5.11 Duality

The critical temperature of the two-dimensional Ising model was found in 1941 by Kramiers and Wannier [Kram41] before the exact solution to the model, which is due to Onsager [Onsa44]. Kramers and Wannier's argument introduces a transformation—called **duality**—which transforms a high-temperature Ising model into an analogous low-temperature model. This transformation is of great importance, and it can be generalized to other statistical models, even though, in the large majority of cases, the transformed model is not identical to the initial one. We will deal with it only in the case of the Ising model in $d = 2$, by means of an argument that is simpler than the one initially formulated by Kramers and Wannier. This argument will also allow us to introduce the high- and low-temperature expansions, which allow us to obtain information about the behavior of models that cannot be solved exactly.

Let us therefore consider the expression of the partition function of a two-dimensional Ising model at temperature T in a vanishing external field:

$$Z(K) = \sum_{\sigma} \exp\left\{\sum_{\langle ij\rangle} K\sigma_i\sigma_j\right\}, \tag{5.63}$$

where the sum over σ runs over all the 2^N spin configurations and that over $\langle ij\rangle$ runs over the $2N$ pairs of nearest-neighbor sites. Since

$$e^{K\sigma\sigma'} = \cosh K + \sigma\sigma' \sinh K = \cosh K(1 + \sigma\sigma' \tanh K), \tag{5.64}$$

Z can also be written

$$Z(K) = (\cosh K)^{2N} \sum_{\sigma} \prod_{\langle ij \rangle} (1 + \sigma_i \sigma_j \tanh K). \tag{5.65}$$

We can now expand the expression on the right-hand-side into a power series of $t = \tanh K$. The terms of this expansion can be represented by diagrams—we associate a factor $t\sigma_i\sigma_j$ to each segment of the diagram that connects site i with site j (where i and j are nearest neighbors). For each pair $\langle ij \rangle$ of nearest-neighbor sites, there can be at most one $t\sigma_i\sigma_j$ factor. We thus obtain

$$Z(K) = (\cosh K)^{2N} \sum_{\sigma} \sum_{\mathcal{G}} t^{|\mathcal{G}|} \prod_{i} \sigma_i^{n_i}. \tag{5.66}$$

where the sum runs over all the $\mathcal{G}$ diagrams that can be drawn on the lattice, $|\mathcal{G}|$ is the number of bonds that appear in diagram $\mathcal{G}$, and for each site i, n_i is the number of bonds that contain the i site in diagram $\mathcal{G}$. When we sum all the spins' configurations, all the diagrams $\mathcal{G}$ in which n_i is odd for one or more i sites will result in a zero contribution; all the other diagrams will result in a contribution equal to $(\cosh K)^{2N} t^{|\mathcal{G}|}$. Therefore,

$$Z(K) = 2^N (\cosh K)^{2N} \sum_{\mathcal{G}}{}' t^{|\mathcal{G}|}. \tag{5.67}$$

where the sum runs over all the diagrams in which the number of bonds that contain each i site is even (in other words, $n_i = 0, 2, 4, \ldots, \forall i$).

These diagrams can be interpreted as the configuration Γ of boundaries between opposite spins in an Ising model defined on a **dual lattice** that results from the original one, and in which a spin variable is associated with each four-spin **plaquette** in the original lattice.

In figure 5.5, the original model is defined in the vertices of the lattice identified by dashed lines. A diagram $\mathcal{G}$ that appears in the expansion of the partition function of this model is drawn with a continuous line. This diagram can be interpreted as a configuration Γ of boundaries in a spin model defined over the dual lattice—the corresponding spin configuration associates an Ising variable with each plaquette.

We thus obtain

$$\sum_{\mathcal{G}}{}' = \sum_{\Gamma} t^{|\Gamma|}, \tag{5.68}$$

where $|\Gamma|$ is the total length of the boundaries that appear in Γ.

We now consider the expression of the partition function of the Ising model over the dual lattice, and we denote its temperature with T'. We impose boundary conditions $+$. One has

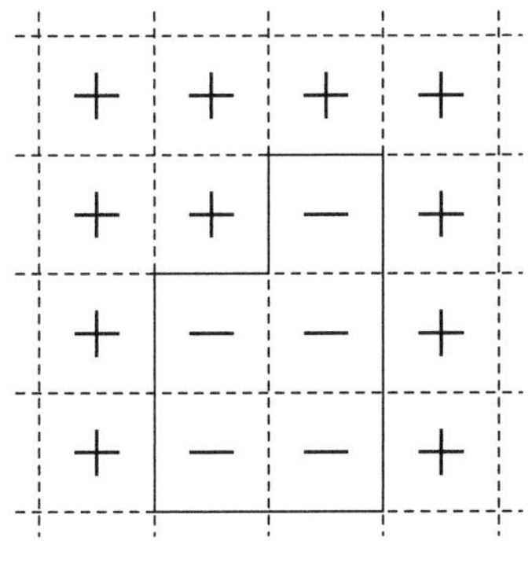

FIGURE 5.5. Diagram $\mathcal{G}$ in the high-temperature expansion of the Ising model interpreted as a configuration Γ of boundaries in an Ising model defined over the dual lattice.

$$Z(K') = \mathrm{e}^{2NK'} \sum_{\Gamma} \mathrm{e}^{-2K'|\Gamma|}, \tag{5.69}$$

where $K' = J/k_{\mathrm{B}}T'$. By comparing this with the expression of $Z(K)$, we obtain the relation

$$t = \tanh K = \mathrm{e}^{-2K'}. \tag{5.70}$$

This is the duality relation for temperatures. It is a duality relation in the sense that if we define $K'(K)$ so that it is satisfied, one gets $K'(K'(K)) = K$—in other words, the $K \longrightarrow K'$ transformation is equal to its inverse. Let us note that if $K \to 0$, one has $K' \to \infty$, and that duality therefore maps high-temperature systems in low-temperature systems, and vice versa. By making use of the relation (5.70), we obtain

$$Z(K') = \mathrm{e}^{2NK'} \frac{1}{2^N} \frac{1}{(\cosh K)^N} Z(K) = \frac{1}{(\sinh 2K)^N} Z(K). \tag{5.71}$$

If the Ising model admits a single transition at $h = 0$, it must occur at a point in which $K' = K$. We thus obtain the critical value K_c:

$$\tanh K_c = \mathrm{e}^{-2K_c}, \tag{5.72}$$

from which one obtains

$$K_c = \frac{1}{2} \ln\left(1 + \sqrt{2}\right) = 0.44069\ldots, \tag{5.73}$$

which corresponds to $k_{\mathrm{B}}T_c \simeq 2.2692J$.

Exercise 5.6 (Duality in the Three-Dimensional Ising Model) By considering the expansion of the partition function of the three-dimensional Ising model as a function of the Γ distribution of boundaries, show that the dual of this model is defined by the Hamiltonian

$$\mathcal{H} = -J \sum_{\mathcal{P}} \prod_{i \in \mathcal{P}} \sigma_i,$$

where the sum runs over all the plaquettes $\mathcal{P}$ composed of the dual lattice's four nearest-neighbor pairs of spin. Find the relation between the temperatures of the Ising model and the corresponding dual model.

Note that, since the dual of the Ising model is not identical to the initial model, this argument does not provide us with information about the critical temperature for $d = 3$.

5.12 Mean-Field Theory

In order to qualitatively calculate the phase diagram for these systems, one resorts to the **mean-field theory**. This approximated theory is extremely useful, and it is the first instrument one resorts to when exploring new models. Obviously, we will present it in the simplest case—in other words, in the Ising model.

Let us suppose that we mentally isolate a spin (which we will denote by a 0) from its environment, in the Ising model. If the spin's state changes, passing for instance from

$\sigma_0 = +1$ to $\sigma_0' = -1$, while all the rest of the environment remains constant, the system's energy varies by

$$\Delta H = 2h + 2 \sum_{i \in \text{p.v.}(0))} J\sigma_i = -\left(h + \sum_{i \in \text{p.v.}(0)} J\sigma_i\right)\Delta\sigma_0, \tag{5.74}$$

where $\Delta\sigma_0 = \sigma_0' - \sigma_0$, and the sum is extended to the nearest-neighbor sites of 0.

This is the same variation that one would obtain for a paramagnet immersed in a field whose value is equal to

$$h_{\text{eff}} = h + \sum_{i \in \text{p.v.}(0)} J\sigma_i. \tag{5.75}$$

The spins close to the one being considered fluctuate, however. As a first approximation, we can assume to describe their action on the spin 0 by this field's mean value—hence, the name **mean-field theory**. We thus obtain

$$h_{\text{eff}} = h + \sum_{i \in \text{p.v.}(0)} J\langle\sigma_i\rangle. \tag{5.76}$$

Let us calculate the mean value of σ for an isolated spin, subjected to an arbitrary external field h. One has

$$\langle\sigma\rangle = \frac{1}{z_0}\sum_{\sigma} \sigma \mathrm{e}^{-h\sigma/k_BT}, \qquad z_0 = \sum_{\sigma} \mathrm{e}^{-h\sigma/k_BT}. \tag{5.77}$$

A simple calculation shows that

$$\langle\sigma\rangle = \tanh\left(\frac{h}{k_BT}\right). \tag{5.78}$$

By exploiting this result, and substituting h with h_{eff}, we obtain

$$\langle\sigma_0\rangle = \tanh\left(\frac{h_{\text{eff}}}{k_BT}\right) = \tanh\left[\frac{1}{k_BT}\left(h + \sum_{i \in \text{p.v.}(0)} J\langle\sigma_i\rangle\right)\right]. \tag{5.79}$$

We can now observe that in the ordered phase, we expect that the mean value of $\langle\sigma_i\rangle$ be the same for all spins. This observation is not as trivial as it seems—if in fact we were to change J's sign, all the observations we made so far would be just as valid, except for this one, because we would instead expect that $\langle\sigma_i\rangle$ assume opposite signs on each of the intertwined sublattices. More generally, the most difficult point of the mean-field theory is that one needs to make a good guess about the type of order one expects. In our case, we posit

$$\langle\sigma_i\rangle = m, \qquad \forall i, \tag{5.80}$$

from which we obtain an equation for m:

$$m = \tanh\left(\frac{h + zJm}{k_BT}\right), \tag{5.81}$$

where ζ is the lattice's coordination number—in other words, the number of nearest-neighbors of a given site.

It is easier to discuss this equation in the form

$$\tanh^{-1} m = \frac{h + zJm}{k_B T}. \tag{5.82}$$

This equation can be solved graphically. Let us consider figure 5.6, in which the curves defined by

$$y = \tanh^{-1} x, \tag{5.83}$$

$$y = \left(\frac{h}{k_B T}\right) + \left(\frac{zJ}{k_B T}\right) x. \tag{5.84}$$

have been traced.

The solution corresponds to the intersection of these two curves. The derivative of $\tanh^{-1} m$ assumes its minimum value in the origin, where it is equal to 1. Therefore, if $k_B T > \zeta J$, the curve and the straight line have only one intersection, and there cannot be finite magnetization (in other words, coexistence). Instead, if $k_B T < \zeta J$, and the external field h is small enough, there will be *three* solutions.

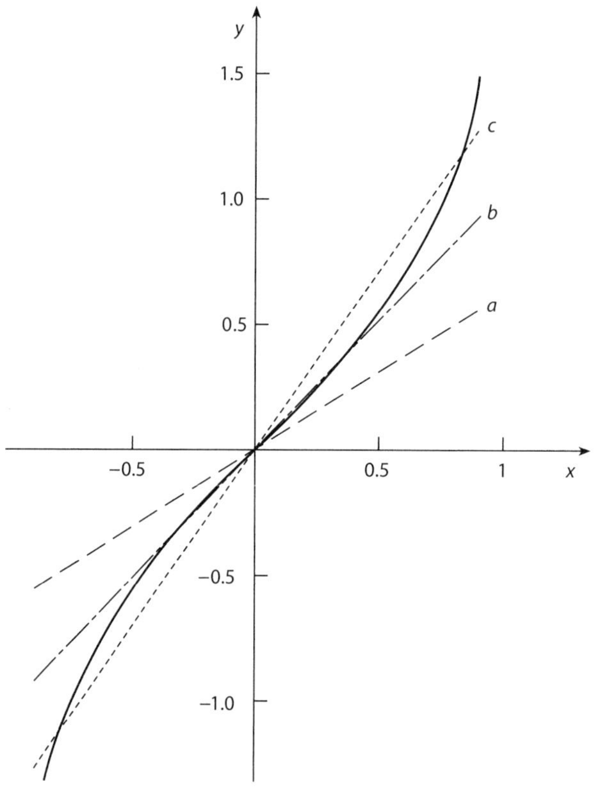

FIGURE 5.6. Graphical solution of the mean-field equation. Continuous line: equation $y = \tanh^{-1} x$. Dashed lines: $y = \zeta J/k_B T$. (a) $T > T_c = \zeta J/k_B$; (b) $T = T_c$; and (c) $T < T_c$.

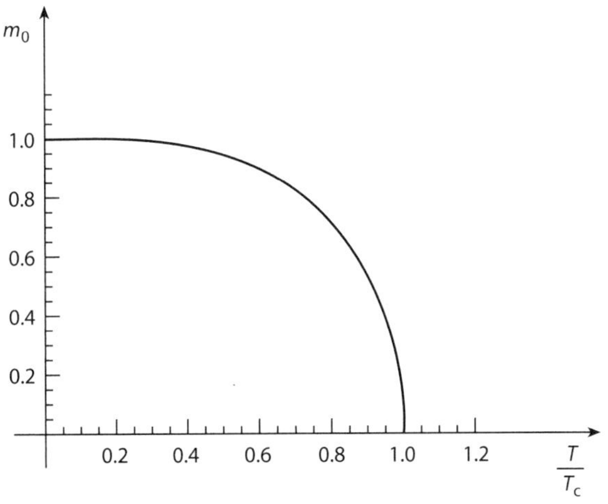

FIGURE 5.7. Spontaneous magnetization for spin m_0 in an Ising model, as a function of $T/T_c = k_B T/\zeta J$, in the mean-field approximation.

Which one is the "physical" solution? Thermodynamics tells us that we must choose the one in which the free energy $E - TS - hM$ is minimal. In our case, we are consistent if we estimate the entropy S from the corresponding valid expression for independent spins:

$$S = -Nk_B\left[\left(\frac{1-m}{2}\right)\ln\left(\frac{1-m}{2}\right) + \left(\frac{1+m}{2}\right)\ln\left(\frac{1+m}{2}\right)\right]. \tag{5.85}$$

By making use of this expression, it is easy to see that the "physical" solution is the one in which m assumes the maximum value (in modulus) of the same sign as the external magnetic field h. We thus obtain the spontaneous magnetization as a function of T, represented by the curve shown in figure 5.7.

In this curve, we have reproduced m as a function of T/T_c, where $k_B T = \zeta J$. We can also calculate the specific heat with a vanishing external field:

$$\begin{aligned} C &= T\left.\frac{\partial S}{\partial T}\right)_h = T\left.\frac{dS}{dm}\frac{\partial m}{\partial T}\right)_h = \\ &= \begin{cases} Nk_B\,(zJm/k_B T)^2\,[1/(1-m)^2 - zJ/k_B T]^{-1}, & \text{for } T < T_c, \\ 0, & \text{for } T > T_c. \end{cases} \end{aligned} \tag{5.86}$$

One should note the specific heat's discontinuity at the critical temperature—in our approximation, the specific heat vanishes above the critical temperature. The phase diagram is particularly simple in the (h, T) plane. The coexistence curve is simply the segment $(0,0) - (0, T_c)$. The magnetization m is subject to a discontinuity along this segment, going from positive h's to negative h's.

5.13 Variational Principle

The reasoning we reproduced in the preceding section someow lacks consistency. Once we had introduced the concept of mean field, we saw that there can be several thermodynamic states (defined by different values of magnetization by spin m) that are solutions of the corresponding equations. To resolve this ambiguity, we recalled the variational principle of free energy, approximating entropy with its expression as obtained for the paramagnet.

In this section, we prove that mean-field theory can be obtained from a variational principle, which allows us to derive both the self-consistency equations we described and the variational principle of free energy in the form we utilized it. This method also allows us to clarify the hypotheses that provide the foundation of the method we used.

The starting point is an inequality valid for each real function $f(x)$ and for each probability distribution:

$$\langle \exp f(x) \rangle \geq \exp \langle f(x) \rangle. \tag{5.87}$$

This inequality is a consequence of the concavity of the exponential function. One can prove this relation by setting

$$\langle e^{f} \rangle = e^{\langle f \rangle} \langle e^{f - \langle f \rangle} \rangle,$$

and using the well-known inequality [equivalent to $x \geq \ln(1 + x)$]

$$e^{x} \geq 1 + x.$$

One thus obtains

$$\langle e^{f} \rangle \geq e^{\langle f \rangle} \langle 1 + f - \langle f \rangle \rangle = e^{\langle f \rangle}.$$

Let us consider, for instance, a variable x that can assume only two values: x_1 with probability p and x_2 with probability $(1 - p)$. As p varies between 0 and 1, $\langle x \rangle$ can assume any value between x_1 and x_2, and $e^{\langle x \rangle}$ will be the value of the corresponding ordinate in the plot of exp (x) (see figure 5.8). On the other hand, the value of $\langle e^{x} \rangle$ will be the value of the corresponding ordinate in the plot of the linear function $pe^{x1} + (1 - p)e^{x2}$. Since this exponential function is concave, the plot of this function always remains above the plot of exp (x) in the interval (x_1, x_2).

This inequality suggests that we can use a variational principle when calculating the free energy. Let us introduce a "trial Hamiltonian" H_0 (arbitrary for the time being). We can then write

$$\begin{aligned} Z &= \sum_{\sigma} \exp\left[-\frac{H(\sigma)}{k_{\mathrm{B}}T}\right] \\ &= Z_0 \left\langle \exp\left[-\frac{H(\sigma) - H_0(\sigma)}{k_{\mathrm{B}}T}\right] \right\rangle_0 \\ &\geq Z_0 \exp\left[-\left\langle \frac{H(\sigma) - H_0(\sigma)}{k_{\mathrm{B}}T} \right\rangle_0\right], \end{aligned} \tag{5.88}$$

where we have defined

$$Z_0 = \sum_{\sigma} \exp\left[-\frac{H_0(\sigma)}{k_{\mathrm{B}}T}\right] = \exp\left[-\frac{(\langle H \rangle_0 - TS_0)}{k_{\mathrm{B}}T}\right], \tag{5.89}$$

$$\langle A(\sigma) \rangle_0 = \frac{1}{Z_0} \sum_{\sigma} A(\sigma) \exp\left[-\frac{H_0(\sigma)}{k_{\mathrm{B}}T}\right]. \tag{5.90}$$

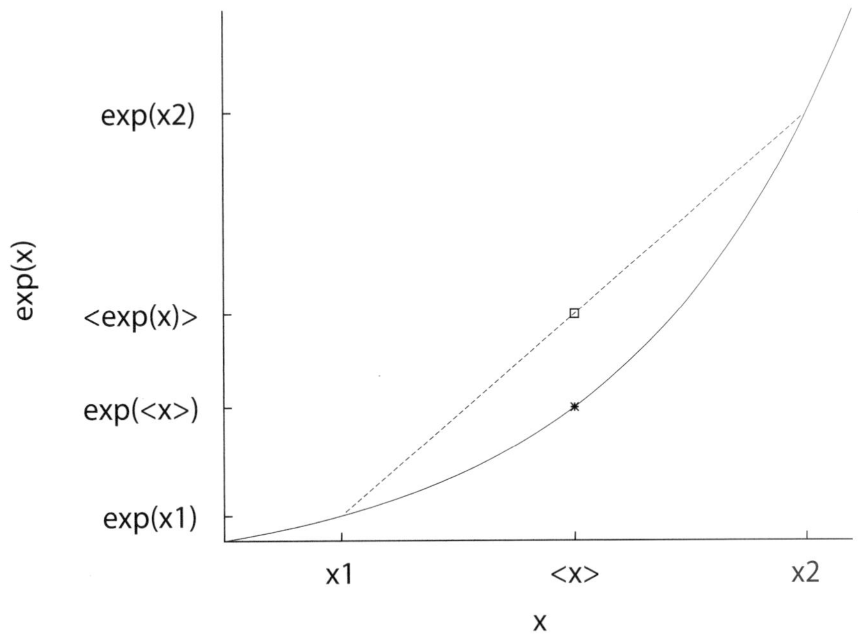

FIGURE 5.8. Concavity of the exponential function. The dashed line represents $\langle \exp x \rangle$ and lies always above the continous curve representing $\exp\langle x \rangle$, when x can take one of two values.

The idea is therefore to look for the maximum of the expression

$$\mathcal{Z} = Z_0 \exp\left[-\left\langle \frac{H - H_0}{k_B T} \right\rangle_0 \right], \tag{5.91}$$

with respect to the possible choices of the trial Hamiltonian H_0. The easiest choice is that of a Hamiltonian with independent spins:

$$H_0 = -\sum_i \lambda \sigma_i. \tag{5.92}$$

Let us now shift to the logarithms, introducing the true free energy $F = -k_B T \ln Z$ and the trial free energy $\mathcal{F} = -k_B T \ln \mathcal{Z}$. The variational principle then takes the form

$$F \leq \mathcal{F}(T, h; \lambda) \equiv \langle H_0 \rangle - TS_0, \tag{5.93}$$

where the internal energy $\langle H \rangle_0$ and the entropy S_0 are evaluated by means of the trial Hamiltonian. Since there is a one-to-one correspondence between the parameter λ and the magnetization (in this simple case!), we can choose m rather than λ as variational parameter. We thus obtain

$$H_0 = -\frac{NzJ}{2} m^2 - Nhm; \tag{5.94}$$

$$\begin{aligned} S_0 = & -Nk_B \left[\left(\frac{1-m}{2} \right) \ln \left(\frac{1-m}{2} \right) \right. \\ & \left. + \left(\frac{1+m}{2} \right) \ln \left(\frac{1+m}{2} \right) \right]. \end{aligned} \tag{5.95}$$

The variational equations are

$$zJm + h = k_B T \tanh^{-1} m, \tag{5.96}$$

which is what we had obtained in the preceding section. This time, however, we additionally obtain the criterion by which to choose this equation's solutions and, even more importantly, the way to generalize it—it will be sufficient to choose some trial Hamiltonians H_0 that are always obtained as sums of Hamiltonians with a few variables independent from the others (so as to allow the calculation of the Z_0's and of the averages $\langle A \rangle_0$'s), but that contain a part of the spin interactions. These generalizations of mean-field theory are called **cluster approximations**.

5.14 Correlation Functions

In the Ising model, the spin variables σ_i relative to different sites are not independent. A measure of their dependence is given by the **correlation function**:

$$C_{ij} \equiv \langle \sigma_i \sigma_j \rangle_c \equiv \langle \sigma_i \sigma_j \rangle - \langle \sigma_i \rangle \langle \sigma_j \rangle. \tag{5.97}$$

This quantity vanishes if σ_i and σ_j are independent.

In this section, we will calculate the C_{ij} in the context of mean-field theory, using a method due to Ornstein and Zernike [Orns14]. In order to do so, let us suppose that we rewrite the partition function, expressing it as a function of an interaction matrix K_{ij}, which is a priori arbitrary (but symmetrical), and of an external field λ_i, which can vary from point to point:

$$Z = \exp\left[\sum_{ij} K_{ij}\sigma_i\sigma_j + \sum_i \lambda_i \sigma_i\right]. \tag{5.98}$$

The physical case corresponds to

$$K_{ij} = \begin{cases} J/k_B T, & \text{if } i \text{ and } j \text{ are nearest neighbors,} \\ 0, & \text{otherwise} \end{cases} \tag{5.99}$$

and $\lambda_i = h/k_B T$.

We observe that

$$\left.\frac{\partial \ln Z}{\partial \lambda_i}\right|_{\text{phys}} = \langle \sigma_i \rangle = m_i, \tag{5.100}$$

$$\left.\frac{\partial^2 \ln Z}{\partial \lambda_i \partial \lambda_j}\right|_{\text{phys}} = \langle \sigma_i \sigma_j \rangle_c. \tag{5.101}$$

The *phys* notation indicates that the derivatives must be calculated with the physical values of K and λ reproduced in equation (5.99) and the line following it. Let us now try to calculate these quantities, using the mean-field expression of the partition function that we derived earlier. We have

$$\ln Z = -S_0/k_B + \sum_{ij} K_{ij} m_i m_j + \sum_i \lambda_i mi, \tag{5.102}$$

where

$$S_0 = -k_B \sum_i \left[\left(\frac{1-m_i}{2}\right) \ln\left(\frac{1-m_i}{2}\right) + \left(\frac{1+m_i}{2}\right) \ln\left(\frac{1+m_i}{2}\right) \right]. \tag{5.103}$$

Taking the derivative of this expression with respect to λ_i, we obtain

$$\frac{\partial \ln Z}{\partial \lambda_i} = \sum_j \frac{\partial \ln Z}{\partial m_j}\bigg|_{\text{phys}} \frac{\partial m_j}{\partial \lambda_i} + m_i, \tag{5.104}$$

where the local magnetization m_i is given by the mean-field equation:

$$0 = \frac{\partial \ln Z}{\partial m_i}\bigg|_{\text{phys}} = -\tanh^{-1} m_i + \sum_{j \in \text{p.v.}(i)} K_{ij} m_j + \lambda_i \big|_{\text{phys}}. \tag{5.105}$$

By taking the derivative of this equation with respect to λ_j, we obtain an equation for the correlation function $C_{ij} = \partial m_i / \partial \lambda_j$:

$$\frac{1}{1-m_i^2} C_{ij} = \sum_{\ell} K_{i\ell} C_{\ell j} + \delta_{ij}. \tag{5.106}$$

In order to solve this equation, it is useful to transform it according to Fourier, taking into account the fact that both K_{ij} and C_{ij} depend only on the vector $\boldsymbol{r}_{ij} = \boldsymbol{r}_j - \boldsymbol{r}_i$, the distance between sites i and j.

We define the Fourier transform according to

$$C(\boldsymbol{k}) = \sum_j \exp(\mathrm{i}\boldsymbol{k} \cdot \boldsymbol{r}_{ij}) C_{ij} \tag{5.107}$$

and analogously for K. Equation (5.106) becomes

$$C(\boldsymbol{k}) = (1 - m^2)[K(\boldsymbol{k}) C(\boldsymbol{k}) + 1], \tag{5.108}$$

and one therefore has

$$C(\boldsymbol{k}) = \frac{1 - m^2}{1 - (1 - m^2) K(\boldsymbol{k})}. \tag{5.109}$$

We can make this expression explicit by keeping in mind that for a simple cubic lattice in d dimensions, one has

$$K(\boldsymbol{k}) = \sum_{j \in \text{p.v.}(i)} \frac{J}{k_{\text{B}} T} \exp(\mathrm{i}\boldsymbol{k} \cdot \boldsymbol{r}_{ij}) = \sum_{\alpha=1}^{d} \frac{J}{k_{\text{B}} T} 2 \cos(k_\alpha a_0), \tag{5.110}$$

where a_0 is the lattice step. For small values of $|\boldsymbol{k}|$, one has

$$K(\boldsymbol{k}) \simeq \frac{J}{k_{\text{B}} T} (\zeta - k^2 a_0^2), \tag{5.111}$$

where $\zeta = 2d$ is the lattice's coordination number. We thus obtain

$$\begin{aligned} C(\boldsymbol{k}) &= (1-m^2)\left[1-(1-m^2)\,(J/k_{\mathrm{B}}T)\sum_{\alpha} 2\cos(k_\alpha a_0)\right]^{-1} \\ &\simeq (1-m^2)\left[1-(1-m^2)\frac{T_{\mathrm{c}}}{T}\left(1-k^2a_0^2\right)\right]^{-1}, \end{aligned} \tag{5.112}$$

in which we have accounted for the fact that the transition temperature is given by $T_{\mathrm{c}} = \zeta J/k_B$.

Let us remark that the Fourier transform of the correlation function, evaluated at $\boldsymbol{k} = 0$, is proportional to the magnetic susceptibility per spin. In fact, one has

$$C(\boldsymbol{k}=0) = \sum_j \frac{\partial\langle\sigma_i\rangle}{\partial\lambda_j} = k_{\mathrm{B}}T\frac{\partial\langle\sigma_i\rangle}{\partial h} = k_{\mathrm{B}}T\chi. \tag{5.113}$$

Calculating this expression for $T > T_{\mathrm{c}}$ (and therefore for $m = 0$), we obtain

$$\chi = \frac{1}{k_{\mathrm{B}}(T-T_{\mathrm{c}})}, \tag{5.114}$$

which diverges when $T \to T_{\mathrm{c}}$. This law is called the **Curie-Weiss law.** On the other hand, it is possible to introduce a length ξ that measures the range of the correlations. Let us set

$$\xi^2 = \frac{1}{2d}\frac{1}{C(\boldsymbol{k}=0)}\sum_j r_{ij}^2 C_{ij}. \tag{5.115}$$

where the $1/2d$ factor has been introduced for convenience's sake. The quantity ξ is called the **coherence** or **correlation length.** It is easy to see that

$$\xi^2 = \sum_\alpha \left.\frac{\partial \ln C(\boldsymbol{k})}{\partial k_\alpha^2}\right|_{\boldsymbol{k}=0} = \frac{Ta_0^2}{2(T-T_{\mathrm{c}})} \propto (T-T_{\mathrm{c}})^{-1}. \tag{5.116}$$

We thus obtain the important result that the correlations' range diverges when $T \to T_{\mathrm{c}}$. This result is not unexpected, given that the divergence of the per spin susceptibility implies that the number of terms different from zero in the expression of the $C(\boldsymbol{k}=0) = \sum C_{ij}$ must diverge, since each term is limited.

It is useful to give an approximate expression of the correlation function in the r space that corresponds to the expressions we derived in the k space. Let us therefore assume that the correlation function in k space is given by the Ornstein-Zernike formula:

$$C(\boldsymbol{k}) \propto \frac{1}{k^2+\xi^{-2}}. \tag{5.117}$$

The Fourier antitransform of this expression is given by

$$C(\boldsymbol{r}) \propto \int \mathrm{d}^d\boldsymbol{k}\,\frac{\mathrm{e}^{-\mathrm{i}\boldsymbol{k}\cdot\boldsymbol{r}}}{k^2+\xi^{-2}}. \tag{5.118}$$

In order to evaluate this expression, let us introduce the identity

$$\frac{1}{x} = \int_0^\infty \mathrm{d}u\,\mathrm{e}^{-ux}, \qquad x > 0. \tag{5.119}$$

We obtain

$$C(\mathbf{r}) \propto \int \mathrm{d}^d \mathbf{k} \int_0^{\infty} \mathrm{d}u\, \mathrm{e}^{-u(k^2+\xi^{-2}) - \mathrm{i}\mathbf{k}\cdot\mathbf{r}}. \quad (5.120)$$

We can now evaluate the Gaussian integral over the k's and obtain

$$C(\mathbf{r}) \propto \int_0^{\infty} \mathrm{d}u\, u^{-d/2} \mathrm{e}^{-u\xi^{-2} - r^2/(4u)}. \quad (5.121)$$

Let us change integration variable, setting

$$w = ur^{-2}. \quad (5.122)$$

We obtain

$$C(\mathbf{r}) \propto r^{2-d} \int_0^{\infty} \mathrm{d}w\, w^{-d/2} \exp\left[-w\left(\frac{r}{\xi}\right)^2 - \frac{1}{4w}\right]. \quad (5.123)$$

If $r \ll \xi$, we can disregard r's dependence in this integral (at least if $d > 2$; otherwise, the integral diverges for large w's). In this manner, we obtain $C(\mathbf{r}) \sim r^{2-d}$. If instead $r \gg \xi$, we can evaluate the integral with the saddle-point method, which one has for $w = w_c = \xi/2r$. We thus obtain

$$C(\mathbf{r}) \sim \mathrm{e}^{-r/\xi}. \quad (5.124)$$

These two behaviors can be summed up by the approximate formula

$$C(\mathbf{r}) \sim \frac{\mathrm{e}^{r/\xi}}{r^{d-2}}, \quad (5.125)$$

which is exact for $d = 3$.

5.15 The Landau Theory

The mean-field theory allows us to obtain an approximate expression for the phase diagram of the system being considered. If, however, we want to aim for the more limited goal of describing the system's behavior in proximity to the critical point, we can resort to a theory that is formally simpler and more easily generalizable—Landau theory.

The mean-field theory requires that we look for the minimum of an effective free energy, defined by

$$\mathcal{F}(h, T; m) = N[f(T; m) - hm], \quad (5.126)$$

expressed as a function of the magnetization per spin m. The trial free energy per spin, expressed as a function of m, has the form

$$\begin{aligned} f(T; m) &= -\frac{zJ}{2} m^2 - Ts(m) \\ &= -\frac{zJ}{2} m^2 + k_B T \left[\left(\frac{1-m}{2}\right) \ln\left(\frac{1-m}{2}\right) \right. \\ &\left. + \left(\frac{1+m}{2}\right) \ln\left(\frac{1+m}{2}\right)\right]. \end{aligned} \quad (5.127)$$

In proximity of the critical point, the equilibrium value of the order parameter m is small, and we can therefore expand this function into a power series:

$$f(T;m) = f(T;0) = \frac{1}{2}a(T)m^2 + \frac{1}{4!}b(T)m^4 + \ldots \tag{5.128}$$

Only the even powers of m appear in this expansion, because the Hamiltonian is symmetrical with respect to the transformation $m \to -m$. It is obviously possible to calculate the coefficients $a(T)$, $b(T)$, and so on explicitly. We have

$$f(T;0) = k_B T \ln 2, \tag{5.129}$$

$$a(T) = k_B T - zJ \propto (T - T_c), \tag{5.130}$$

$$b(T) = 10 k_B T. \tag{5.131}$$

We see that the coefficient a changes sign at the critical temperature, while the coefficient b remains positive. In effect, when we look for $\mathcal{F}$'s minimum with respect to m, we expect to find a vanishing solution for $T > T_c$ and two opposite minima, different from zero, for $T < T_c$. The behavior of $f(T;m)$ for small values of m, and for temperatures close to T_c, is shown in figure 5.9.

Therefore, continuing to follow our goal of studying T_c's neighborhood, we can set:

$$a(T) \simeq a'(T - T_c), \quad a' > 0, \tag{5.132}$$

$$b(T) \simeq b(T_c) = b > 0. \tag{5.133}$$

The equations for $\mathcal{F}$'s minimum take the form

$$a'(T - T_c)m + \frac{b}{3!}m^3 = h. \tag{5.134}$$

For $T > T_c$ we have a real solution:

$$m \simeq \frac{1}{a'(T - T_c)}h = \chi h, \tag{5.135}$$

which confirms the result we had already arrived at:

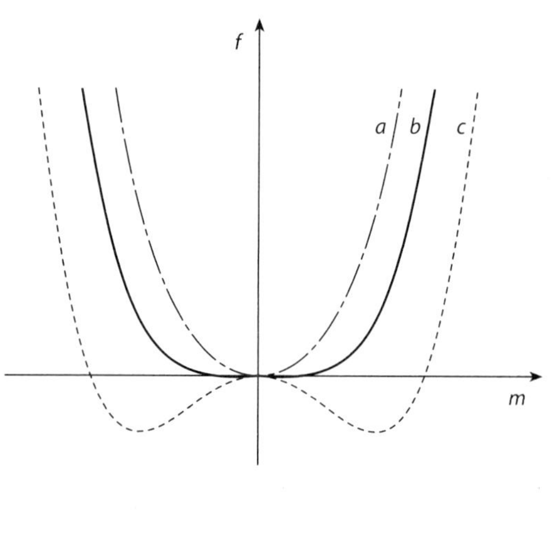

FIGURE 5.9. Landau free energy $f = f(T;m)$ as a function of magnetization m near the critical temperature. (a) $T > T_c$; (b) $T = T_c$; and (c) $T < T_c$.

$$\chi \propto (T - T_c)^{-1}. \tag{5.136}$$

On the other hand, for $T < T_c$, we have *three* solutions—

$$m \simeq \frac{1}{a'(T - T_c)} h, \tag{5.137}$$

but there is one which corresponds to a *maximum* of $\mathcal{F}$ (as is also obvious from figure 5.9). The other two solutions correspond to minima and are given approximately by

$$\frac{b}{3!} m^2 = a' |T - T_c|, \tag{5.138}$$

with additional contributions that vanish for $h \to 0$. We thus obtain the behavior for the spontaneous magnetization close to T_c:

$$m_0(T) = \lim_{h \to 0^+} m(T, h) = \left(\frac{3! a' |T - T_c|}{b} \right)^{1/2} \propto |T - T_c|^{1/2}. \tag{5.139}$$

It is also interesting to calculate m's behavior as a function of h exactly at the critical temperature—we have

$$\frac{b}{3!} m^3 = h, \tag{5.140}$$

and as result, the magnetization is proportional to $h^{1/3}$.

One can also calculate the specific heat. In fact, since

$$S = -\frac{\partial F}{\partial T}, \tag{5.141}$$

one obtains

$$C_h = -\left.\frac{\partial^2 F}{\partial T^2}\right|_h = \begin{cases} 0, & \text{for } T > T_c, \\ \left(\frac{3}{4} a'\right)^2 & \text{for } T < T_c. \end{cases} \tag{5.142}$$

The specific heat is therefore discontinuous at the transition.

All the results for the critical behavior of thermodynamic quantities derive from the hypothesis that the thermodynamic value of the order parameter can be obtained by means of a variational principle, in which a function $f(T; m)$ appears that is (1) analytic and (2) symmetrical with respect to a certain transformation group of m. In our case, this group is composed of the identity and the inversion $m \to -m$.

One can consider more general cases, in which the symmetry group is more complicated. In this case, in the expansion of $f(T; m)$, only the invariants that can be built starting from m can appear. Let us suppose for example that the order parameter is a vector $\mathbf{m} = (m_\alpha)$ in n dimensions and that it transforms according to the rotation group in n dimensions, $O(n)$. The only invariant we can construct is $m^2 = \sum_\alpha m_\alpha^2$. Therefore, $f(T; m)$ is in fact only a function of m^2, and its expansion into a Taylor series has the same form as in the Ising model. The results we obtained are therefore still essentially valid.[1]

[1] There are, however, some effects related to the vectorial nature of the order parameter, such as the existence of a transverse susceptibility, that diverges for $h \to 0$ below T_c.

On the other hand, in more complicated situations, there can be cubic invariants, or invariants proportional to higher odd powers of m. When this happens, the Landau theory predicts that the transition becomes discontinuous. In effect, exactly at the critical temperature, the f has to admit a minimum in $m = 0$, but this is impossible if the term proportional to m^3 does not vanish. Cubic invariants of this type are always present at the fluid–solid transition—therefore, the order parameter of the fluid–solid transition cannot vanish continuously (at least within the context of mean-field theory).

The analyticity of f is due to the very nature of mean-field approximation—the effect of the rest of the sample on a degree of freedom (or on a small number of degrees of freedom) is represented by effective interaction parameters, such as the "effective" magnetic field λ. The trial free energy is then obtained by calculating the partition sum over these few degrees of freedom—it is obvious that this procedure cannot produce nonanalytic dependencies in the order parameter.

It is possible, in principle, to improve the mean-field approximation by considering a larger, but finite, number of degrees of freedom—this improvement, however, cannot modify those results of the mean-field theory that depend only on the analyticity of the trial free energy.

5.16 Critical Exponents

We have thus obtained the behaviors of the various thermodynamic quantities in T_c's neighborhood. These behaviors are described by power-laws, whose exponents are called **critical exponents**. The results we obtained are summarized in table 5.1. These values of the critical exponents follow from the analyticity hypothesis and are called the **classical values**. Given the generality of the hypotheses underlying Landau theory, it was a surprise to see that the experimentally observed values of the critical exponents are not the same as the classical ones—physical systems can be grouped into broad classes, called **universality classes**, within which the exponents' values are constant, and it is actually possible to map the critical behavior of one system into that of another by means of some simple transformations.

Table 5.1 Classical Critical Exponents

Quantity	Behavior	Region	Exponent
C_h	$\|T - T_c\|^{-\alpha}$	$h = 0$	$\alpha = 0$ (discont.)
χ_T	$\|T - T_c\|^{-\gamma}$	$h = 0$	$\gamma = 1$
M	$\|T - T_c\|^{-\beta}$	$h = 0$	$\beta = 1/2$
	$h^{1/\delta}$	$T = T_c$	$\delta = 3$
ξ	$\|T - T_c\|^{-\nu}$	$h = 0$	$\nu = 1/2$

Two challenges remain:

1. Identify the reason for the Landau theory's lack of success.
2. Identify the reason for the existence of the universality classes, and if possible, provide a method for calculating the critical behavior within each class.

Let us observe that the hypothesis that the order parameter has a well-specified value, obtained from free energy's minimum condition, corresponds to the hypothesis that it is possible to disregard its fluctuations. We will see further on that this allows us to define a criterion (called the **Ginzburg criterion**) to evaluate the region of validity for mean-field theory. In order to make this criterion explicit, however, it is necessary to describe the behavior of the critical fluctuations.

5.17 The Einstein Theory of Fluctuations

In order to understand the critical behavior, it is necessary to treat the fluctuations of thermodynamic quantities in small, but macroscopic, systems. This theory of fluctuations is essentially due to Einstein.

In order to clarify our ideas, let us consider a system described by the Ising model, at a given value of the temperature T and magnetic field h. Within this system, let us focus our attention on a small but macroscopic region S. We want to evaluate the probability that the energy and the magnetizations of the spins contained in this region have respectively the values E_S and M_S.

To evaluate this probability, let us consider those microstates ν_S of the subsystem S such that the internal energy of the subsystem is equal to E_S and its magnetization is equal to M_S. The probability that the subsystem is in any microstate of this kind is proportional to the Boltzmann factor $\exp[-(E_S - hM_S)/k_B T]$, with the given values of T and h. Thus, the probability that the subsystem S has these values of internal energy and magnetization is obtained by summing this probability over all the microstates satisfying the condition just mentioned:

$$p(E_S, M_S) = \frac{1}{Z}\sum_{\nu_S} \delta[E_S - E(\nu_S)]\delta[M_S - M(\nu_S)] \times \exp\left(-\frac{E_S - hM_S}{k_B T}\right). \quad (5.143)$$

In this expression, Z is the partition function, which imposes the normalization condition on the probability distribution of the microstates of S, and the delta functions impose the condition on the values of E_S and M_S.

The sum over all microscopic states gives the exponential of the entropy, expressed as a function of the internal energy and the magnetization:

$$\sum_{\nu_S} \delta[E_S - E(\nu_S)]\delta[M_S - M(\nu_S)] = \exp\left[\frac{S_S(E_S, M_S)}{k_B}\right]. \quad (5.144)$$

Therefore,

$$p(E_S, M_S) = \frac{1}{Z}\exp\left[\frac{E_S - TS_S(E_S, M_S) - hM_S}{k_B T}\right] = \exp\left[-\frac{\Delta\mathcal{F}(E_S, M_S; T, h)}{k_B T}\right], \tag{5.145}$$

where

$$\Delta\mathcal{F}(E_S, M_S; T, h) = E_S - TS_S(E_S, M_S) - hM_S - F(T, h), \tag{5.146}$$

is the **availability**. A normalization constant is understood in this expression, and $F(T, h) = k_B T \ln Z$ is the Helmholtz free energy, expressed as a function of the temperature and magnetic field. Note that the availability vanishes for the equilibrium values of energy and magnetization by virtue of the variational principle of the Legendre transform. Availability is often called *free energy difference*—let us remember, however, that unlike actual free energy, it is a function both of intensive variables (in our case, the magnetic field and the temperature) and of fluctuating extensive variables.

Let us observe that we have obtained an apparently nonnormalized probability distribution. Indeed, the quantities that appear explicitly in equation (5.146) are extensive, and are defined up to subextensive corrections. One can easily check that, since the normalization factor behaves like a power of the subsystem's size, imposing the normalization also leads to subextensive corrections, which are therefore negligible.

If, as usually occurs, entropy is a regular function of its arguments around the equilibrium values, then it is possible to deduce a certain number of important relations for the fluctuations.

Let us first consider a system enclosed by nonadiabatic walls, in which the internal energy can fluctuate. We obtain

$$\langle \Delta E^2 \rangle = -k_B \left.\frac{\partial^2 S}{\partial E^2}\right|_V^{-1} = -k_B \left.\frac{\partial (1/T)}{\partial E}\right|_V^{-1} = k_B T^2 \left.\frac{\partial E}{\partial T}\right|_V = k_B T^2 C_V. \tag{5.147}$$

This result obviously coincides with the one obtained for the fluctuations in the canonical ensemble.

Let us now focus our attention on a certain collection (X_i), $i = 1, 2, \ldots, r$ of *extensive* quantities. Let us denote the relative equilibrium values with X_i^0. The first nonvanishing term of the Taylor expansion of availability will be the second one:

$$\frac{\Delta\mathcal{F}}{k_B T} \simeq \frac{1}{2k_B}\sum_{ij}\frac{\partial^2 S}{\partial X_i \partial X_j}\Delta X_i \Delta X_j, \tag{5.148}$$

where $\Delta X_i = X_i - X_i^0$. The linear terms of the expansion vanish due to the variational principle of the Legendre transform. Because of entropy's convexity, this quadratic form must be positive-semidefinite. More particularly, if it is positive-definite, and if it is possible to disregard the further terms of the Taylor expansion, we deduce from equation (5.148) that the distribution of fluctuations of the (X_i)'s is a Gaussian. The correlation matrix is the inverse of the one that appears in this formula:

$$C_{ij}^{-1} = -\frac{1}{k_B}\frac{\partial^2 S}{\partial X_i \partial X_j^0} = \frac{1}{k_B}\left.\frac{\partial (f_j/T)}{\partial X_i^0}\right)_X. \tag{5.149}$$

We have made explicit the fact that the derivative is taken by keeping the average values of the other extensive variables constant. This result coincides with what can be obtained by considering the corresponding generalized ensemble:

$$C_{ij} = \langle \Delta X_i \Delta X_j \rangle = \frac{\partial^2 \ln Z}{\partial (f_i/k_B T)\, \partial (f_j/k_B T)} = k_B \left.\frac{\partial X_i^0}{\partial (f_j/T)}\right)_{f/T}. \tag{5.150}$$

In this case, the derivative is taken by keeping the values of the ratios f/T constant. Let me remind you that the force conjugated with internal energy is $1/T$. These expressions are equivalent, since in effect one has

$$\sum_\ell \left.\frac{\partial X_i^0}{\partial (f_\ell/T)}\right)_{f/T} \left.\frac{\partial (f_\ell/T)}{\partial X_j^0}\right)_X = \delta_{ij}. \tag{5.151}$$

This approach to the statistics of fluctuations is not limited to just the usual thermodynamic quantities. More specifically, it can be generalized to the study of the spatial distribution of magnetization density.

We can, for instance, evaluate the probability that a certain Fourier component of magnetization density has a given value ϕ_k. The component of the magnetization density is an observable, because it is expressed by

$$\phi_k = \sum_j e^{-i k \cdot r_i} \sigma_j. \tag{5.152}$$

This probability is proportional to $\exp(-\Delta \mathcal{F}(\phi_k)/k_B T)$, where, for small fluctuations, one has

$$\frac{\Delta \mathcal{F}(\phi_k)}{k_B T} \simeq \frac{1}{2N}\frac{|\phi_k|^2}{C(\boldsymbol{k})}. \tag{5.153}$$

In this expression, $C(\boldsymbol{k})$ is the Fourier transform of the correlation function:

$$C(\boldsymbol{k}) = \frac{1}{N}\langle |\phi_k|^2 \rangle = \sum_j e^{i k \cdot r_{ij}} \langle \sigma_i \sigma_j \rangle_c. \tag{5.154}$$

For small values of $|\boldsymbol{k}|$, one can approximate $C(\boldsymbol{k})$ by means of its Ornstein-Zernike expression:

$$C(\boldsymbol{k}) = \frac{k_B T \chi}{1 + k^2 \xi^2}, \tag{5.155}$$

where ξ is the coherence length, and χ the susceptibility per spin. The expressions of $\chi$and ξ are not necessarily given correctly by mean-field theory. In order for this expression to be valid, it is sufficient that $C^{-1}(\boldsymbol{k})$ can be expanded into a Taylor series as a function of k^2—in other words, basically that the correlations be of finite range.

We have so far considered the fluctuations of extensive quantities in a finite region of the system, which are in contact with a much larger system, and therefore characterized by well-defined values of the corresponding intensive quantities (generalized forces and temperature). If we consider an isolated system, we can ask ourselves whether it is possible to observe analogous fluctuations in the intensive quantities. This problem is still the subject of discussion. One can formally associate corresponding fluctuations of the intensive variables with the fluctuations of the extensive variables by means of the equations of state. This procedure is fairly arbitrary however and, above all, does not allow one to obtain further physical insights.

5.18 Ginzburg Criterion

Let us now suppose that we are considering the fluctuations of the magnetization as we approach the critical temperature. We can estimate the availability by using the results of mean-field theory (or, equivalently, of the Landau theory):

$$\Delta\mathcal{F} \simeq \frac{N}{2}a'(T-T_c)\,m^2 + \frac{Nb}{4!}m^4. \tag{5.156}$$

As long as $T > T_c$, the first term dominates the second one when $N \to \infty$. The magnetization fluctuations are therefore Gaussian and $O(N^{-1/2})$. The variance diverges as we get closer to the critical temperature, and exactly at T_c, the fluctuations are no longer Gaussian:

$$p(m) \propto \exp\left(-\frac{Nb}{4!k_BT_c}m^4\right). \tag{5.157}$$

Close to the critical temperature, the fluctuations become very intense and are spatially correlated over very large distances—we have in fact seen that the coherence length diverges for $T \to T_c$. It is therefore essential to understand under which conditions it is possible to neglect the consequences of the fluctuations, and therefore the mean-field's theory's predictions can be considered valid, and when they become unreliable.

In order to estimate the fluctuations' relative importance, we employ a criterion that is due to Ginzburg [Ginz60]. The idea is to compare the mean value of the order parameter as predicted by mean-field theory with the fluctuations predicted by Ornstein-Zernicke's theory.

Let us consider the expression (5.155) of the Fourier transform of the correlation function, which we will rewrite (up to a constant factor) in the form

$$C(\mathbf{k}) = \frac{1}{k^2+t}, \tag{5.158}$$

where $t \propto |T-T_c|$.

By evaluating the Fourier antitransform of this expression, we obtain an estimate of the local spin fluctuations:

$$\langle\Delta\sigma_i^2\rangle = \langle\sigma_i^2\rangle - \langle\sigma_i\rangle^2 \propto \int \mathrm{d}^d\mathbf{k}\,C(\mathbf{k}) \propto t^{(d-2)/2} + \text{regular terms}. \tag{5.159}$$

The "regular" terms come from larger values of $|\mathbf{k}|$, for which the expression (5.155) is no longer applicable. They do not have consequences for the critical behavior (at least if $d > 2$), and we can forget them as far as our argument is concerned.

The order of magnitude of the spin's critical fluctuations is therefore given by

$$\Delta\sigma \propto \sqrt{\langle \Delta\sigma_i^2 \rangle} \propto t^{(d-2)/4}. \tag{5.160}$$

We want to compare this fluctuation with the mean value of the order parameter, and we therefore have

$$\langle \sigma \rangle \propto t^{1/2}. \tag{5.161}$$

As long as $\Delta\sigma \ll \langle\sigma\rangle$ the mean-field theory's predictions remain valid. This also occurs for $t \to 0$, as long as $d > 4$. We can therefore expect some deviations from mean-field behavior when $d \leq 4$.

This argument can be made more quantitative by reinstating the coefficients we gave as understood, and by evaluating them sufficiently far from the critical point. It can also be formulated in a more rigorous manner, by introducing an approximation scheme that reproduces mean-field theory at the lowest order and provides corrections to it as a systematic power series expansion of a fictitious parameter. The Ginzburg criterion then follows from the comparison of the second and first terms of this expansion [Amit84, p. 105].

In conclusion, we can expect that in two and three dimensions, the asymptotic behavior in proximity of the critical point will not be well described by mean-field theory.

5.19 Universality and Scaling

The values of the critical exponents we calculated on the rare occasions in which it is possible to do so exactly, or that were obtained from real or numerical experiments, are in effect not the same as those of the classical exponents. In table 5.2, we report the values of the critical exponents for a certain number of physical systems. The exponent η characterizes the behavior of the correlation function exactly at the critical temperature—in Fourier space, one has

$$C(\mathbf{k}, T = T_c) \propto |\mathbf{k}|^{-2+\eta}. \tag{5.162}$$

These values are obviously not equal to the classical values. One can notice, however, that the exponents of the first three systems are mutually compatible, and so are the exponents of the last two—the exponents seem to depend only on the dimensionality of the order parameter. This conjecture is corroborated by considering the exponents obtained numerically (with different methods) for some statistical mechanics models on a lattice, which are shown in table 5.3. Analogously, the measurable exponents of the λ transition of helium correspond to those of the planar model, while the ferromagnetic or antiferromagnetic transitions characterized by a vectorial order parameter in three dimensions are described by the same exponents. These observations have led to the formulation of the **universality hypothesis**—phase transitions can be sorted into a small number of

Table 5.2 Critical Exponents of Different Physical Systems

	Xe	Bin. mixt.	β-brass	^{4}He	Fe	Ni
n	1	1	1	2	3	3
α	<0.2	$0.113\pm.005$	$0.05\pm.06$	-0.014 ± 0.16	$-0.03\pm.12$	$0.04\pm.12$
β	$0.35\pm.15$	$0.322\pm.002$	$0.305\pm.005$		$0.37\pm.01$	$0.358\pm.003$
γ	$1.3\pm^{.1}_{.2}$	$1.239\pm.002$	$1.25\pm.02$		$1.33\pm.015$	$1.33\pm.02$
δ	$4.2\pm^{.6}_{.3}$				4.3 ± 1	$4.29\pm.05$
η	$0.1\pm.1$	$.0017\pm.015$	$0.08\pm.07$		$0.07\pm.04$	$0.041\pm.01$
ν	≈0.57	$0.615\pm.006$	$0.65\pm.02$	$0.672\pm.001$	$0.69\pm.02$	$0.64\pm.1$

Note: The first three columns are systems with a scalar order parameter ($n=1$): Xe at its critical point, different binary mixtures at their consolution point, and β-brass, a form of copper and tin alloy. The transition $\lambda(\text{He}^{\text{I}}\text{-}\text{He}^{\text{II}})$ corresponds to a planar order parameter ($n=2$), while Fe and Ni are Heisenberg ferromagnets ($n=3$) with weak anisotropy. I have included only exponents that have been actually measured. More specifically, in the case of the λ transition of helium, the order parameter is not accessible, and therefore the exponents β, γ, δ, and η cannot be measured.

Table 5.3 Critical Exponents for Different Models as a Function of the Dimensionality *n* of the Order Parameter and the Dimensionality *d* of Space.

	Mean-field	Ising $d=2$	Ising $d=3$	Heisenberg	Spherical
(n, d)		(1,2)	(1.3)	(3,3)	$(\infty, 3)$
α	0 (disc.)	0 (log)	$0.119\pm.006$	$-0.08\pm.04$	-1
β	1/2	1/8	$0.326\pm.004$	$0.38\pm.03$	1/2
γ	1	7/4	$1.239\pm.003$	$1.38\pm.02$	2
δ	3	(15)	$4.80\pm.05$	$4.65\pm.29$	5
η	0	1/4	$0.024\pm.007$	$0.07\pm.06$	0
ν	1/2	1	$0.627\pm.002$	$0.715\pm.02$	1

Note: The exponent δ of the two-dimensional Ising model is conjectured; the others are obtained from the exact solution. Exponents obtained from an exact solution of the spherical model, which corresponds to $n\to\infty$, are also shown.

universality classes, characterized by the dimensions of the system and of the order parameter, and the critical exponents are equal each class.

In fact, one can observe a stronger relation. Let us consider the equation of state for the order parameter, expressed as a function of h and of

$$t \equiv \frac{T - T_c}{T_c} \tag{5.163}$$

for two different systems:

$$m^{(1)} = m^{(1)}(t, h), \tag{5.164}$$

$$m^{(2)} = m^{(2)}(t, h). \tag{5.165}$$

It is then possible to map the two equations of state onto each other by rescaling m, h, and t by an appropriate positive factor:

$$m^{(1)}(t, h) = \lambda_m m^{(2)}(\lambda_t t, \lambda_h h). \tag{5.166}$$

A relation of this kind, on the other hand, must also be valid if one is discussing the *same system*—by arbitrarily choosing one of the scale factors, we can choose the other two so that

$$m(t, h) = \lambda_m m(\lambda_t t, \lambda_h h). \tag{5.167}$$

Let us consider what occurs at $h = 0$, $t < 0$. On the one hand, we obtain

$$m \propto |t|^{\beta}, \tag{5.168}$$

and on the other,

$$m(t, h = 0) = \lambda_m m(\lambda_t t, h = 0). \tag{5.169}$$

By choosing $\lambda_t = |t|^{-1}$, we obtain

$$\lambda_m = \lambda_t^{-\beta}. \tag{5.170}$$

Following the same reasoning at $t = 0$, we obtain

$$\lambda_m = \lambda_h^{-1/\delta}. \tag{5.171}$$

The magnetization is therefore a general homogeneous function of its arguments:

$$m(\lambda t, \lambda^{\beta\delta} h) = \lambda^{\beta} m(t, h). \tag{5.172}$$

Since the magnetization is the derivative of the free energy with respect to h, an analogous relation also exists for free energy density (or, more exactly, for its singular part):

$$f(\lambda t, \lambda^{\beta\delta} h) = \lambda^{a} f(t, h). \tag{5.173}$$

We can determine the exponent a from the behavior of specific heat. We know that

$$C_h \propto \frac{\partial^2 f}{\partial t^2} \tag{5.174}$$

One therefore has

$$C_h(\lambda t, 0) = \lambda^{a-2} C_h(t, 0), \tag{5.175}$$

from which we obtain

$$a - 2 = -\alpha. \tag{5.176}$$

The free energy density therefore satisfies the general homogeneity relation:

$$f(\lambda t, \lambda^{\beta\delta} h) = \lambda^{2-\alpha} f(t, h). \tag{5.177}$$

This expression vanishes at the critical point—obviously, this does not mean that *all* the free energy vanishes at the critical point, but only its singular part.

By taking the derivative of this relation, one can express the exponents of the thermodynamic quantities as a function of only two exponents, α and $\beta\delta$, which appear in the expression of the free energy. By taking its derivative with respect to h, for example, one obtains the expression of the magnetization:

$$\lambda^{\beta\delta} m(\lambda t, \lambda^{\beta\delta} h) = \lambda^{2-\alpha} m(t, h). \tag{5.178}$$

One therefore has

$$\beta = 2 - \alpha - \beta\delta. \tag{5.179}$$

By taking once more the derivative, one obtains the relation for susceptibility:

$$\lambda^{2\beta\delta} \chi(\lambda t, \lambda^{\beta\delta} h) = \lambda^{2-\alpha} \chi(t, h), \tag{5.180}$$

from which we obtain the exponent γ:

$$-\gamma = 2 - \alpha - \beta\delta. \tag{5.181}$$

These two expressions imply a relation between the first three exponents:

$$\alpha + 2\beta + \gamma = 2. \tag{5.182}$$

Relations of this type (known as **scaling laws**) can be written between any three exponents, and follow from the homogeneity of the free energy and from the fact that only two independent exponents appear in it. They are well satisfied by the experimentally measured exponents, and currently there are few doubts that they are valid for the asymptotic exponents.

Another relation of this type connects β, γ, and δ:

$$\gamma = \beta(\delta - 1). \tag{5.183}$$

It is enlightening to consider the implication of these relations for the coherence length and for the correlation function. The lines of reasoning we have followed show that the coherence length must itself also satisfy a general homogeneity relation:

$$\xi(\lambda t, \lambda^{\beta\delta} h) = \lambda^{-\nu} \xi(t, h). \tag{5.184}$$

An analogous relation must also hold for the correlation function, where one will, however, also have to rescale the wave vector:

$$C(\lambda^{b} k, \lambda t, \lambda^{\beta\delta} h) = \lambda^{c} C(k, t, h). \tag{5.185}$$

In order to determine the exponents b and c, let us observe that for $k = 0$, the correlation function is proportional to the susceptibility. We thus obtain

$$c = -\gamma. \tag{5.186}$$

On the other hand, at $t = h = 0$, we must have

$$C(k) \propto k^{-2+\eta}. \tag{5.187}$$

Therefore,

$$\frac{\gamma}{b} = 2 - \eta. \tag{5.188}$$

Then, recalling that the coherence length is defined by

$$\xi^2 \propto \frac{1}{C(0)} \left.\frac{\partial C(k)}{\partial k^2}\right|_{k=0}, \tag{5.189}$$

and comparing with the relation for the ξ, we obtain

$$b = \nu. \tag{5.190}$$

We have thus obtained another scaling law:

$$\frac{\gamma}{\nu} = 2 - \eta. \tag{5.191}$$

And last, we want to connect the exponent ν to the other thermodynamic variables. To do this, we observe that due to Einstein's theory of fluctuations, we expect that the free energy of a fluctuation (which, as we know, is correlated over distances on the order of the coherence length) should be of the order of magnitude of $k_B T_c$—in other words, a constant. When the coherence length is multiplied by a factor $\ell = \lambda^{-\nu}$, the contribution of the fluctuations to the free energy density is multiplied by a factor $\ell^{-d} = \lambda^{d\nu}$, where d is the system's dimensionality. We thus obtain the scaling law:

$$2 - \alpha = d\nu. \tag{5.192}$$

5.20 Partition Function of the Two-Dimensional Ising Model

In this section, I report the solution of the two-dimensional Ising model in zero magnetic field, as reported by Vdovichenko [Vdov64 and Vdov65]. The model was first solved by Lars Onsager in 1944 [Onsa44] by a mathematical *tour de force*. Simpler derivations were found later by Kac and Ward [Kac52]; Schulz, Mattis, and Lieb [Schu64]; and others. The present solution was inspired by Kac and Ward's solution, but is simpler. It became widely known due to Landau and Lifshitz's treatise on theoretical physics [Land80, p. 4798ff] .

We consider a system of $N = L^2$ Ising spins placed on a square lattice. Thus the spin placed at the (k, ℓ) lattice point is denoted by $\sigma_{k\ell}$ and one has $\sigma_{k\ell} = \pm 1$, $k, \ell \in \{1, \dots, L\}$. The Hamiltonian $H(\{\sigma\})$ is given by

$$H(\{\sigma\}) = -\sum_{k\ell} \left[J(\sigma_{k\ell}\sigma_{k,\ell+1} + \sigma_{k\ell}\sigma_{k+1,\ell}) + h\sigma_{k\ell} \right], \tag{5.193}$$

where we have assumed periodic boundary conditions:

$$\sigma_{k+L,\ell} = \sigma_{k,\ell+L} = \sigma_{k\ell}, \qquad \forall k, \ell. \tag{5.194}$$

We set $h = 0$ from now on. Then, we have seen in section 5.11 that the partition function can we written

$$Z(K) = \left(\frac{2}{1-t^2}\right)^N \sum_{\mathcal{G}}{}' t^{|\mathcal{G}|}, \tag{5.195}$$

where

$$t = \tanh \frac{J}{k_B T}, \tag{5.196}$$

and the sum runs over all diagrams $\mathcal{G}$ that can be drawn on the lattice, such that (1) each bond appears at most once, and (2) at each vertex, only an even number of bonds (zero, two, or four) can meet. In this expression, $|\mathcal{G}|$ is the number of bonds that appear in the diagram $\mathcal{G}$. Then, this expression can be written in the form

$$S = \sum_{\mathcal{G}}{}' t^{|\mathcal{G}|} = \sum_r t^r g_r, \tag{5.197}$$

where g_r is the total number of diagrams satisfying the two preceding rules and containing exactly r bonds.

We will now evaluate this expression by transforming it into a sum over loops. The resulting expression will then be evaluated by reducing it to a random-walk problem.

A generic diagram $\mathcal{G}$ can be considered as a collection of **loops**. A loop is the trajectory of a walk that starts and ends on the same site. However, the decomposition of a diagram into loops is ambiguous if there are self-intersections, that is, if there are vertices where four bonds meet. Let us consider, for example, the diagram in figure 5.10. It can be considered as the collection of two loops that meet at one vertex (a), or as a single loop whose path does intersect itself (c) or does not (b). In order to obtain a nonambiguous sum, we assign to each diagram a factor $(-1)^n$, where n is the number of intersections. In this situation, the contribution of case (c) will be opposite to that of case (b), and they cancel out, leaving only the contribution of case (a). One can easily realize, then, that with this convention, the contribution of diagrams in which three bonds meet at a vertex identically vanishes, as can be seen in figure 5.11. In this way, the sum over all diagrams $\mathcal{G}$ is reduced to a sum over all loops, in which each loop appears with a weight proportional to $(-1)^n$, where n is the number of self-intersections. Notice that we do not allow vertices connected to only one bond and, therefore, the possibility that a walker gets back in its steps.

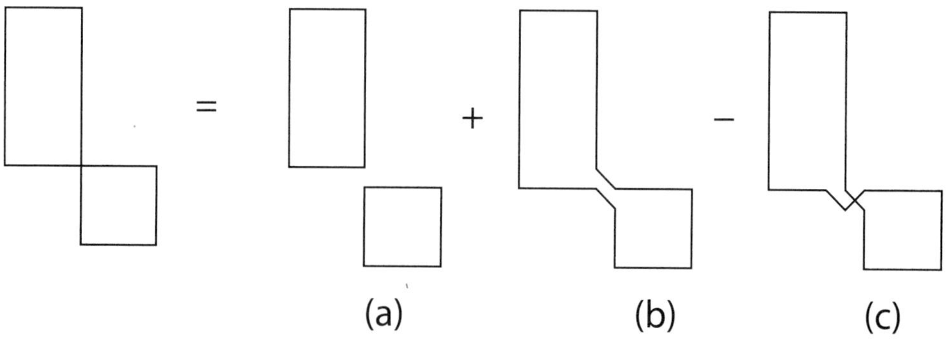

FIGURE 5.10. A diagram with self-intersections can be decomposed in several different ways into loops.

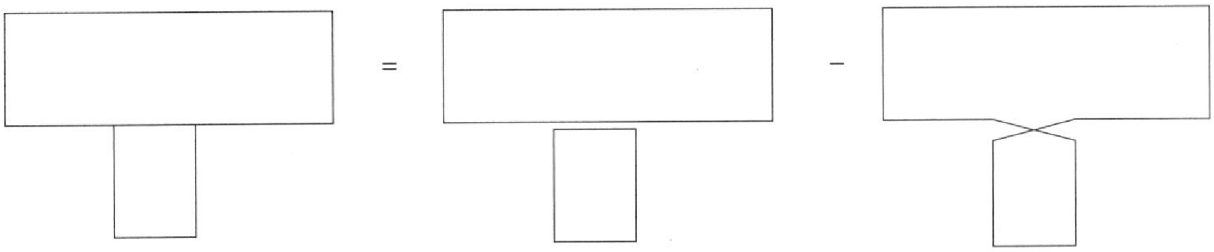

FIGURE 5.11. A diagram with a three-bond vertex can be obtained as the sum of two diagrams with numbers of intersections that differ by one. Their contributions cancel out.

Now, we can express the number of self-intersections of a loop by means of the following trick. It is well known that the total angle through which the tangent angle to the trajectory of a walker performing a loop turns around is given by $2\pi(\ell+1)$, where the parity of ℓ is equal to the parity of the number of intersections n. Thus, if we assign a factor $e^{i\phi/2}$ to each lattice point with turning angle ϕ, then we will have at the end of the loop a factor $(-1)^{\ell+1}=(-1)^{n+1}$, where n is the number of intersections. With this counting, each diagram made up of s loops will give a contribution proportional to $(-1)^{s+n}$.Thus we have to multiply this contribution by $(-1)^s$ in order to have the required sign in equation (5.197).

In order to evaluate the angle ϕ, it is convenient to deal with directed loops. Let us denote by f_r the sum over all undirected loops consisting of r bonds (taking into account the factors t^r and $e^{i\phi n/2}$). Then, the sum over all double loops of ℓ bonds will be given by

$$\frac{1}{2}\sum_{r_1+r_2=\ell} f_{r_1}f_{r_2},$$

taking into account the possible permutations of the loops. Thus we have in general

$$S=\sum_{s=1}^{\infty}(-1)^s\frac{1}{s!}\sum_{r_1,r_2,\dots,r_s=1} f_{r_1}f_{r_2}\cdots f_{r_s}=\exp\left\{-\sum_{r=1}^{\infty}f_r\right\}. \tag{5.198}$$

In going from undirected to directed loops, each loop is encountered twice, and thus if we denote by ν_r the sum of the contributions of directed loops with r bonds, we have

$$S=\exp\left\{-\frac{1}{2}\sum_{r=1}^{\infty}\nu_r\right\}. \tag{5.199}$$

We will now evaluate ν_r. Given a lattice point (k,ℓ), let denote the possible directions as follows:

$$\begin{aligned}
&N\colon\ (k,\ell)\longrightarrow(k,\ell+1),\\
&E\colon\ (k,\ell)\longrightarrow(k+1,\ell),\\
&S\colon\ (k,\ell)\longrightarrow(k,\ell-1),\\
&W\colon\ (k,\ell)\longrightarrow(k-1,\ell).
\end{aligned}$$

Let us denote by $W_r(k\ell\nu|k_0\ell_0\nu_0)$ the sum of all contributions of r-bond diagrams starting from lattice point $k_0\ell_0$ in the direction $\nu_0\in\{\mathrm{N},\mathrm{E},\mathrm{S},\mathrm{W}\}$ and ending in lattice point (k,ℓ) in the direction ν. Each bond occurs with a factor $te^{i\phi/2}$, where ϕ is the change of direction in going to the next bond.

Then, it is possible to write a linear recursion relation for W_r:

$$W_{r+1}(k\ell\nu|k_0\ell_0\nu_0) = \sum_{k'\ell'\nu'} T_{k\ell\nu,k'\ell'\nu'} W(k'\ell'\nu'|k_0\ell_0\nu_0). \tag{5.200}$$

The transition matrix $\mathsf{T} = (T_{k\ell\nu,k'\ell'\nu'})$ has the expression

$$T_{k\ell\nu,k'\ell'\nu'} = tA_{\nu\nu'}\delta_{k',k+\alpha(\nu')}\delta_{\ell',\ell+\beta(\nu')}, \tag{5.201}$$

where

$$\alpha(\mathrm{N}) = 0, \qquad \beta(\mathrm{N}) = -1,$$
$$\alpha(\mathrm{E}) = -1, \qquad \beta(\mathrm{E}) = 0,$$
$$\alpha(\mathrm{S}) = 0, \qquad \beta(\mathrm{S}) = +1,$$
$$\alpha(\mathrm{W}) = +1, \qquad \beta(\mathrm{N}) = 0.$$

The matrix $\mathsf{A} = (A_{\nu\nu'})$ (where $\nu, \nu' = \mathrm{N, E, S, W}$) is given by

$$\mathsf{A} = \begin{pmatrix} 1, \omega, 0, \omega^* \\ \omega^*, 1, \omega, 0 \\ 0, \omega^*, 1, \omega \\ \omega, 0, \omega^*, 1 \end{pmatrix}, \tag{5.202}$$

where

$$\omega = e^{i\pi/4}, \tag{5.203}$$

and ω^* is the complex conjugate of ω.

The connection between the weights W and the loop contributions v_r is given by

$$v_r = \frac{b_r}{r}, \tag{5.204}$$

where

$$b_r = \mathrm{Tr}\,\mathsf{W}_r = \sum_{k\ell\nu} W(k\ell\nu|k\ell\nu). \tag{5.205}$$

The factor $1/r$ comes from the fact that a single diagram with r bonds can be obtained from r different walks, with different the starting points. Now, from equation (5.200), we have

$$\mathrm{Tr}\,\mathsf{W}_r = \mathrm{Tr}\,\mathsf{T}^r = \sum_i \lambda_i^r, \tag{5.206}$$

where λ_i are the eigenvalues of the matrix T. From this equation, taking into account equations (5.199) and (5.204), we obtain

$$S = \exp\left\{-\frac{1}{2}\sum_{r,i}\frac{1}{r}\lambda_i^r\right\} = \exp\left\{\frac{1}{2}\sum_i \ln(1-\lambda_i)\right\} = \prod_i \sqrt{1-\lambda_i}. \tag{5.207}$$

Thus the problem boils down to the diagonalization of the matrix T. One can see from equation (5.201) that T depends only on the differences in the indices k, ℓ. It can thus be diagonalized by a Fourier transformation. We set

$$T_{\nu\nu'}(m,n) = \sum_{k\ell} \mathrm{e}^{-2\pi\mathrm{i}(mk+n\ell)}\, T_{k\ell\nu,00\nu'}. \tag{5.208}$$

We then find that

$$T_{\nu\nu'}(m,n) = t \begin{pmatrix} \gamma^*(n), & \omega\gamma^*(m), & 0, & \omega^*\gamma(m) \\ \omega^*\gamma^*(n), & \gamma^*(m), & \omega\gamma(n), & 0 \\ 0, & \omega^*\gamma^*(m), & \gamma(n), & \omega\gamma(m) \\ \omega\gamma^*(n), & 0, & \omega^*\gamma(n), & \gamma(m) \end{pmatrix}, \tag{5.209}$$

where

$$\gamma(m) = \mathrm{e}^{2\pi\mathrm{i}m/L)}. \tag{5.210}$$

Thus, for given values of (m, n), we have

$$\begin{aligned} \prod_{i=1}^{4} \{1 - \lambda_i(m,n)\} &= \det[\mathsf{I} - \mathsf{T}(m,n)] \\ &= (1+t^2)^2 - 2t(1-t^2)\left(\cos\frac{2\pi m}{L} + \cos\frac{2\pi n}{L}\right). \end{aligned} \tag{5.211}$$

Thus we obtain

$$Z = 2^N(1-t^2)^{-N} \prod_{mn}\left[(1+t^2)^2 - 2t(1-t^2)\left(\cos\frac{2\pi m}{L} + \cos\frac{2\pi n}{L}\right)\right]^{1/2}, \tag{5.212}$$

where the product runs over L consecutive values of m and of n.

Setting $p = 2\pi m/L$ and $q = 2\pi n/L$, the Helmholtz free energy is given by

$$\begin{aligned} F(T) &= -Nk_{\mathrm{B}}T\Big\{\ln 2 - \ln(1-t^2) \\ &\quad + \frac{1}{2}\int_{-\pi}^{+\pi} \frac{\mathrm{d}p\mathrm{d}q}{(2\pi)^2} \ln\big[(1+t^2)^2 - 2t(1-t^2)(\cos p + \cos q)\big]\Big\}. \end{aligned} \tag{5.213}$$

Let us consider the contribution of the integral. The minimum value of the integrand is reached for $p = q = 0$ and is given by

$$\ln\big[(1+t^2)^2 - 4t(1-t^2)\big] = \ln[t^2 + 2t - 1]^2.$$

The argument of the logarithm vanishes for

$$t = t_{\mathrm{c}} = \sqrt{2} - 1,$$

which corresponds to the transition temperature T_{c} given by equation (5.72):

$$\frac{J}{k_{\mathrm{B}}T_{\mathrm{c}}} = \frac{1}{2}\ln\left(1 + \sqrt{2}\right).$$

In order to understand the behavior of F in the neighborhood of this temperature, let us introduce $\tau = t - t_{\mathrm{c}}$ and expand the integrand for small values of τ and of p, q. One has

$$F(T) = \frac{1}{2}\int_{-\pi}^{+\pi} \frac{\mathrm{d}p\mathrm{d}q}{(2\pi)^2} \ln[c_1\tau^2 + c_2(p^2+q^2)] + \text{regular terms},$$

where c_1 and c_2 are constants. Integrating, one obtains

$$F(T) = -a\tau^2 \ln|\tau| + \text{regular terms},$$

where $a > 0$ is a constant. The specific heat C is proportional to $-\mathrm{d}^2F/\mathrm{d}\tau^2$. Thus we have

$$C \simeq a \ln|\tau| + \text{regular terms}, \tag{5.214}$$

indicating that the specific heat exhibits a logarithmic divergence at the critical temperature.

The evaluation of the spontaneous magnetization $m_0 = \langle\sigma\rangle$ proceeds in a similar way [Vdov65], starting, e.g., from the relation

$$m_0^2 = \lim_{k\to\infty} \langle \sigma_{1\ell}\sigma_{1,\ell+k} \rangle.$$

One obtains $m_0 = 0$ for $t < t_c$ given earlier, and

$$m_0 = \left\{1 - \left(\frac{t^{-1} - t}{2}\right)^4\right\}^{1/8} \tag{5.215}$$

for $t > t_c$, i.e., below the transition temperature. Thus for small positive values of $\tau = t - t_c$, one has

$$m_0 \propto \tau^{\beta}, \tag{5.216}$$

where the exponent β is given by

$$\beta = \frac{1}{8}. \tag{5.217}$$

The connection between the Ising model and the statistics of loops can be interpreted more deeply as the equivalence between the model and a system of noninteracting fermions. This correspondence is exploited by Schulz, Mattis and Lieb [Schu64] in their solution of the Ising model.

Recommended Reading

The phenomenology of phase transitions is explained in H. E. Stanley, *Introduction to Phase Transitions and Critical Phenomena*, Oxford, UK: Oxford University Press, 1971. The books devoted to phase transitions are now legion: it is sufficient to quote the series of volumes edited by C. Domb and M. Green (and later by C. Domb and J. L. Lebowitz), *Phase Transitions and Critical Phenomena*, London: Academic Press, 1971–present. The mean-field and Landau theories are discussed in detail in L. D. Landau and E. M. Lifshitz, *Statistical Physics*, 3rd ed., part 1, Oxford, UK: Pergamon Press, 1980. A simple presentation of the different ordered phases in condensed matter can be found in the excellent volume by P. M. Chaikin and T. C. Lubensky, *Principles of Condensed Matter Physics*, Cambridge, UK: Cambridge University Press, 1995. One book is entirely devoted to Landau phase transition theory in complex systems: J.-C.Tolédano and P. Tolédano, *The Landau Theory of Phase Transitions: Application to Structural, Incommensurate, Magnetic, and Liquid Crystal Systems*, Singapore: World Scientific, 1987.

6 Renormalization Group

Never do a calculation before you know the answer.
— J. A. Wheeler

6.1 Block Transformation

The results of the preceding chapter imply that the statistics of fluctuations close to the critical point are very similar for the same system at different values of the temperature and magnetic field (or, more generally, of those quantities that characterize distance from the critical point), provided one performs a rescaling of the lengths and of the order parameter. In 1966, Leo P. Kadanoff put forth an argument to explain this resemblance in the fluctuations' properties [Kada66]. The ideas described in that work were reformulated by Kenneth G. Wilson in several papers of 1971 and reviewed in [Wils72]. They form the basis of the modern theory of phase transitions. Unfortunately, this theory goes by the quite awkward name of *renormalization group*.

Let us consider a two-dimensional Ising model in the proximity of the critical point. In the upper part of figure 6.1, you can see one of its configurations (the black squares represent an "up" spin). Since the distance from the critical point is small, the coherence length is large. We now define a new configuration using the majority rule: we group the spins together in 3×3 cells and assign an up spin to each cell if the majority of the spins that belong to it are up. We thus obtain a new configuration, which is represented in the lower portion of figure 6.1.

Kadanoff's idea is that this new configuration is a typical configuration of a two--dimensional Ising model, at a different (and, as we shall see, larger) distance from the critical point. To simplify, let us assume that the magnetic field vanishes. For reasons of symmetry, the magnetic field of the new configuration will also vanish. In this manner, we will be dealing only with the t variable, which represents the distance from the critical temperature. We will denote the value of this variable in the first configuration by t_0, and the corresponding value in the second by t_1. It is obvious that a certain value of t_1 corresponds to each value of t_0. We have thus defined a transformation:

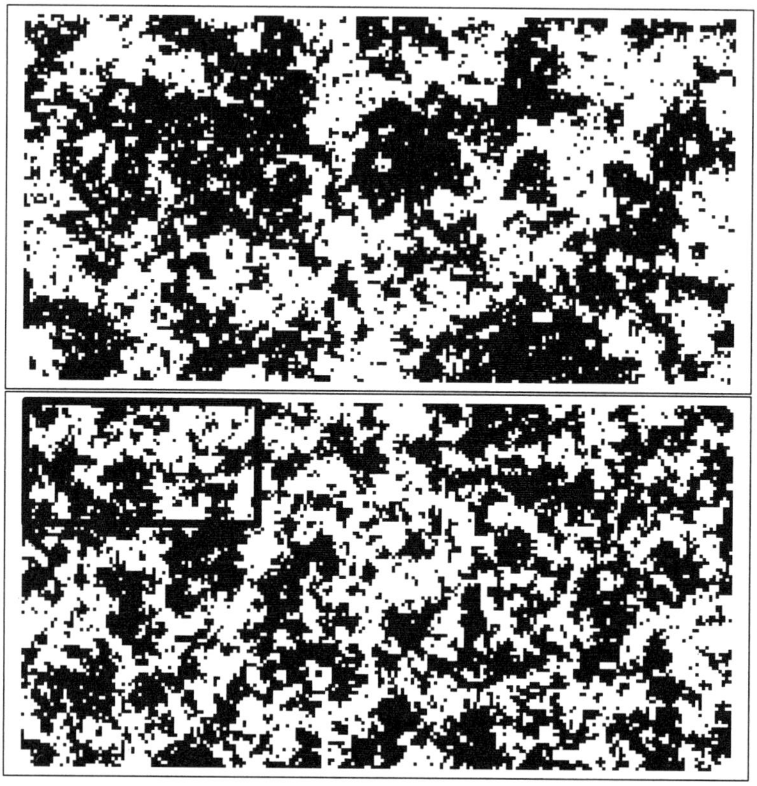

FIGURE 6.1. Block transformation The upper image represents a typical configuration of an Ising model at temperature $T = 1{,}1\ T_c$. From this configuration, one obtains the configuration in the lower image by grouping the spins in the first configuration into 3×3 blocks and assigning the spin value of the majority of the spins that compose it to each block. The box in the bottom panel highlights the spins corresponding to those of the top panel. Adapted from [Wils79].

$$t_1 = T(t_0). \tag{6.1}$$

By carrying out this transformation, we have modified the coherence length. So if in the original model the spins were correlated up to a distance $\ell = \xi(t_0)$ (measured in lattice step units), in the transformed model this distance will be equal to $\ell/3$, since each spin of the "new" model corresponds to 3^2 spins of the "old," and the new lattice step is therefore $b = 3$ times greater. We are able in general to define more general transformations, with other values of b. The relation

$$\xi(t_1) = \frac{1}{b}\xi(t_0). \tag{6.2}$$

is therefore valid.

Since the new coherence length is smaller than the old, the new model is further away from the critical point than the old, as we had anticipated. The only exception occurs if we are exactly at the critical point, because in that case, the coherence length is infinite, and it must remain infinite also for the new model. The transformation T must therefore leave the critical point unchanged:

$$T(0) = 0. \tag{6.3}$$

We observe, moreover, that the transformation T is exclusively local, and that it requires considering the effects of a finite number of degrees of freedom each time. One can therefore expect it to be regular in its argument, and that for small values of t, one will have

$$T(t) \simeq \kappa t + O(t^2), \tag{6.4}$$

where κ is a certain constant larger than 1.

We can repeat the same remarks by placing ourselves at the critical temperature and in a nonvanishing magnetic field, and therefore obtain the following expression for the transformation of h:

$$T(h) \simeq \kappa_h h + O(h^3). \tag{6.5}$$

For reasons of symmetry, only odd powers of h may appear. By combining everything, we obtain

$$T(t, h) \simeq \kappa t + \kappa_h h, \tag{6.6}$$

where $t' = \kappa t$, $h' = \kappa_h h$, while the corresponding transformation of the coherence length is given by

$$\xi(T(t, h)) = \frac{1}{b}\xi(t, h). \tag{6.7}$$

These results imply the homogeneity of the thermodynamic functions and the scaling laws.

In order to clarify our ideas, let us place ourselves once more in a vanishing magnetic field, and let us assume that we let t tend to zero. By repeatedly applying the transformation, we can express the model's coherence length (and the other thermodynamic quantities) in terms of their values for a corresponding model in which t is kept fixed. In other words, we can apply the transformation n times so as to obtain

$$\xi(\kappa^{-n}t, h = 0) = b^n \xi(t, h = 0). \tag{6.8}$$

This relation has the form of the generalized homogeneity relations that we discussed in chapter 5 if we perform the identification

$$\lambda = \kappa^{-n}, \qquad \lambda^{-\nu} = b^n. \tag{6.9}$$

The fact that λ was arbitrary in section 5.19, while our factors can assume only discrete values, is not essential. As a result of this comparison, we can express the critical exponent ν as a function of the constant κ:

$$\frac{1}{\nu} = \frac{\ln \kappa}{\ln b}. \tag{6.10}$$

Analogously, by considering the transformation at the critical temperature and in a vanishing field, we obtain

$$\frac{\beta\delta}{\nu} = \frac{\ln \kappa_h}{\ln b}. \tag{6.11}$$

These results were derived by Kadanoff.

In order to transform these observations into a theory that will allow for the effective calculation of the critical behavior, we must try to calculate the transformation T. We then realize that the hypothesis that the new model is still an Ising model like the old one (although at a different distance from the critical point) is simplistic. If we want to describe the statistics of the fluctuations of the collective cell variables by means of a Hamiltonian, we will be forced in general to introduce new interactions, such as interactions with several spins, or with next-to-nearest neighbors, and so on. It will therefore be necessary to consider the transformation T in a sufficiently large space of Hamiltonian functions. The key point is that since the transformation is, in principle, regular in the parameters that define the Hamiltonian, we will be able to find effective approximation schemes. We will devote the next sections to the definition of some of these schemes.

6.2 Decimation in the One-Dimensional Ising Model

In this section, following the discussion in [Chan87, pp. 139 ff], we will provide an exact definition of the Kadanoff transformation T (or, rather, of one of its analogs) for the one-dimensional Ising model.

Let us consider a model defined by the Hamiltonian

$$H = -\sum_i (J\sigma_i\sigma_{i+1} + h\sigma_i). \tag{6.12}$$

It is useful to consider the expression of the partition function:

$$Z = \sum_\sigma \exp\left[\sum_i (K\sigma_i\sigma_{i+1} + \mu\sigma_i)\right], \tag{6.13}$$

where we have introduced the shorthand

$$K = \frac{J}{k_B T}, \qquad \mu = \frac{h}{k_B T}. \tag{6.14}$$

Let us now try to define a transformation in the spirit of the Kadanoff transformation **T** described in the preceding section. This transformation must allow one to pass from a state identified by certain values (K, μ) to another defined by (K', μ') in a calculable manner and so that in the course of this transformation the fluctuations' statistics are transformed in a known fashion, and more specifically, so that the coherence length is divided by a known b factor.

One transformation that allows us to obtain these results is **decimation**. We sum over the fluctuations of spins with an even index, and we obtain a model in which only spins with odd index appear—this is a new one-dimensional Ising model, with nearest-neighbor interactions, but in which the coherence length (measured in units equal to the lattice step) has been halved.

In order to perform this transformation explicitly, let us consider a spin with an even index i and its interaction with neighboring spins. For symmetry's sake, we assign half of each spin's magnetic field term to its bond with the spin that precedes it, and half to that with the spin that follows it. We can then express the partition function as follows:

$$Z = \sum_{\{\sigma_{\text{odd}}\}} \prod_i \sum_{\sigma_{2i}} \exp\left[K(\sigma_{2i-1} + \sigma_{2i+1})\sigma_{2i} + \frac{\mu}{2}(\sigma_{2i-1} + 2\sigma_{2i} + \sigma_{2i+1})\right]. \tag{6.15}$$

By performing the sum over the even index spins, we easily obtain:

$$Z = \sum_{\{\sigma_{\text{odd}}\}} \exp\left\{\sum_i \left[\frac{\mu}{2}(\sigma_{2i-1} + \sigma_{2i+1}) + \ln 2\cosh\left[K(\sigma_{2i-1} + \sigma_{2i+1}) + \mu\right]\right]\right\}. \tag{6.16}$$

It does not appear to be, but this is the partition function of a one-dimensional Ising model. In order to see this, let us first redefine the index:

$$2i - 1 \to i, \qquad 2i + 1 \to i + 1. \tag{6.17}$$

We also set

$$2\cosh\left[K(\sigma_i + \sigma_{i+1}) + \mu\right] = \zeta_0 \exp\left[K'\sigma_i\sigma_{i+1} + \frac{\delta\mu}{2}(\sigma_i + \sigma_{i+1})\right]. \tag{6.18}$$

The left-hand side can assume three values, depending on the values assumed by the two spins being considered; the right-hand side contains three parameters that have to be determined—in other words ζ_0, K', and μ. A solution can therefore exist. In fact, with a little algebra, one obtains:

$$\exp(4K') = \cosh(\mu + 2K)\cosh(\mu - 2K)(\cosh^2\mu)^{-1}, \tag{6.19}$$

$$\exp(2\delta\mu) = \cosh(\mu + 2K)\left[\cosh(\mu - 2K)\right]^{-1}, \tag{6.20}$$

$$\zeta_0 = 2\cosh\mu \exp(K'). \tag{6.21}$$

We can set $\mu' = \mu + \delta\mu$ and thus obtain the transformation of the parameters that define the interaction. As a result of this transformation, the partition function has been multiplied by $\zeta_0^{N/2}$. We can therefore write the effect of this transformation for $f = \ln Z/N$, which is proportional to the free energy per site:

$$f(K,\mu) = \frac{1}{2}\ln\zeta_0 + \frac{1}{2}f(K',\mu'). \tag{6.22}$$

On the other hand, the new model's lattice step is twice that of the old, and the b factor is therefore equal to 2:

$$\xi(K',\mu') = \frac{1}{2}\xi(K,\mu). \tag{6.23}$$

In order to discuss the transformation's effect, let us place ourselves in a vanishing magnetic field ($\mu = 0$). One then has $\mu' = 0$, as could be expected from symmetry considerations. The transformation of K takes the form

$$K' = \frac{1}{2}\ln\cosh 2K, \tag{6.24}$$

and one has

$$\zeta_0 = 2e^{K'} = 2\sqrt{\cosh 2K}. \tag{6.25}$$

It is easy to see that $K' \leq K$—one has equality only for $K = 0$. Therefore, starting from any *finite* value of K, the transformation's interaction takes us toward the fixed value $K = 0$, which corresponds to infinite temperature (and vanishing coherence length). There is another value of K that remains unchanged under the transformation—in other words, $K = \infty$ (which corresponds to vanishing temperature and an infinite coherence length). The transformation tends to take the system away from this **fixed point**.

This behavior of the interactions under the transformation shows that there is no phase transition at a finite temperature. (Strictly speaking, one can talk of a transition at vanishing temperature.) We now show that we can use this transformation in order to evaluate the free energy.

In order to have an effective calculation scheme, it is useful to express the free energy calculated in K as a function of that calculated in K', by making use of the equation

$$f(K) = \frac{1}{2}f(K') + \frac{1}{2}\ln 2 + \frac{K'}{2}, \tag{6.26}$$

which derives from the expressions of ζ_0 and K' obtained earlier. We then have to invert the relation between K and K' we obtained previously, so as to have

$$K = \mathcal{T}^{-1}(K') = \frac{1}{2}\cosh^{-1}(e^{2K'}). \tag{6.27}$$

In this manner, possible errors made when estimating $f(K')$ will be compressed, since they are multiplied by a factor of $1/2$ at each iteration.

Our strategy is to start from a specific value K_0 of K, and from a corresponding initial estimate of $f(K_0)$. By making use of equation (6.27) we obtain a sequence $K_1, K_2, \dots$, where $K_n = \mathcal{T}^{-1}(K_{n-1})$. The relation (6.26) then gives us the corresponding value of $f(K_n)$. Since K grows at each iteration, we can start with a very small value—for example, 0.01—where we can estimate $f(0.01) \simeq \ln 2$, which is the value for $K = 0$. By iterating, we obtain table 6.1, in which the first column contains successive values of K, the second column contains the corresponding values of $f(K)$ estimated by means of the transformation, and the third column contains the values calculated from the exact solution. We can see that a moderate computational investment allows us to obtain almost exact free energy values.

Table 6.1 Decimation Results for the Free Energy in the One-Dimensional Ising Model

K	$f(K)$ (transformation)	$f(K)$ exact
0.01	$\ln 2 = 0.693\,147$	0.693 197
0.100 334	0.698 147	0.698 172
0.327 447	0.745 814	0.745 827
0.636 247	0.883 204	0.883 210
0.972 710	1.106 299	1.106 302
1.316 710	1.386 078	1.386 080
1.662 637	1.697 968	1.697 968
2.009 049	2.026 876	2.026 877
2.355 582	2.364 536	2.236 537
2.702 146	2.706 633	2.706 634

Note: In the first column, $K = J/k_BT$ for successive iterations, and in the second column, the corresponding value of free energy per spin $f(K)$, calculated by means of equation (6.26), and compared with the exact value $f(K) = \ln 2 \cosh K$.

Source: Chan87, p. 141.

6.3 Two-Dimensional Ising Model

Let us now consider a two-dimensional Ising model in a vanishing magnetic field, defined over a square lattice. As with the one-dimensional model, let us introduce the notation $K = J/k_BT$.

The spins that belong to the model are arranged on two intertwined sublattices, such that each site that belongs to one sublattice is surrounded by spins belonging to the other sublattice. The sublattices are distinguished by the sum of the indices corresponding to the two coordinates—if this sum is even, the site belongs to the "even" sublattice; otherwise, it belongs to the "odd" one. The new model is also defined over a square lattice (rotated by 45 deg), while the lattice step is increased by a factor $b = \sqrt{2}$.

Let us focus our attention on one element (figure 6.2), composed by a spin (which we will denote by σ_0) and by its four neighbors. The contribution of this element to the partition function is given by

$$\sum_{\sigma_0} \exp[K\sigma_0(\sigma_1 + \sigma_2 + \sigma_3 + \sigma_4)] = 2\cosh[K(\sigma_1 + \sigma_2 + \sigma_3 + \sigma_4)]. \tag{6.28}$$

We would want to be able to write this expression in the form

$$\exp\left[\frac{K'}{2}(\sigma_1\sigma_2 + \sigma_2\sigma_3 + \sigma_3\sigma_4 + \sigma_4\sigma_1)\right], \tag{6.29}$$

eventually up to a constant factor. The factor $1/2$ is due to the fact that the 1–2, 2–3, 3–4, 4–1 bonds appear twice in the new model—once in the cell we are considering, and a

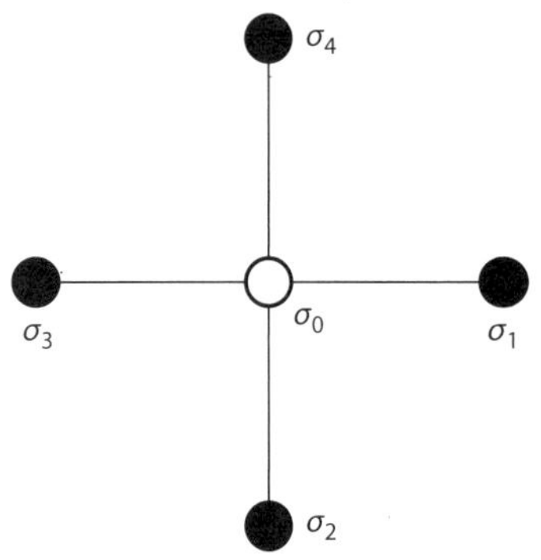

FIGURE 6.2. The "decimating" spin σ_0 and its four neighbors.

second time in each of the neighboring cells. One cannot, however, obtain this expression by means of a single interaction constant. We should actually represent the three possible values of cosh $[K(\sigma_1+\sigma_2+\sigma_3+\sigma_4)]$ using only two parameters, not to speak of the fact that the last expression is symmetrical with respect to all the permutations (1234), while the interaction among nearest neighbors is not.

It is therefore necessary to introduce other interactions into the new model—in our case, we will need the next-to-nearest-neighbor interactions ($\sigma_1-\sigma_3$ and $\sigma_2-\sigma_4$, for example), represented by a parameter K_1, and a four-spin interaction, represented by a parameter L. We therefore set

$$\begin{aligned}\cosh[K(\sigma_1+\sigma_2+\sigma_3+\sigma_4)] &= \zeta_0(K)\\ &\times \exp\left[\frac{K'}{2}(\sigma_1\sigma_2+\sigma_2\sigma_3+\sigma_3\sigma_4+\sigma_4\sigma_1)\right.\\ &\left.+K_1(\sigma_1\sigma_3+\sigma_2\sigma_4)+L\sigma_1\sigma_2\sigma_3\sigma_4\right].\end{aligned} \tag{6.30}$$

One thus obtains the equations

$$2\cosh 4K = \zeta_0(K)\exp(2K'+2K_1+L), \tag{6.31}$$

$$2\cosh 2K = \zeta_0(K)\exp(-L), \tag{6.32}$$

$$2 = \zeta_0(K)\exp(-2K_1+L), \tag{6.33}$$

$$2 = \zeta_0(K)\exp(-2K'+2K_1+L). \tag{6.34}$$

It is a simple algebra exercise to show that

$$K' = \frac{1}{4}\ln\cosh 4K, \tag{6.35}$$

$$K_1 = \frac{1}{8}\ln\cosh 4K, \tag{6.36}$$

$$L = \frac{1}{8}\ln\cosh 4K - \frac{1}{2}\ln\cosh 2K, \tag{6.37}$$

$$\zeta_0(K) = 2(\cosh 2K)^{1/2}(\cosh 4K)^{1/8}. \tag{6.38}$$

Even if we start from a model in which only nearest-neighbor interactions are present, as a result of the first decimation, we also obtain next-nearest-neighbor and four-spin interactions. A second decimation would introduce even more numerous interactions. In order to be able to implement our program, we need to simplify the result so that it can be represented by a smaller number of interaction parameters.

One possibility is to "project" the transformation only onto the K axis—in other words, to disregard all other interactions. The resulting equation is

$$K' = \frac{1}{4} \ln \cosh 4K, \tag{6.39}$$

which however admits only the trivial fixed points $K = 0$ and $K = \infty$ and does not describe a transition.

In order to improve our approximation, we must in some fashion also take the other interactions into account. We are not able to iterate the transformation if other interactions appear in our Hamiltonian, such as the next-nearest-neighbor ones, for instance. We must therefore find a way to project the transformation (6.35) to (6.37) over the nearest-neighbor interactions axis, but take the effect of the other interactions at least partially into account. There is no systematic way of proceeding, and the procedure will thus be validated by the results we find. One choice that has been found acceptable is to neglect the four-body interaction and to choose the nearest-neighbor interaction coefficient so that the ground state's energy will be equal to that due to K' and K_1:

$$K' \sum_{\langle ij \rangle} \sigma_i \sigma_j + K_1 \sum_{|\ell m|} \sigma_\ell \sigma_m \simeq \overline{K'} \sum_{\langle ij \rangle} \sigma_i \sigma_j . \tag{6.40}$$

The first sum runs over the pairs of nearest neighbors, and so does that for the right-hand side; the second sum on the left-hand side runs over the pairs of next-nearest neighbors.

By comparing the two sides of this equation in the state in which all spins are parallel, we obtain

$$NK' + NK_1 \simeq N\overline{K'}, \tag{6.41}$$

and therefore,

$$\overline{K'} \simeq K' + K_1 = \frac{3}{8} \ln \cosh 4K. \tag{6.42}$$

The approximate transformation has the expression (with an obvious change in notation):

$$f(K') = 2f(K) - \ln\left[2(\cosh 2K)^{1/2}(\cosh 4K)^{1/8}\right], \tag{6.43}$$

$$K' = \frac{3}{8} \ln \cosh 4K. \tag{6.44}$$

In figure 6.3, we show the curve K' as a function of K, as can be seen from the preceding equation. This equation admits one fixed point, defined by the equation

$$K^* = \frac{3}{8} \ln \cosh 4K^*, \tag{6.45}$$

and therefore,

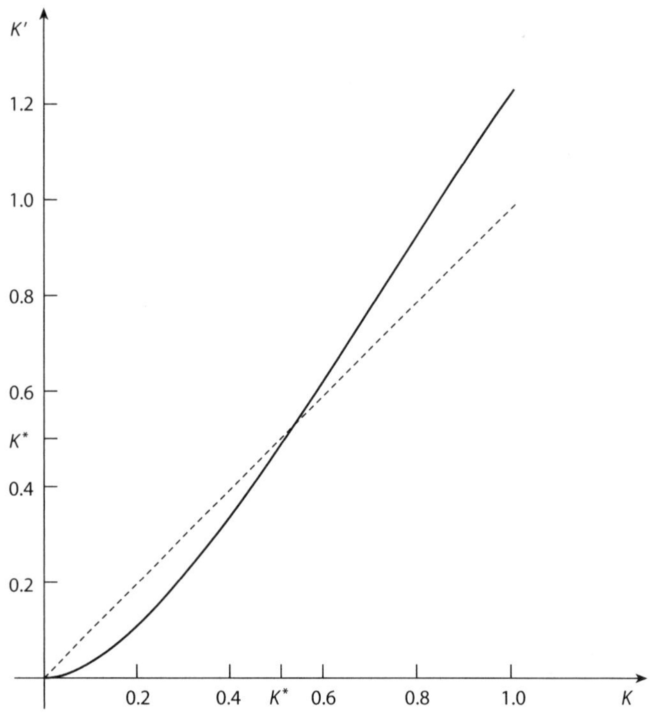

FIGURE 6.3. Recurrence relation K' as a function of K for the two-dimensional Ising model, according to equation (6.44). Note the fixed point at $K^* = 0.50689$.

$$K^* = 0.50689. \tag{6.46}$$

This fixed point is *unstable*. Let us suppose that we take

$$K = K^* + \delta K, \tag{6.47}$$

and then, for small values of δK, we obtain

$$K' = K^* + \left.\frac{\mathrm{d}K'}{\mathrm{d}K}\right|_{K^*} \delta K, \tag{6.48}$$

where

$$\left.\frac{\mathrm{d}K'}{\mathrm{d}K}\right|_{K^*} = 1.44892. \tag{6.49}$$

Since this quantity is greater than one, the iterates will move away from the fixed point, just as we had supposed in Kadanoff's argument.

For $K = K^*$, we remain at the fixed point, and we can therefore assume that δK is proportional to the distance from the transition point, and that it thus plays the role of t in Kadanoff's argument. In this fashion, we have realized, if only approximately, Kadanoff's program for the two-dimensional Ising model, and we have obtained an estimate of the exponent ν:

$$\frac{1}{\nu} = \frac{1}{\ln b} \ln \left.\frac{dK'}{dK}\right|_{K^*} = \frac{2}{\ln 2} \ln \left.\frac{dK'}{dK}\right|_{K^*} \simeq 1.06996, \tag{6.50}$$

which is an excellent approximation of the exact result $1/\nu = 1$, especially if compared to the classical result $1/\nu = 2$. We have expressed the results as a function of $1/\nu$, because as one can see from the equation reproduced, this is the quantity that can be obtained most directly from the calculation.

6.4 Relevant and Irrelevant Operators

As we have seen, the transformation from the original model to the model with collective variables introduces in general new interactions. At least in principle, we must therefore define the Kadanoff transformation T in a space that contains all the possible interactions.

Let us suppose that we want to deal with a model defined over a cubic lattice in d dimensions, and in which a real variable ϕ_i is associated with each lattice point i. In general, the partition function will have the expression

$$Z = \int \prod_i d\phi_i \exp(-\mathcal{H}\{\sigma_i\}), \tag{6.51}$$

where

$$-\mathcal{H}\{\sigma_i\} = \sum_{n=1}^{\infty} \frac{1}{n!} K^{(n)}_{\{i_1,\ldots,i_n\}} \phi_{i_1} \cdots \phi_{i_n}. \tag{6.52}$$

We will consider in the following, except when stated explicitly otherwise, only Hamiltonians that are invariant with respect to the transformation $\phi \to -\phi$. Therefore, only even values of n appear in this expansion. Since the transformation T maintains this symmetry, this will be true also for the transformed Hamiltonians.

The set of quantities $K^{(n)}_{\{i_1,\ldots,i_n\}}$ therefore allows one to represent the model. Let us consider a space (with infinite dimensions), identified by the coordinates $K^{(n)}_{\{i_1,\ldots,i_n\}}$. In this space, we can define the transformation T in the following manner:

1. Group b^d variables (which belong to one of the original lattice's cells) together into a single collective variable $\Phi_I = \sum_{i \in I} \phi_i$.
2. Express the interactions of the original model in terms of effective interactions for the collective variables $\{\Phi_I\}$ according to the expression

$$\begin{aligned} Z &= \int \prod_i d\phi_i \exp(-\mathcal{H}\{\sigma_i\}) \\ &= \int \prod_I d\Phi_I \int \prod_i d\phi_i \delta\Big(\Phi_I - \sum_{i \in I} \phi_i\Big) \exp(-\mathcal{H}\{\sigma_i\}) \\ &= \int \prod_I (\zeta_0 d\Phi_I) \exp(-\overline{\mathcal{H}}\{\Phi_I\}). \end{aligned} \tag{6.53}$$

3. Redefine the lattice step, so as to have a variable Φ_I for each point of the lattice—the distances will be therefore multiplied by a factor of $1/b$.

4. Define the new variable $\phi'_I = \zeta\Phi_I$ (by suitably rescaling the collective variable Φ_I). The ζ factor is chosen so that the transformation has a nontrivial fixed point—practically speaking, it is sufficient to require that the interactions one obtains decrease fairly rapidly with the distance.

In this fashion, we have defined a transformation $\boldsymbol{T}$ between the Hamiltonians

$$\mathcal{H}'\{\phi'_i\} = T\mathcal{H}\{\phi_i\}. \tag{6.54}$$

By defining $f(\mathcal{H}) = \ln Z(\mathcal{H})/N$, the transformation's properties can be summarized by the relations

$$f(\mathcal{H}) = (N/b^d)[\ln \zeta_0(\mathcal{H}') + f(\mathcal{H}')], \tag{6.55}$$

$$\xi(\mathcal{H}) = b \cdot \xi(\mathcal{H}'), \tag{6.56}$$

$$\langle \phi_0 \phi_r \rangle_{\mathcal{H}} = b^{2d} \zeta^{-2} \langle \phi'_0 \phi'_{r/b} \rangle_{\mathcal{H}'}. \tag{6.57}$$

In the third equation, we have taken into account the fact that the new variables ϕ' are proportional to the sum of b^d old variables.

In the space of the Hamiltonians, there will exist a **critical surface** S_c defined by $\xi(\mathcal{H}) = \infty$, $\mathcal{H} \in S_c$. This surface is invariant with respect to $\boldsymbol{T}$, as one can see from equation (6.56). We can assume that, starting from any point on this surface and iterating the transformation, we will end up on one of the transformation's **fixed points**, which satisfy the equation

$$\mathcal{H}^* = T\mathcal{H}^*. \tag{6.58}$$

Keeping equation (6.57) in mind, we see that for each Hamiltonian belonging to the critical surface, the correlation functions relative to fairly large distances are proportional to those characteristic of $\mathcal{H}^*$. This is the microscopic justification of the universality hypothesis: the properties of the critical fluctuations are determined by $\mathcal{H}^*$—in other words, by the transformation $\boldsymbol{T}$—rather than by the details of the system's Hamiltonian (which only plays the role of the initial condition from which one begins to iterate the transformation $\boldsymbol{T}$).

Let us now consider the problem from a slightly more abstract point of view. In the space of the Hamiltonians, we identified the critical surface S_c, which has codimension 1 because it is defined by a single equation. For a well-defined model (for example, for our initial Ising model), the Hamiltonian, as the temperature varies, runs over a line S_0 (if we can make several parameters vary, it will generally be a surface instead of a line), which we can call the **elementary surface**. Generically, the elementary surface meets the critical surface in a point $\mathcal{H}_c$, which represents the critical Hamiltonian. This situation is schematically represented in figure 6.4.

Starting from $\mathcal{H}_c$ and iterating the transformation, we reach the fixed point $\mathcal{H}^*$. Iterating the transformation at this point will have no effect on the Hamiltonian. We can now express the effect of the transformation on the correlation functions:

$$\langle \phi_0 \phi_r \rangle_{\mathcal{H}^*} = b^{-2d} \zeta^{-2} \langle \phi_0 \phi_{r/b} \rangle_{\mathcal{H}^*}. \tag{6.59}$$

By transforming according to Fourier, we obtain:

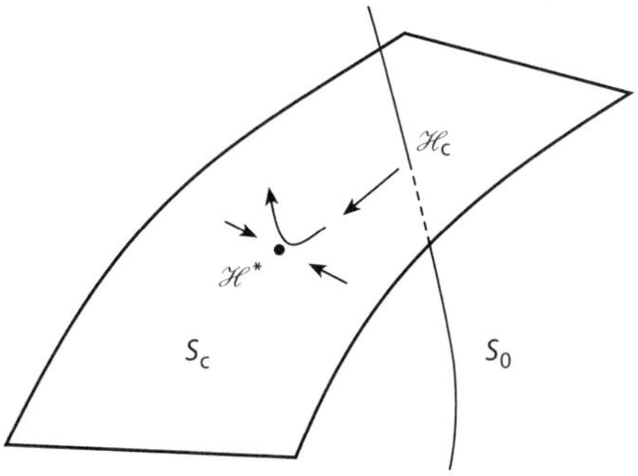

FIGURE 6.4. Scheme of the critical surface (S_c) and the elementary surface (S_0) in the space of the Hamiltonians. The critical Hamiltonian $\mathcal{H}_c$ is at the intersection of S_0 with S_c. The fixed point $\mathcal{H}^*$ of the Kadanoff transformation $\boldsymbol{T}$ is also shown. The arrows point to the direction of the flux under the iteration of the Kadanoff transformation.

$$C(\boldsymbol{k}, \mathcal{H}^*) = b^{-d}\zeta^{-2}C(\boldsymbol{k}b, \mathcal{H}^*). \tag{6.60}$$

This is a generalized homogeneity relation, from which we obtain

$$d + 2 - \eta = 2\frac{\ln\zeta}{\ln b}. \tag{6.61}$$

This result also shows us that we need to introduce a nontrivial ζ factor. If we had chosen another value of ζ, we would not have obtained a fixed point in the space of the Hamiltonians we are considering.

Let us now consider our model at a temperature that is not the critical one. In other words, let us set

$$\mathcal{H}_0 = \mathcal{H}_c + \delta\mathcal{H}, \tag{6.62}$$

where $\delta\mathcal{H}$ is proportional to the distance from the critical temperature. By continuity, we expect that the iteration of the transformation $\boldsymbol{T}$ will at first bring us into proximity to the fixed point $\mathcal{H}^*$—in other words, we will have, for a fixed number of iterations n_0,

$$\boldsymbol{T}^{n_0}\mathcal{H}_0 = \mathcal{H}^* + \delta\mathcal{H}^*, \tag{6.63}$$

where $\delta\mathcal{H}^*$ is small. Let us apply the transformation once more:

$$\boldsymbol{T}(\mathcal{H}^* + \delta\mathcal{H}^*) \simeq \mathcal{H}^* + \mathcal{L}\delta\mathcal{H}^*, \tag{6.64}$$

where $\mathcal{L}$ is a linear operator. We can decompose $\delta\mathcal{H}^*$ into the right eigenvectors of $\mathcal{L}$:

$$\delta\mathcal{H}^* = \sum_{\ell} \delta\mathcal{H}^{(\ell)}, \tag{6.65}$$

where

$$\mathcal{L}\delta\mathcal{H}^{(\ell)} = \lambda^{(\ell)}\mathcal{H}^{(\ell)}. \tag{6.66}$$

It follows from our hypotheses that $\mathcal{L}$ admits only one eigenvector greater than 1. Before "proving" this, let us suppose that it is true and examine its consequences. If

$$\delta\mathcal{H}^{*} = \begin{pmatrix} \delta\mathcal{H}^{(1)} \\ \delta\mathcal{H}^{(2)} \\ \vdots \end{pmatrix}, \tag{6.67}$$

we will have

$$\mathcal{L}\delta\mathcal{H}^{*} = \begin{pmatrix} \lambda^{(1)}\delta\mathcal{H}^{(1)} \\ \lambda^{(2)}\delta\mathcal{H}^{(2)} \\ \vdots \end{pmatrix}, \tag{6.68}$$

where all the elements (except the first) are smaller than in $\delta\mathcal{H}^{*}$. In order for $\mathcal{H}_c$ to be reached from $\mathcal{H}^{*}$, it is necessary that $\delta\mathcal{H}^{(1)}$ vanish at the critical temperature. In other words, we must have

$$\delta\mathcal{H}^{(1)} \propto (T - T_c). \tag{6.69}$$

While successive iterations make $\delta\mathcal{H}^{(1)}$ increase, they also make all the other perturbations decrease, which are consequently called **irrelevant** perturbations. Other models whose fundamental surface crosses the same critical surface are governed by the same fixed point (and, as we shall see, have the same critical exponents). The irrelevant perturbations, therefore, do not influence the asymptotic critical behavior.

As a first approximation, we can disregard the irrelevant perturbations and assume that the thermodynamic properties of the Hamiltonian are determined only by $\delta\mathcal{H}^{(1)}$. We thus obtain

$$\xi(\delta\mathcal{H}^{(1)}) = b\xi(\lambda^{(1)}\delta\mathcal{H}^{(1)}). \tag{6.70}$$

When we recall that $\delta\mathcal{H}^{(1)}$ is proportional to the distance from the critical temperature, we obtain the expression of the exponent $1/\nu$:

$$\frac{1}{\nu} = \frac{\ln \lambda^{(1)}}{\ln b}. \tag{6.71}$$

$\mathcal{L}$'s other eigenvalues govern the approach to the fixed point, and therefore corrections to the asymptotic behavior.

By considering the fixed point $\mathcal{H}^{*}$, we have thus defined $\mathcal{H}^{*}$'s **basin of attraction** as the locus of the Hamiltonians that, under the iteration of the transformation T, tend toward $\mathcal{H}^{*}$—in our case, it is composed of the critical surface S_c. Let us now consider a generic Hamiltonian close to $\mathcal{H}^{*}$, where the perturbation $\delta\mathcal{H}^{(1)}$ is expanded as in equation (6.65). In this expansion, some terms grow on the application of the transformation T: they are called the **relevant perturbations** (or **operators**), because they take the iterates away from the fixed point $\mathcal{H}^{*}$. When these are present, the asymptotic critical behavior is no more described by $\mathcal{H}^{*}$. In our case, there is only $\delta\mathcal{H}^{(1)}$. However, if one adds to the Hamiltonian a perturbation that explicitly breaks the symmetry $\rho \to -\rho$, this will introduce at least one additional relevant operator. As for the terms that grow smaller as the transformation T is applied, they do not modify the asymptotic critical behavior, which justifies the term **irrelevant perturbations** that we have applied to them. We can also see that the codimension of the basin of attraction of $\mathcal{H}^*$, and therefore of S_c, is equal to the number of relevant operators in the expansion appearing in equation (6.65). Thus, if this codimension is equal to one, as we have assumed, there can be one, and only one, relevant operator, as we had anticipated.

6.5 Finite Lattice Method

In this section, we discuss a technique that can be used (in principle) to obtain a succession of approximations to the exact Kadanoff transformation T. For simplicity's sake, and following the authors of the method [Niem76], we will consider a two-dimensional Ising model over a triangular lattice. Universality (and the exact solution) will guarantee that the critical exponents of this model are the same as those of the usual Ising model over a square lattice.

We want to define the Kadanoff transformation in the following manner: one groups the spins from the original model together into groups of three, according to the sketch shown in figure 6.5. One then associates a block-spin μ_I with each block I. These spins are also arranged over a triangular lattice, with a lattice step equal to √3 times the original constant.

Given the Ising Hamiltonian of the original model:

$$\mathcal{H}(\sigma, K, h) = -K\sum_{\langle ij \rangle}\sigma_i\sigma_j - h\sum_i \sigma_i, \tag{6.72}$$

we want to obtain the new Hamiltonian $\mathcal{H}'$ by summing over the spins σ_i according to the following rule:

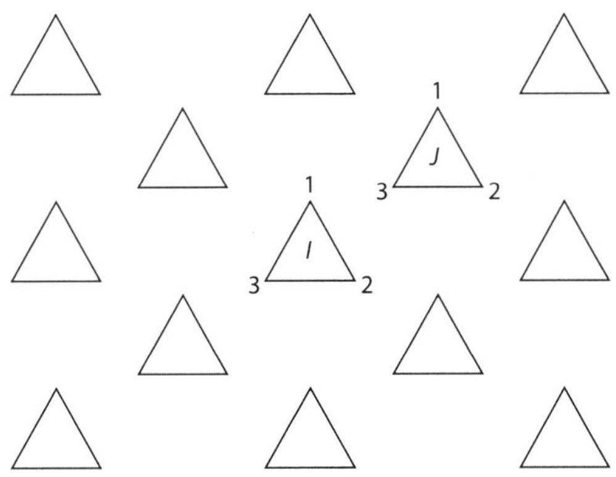

FIGURE 6.5. Subdivision of a triangular lattice into blocks of three spins. The numbers identify the spins from the original model; the letters identify the blocks.

$$e^{-Nf_0(K,h) - \mathcal{H}'(\mu,K',h')} = \sum_{\sigma} \mathcal{P}(\sigma,\mu)\, e^{\mathcal{H}(\sigma,K,h)}. \tag{6.73}$$

In this equation, the *projector* $\mathcal{P}$ represents the local relation that holds between the σ's and the μ's and is chosen in order to satisfy the relation

$$\sum_{\mu} \mathcal{P}(\sigma,\mu) = 1, \qquad \forall \sigma, \tag{6.74}$$

so as to guarantee the invariance of the partition function under the transformation (6.73).

Choosing the projector $\mathcal{P}$ is equivalent to choosing the relation between the σ's and the μ's. One possible choice is the *majority rule*:

$$\mu_i = \phi_i(\sigma) = \operatorname{sign} \sum_{i \in I} \sigma_i. \tag{6.75}$$

In this manner, $\mathcal{P}$ admits the expression

$$P(\sigma,\mu) = \prod_{I} \delta_{\mu_I,\phi_I(\sigma)}. \tag{6.76}$$

Expression (6.73), together with (6.76), is explicit, but this does not imply that it is actually calculable. We will discuss one of many possible methods, in which one considers explicitly a finite lattice, for which the sum (6.73) can be calculated exactly. Let us therefore consider two blocks of three spins, I and J, nearest neighbors of one another, as shown in figure 6.6.

To simplify, let us assume that we are in a vanishing magnetic field. The recurrence formula takes the form

$$e^{-f_0(K) + K'_{\mu_I \mu_J}} = \sum_{\sigma} \mathcal{P}(\sigma,\mu)\, e^{\mathcal{H}}. \tag{6.77}$$

By making the dependence on μ explicit, we obtain

$$e^{-f_0(K) + K'} = \sum_{\{\sigma\}} \delta_{1,\phi_I(\sigma)}\, \delta_{1,\phi_J(\sigma)}\, e^{-\mathcal{H}(\sigma)}; \tag{6.78}$$

$$e^{-f_0(K) - K'} = \sum_{\{\sigma\}} \delta_{1,\phi_I(\sigma)}\, \delta_{-1,\phi_J(\sigma)}\, e^{-\mathcal{H}(\sigma)}. \tag{6.79}$$

We can calculate these expressions by brute force, summing over all the configurations of spins allowed by the delta function. They reduce the potential number $2^6 = 64$ configurations to a more manageable 16. One thus obtains

$$e^{-f_0(K) + K'} = e^{8K} + 3e^{4K} + 2e^{2K} + 3 + 6e^{-2K} + e^{-4K}, \tag{6.80}$$

$$e^{-f_0(K) - K'} = 2e^{4K} + 2e^{3K} + 4 + 6e^{-2K} + 2e^{-4K}. \tag{6.81}$$

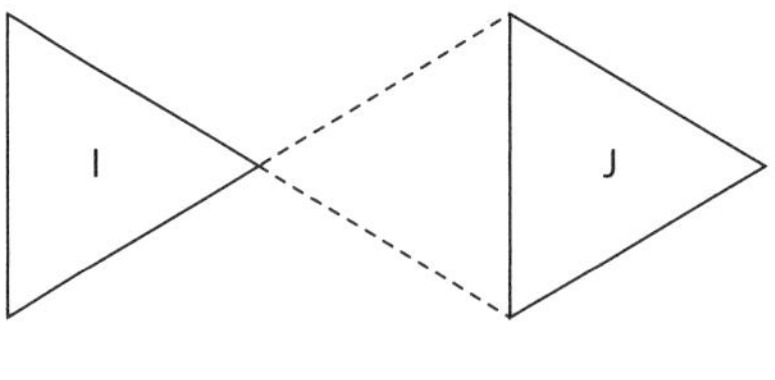

FIGURE 6.6. A finite lattice with six spins.

This implies that

$$K' = \frac{1}{2}\ln\frac{e^{8K} + 3e^{4K} + 2e^{2K} + 3 + 6e^{-2K} + e^{-4K}}{2e^{4K} + 2e^{3K} + 4 + 6e^{-2K} + 2e^{-4K}}. \tag{6.82}$$

This recurrence relation admits one fixed point $K^* = 0.3653$ and yields the exponent $1/\nu = 0.7922$. Although these results are not very good, one can systematically improve the approximation by considering larger lattices [Niem76]. In order to do this, *all* the interactions compatible with the problem's symmetry will come into play—it will therefore be necessary to broaden the space of the interactions being considered.

6.6 Renormalization in Fourier Space

6.6.1 Introduction

In the following sections, we will describe how one can obtain the universal critical properties by means of a version of the Kadanoff transformation in Fourier space. These developments closely follow Wilson's original formulation, which is discussed in [Wils72]. It may also be useful to consult Amit [Amit84].

This approach's point of view is quite different from that in the previous chapter. Instead of considering specific models, we take advantage of universality in order to define some Hamiltonians that possess only those properties we consider essential in determining the critical behavior. This approach leads to approximate expressions of the critical exponents, in the form of an expansion in which only the model's essential properties make their appearance: the dimensionality of space, d, and that of the order parameter, n.

Let us consider an n-component generalization of the Ising model, defined by the Hamiltonian

$$\mathcal{H} = -\frac{1}{2}\sum_{ij} K_{ij}\boldsymbol{\phi}_i \cdot \boldsymbol{\phi}_j - \sum_i \boldsymbol{h}\cdot\boldsymbol{\phi}_i. \tag{6.83}$$

For convenience's sake, we have included the factor $1/k_B T$ in the interaction constants K and h, and the Boltzmann factor is therefore given simply by $e^{-\mathcal{H}}/Z$, where

$$Z = \mathrm{Tr}\, e^{-\mathcal{H}}. \tag{6.84}$$

The interaction matrix K_{ij} depends only on the distance $\boldsymbol{r}_{ij} = \boldsymbol{r}_j - \boldsymbol{r}_i$ between the two lattice sites i and j:

$$K_{ij} = K(\boldsymbol{r}_{ij}) = \begin{cases} K_0, & \text{if } i \text{ and } j \text{ are nearest neighbors,} \\ 0, & \text{otherwise.} \end{cases} \tag{6.85}$$

The $\boldsymbol{\phi}$ variables are n-component unit vectors:

$$\boldsymbol{\phi}_i = (\phi_i^1, \phi_i^2, \ldots, \phi_i^n), \qquad \sum_{\alpha=1}^{n} (\phi_i^\alpha)^2 = 1. \tag{6.86}$$

In this manner, the trace Tr in equation (6.83) is defined by

$$\mathrm{Tr} \equiv \int \prod_i \prod_{\alpha=1}^{n} \mathrm{d}\phi_i^\alpha \, \delta\left(\sum_{\alpha=1}^{n} (\phi_i^\alpha)^2 - 1 \right). \tag{6.87}$$

It is often useful to eliminate the constraint over the spin magnitude and consider a more general formulation in which

$$\mathrm{Tr} = \int \prod_i \prod_{\alpha=1}^{n} \mathrm{d}\phi_i^\alpha \, W(\phi_i), \tag{6.88}$$

where $W(\phi)$ is a given function. In the case we have considered so far, one has

$$W(\phi) = \delta\left(\sum_{\alpha=1}^{n} (\phi^\alpha)^2 - 1 \right). \tag{6.89}$$

Another possible choice of W is the following:

$$W(\phi) = \mathcal{N} \mathrm{e}^{-u(|\phi|^2 - 1)^2}, \tag{6.90}$$

where $\mathcal{N}$ is a normalization constant. This function has a maximum for $|\phi| = 1$ and decreases all the more rapidly the larger u becomes. Expression (6.89) therefore can be recovered in the limit $u \to \infty$. On the basis of universality however, we expect to obtain the same results even if we keep u finite.

At this point, we can reformulate the problem in Fourier space, defining ϕ_k by means of

$$\phi_i = \sum_i \phi_k \mathrm{e}^{\mathrm{i}k \cdot r_i}. \tag{6.91}$$

Since the Hamiltonian (6.83) is quadratic in the ϕ, its expression in Fourier space is simple:

$$\mathcal{H} = -\frac{1}{2N} \sum_k K(\boldsymbol{k}) \, \phi_{-k} \cdot \phi_k - \boldsymbol{h} \cdot \phi_{k=0}. \tag{6.92}$$

In this expression, $K(\boldsymbol{k})$ is the Fourier transform of the interaction matrix:

$$K(\boldsymbol{k}) = \sum_j K(\boldsymbol{r}_{ij}) \, \mathrm{e}^{-\mathrm{i}k \cdot r_{ij}}, \tag{6.93}$$

and the sum is extended to the first Brillouin zone. For a simple cubic lattice, as we saw in chapter 2, one has

$$K(\boldsymbol{k}) = 2K_0 \sum_{j=1}^{d} \cos k_j a_0 \simeq 2K_0 \left(d - k^2 a_0^2 / 2 \right), \tag{6.94}$$

where a_0 is the lattice constant, and the first Brillouin zone is formed by those $\boldsymbol{k}$ whose k_j components satisfy

$$|k_j| \leq \frac{\pi}{a_0} = \Lambda, \quad j = 1, \ldots, d, \tag{6.95}$$

We refer to $\Lambda = \pi/a_0$ as the "cutoff." One should note that if the K_{ij} interaction is short-range, its Fourier transform $K(\mathbf{k})$ is analytic in $\mathbf{k}$, and therefore admits an expansion in powers of $\mathbf{k}$ around $k = 0$. In the case of the simple cubic lattice, this expansion will contain only even powers of $\mathbf{k}$ and contains only k^2 to the first order (as we saw), while the k^4 order term will contain a combination of $(k^2)^2$ and of $\sum_j k_j^4$.

By introducing the *weight* $W(\phi)$, we thus obtain the following expression of the partition function:

$$Z = \int \prod_{\mathbf{k}} \mathrm{d}\phi_{\mathbf{k}}\, W(\{\phi_{\mathbf{k}}\})\, \mathrm{e}^{-\mathcal{H}}, \tag{6.96}$$

where $\mathcal{H}$ is given by equation (6.92) and $W(\{\phi_{\mathbf{k}}\})$ is the weight expressed as a function of the Fourier components $\phi_{\mathbf{k}}$.

6.6.2 *Gaussian Model*

The **Gaussian model** is obtained by considering a Gaussian weight $W(\phi)$:

$$W(\phi) = \mathcal{N}\mathrm{e}^{-(1/2)\lambda|\phi_i|^2}. \tag{6.97}$$

For $\lambda > 0$, one obtains a distribution centered around $|\phi| = 0$. At this point, the functional $W(\{\phi_{\mathbf{k}}\})\mathrm{e}^{-\mathcal{H}}$ is Gaussian, and the partition function can be calculated directly. If for simplicity's sake we set $\mathbf{h} = 0$, we have

$$Z \propto \int \prod_{\mathbf{k}} \mathrm{d}\phi_{\mathbf{k}} \exp\left\{-\frac{1}{2N}\sum_{\mathbf{k}} [\lambda + K(\mathbf{k})]\,\phi_{-\mathbf{k}} \cdot \phi_{\mathbf{k}}\right\}. \tag{6.98}$$

The value of the Fourier components $\phi_{\mathbf{k}}$ depends on the number N of spins that make up the system, as one can see from equation (6.91). In order to take the thermodynamic limit, it is convenient to redefine them to make their value independent of N, which can be accomplished by rescaling them by a factor $1/\sqrt{N}$. The preceding expression then takes the form

$$Z \propto \int \prod_{\mathbf{k}} \mathrm{d}\phi_{\mathbf{k}} \exp\left\{-\frac{1}{2}\sum_{\mathbf{k}} \Delta(\mathbf{k})\,\phi_{-\mathbf{k}} \cdot \phi_{\mathbf{k}}\right\}, \tag{6.99}$$

where

$$\Delta(\mathbf{k}) = r + ck^2 + O(k^4), \tag{6.100}$$

$$r = \lambda - 2dK_0, \tag{6.101}$$

$$c = K_0 a_0^2, \tag{6.102}$$

and one can easily evaluate the next terms in the expansion of $\mathbf{k}$. We can therefore integrate over the wave vectors $\mathbf{k}$ by employing the usual recipe:

$$\sum_{\mathbf{k}} \longrightarrow V\int \frac{\mathrm{d}^d\mathbf{k}}{(2\pi)^d} \equiv V\int_{\mathbf{k}}.$$

The Hamiltonian *density* (which we will henceforth identify with $\mathcal{H}$) is therefore given by

$$\mathcal{H} = \frac{1}{2} \int_k \Delta(\boldsymbol{k})\, \phi_{-k} \cdot \phi_k. \tag{6.103}$$

The Gaussian model can be resolved exactly. It is, however, more instructive to analyze it by means of the Kadanoff transformation in the following manner. First, we exploit universality so as to redefine the first Brillouin zone. In this fashion, we postulate that our field ϕ_i has Fourier components ϕ_k for $\boldsymbol{k}$ included in a sphere of d dimensions and of radius equal to the "cutoff" Λ. We then define the Kadanoff transformation as the sequential application of the following three operations:

1. We divide the *field* ϕ_i into two components, $\phi_i^<$ and $\phi_i^>$, where $\phi_i^<$ has Fourier components $\phi_k^<$ for $|\boldsymbol{k}| \leq \Lambda/b$ (where $b > 1$ is the scaling factor), while $\phi_i^>$ has them for $\boldsymbol{k}$ belonging to the spherical *shell* $\Lambda/b < |\boldsymbol{k}| < \Lambda$. We then integrate over $\phi^>$ in the expression of Z given by equation (6.99). We thus obtain the following expression of the partition function Z:

$$Z = Z_0 \int \prod_k \mathrm{d}\phi_k^< \, \mathrm{e}^{-V\mathcal{H}^<}, \tag{6.104}$$

where Z_0 is a constant (which contributes to the free energy), and the Hamiltonian density $\mathcal{H}^<$ is given by

$$\mathcal{H}^< = \frac{1}{2} \int_k^< \Delta(\boldsymbol{k})\, \phi_{-k} \cdot \phi_k, \tag{6.105}$$

2. At this point, the maximum value of $|\boldsymbol{k}|$ is equal to $\Lambda/b = \pi/(a_0 b)$—this corresponds to a model in which the lattice step is $a_0 b$ rather than a_0. In order to return to the preceding value of the step, we change variables in the expression of Hamiltonian density:

$$\boldsymbol{k} = \boldsymbol{k}'/b, \tag{6.106}$$

and we redefine $\phi_k^< = \bar{\phi}_{bk}$. The Hamiltonian density becomes

$$\begin{aligned} \bar{\mathcal{H}} &= \frac{1}{2} \int_{k'} b^{-d} \Delta(\boldsymbol{k}'/b)\, \bar{\phi}_{-k'} \cdot \bar{\phi}_{k'}, \\ &= \frac{1}{2} \int_{k'} \tilde{\Delta}(\boldsymbol{k}')\, \bar{\phi}_{-k'} \cdot \bar{\phi}_{k'}, \end{aligned} \tag{6.107}$$

where

$$\tilde{\Delta}(k') = r b^{-d} + c b^{-(d+2)} k'^2 + O(k'^4). \tag{6.108}$$

3. We now rescale the variables $\bar{\phi}_k$ by a factor ζ, which will be chosen appropriately later:

$$\bar{\phi}_k = \zeta \phi_k. \tag{6.109}$$

The final expression of the renormalized Hamiltonian density is therefore:

$$\mathcal{H}' = \frac{1}{2} \int_k \Delta'(\boldsymbol{k})\, \phi_{-k} \cdot \phi_k, \tag{6.110}$$

where

$$\Delta'(\mathbf{k}) = r' + c'k^2 + O(k^4), \tag{6.111}$$

$$r' = rb^{-d}\zeta^2, \tag{6.112}$$

$$c' = cb^{-(d+2)}\zeta^2. \tag{6.113}$$

We can now choose the scaling factor ζ so as to have some fixed points. By choosing $\zeta = b^{d/2}$, one has $r' = r$, and therefore the point in which $r \neq 0$, $c = 0$ (and all the coefficients of k^n with $n \geq 2$ vanish) is a fixed point. It is easy to see that it is a *stable* fixed point. In fact, with this choice of ζ, one has $c' = cb^{-2} < c$, and c is therefore an irrelevant parameter.

Exercise 6.1 Show that all the coefficients of κ^n with $n \neq 0$ are irrelevant parameters around this fixed point.

This fixed point corresponds to infinite temperature. Since the Fourier transform $K(\mathbf{k})$ of the interaction is constant, its antitransform K_{ij} is local—in other words, it vanishes whenever $i \neq j$. The system is therefore made of noninteracting local variables ϕ_i—this is the situation one approaches by increasing the temperature in a system of variables interacting at short range. We will therefore call this fixed point ($r \neq 0$, $c = 0$ etc.) the **infinite temperature fixed point.**

We can now choose $\zeta = b^{(d+2)/2}$. In this case, one has $c' = c$, but $r' = b^2 r$. In order to obtain the fixed point, we have to set $r = 0$, $c \neq 0$ and put all k^n coefficients with $n \geq 4$ to zero. Now, it is easy to see that r becomes the only relevant operator.

Exercise 6.2 Show that all the coefficients of k^n with $n \geq 4$ are irrelevant with this choice of ζ around this fixed point.

This fixed point describes the critical point for the Gaussian model. In fact, let us calculate the correlation function

$$\langle \phi_{-k} \cdot \phi_k \rangle = \frac{1}{Z} \int \prod_{k'} \mathrm{d}\phi_{k'} \phi_{-k} \cdot \phi_k \exp\left\{ -\frac{1}{2} \int_{k'} \Delta(\mathbf{k}')\, \phi_{-k'} \cdot \phi_{k'} \right\}. \tag{6.114}$$

The integral is Gaussian and can be factorized into independent integrals, each relative to a pair $(\mathbf{k}', -\mathbf{k}')$. These integrals cancel out with the analogous denominator integrals, except for the one relative to the pair $(\mathbf{k}, -\mathbf{k})$, which gives

$$G(\mathbf{k}) \equiv \langle \phi_{-k} \phi_k \rangle = \frac{1}{\Delta(\mathbf{k})} \simeq \frac{1}{r + ck^2}, \tag{6.115}$$

for small values of $\mathbf{k}$. The correlation function therefore has the Ornstein-Zernike form, and the coherence length ξ is proportional (for small values of r) to $r^{-1/2}$. It becomes infinite for $r = 0$, which therefore corresponds to the critical point. We can also obtain the value of the exponent ν from this argument:

$$\nu = \frac{1}{2}. \tag{6.116}$$

Let us remark that the same result can be obtained from the recurrence relations (6.111):

$$\frac{1}{\nu} = \frac{1}{\ln b} \ln \left(\left. \frac{dr'}{dr} \right|_{r=r^*} \right) = 2. \tag{6.117}$$

It is not difficult to calculate the other exponents—in this fashion, one obtains, for example,

$$\alpha = \frac{d}{2}, \qquad \gamma = 1. \tag{6.118}$$

Since for $r = 0$, the correlation function is proportional to k^{-2}, one has $\eta = 0$. The exponent β is not defined in the Gaussian model. In fact, the region $T < T_c$ should be defined by $r < 0$. In this case, however, the partition function Z is not defined, because the integral over the ϕ is divergent. Analogously, the exponent δ is not defined.

Let us note that all these exponents satisfy the scaling laws, and become equal to the classical exponents when $d = 4$.

6.6.3 The ϕ^4 Model

Let us now express the weight $W(\{\phi\}) = \Pi_i W(\phi_i)$, where $W(\phi)$ is defined by equation (6.90), in terms of the Fourier components ϕ_k. We obtain

$$W(\{\phi\}) \propto \exp\left\{ \frac{\lambda}{2N} \sum_k \phi_{-k} \cdot \phi_k - \frac{u}{N^2} \sum_{k_1 k_2 k_3} (\phi_{k_1} \cdot \phi_{k_2}) (\phi_{k_3} \cdot \phi_{-k_1-k_2-k_3}) \right\}, \tag{6.119}$$

where λ is a coefficient that we do not need to make explicit. The Hamiltonian density in the Fourier space therefore has the following expression:

$$\begin{aligned} \mathcal{H} &= \frac{1}{2} \int_k \Delta(k)\, \phi_{-k} \cdot \phi_k \\ &\quad + u \int_{k_1} \int_{k_2} \int_{k_3} (\phi_{k_1} \cdot \phi_{k_2}) (\phi_{k_3} \cdot \phi_{-k_1-k_2-k_3}) + \cdots \end{aligned} \tag{6.120}$$

where we have redefined the fields ϕ and the coefficient u and understood the higher-order terms in the ϕ variables. The model defined in this manner is known in field theory as the ϕ^4 **model**. We can obtain the same model as a starting point from the Einstein theory of fluctuations, assuming the Landau expression for the availability. The idea is to consider only those fluctuations whose wavelength is greater than a given elementary length, which we will denote by a_0. The relative availability must be a regular function of the ϕ_k, since it results from the integration over local degrees of freedom, and therefore Landau's expression can legitimately be assumed.

The model described by this Hamiltonian is no longer exactly solvable, and it is necessary to resort to some approximations. We rewrite the Hamiltonian in the following form:

$$\mathcal{H} = \mathcal{H}_0 + \mathcal{H}_1, \tag{6.121}$$

where

$$\mathcal{H}_0 = \frac{1}{2}\int_{k}\Delta(k)\,\phi_{-k}\cdot\phi_k, \tag{6.122}$$

$$\mathcal{H}_1 = u\int_{k_1}\int_{k_2}\int_{k_3}(\phi_{k_1}\cdot\phi_{k_2})(\phi_{k_3}\cdot\phi_{-k_1-k_2-k_3}) + \cdots \tag{6.123}$$

Let us now consider step 1 of the Kadanoff transformation. We have

$$\mathrm{e}^{-\mathcal{H}^<} = \int\prod_{\Lambda/b<k<\Lambda}\mathrm{d}\phi_k^>\,\mathrm{e}^{-(\mathcal{H}_0+\mathcal{H}_1)}. \tag{6.124}$$

We are not able to evaluate this integral exactly. We can, however, expand $\mathrm{e}^{\mathcal{H}_1}$ in the form

$$\mathrm{e}^{-\mathcal{H}_1} = 1-\mathcal{H}_1+\frac{1}{2}\mathcal{H}_1^2+\cdots \tag{6.125}$$

By substituting this expression in equation (6.124), we obtain

$$\mathrm{e}^{-\mathcal{H}^<} = Z_0\exp\left[-\mathcal{H}_0^< - \langle\mathcal{H}_1\rangle_0 + \frac{1}{2}\left(\langle\mathcal{H}_1^2\rangle_0 - \langle\mathcal{H}_1\rangle_0^2\right)+\cdots\right], \tag{6.126}$$

where

$$\langle A\rangle_0 \equiv \frac{1}{Z_0}\left[\int\prod_{\Lambda/b<k<\Lambda}\mathrm{d}\phi_k^>\,\mathrm{e}^{-\mathcal{H}_0}A\right], \tag{6.127}$$

with Z_0 a normalization factor. One therefore has

$$\mathcal{H}^< = \mathcal{H}_0^< + \langle\mathcal{H}_1\rangle_0 - \frac{1}{2}\left(\langle\mathcal{H}_1^2\rangle_0 - \langle\mathcal{H}_1\rangle_0^2\right)+\cdots \tag{6.128}$$

This expression is a **cumulant expansion** of the renormalized Hamiltonian $\mathcal{H}^<$. At zero order in $\mathcal{H}_1$, we obtain again the Gaussian Hamiltonian that we saw in the preceding section.

Let us consider the first nontrivial term, proportional to $\langle\mathcal{H}_1\rangle_0$. It is defined by

$$-u\left(\frac{1}{Z_0}\right)\int_{k_1}\int_{k_2}\int_{k_3}\int\prod_{k}\mathrm{d}\phi_k^>\,\mathrm{e}^{-\mathcal{H}_0}\left(\phi_{k_1}\cdot\phi_{k_2}\right)\left(\phi_{k_3}\cdot\phi_{-k_1-k_2-k_3}\right).$$

The terms that contribute to this expression can be sorted into the following classes:

1. Terms in which all the wave vectors k_i, $i = 1, \ldots, 4$ (where we have defined $k_4 = -\sum_{j=1}^{3}k_j$) have a magnitude greater than Λ/b. The contribution of these wave vectors is a constant that modifies the free energy, but not the interaction itself.

2. Terms in which only one of the wave vectors (or three of them) have a magnitude greater than Λ/b vanish because of symmetry when one integrates over $\phi_k^>$.

3. Terms in which the two vectors are larger than Λ/b. We can distinguish two cases:

   - $k_1 = -k_2$, $k_3 = -k_4$. Let us assume that $|k_3| > \Lambda/b$. We obtain

     $$n(\phi_{-k_1}\cdot\phi_{k_1})\int_{k_3}^{>}G(k_3), \tag{6.129}$$

     as a contribution, where $G(k)$ is the correlation function of the Gaussian model

$$G(\mathbf{k}) = \frac{1}{\Delta(\mathbf{k})}. \tag{6.130}$$

The factor n is the result of the sum over the n components of the ϕ fields. There are two terms of this type. We have denoted the integral extended to the spherical shell $\Lambda/b < k < \Lambda$ by $\int_{k}^{>}$.

- $\mathbf{k}_1 = -\mathbf{k}_3$, $\mathbf{k}_2 = -\mathbf{k}_4$ (or $\mathbf{k}_1 = -\mathbf{k}_4$, $\mathbf{k}_2 = -\mathbf{k}_3$). If we choose the first possibility and assume that $|\mathbf{k}_2| > \Lambda/b$, we obtain a contribution of the form

$$(\phi_{-\mathbf{k}_1} \cdot \phi_{\mathbf{k}_1}) \int_{\mathbf{k}_2}^{>} G(\mathbf{k}_2). \tag{6.131}$$

There are four terms of this type.

4. The terms whose wave vectors are smaller than Λ/b in magnitude: these are not modified. The result is therefore

$$\begin{aligned} \langle \mathcal{H}_1 \rangle_0 &= \text{const.} + u \int_{\mathbf{k}_1}^{<} (2n+4)(\phi_{-\mathbf{k}_1} \cdot \phi_{\mathbf{k}_1}) \int_{\mathbf{k}}^{>} G(\mathbf{k}) \\ &\quad + u \int_{\mathbf{k}_1}^{<} \int_{\mathbf{k}_2}^{<} \int_{\mathbf{k}_3}^{<} (\phi_{\mathbf{k}_1} \cdot \phi_{\mathbf{k}_2})(\phi_{\mathbf{k}_3} \cdot \phi_{-\mathbf{k}_1-\mathbf{k}_2-\mathbf{k}_3}). \end{aligned} \tag{6.132}$$

Remember that $\int_{\mathbf{k}}^{<}$ refers to the integrals extended to the new *Brillouin zone* $|\mathbf{k}| < \Lambda/b$, while $\int_{\mathbf{k}}^{>}$ refers to those extended to the spherical shell $\Lambda/b < |\mathbf{k}| < \Lambda$.

Starting from this expression, we perform steps 2 and 3 of the Kadanoff transformation. After some further reordering, we obtain the recurrence formulas:

$$r' = b^{-d}\zeta^2 \left[r + 2(2n+4)\, u \int_{\mathbf{k}}^{>} G(\mathbf{k}) \right], \tag{6.133}$$

$$c' = cb^{-(d+2)}\zeta^2, \tag{6.134}$$

$$u' = ub^{-3d}\zeta^4. \tag{6.135}$$

Let us consider equation (6.135). It implies that $u = 0$, $r = 0$ is a fixed point in the transformation, the Gaussian fixed point we studied in the preceding section. If we choose $\zeta = b^{(d+2)/2}$, we obtain $u' = ub^{4-d}$. Therefore, if $d > 4$, $u' < u$, and u is in this case an irrelevant perturbation with respect to the Gaussian fixed point. On the other hand, if $d < 4$, one has $u' > u$, and u is a relevant perturbation. We can therefore expect that the Gaussian model provides a good description of the critical behavior (at least above T_c) for $d > 4$.

In order to study the system's behavior for $d < 4$, we must evaluate the transformations (6.133) to (6.135) and especially the value of u' at the next order in u. Actually, as we will see immediately, it will be sufficient to calculate the recurrence formula for the coefficient u only, and it comes from the term $\langle \mathcal{H}_1^2 \rangle_0 - \langle \mathcal{H}_1 \rangle_0^2$. In this fashion, we will obtain an expression of the form

$$u' = ub^{4-d} - u^2 w(b, r, n), \tag{6.136}$$

which implies the existence of a fixed point value of u^* given by

$$u^* = \frac{1 - b^{4-d}}{w(b, r^*, n)}, \tag{6.137}$$

where r^* is the value of r corresponding to the critical fixed point. Obviously, our approximations will be acceptable if u^* is small. Let us introduce the quantity $\epsilon = 4 - d$ and assume that we can treat it as a small quantity. We obtain

$$u^* \simeq \epsilon \frac{\ln b}{w(b,r,n)}. \tag{6.138}$$

It is therefore reasonable to calculate the renormalization transformations by means of perturbation techniques in the context of an expansion in ϵ.

6.6.4 Critical Exponents at the First Order in ϵ

We must therefore evaluate the expression

$$\langle \mathcal{H}_1^2 \rangle_0 - \langle \mathcal{H}_1 \rangle_0^2.$$

The first term contains a product of eight ϕ fields. Let us write $\phi = \phi^< + \phi^>$, and evaluate the average over $\phi^>$ with respect to the distribution e^{-H_0}. The terms with an odd number of $\phi^>$ vanish because of symmetry. Let us consider the other terms one by one:

1. The terms with eight $\phi^>$ factors are constant and contribute to free energy, but not to $\mathcal{H}^<$.
2. The terms with six $\phi^>$ factors are proportional to $(\phi^<)^2$ and therefore contribute to the renormalization of r and c. Since they are proportional to u^2, they are of an order higher than the ones we already evaluated, and we can disregard them for the time being.
3. The terms with four $\phi^>$ factors are of the form $\phi^<_{k_1} \cdots \phi^<_{k_4} \langle \phi^>_{q_1} \cdots \phi^>_{q_4} \rangle_0$. Since the average is evaluated with a Gaussian distribution, the average of the four $\phi^>$ fields is broken into the sums of the products of two averages of the $\langle \phi^>_{q_1} \phi^>_{q_2} \rangle_0$ type. We thus obtain different terms, which can be sorted into two categories:

   - Those in which the $\phi^>$ factors that appear in each of these averages belong to the same factor $\mathcal{H}_1$. These terms contribute to $\langle \mathcal{H}_1 \rangle_0^2$ and are therefore cancelled out by the second term of our expression.
   - Those in which the $\phi^>$ factors in each of these two averages belong to two different factors $\mathcal{H}_1$. These are the contributions we must take into account. They are of the form

   $$\phi^<_{k_1} \phi^<_{k_2} \phi^<_{k_3} \phi^<_{k_4} \int_{q_1}^> \int_{q_2}^> \int_{q_3}^> \int_{q_4}^>$$
   $$\times \delta(k_1 + k_2 + q_1 + q_3)\, \delta(k_3 + k_4 + q_2 + q_4)$$
   $$\times \langle \phi^>_{q_1} \phi^>_{q_2} \rangle_0 \langle \phi^>_{q_3} \phi^>_{q_4} \rangle_0.$$

   The delta functions come from the constraints imposed on the wave numbers in the expression of $\mathcal{H}_1$.

   The coefficient of $\phi^<_{k_1} \cdots \phi^<_{k_4}$ in this expression depends on the wave numbers $\boldsymbol{k}$, which means that the resulting renormalized interaction will not be purely local. As long as it is short-range, however (and therefore analytic as a function of $\boldsymbol{k}$), we can consider expanding it in series around $\boldsymbol{k} = 0$. As we will see shortly, the terms of this expansion proportional to

k^2 and so on are irrelevant and can be disregarded (at least at this order). We can therefore evaluate this coefficient for $\boldsymbol{k}_i = 0$. We obtain

$$\int_{q_1}^{>}\int_{q_2}^{>}\langle\phi_{q_1}^{>}\phi_{q_2}^{>}\rangle_0\langle\phi_{-q_1}^{>}\phi_{-q_2}^{>}\rangle_0.$$

Now translation invariance imposes that $\langle\phi_{q_1}^{>}\phi_{q_2}^{>}\rangle_0$ is proportional to $\delta(q_1+q_2)$, and this expression is therefore reduced to

$$\int_{q}^{>}\langle\phi_{q}^{>}\phi_{-q}^{>}\rangle_0^2 = \int_{q}^{>} G(q)^2.$$

We now have to calculate how many terms of this type are present, accounting for the sums over the ϕ fields' different components, among other things. This calculation is slightly technical and is discussed in section 6.6.5. The result is $8(n+8)$, which, if we take into account the factor $1/2$ that appears in the cumulant expansion, gives us a coefficient $4(n+8)$.

4. The terms with two $\phi^{>}$ factors provide a contribution to the renormalization interaction, proportional to ϕ^6. Therefore, although our initial interaction contained only local terms in ϕ^4, we have obtained an interaction containing nonlocal terms (but short-range ones) in ϕ^4, as well as terms proportional to higher powers of ϕ. We will, however, be able to see that these terms are irrelevant and can be disregarded, at least at the lowest order in ϵ.
5. The terms with no $\phi^{>}$ factors are contributions to $\langle\mathcal{H}_1\rangle^2$ and are cancelled out by our expression's second term.

Taking these new contributions into account, and applying the other two steps of the Kadanoff transformations, we obtain the recurrence formulas:

$$r' = b^{-d}\zeta^2\left[r + 4(n+2)u\int_q^{>} G(q)\right], \tag{6.139}$$

$$c' = cb^{-(d+2)}\zeta^2, \tag{6.140}$$

$$u' = b^{-3d}\zeta^4\left[u - 4(n+8)u^2\int_q^{>} G^2(q)\right]. \tag{6. 141}$$

We see that in order to maintain $c' = c$, one needs to set $\zeta = b^{(d+2)/2}$, as in the Gaussian case. Let us now consider interaction terms proportional to ϕ^p, with $p \geq 4$. The relevant recurrence formula gives

$$v_p' = v_p b^{-(p-1)d}\zeta^p\left[1 + O(u^2)\right]. \tag{6.142}$$

With our choice of ζ, we obtain $v_p' \simeq b^{p(-d/2+1)+d}v_p$, and therefore, v_n is irrelevant if $d \simeq 4$ and u is small at the fixed point. Analogously, if we consider some four-field interaction terms, which are proportional to k^p (with $p = 2, 4, \ldots$), we see that the relevant coefficients are rescaled by a factor b^{4-d-p}, up to contributions of the order of u^2. These fields are therefore irrelevant, as long as ϵ and u are small.

We can now suppose $b = 1 + \delta\ell$, with $\delta\ell$ very small, and evaluate the wave numbers in units of Λ. More specifically, this allows us to evaluate the integrals that appear in these expressions in a very simple fashion:

$$\int_q^> G(q) \simeq K_d \delta\ell \frac{1}{r+c'} \tag{6.143}$$

$$\int_q^> G^2(q) \simeq K_d \delta\ell \frac{1}{(r+c)^2}, \tag{6.144}$$

where $K_d = S_d/(2\pi)^d$ (S_d is the surface of the unit sphere in d dimensions) is a factor that comes from the integration, and which we can get rid of by redefining u appropriately. With $\zeta = b^{(d+2)/2}$ (as we shall from now on), we obtain

$$r' = (1+\delta\ell)^2 \left[r + 4(n+2)\,\delta\ell \frac{u}{r+c} \right], \tag{6.145}$$

$$c' = c, \tag{6.146}$$

$$u' = (1+\delta\ell)^{\epsilon} \left[1 - 4(n+8) u \delta\ell \frac{u}{(r+c)^2} \right]. \tag{6.147}$$

We can therefore set $c = 1$ and write these relations as differential equations:

$$\frac{dr}{d\ell} = 2r(\ell) + 4(n+2) \frac{u(\ell)}{1+r(\ell)}, \tag{6.148}$$

$$\frac{du}{d\ell} = \epsilon u(\ell) - 4(n+8) \frac{u^2(\ell)}{(1+r(\ell))^2}. \tag{6.149}$$

The flow corresponding to these differential equations is shown in figure 6.7 in the two cases $d > 4$ (when the fixed Gaussian point is stable) and $d < 4$—in this case, the fixed Gaussian point becomes unstable, and a new stable fixed point appears with $u^* > 0$. This fixed point (r^*, u^*) satisfies (at first order in ϵ)

$$r^* = -\frac{1}{2}\frac{n+2}{n+8}\epsilon, \tag{6.150}$$

$$u^* = \frac{\epsilon}{4(n+8)}. \tag{6.151}$$

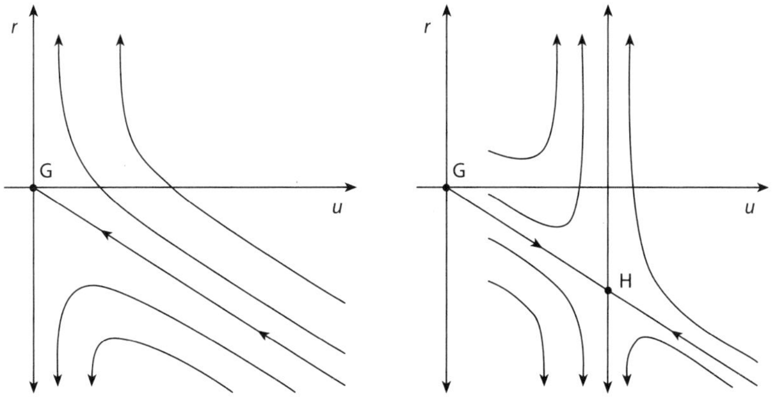

FIGURE 6.7. Renormalization flux in the plane (u, r) for the ϕ^4 model. Left: $d > 4$. The G (Gaussian) point is stable for the entire critical line $r = r_c(u)$. Right: $d < 4$. The H (Heisenberg) fixed point appears, which is stable for the critical surface in which $u > 0$.

This fixed point describes the critical behavior of the n-component Heisenberg model for $d < 4$. We will therefore call it the **Heisenberg fixed point**.

Let us now consider the linearized flow equations around this fixed point:

$$\frac{\mathrm{d}}{\mathrm{d}t}\begin{pmatrix} r \\ u \end{pmatrix} = \begin{pmatrix} 2 - [(n+2)/(n+8)]\epsilon, & 4(n+2) \\ 0, & -\epsilon \end{pmatrix}\begin{pmatrix} r - r^* \\ u - u^* \end{pmatrix}. \tag{6.152}$$

The eigenvalues of the matrix that defines the linearized flow are

$$\lambda_r = \frac{1}{\nu} = 2 - \frac{n+2}{n+8}\epsilon, \tag{6.153}$$

$$\lambda_u = -\epsilon. \tag{6.154}$$

The second eigenvalue controls the approach to the fixed point. The first eigenvalue determines the value of the exponent ν:

$$\nu = \frac{1}{2} + \frac{n+2}{2(n+8)}\epsilon + O(\epsilon^2). \tag{6.155}$$

The second independent critical exponent, η, can be obtained from the expression of ζ. When the lengths are rescaled by a factor $1/b$ at the fixed point, the correlation function in real space is rescaled by a factor ζ^2. Therefore,

$$G(r/b) = \zeta^2 G(r) = b^{d+2} G(r). \tag{6.156}$$

This implies that $G(r) \sim r^{-(d+2)}$, which should be compared with the definition $G(r) \sim r^{-(d+2-\eta)}$, which defines the exponent η. We thus obtain

$$\eta = 0 + O(\epsilon^2). \tag{6.157}$$

The other exponents can be obtained from the scaling laws, stopping of course at the first order in ϵ. One thus has

$$\alpha = \epsilon\left(\frac{1}{2} - \frac{n+2}{n+8}\right) + O(\epsilon^2), \tag{6.158}$$

$$\beta = \frac{1}{2} - \frac{3\epsilon}{2(n+8)} + O(\epsilon^2), \tag{6.159}$$

$$\gamma = 1 + \frac{n+2}{2(n+8)}\epsilon + O(\epsilon^2), \tag{6.160}$$

$$\delta = 3 + \epsilon + O(\epsilon^2), \tag{6.161}$$

$$\eta = O(\epsilon^2). \tag{6.162}$$

By substituting $\epsilon = 1$ in these expressions, one does not obtain particularly good results—the interest of this approach lies in the fact that it can be systematically improved (calculations up to the fifth order in ϵ have been performed) and that it can provide general qualitative information, about the dependence in n, for instance. Most of all, this method allows us to systematically control universality properties in the most general fashion.

6.6.5 Multiplicity of the Contributions to Renormalization of u

In this section, I show how one obtains the coefficient for multiplicity $4(n+8)$ in the equation (6.141).

By writing all the sums, both over the vectors and over the field components, $\langle \mathcal{H}_1^2 \rangle_0$ can be written

$$\begin{aligned}\langle \mathcal{H}_1^2 \rangle_0 &= \frac{1}{Z_0} \int \prod_q \mathrm{d}\phi_q^> \, \mathrm{e}^{-\mathcal{H}_0} u^2 \sum_{\alpha_1\alpha_2} \sum_{\beta_1\beta_2} \int_{k_1} \cdots \int_{k_4} \int_{q_1} \cdots \int_{q_4} \\ &\times \delta(k_1+k_2+k_3+k_4)\, \delta(q_1+q_2+q_3+q_4) \\ &\times \left(\phi_{k_1}^{\alpha_1} \phi_{k_2}^{\alpha_1} \phi_{k_3}^{\alpha_2} \phi_{k_4}^{\alpha_2}\right) \left(\phi_{q_1}^{\beta_1} \phi_{q_2}^{\beta_1} \phi_{q_3}^{\beta_2} \phi_{q_4}^{\beta_2}\right).\end{aligned} \tag{6.163}$$

The terms we are considering are those in which two fields $\phi^>$ belonging to the first factor [the one referred to by $(\boldsymbol{k}, \alpha)$] are paired each with a field belonging to the second factor $(\boldsymbol{q}, \beta)$. One thus obtains two factors $\langle \phi_{k\alpha}^> \phi_{k\beta}^> \rangle_0$ that vanish, unless one has $\boldsymbol{k} = -\boldsymbol{q}$ and $\alpha = \beta$.

The terms one obtains are of three types:

1. Terms of the type

   $$\langle \phi_{k_1\alpha_1}^> \phi_{q_1\beta_1}^> \rangle_0 \langle \phi_{k_2\alpha_1}^> \phi_{q_2\beta_1}^> \rangle_0 \phi_{k_3\alpha_2}^< \phi_{k_4\alpha_2}^< \phi_{q_3\beta_2}^< \phi_{q_4\beta_2}^<.$$

   In this case, one can freely add over $\alpha_1 = \beta_1$, and one therefore obtains a contribution that is proportional to the number n of components of the $\phi^>$ field. It is easy to see that there are eight contributions of this sort: One can choose α_1 or α_2 and connect it with β_1 or β_2, and there are two ways to make the connection.

2. Terms of the type

   $$\langle \phi_{k_1\alpha_1}^> \phi_{q_1\beta_1}^> \rangle_0 \langle \phi_{k_3\alpha_2}^> \phi_{q_3\beta_2}^> \rangle_0 \phi_{k_2\alpha_1}^< \phi_{k_4\alpha_2}^< \phi_{q_2\beta_1}^< \phi_{q_4\beta_2}^<.$$

   In this case, one must have $\alpha_1 = \beta_1$ and $\alpha_2 = \beta_2$, and there are therefore no free components of the $\phi^>$ field over which to sum. The result therefore does not depend on n. There are 32 terms of this type: one can connect any sort of α field with any of the four β fields (16 choices), and one then needs to connect an α field that contains a different component to one of the β fields with a component that differs from the one previously connected (4 choices). One then needs to divide by 2 because the same result can be obtained by making one's choices in two α fields in the reverse order.

3. Terms of the type

   $$\langle \phi_{k_1\alpha_1}^> \phi_{q_1\beta_1}^> \rangle_0 \langle \phi_{k_3\alpha_2}^> \phi_{q_2\beta_1}^> \rangle_0 \phi_{k_2\alpha_1}^< \phi_{k_4\alpha_2}^< \phi_{q_3\beta_2}^< \phi_{q_4\beta_2}^<.$$

   In this case, one must have $\alpha_1 = \beta_1 = \alpha_2$, and there are therefore once again no free components of the $\phi^>$ field. Using the same type of reasoning as earlier, we see that there are 32 components of this type.

The multiplicity of contributions is therefore $8n + 64$. One can check this result by setting $n = 1$. One obtains a multiplicity equal to 72. On the other hand, for $n = 1$, one has $(4 \times 4 \times 3 \times 3)/2 = 72$ terms (4 possible α's with 4 possible β's, then one of the remaining

3 α's with one of the remaining β's, and then dividing by 2 because one obtains the same result by switching the two pairs).

Taking the factor 1/2 that comes from the cumulant expansion into account, we obtain the coefficient $4(n+8)$ in equation (6.141).

These evaluations are greatly facilitated if diagrammatic techniques are employed, and these are presented in almost all the books on the topic.

6.7 Quadratic Anisotropy and Crossover

Let us now consider a Hamiltonian in which the different components of the order parameter interact with different intensities. If the order parameter ϕ is an n-component unit vector, for instance, we can consider the following generalization of Heisenberg's Hamiltonian:

$$\mathcal{H} = -\frac{1}{2}\sum_{ij} K_{ij} \sum_{\alpha=1}^{n} (1+\Delta_\alpha)\,\phi_i^\alpha \phi_j^\alpha. \qquad (6.164)$$

Let us assume that $\Delta_\alpha = \Delta > 0$ for $\alpha = 1, \dots, m$ and $\Delta_\alpha = 0$ for $\alpha = m+1, \dots, n$. In this case, the first m components of ϕ will interact more strongly than the others, and we can expect the symmetry breaking to privilege them: the critical behavior of the model will be the one of a model with an m-dimensional order parameter. If Δ is very small, on the other hand, we expect the model's behavior for T not too close to T_c (or for $T = T_c$, but for fairly small distances) to be similar to that of the symmetrical model.

Let us see how one can deal with the effect of this type of perturbation at the first order in ϵ.

By switching to continuous fields, it is easy to see that the effect of perturbation Δ will modify the quadratic part of the Hamiltonian density. A term of this type is therefore called a **quadratic anisotropy** term. Let us attempt to describe its effect in the simplest fashion. Let us therefore consider a Hamiltonian density of the form

$$\mathcal{H} = \mathcal{H}_0 + \mathcal{H}_1, \qquad (6.165)$$

where $\mathcal{H}_1$ has the form seen in the preceding paragraph, while

$$\mathcal{H}_0 = \frac{1}{2}\int_k \sum_{\alpha=1}^{n} (r_\alpha + k^2)\,\phi_{-k}^\alpha \phi_k^\alpha. \qquad (6.166)$$

It is easy to see that, at the lowest order in u, the recurrence relations for the r_α are given by

$$\frac{dr_\alpha}{d\ell} = 2r_\alpha + 4u\left(\frac{2}{1+r_\alpha} + \sum_{\beta=1}^{n}\frac{1}{1+r_\alpha}\right). \qquad (6.167)$$

By linearizing around the Heisenberg fixed point and setting $\delta r_\alpha = (r_\alpha - r^*)$, $\delta r = \sum_\alpha \delta r_\alpha$ and $\Delta_\alpha = \delta r_\alpha - n^{-1}\delta r$, we obtain

$$\frac{d\delta r}{d\ell} = [2 - 4u^*(n+2)]\delta r, \tag{6.168}$$

$$\frac{d\Delta_\alpha}{d\ell} = 2(1-4u^*)\Delta_\alpha. \tag{6.169}$$

The first equation governs the behavior of the correlation length around the Heisenberg fixed point and shows that the exponent ν is not modified. The second equation determines the critical behavior of the system with a small anisotropy.

Let us, in fact, consider the system at the critical point (therefore, $\delta r = 0$) and with, for example, $\Delta_1 > 0$, $\Delta_\alpha = 0$ ($\alpha = 2, \ldots, n$). By solving the recurrence relations, we will have $\Delta_1(\ell) = \Delta_1(0)e^{\psi\ell}$, where the exponent ψ is given by

$$\psi = 2(1-4u^*) = 2 - \frac{2\epsilon}{n+8} + O(\epsilon^2). \tag{6.170}$$

Let us recall that the length rescaling factor b is given by $b = e^\ell$. The correlation function $G(\mathbf{r}, \Delta)$, where $\Delta = \Delta_1(0)$, satisfies the relation

$$G(\mathbf{r},\Delta) = b^{-(d+2)}G(\mathbf{r}/b, \Delta b^\psi). \tag{6.171}$$

By choosing $b = |\mathbf{r}|$, we obtain $G(\mathbf{r},\Delta) = |\mathbf{r}|^{-(d+2)} f(r/\Delta^{-1/\psi})$, where $f(x)$ is a certain function. We see, therefore, that there is a certain characteristic length $\ell_\Delta \sim \Delta^{-1/\psi}$ that governs the effects of anisotropy at the critical point. For $|\mathbf{r}| \ll \ell_\Delta$, the behavior is similar to that with $\Delta = 0$ (isotropic model), while for $|\mathbf{r}| \gg \ell_\Delta$, the behavior corresponds to that of large values of Δ, and therefore to a model with a smaller number of dimensions of the order parameter.

If we instead consider $T \neq T_c$, we will have to compare the correlation length $\xi \sim |T - T_c|^{-\nu}$ with ℓ_Δ. For $\mathbf{k} = 0$, therefore, the change in behavior will occur for $|T - T_c| \sim \ell^{-1/\phi}$, where $\phi = \psi\nu$ is called the **crossover exponent**:

$$\phi = \psi\nu = 1 + \frac{n}{2(n+8)} + O(\epsilon^2). \tag{6.172}$$

More particularly, this result implies that $T_c(\Delta) = T_c(0) + O(\Delta^{1/\phi})$. Since $\phi > 1$, the line of critical points exhibits a cusp in $\Delta = 0$.

In this way, the critical behavior of the model is described by one fixed point further from the critical point, and by another one closer to it. This phenomenon is called **crossover**.

6.8 Critical Crossover

The most common case of crossover is obviously that by means of which a system's behavior is described by a Gaussian fixed point when it is far from the critical temperature, and by a Heisenberg fixed point close to the critical temperature—the transition from one behavior to the other occurs at a temperature determined by the Ginzburg criterion. In this section, we show how the equations (6.148) and (6.149) describing the renormalization flow at the first order in ϵ allow one to obtain the crossover behavior of the susceptibility.

Let us rewrite equations (6.148) and (6.149) in the form

$$\frac{dr}{d\ell} = 2r + A\frac{u}{1+r}, \tag{6.173}$$

$$\frac{du}{d\ell} = \epsilon u - B\frac{u^2}{(1+r)^2}, \tag{6.174}$$

where $A = 4(n+2)$ and $B = 4(n+8)$. If $[r(\ell), u(\ell)]$ is a solution of the flow equations, the correlation function $C(\mathbf{k}, r, u) = \langle \phi_{-k}\phi_k \rangle$ satisfies the relation

$$C(\mathbf{k}, r, u) = \mathrm{e}^{(2-\eta)\ell} C(\mathrm{e}^{\ell}\mathbf{k}, r(\ell), u(\ell)). \tag{6.175}$$

(More generally, the η exponent obtained by rescaling the fields, will also depend on ℓ, and η should be substituted by $\int_0^{\ell} d\ell' \eta(\ell')$ in the first factor. At the first order in ϵ, however, $\eta = 0$, and we can disregard this complication.) We can use this relation to express $C(\mathbf{k}, r, u)$ for small r's and k's (in the proximity of the critical point) in terms of $C(\mathrm{e}^{\ell}\mathbf{k}, r(\ell), u(\ell))$, where $r(\ell)$ and $\mathrm{e}^{\ell}\mathbf{k}$ are of order one, which should allow us to evaluate C in perturbation theory with ease.

We will soon see that, at the first order in ϵ, we can set $r = 0$ in the equation for u, so that equation (6.174) takes the form

$$\frac{du}{d\ell} = \epsilon u - Bu^2. \tag{6.176}$$

This equation can be solved exactly, and has the solution

$$u(\ell) = u_0 \frac{\mathrm{e}^{\epsilon\ell}}{Q(\ell)}, \tag{6.177}$$

where $u_0 = u(\ell = 0)$ and

$$Q(\ell) = 1 + \frac{Bu_0}{\epsilon}(\mathrm{e}^{\epsilon\ell} - 1). \tag{6.178}$$

It is clear that

$$u(\ell) \to \begin{cases} u_0, & \text{for } \ell \to 0, \\ \epsilon/B = u^*, & \text{for } \ell \to \infty. \end{cases} \tag{6.179}$$

The transition from one value to the other occurs for $(Bu_0/\epsilon)\mathrm{e}^{\epsilon\ell} \sim 1$. We will always assume that u_0 is on the order of ϵ.

We will denote the solution we just found by $u_1(\ell)$ and the solution of equation (6.174) by $u(\ell)$. Let us set $u(\ell) = f(\ell)u_1(\ell)$. Then,

$$u_1\frac{df}{d\ell} = -Bu_1^2(f^2 - f) + f^2u_1^2\left[\frac{1}{(1+r)^2} - 1\right]. \tag{6.180}$$

Since u_1 is of order ϵ, and $f \to 1$ for $\ell \to 0$, one has $f = 1 + O(\epsilon)$—we can therefore set $u = u_1$ at the first order in ϵ.

The solution of equation (6.173) is a more delicate matter, especially since we must consider the possibility that r becomes of order 1. Let us introduce the quantity

$$t(\ell) = r(\ell) + \frac{1}{2} Au(\ell) - \frac{1}{2} Au(\ell)\, r(\ell) \ln[1 + r(\ell)], \tag{6.181}$$

and take into account the fact that $du/d\ell = O(\epsilon^2)$. By neglecting contributions of order ϵ^2, one then has

$$\begin{aligned} \frac{dt}{d\ell} &= \frac{dr(\ell)}{d\ell}\left[1 - \frac{1}{2} Au(\ell) \ln[1 + r(\ell)] - \frac{1}{2} Au(\ell) \frac{r(\ell)}{1 + r(\ell)}\right] \\ &= \left[2r(\ell) + A\frac{u(\ell)}{1 + r(\ell)}\right]\left[1 - \frac{1}{2} Au(\ell) \ln[1 + r(\ell)] - \frac{1}{2} Au(\ell) \frac{r(\ell)}{1 + r(\ell)}\right] \\ &= 2r(\ell) + Au(\ell) - Au(\ell)\, r(\ell) \ln[1 + r(\ell)] + Au(\ell)\, r(\ell) \\ &= [2 - Au(\ell)]\, t(\ell) + O(\epsilon^2). \end{aligned} \tag{6.182}$$

Then,

$$t(\ell) = e^{S(\ell)}\, t(0), \tag{6.183}$$

where

$$S(\ell) = 2\ell - A \int_0^{\ell} d\ell' u(\ell'). \tag{6.184}$$

We see that, if $t(0) = 0$, then $t(\ell) = 0$, $\forall \ell$. We can therefore identify t with the distance from the critical surface. The transition therefore occurs for $t(0) = 0$—in other words, for $r(0) = -1/2Au(0) + O(\epsilon^2)$.

It is possible to evaluate $S(\ell)$ explicitly:

$$\begin{aligned} S(\ell) &= 2\ell - A \int_0^{\ell} d\ell' u(\ell') = 2\ell - Au_0 \int_0^{\ell} d\ell' \frac{e^{\epsilon \ell'}}{1 + (Bu_0/\epsilon)(e^{\epsilon \ell} - 1)} \\ &= 2\ell - \frac{A}{B} \ln Q(\ell). \end{aligned} \tag{6.185}$$

Therefore,

$$t(\ell) = e^{2\ell}[Q(\ell)]^{-A/B}\, t(0) \tag{6.186}$$

$$\simeq \begin{cases} e^{2\ell}\, t(0), & \text{for } \ell \to 0; \\ e^{(2 - A\epsilon/B)\ell} (Bu_0/\epsilon)\, t(0), & \text{for } \ell \to \infty. \end{cases} \tag{6.187}$$

This is the crossover we were looking for. The mean-field behavior is valid as long as $(Bu_0/\epsilon)\, e^{\epsilon \ell} \ll 1$. For $(Bu_0/\epsilon)\, e^{\epsilon \ell} \gg 1$ the behavior is dictated by the Heisenberg fixed point, with $t(\ell) \simeq e^{\ell/\nu} t(0)$, where $1/\nu = 2 - [(n+2)/(n+8)]\epsilon$. The crossover occurs at the temperature defined by $t = (Bu_0/\epsilon)^{2/\epsilon}$.

We must now relate the physical quantities to the solution obtained for $t(\ell)$. Let us choose ℓ^* so that

$$t(\ell^*) = 1. \tag{6.188}$$

At this point, it is possible to use perturbation theory to evaluate the correlation functions. The susceptibility χ is proportional to the correlation function for $\boldsymbol{k} = 0$. In perturbation theory, it is more convenient to evaluate the *inverse* of $C(\mathbf{k}, r, u)$. In effect, at the lowest order in u, one has

$$C^{-1}(\mathbf{k}, r, u) = r + k^2 + 4u\,(n+2)\int_{\mathbf{q}} G^2(\mathbf{q}, r), \tag{6.189}$$

where $G(\mathbf{q}, r) = 1/(r + q^2)$ is the correlation function of the Gaussian model. This relation is derived at the end of this section. In $d = 4$, one has, at the lowest order in the perturbation,

$$\begin{aligned} C^{-1}(0, r, u) &= r + 4\,(n+2)\int_{\mathbf{q}} \frac{1}{(1+r)^2} \\ &= r + \frac{1}{2}Au. - \frac{1}{2}Ar\,[\ln(1+r) - \ln r] \\ &\simeq t + \frac{1}{2}u\,\ln t. \end{aligned} \tag{6.190}$$

Therefore, for $\ell = \ell^*$, one has $\chi(r(\ell^*), u(\ell^*)) = 1/t(\ell^*) = 1$. By using equation (6.175) with $\eta = 0$, we obtain

$$\chi(r, u) = \mathrm{e}^{2\ell^*}\chi(r(\ell^*), u(\ell^*)) = \mathrm{e}^{2\ell^*}, \tag{6.191}$$

where ℓ^* satisfies the equation

$$\mathrm{e}^{2\ell^*}\left[1 + \frac{Bu_0}{\epsilon}\left(\mathrm{e}^{\epsilon\ell^*} - 1\right)\right]^{-A/B} t = 1, \tag{6.192}$$

in which we have set $t = t(0)$. An approximate solution of this equation is given by

$$\mathrm{e}^{\ell^*} = t^{-1/2}\left[1 + B\frac{u_0}{\epsilon t^{\epsilon/2}}\right]^{A/2B}. \tag{6.193}$$

We thereby have

$$\begin{aligned} \chi &= t^{-1}\left[1 + B\frac{u_0}{\epsilon t^{\epsilon/2}}\right]^{A/2B} \\ &\sim \begin{cases} t^{-1}, & \text{for } (Bu_0/\epsilon) \ll t^{\epsilon/2}, \\ t^{1-(\epsilon/2)A/B}, & \text{for } (Bu_0/\epsilon) \gg t^{\epsilon/2}. \end{cases} \end{aligned} \tag{6.194}$$

The crossover can be visualized by introducing the effective exponent γ_{eff}, defined by

$$\gamma_{\mathrm{eff}} = -\frac{\mathrm{d}\ln\chi}{\mathrm{d}\ln t}. \tag{6.195}$$

This quantity can be evaluated directly by taking the derivative of relation (6.192) with respect to t, and taking (6.191) into account.

Figure 6.8 shows $\gamma_{\mathrm{eff}}(t)$ evaluated with $n = 1$, $\epsilon = 2$, and $Bu_0 = 10^{-1}$.

6.8.1 Derivation of Equation (6.189)

It is convenient to start from the expression of the average of ϕ_i^α:

$$\langle\phi_i^\alpha\rangle = \mathrm{Tr}\,\phi_i^\alpha\,\mathrm{e}^{-\mathcal{H}(\phi)}/\mathrm{Tr}\,\mathrm{e}^{-\mathcal{H}(\phi)}, \tag{6.196}$$

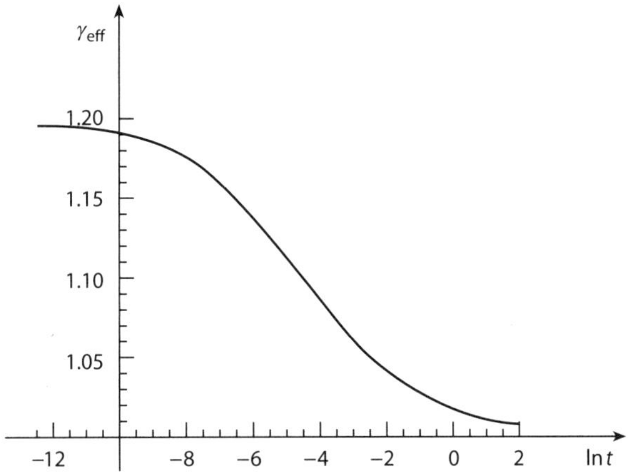

FIGURE 6.8. Effective critical exponent γ_{eff} as a function of $\ln t$, which describes the crossover from Gaussian to Heisenberg critical behavior. The plot corresponds to the case in which $n = 1$, $\epsilon = 1$, and $Bu_0 = 0.1$.

where

$$\mathrm{Tr} = \int \prod_{i\alpha} \mathrm{d}\phi_i^\alpha, \tag{6.197}$$

and

$$\mathcal{H} = \mathcal{H}_0 + \mathcal{H}_1, \tag{6.198}$$

$$\mathcal{H}_0 = \frac{1}{2} \sum_{ij} \sum_{\alpha} \phi_i^\alpha G_{ij}^{(-1)} \phi_j^\alpha, \tag{6.199}$$

$$\mathcal{H}_1 = u \sum_i \left(\sum_\alpha \phi_i^\alpha \right)^2. \tag{6.200}$$

In expression (6.196), ϕ appears as a dummy integration variable. The result does not change, therefore, if one performs the transformation

$$\phi_i^\alpha \longrightarrow \phi_i^\alpha + c, \qquad \forall i, \tag{6.201}$$

in the integral appearing in the numerator. By equating the derivative with respect to c of equation (6.196) to zero, we obtain

$$\sum_j G_{ij}^{-1} \langle \phi_i^\alpha \phi_j^\alpha \rangle + 4u \sum_j \sum_\beta \left\langle \phi_i^\alpha \phi_j^\alpha \left(\phi_j^\beta \right)^2 \right\rangle = 1. \tag{6.202}$$

At the lowest order, we can set

$$\sum_\beta \left\langle \phi_i^\alpha \phi_j^\alpha \left(\phi_j^\beta \right)^2 \right\rangle = (n+2) \langle \phi_i^\alpha \phi_j^\alpha \rangle \left\langle \left(\phi_j^\beta \right)^2 \right\rangle, \tag{6.203}$$

(where one no longer sums over the indexes) and evaluate the average of ϕ^2 in the Gaussian model. Multiplying by C_{jk}^{-1} and summing over j, we thus obtain

$$C_{ik}^{-1} = G_{ik}^{-1} + 4u\,(n+2)\left\langle\left(\phi_j^{\beta}\right)^2\right\rangle_0. \tag{6.204}$$

By passing to the Fourier transforms, we arrive at equation (6.189).

6.9 Cubic Anisotropy

Perturbations that explicitly break the rotation symmetry $O(n)$ of the order parameter give rise to an interesting structure of fixed points and flow over the critical surface. We describe these effects in the ϕ^4 model by introducing an interaction term of the form:

$$\mathcal{H}_{\text{anis}} = v\int \mathrm{d}\mathbf{r}\sum_{\alpha=1}^{n}\phi_\alpha^4. \tag{6.205}$$

For $u = 0$, the n-component model is reduced to a collection of n independent Ising models. It is easy to see that for $d > 4$, the new term is irrelevant. By using the techniques we have developed, one can obtain the first-order flow equations in ϵ:

$$\frac{\mathrm{d}r}{\mathrm{d}\ell} = 2r + 4[(n+2)\,u + 3v]\frac{1}{1+r}, \tag{6.206}$$

$$\frac{\mathrm{d}u}{\mathrm{d}\ell} = \epsilon u - 4[(n+8)u^2 + 6uv]\frac{1}{(1+r)^2}, \tag{6.207}$$

$$\frac{\mathrm{d}v}{\mathrm{d}\ell} = \epsilon v - 4[12uv + 9v^2]\frac{1}{(1+r)^2}. \tag{6.208}$$

Exercise 6.3 Obtain these flux equations.

We have proceeded to an appropriate redefinition of u and v in order to get rid of the K_d factor. These equations have four fixed points.

$$\text{Gaussian:}\quad u_{\mathrm{G}}^* = v_{\mathrm{G}}^* = 0, \tag{6.209}$$

$$\text{Heisenberg:}\quad u_{\mathrm{H}}^* = \epsilon/[4(n+8)],\quad v_{\mathrm{H}}^* = 0, \tag{6.210}$$

$$\text{Ising:}\quad u_{\mathrm{I}}^* = 0,\quad v_{\mathrm{I}}^* = \epsilon/36, \tag{6.211}$$

$$\text{Cubic:}\quad u_{\mathrm{C}}^* = \epsilon/(12n),\quad v_{\mathrm{C}}^* = [(n-4)\,\epsilon]/(12n). \tag{6.212}$$

The Ising fixed point has this name because it corresponds to n independent Ising models.

The linearized flow equations in the proximity of the fixed points have the form

$$\frac{\mathrm{d}\delta u}{\mathrm{d}\ell} = \left\{\epsilon - 4[2(n+8)\,u^* + 6v^*]\right\}\delta u - 24u^*\delta v, \tag{6.213}$$

$$\frac{\mathrm{d}\delta v}{\mathrm{d}\ell} = \left\{\epsilon - 4[12u^* + 18v^*]\right\}\delta u - 48v^*\delta v. \tag{6.214}$$

The corresponding stability exponents are

$$\text{Gaussian:}\quad \lambda_{\mathrm{G}}^{u} = \lambda_{\mathrm{G}}^{v} = \epsilon, \tag{6.215}$$

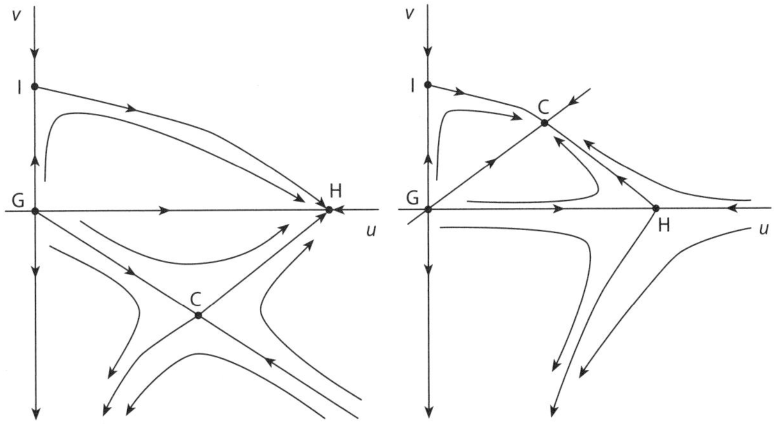

FIGURE 6.9. Renormalization flux in the plane (u, v) for the ϕ^4 model with cubic anisotropy. The following fixed points are shown: (G) Gaussian; (H) Heisenberg; (I) Ising; and (C) cubic. Left: $n < n_c$. Right: $n > n_c$.

$$\text{Heisenberg:} \quad \lambda_{\mathrm{H}}^{u} = -\epsilon, \quad \lambda_{\mathrm{H}}^{v} = \frac{n-4}{n}\epsilon, \tag{6.216}$$

$$\text{Ising:} \quad \lambda_{\mathrm{I}}^{u} - \frac{\epsilon}{3}, \quad \lambda_{\mathrm{I}}^{v} = -\epsilon, \tag{6.217}$$

$$\text{Cubic:} \quad \lambda_{\mathrm{C}}^{(1)} = -\epsilon, \lambda_{\mathrm{C}}^{(2)} = \frac{4-n}{3n}\epsilon. \tag{6.218}$$

One should note that the Heisenberg fixed point is unstable and the cubic fixed point is stable for $n > n_c = 4 + O(\epsilon)$, while the opposite is true for $n < n_c$. These equations imply the flow diagrams shown in figure 6.9—for all values of n, there is a fixed point that is locally stable. There exist regions of the plane (u, v), however, that do not go to a fixed point—the entire region $v < 0$ for $n > n_c$, for example. This phenomenon is known as **runaway**. In order to define the behavior of a system in which runaway appears, further analysis is necessary. In our case, it is possible to see that the transition becomes discontinuous.

In this manner, we can evaluate the exponent ν around the new cubic fixed point:

$$\nu_{\mathrm{C}} = \frac{1}{2} + \frac{n-1}{6n}\epsilon + O(\epsilon^2). \tag{6.219}$$

6.10 Limit $n \to \infty$

In this section, we will consider a system with an n-component vector order parameter in the limit $n \to \infty$, and we will show that it can be resolved exactly, and that the corresponding critical exponents are not classical.

Let us therefore consider the system defined by the Hamiltonian (6.83), which we rewrite as

$$\mathcal{H} = -\frac{1}{2}\sum_{ij} K_{ij}\boldsymbol{\phi}_i \cdot \boldsymbol{\phi}_j,$$

where the interaction matrix $\mathsf{K} = (\mathsf{K}_{ij})$ is defined by

$$K_{ij} = \begin{cases} K = J/k_B T, & \text{if } i \text{ and } j \text{ are nearest neighbors,} \\ 0, & \text{otherwise,} \end{cases}$$

and where the $\boldsymbol{\phi}$ are n-component vectors ($\boldsymbol{\phi}_i = \phi_i^1, \phi_i^2, \ldots, \phi_i^n$) with a magnitude equal to $\sqrt{n}$:

$$\sum_{\alpha=1}^{n} (\phi_i^\alpha)^2 = n, \qquad \forall i. \tag{6.220}$$

The system's partition function in the presence of the external magnetic field h (which we assume to be oriented in direction 1) is given by

$$\begin{aligned} Z(h) &= \int \prod_i \left\{ \prod_\alpha \mathrm{d}\phi_i \delta\left(\sum_\alpha (\phi_i^\alpha)^2 - n \right) \right\} \\ &\times \exp\left\{ \frac{1}{2}\sum_{ij} K_{ij} \sum_\alpha \phi_i^\alpha \phi_j^\alpha + \sum_i h\phi_i^1 \right\}. \end{aligned} \tag{6.221}$$

In order to impose the constraint on the magnitude of $\boldsymbol{\phi}_i$'s, we use the identity

$$\delta(x) = \int_{-\mathrm{i}\infty}^{\mathrm{i}\infty} \frac{\mathrm{d}y}{2\pi \mathrm{i}}\, \mathrm{e}^{-xy}. \tag{6. 222}$$

One therefore has

$$\delta\left(\sum_\alpha (\phi_i^\alpha)^2 - n\right) = \int_{-\mathrm{i}\infty}^{\mathrm{i}\infty} \frac{\mathrm{d}\lambda_i}{2\pi \mathrm{i}} \exp\left[-\lambda_i \left(\sum_\alpha (\phi_i^\alpha)^2 - n \right) \right]. \tag{6.223}$$

In this manner, we can express the partition function (up to constant factors) in the following form:

$$Z(h) = \int \prod_i \left(\mathrm{d}\lambda_i \prod_\alpha \mathrm{d}\phi_i^\alpha \right) \exp\left\{ -\frac{1}{2} \sum_\alpha \sum_{ij} [2\lambda_i \delta_{ij} - K_{ij}] \phi_i^\alpha \phi_j^\alpha + n \sum_i \lambda_i \right\}. \tag{6.224}$$

Since ϕ_i^α appears quadratically in this expression, it is possible to evaluate the corresponding functional integral exactly. Actually, it is convenient to integrate over all components of α, except for the first ϕ_i^1, which we shall denote by σ_i. This integral converges if the matrix $[2\lambda_i \delta_{ij} - K_{ij}]$ is positive definite. One thus obtains

$$\begin{aligned} Z(h) &= \int \prod_i (\mathrm{d}\lambda_i \mathrm{d}\sigma_i) \exp\left\{ \frac{1}{2} \sum_{ij} [K_{ij} - 2\lambda_i \delta_{ij}] \sigma_i \sigma_j \right. \\ &\left. + h\sum_i \sigma_i + n \sum_i \lambda_i - \frac{n-1}{2} \mathrm{Tr} \log[2\lambda_i \delta_{ij} - K_{ij}] \right\}. \end{aligned} \tag{6.225}$$

The term in Tr log comes from the determinant that one obtains from the Gaussian integral. This term can be evaluated more easily by shifting to the Fourier transforms. Let us introduce the Fourier transform of K_{ij}:

$$K\omega(\boldsymbol{k}) = \sum_j e^{i\boldsymbol{k}\cdot(\boldsymbol{r}_j - \boldsymbol{r}_i)} K_{ij}. \tag{6.226}$$

For an d-dimensional simple cubic lattice with an a_0 constant, one has (as we have already seen)

$$\omega(\boldsymbol{k}) = 2\sum_{\alpha=1}^{d} \cos k_\alpha a_0, \tag{6.227}$$

and therefore $\omega(0) = 2d$, while for $ka_0 \ll 1$, one has

$$\omega(\boldsymbol{k}) \simeq 2d - k^2 a_0^2. \tag{6.228}$$

The logarithm $\Delta(\lambda)$ of the determinant of the matrix $(2\lambda\delta_{ij} - K_{ij})$ can be expressed as follows:

$$\begin{aligned} \Delta(\lambda) &= \mathrm{Tr}\log[2\lambda\delta_{ij} - K_{ij}] = \sum_{\boldsymbol{k}} \log[2\lambda - K\omega(\boldsymbol{k})] \\ &= \frac{Na_0^d}{(2\pi)^d}\int_{\text{Z.B.}} \mathrm{d}\boldsymbol{k}\, \log[2\lambda - K\omega(\boldsymbol{k})], \end{aligned} \tag{6.229}$$

where the integral over $\boldsymbol{k}$ is extended to the first Brillouin zone.

In the limit $n \to \infty$, we can set $n - 1 \approx n$ and evaluate the remaining integrals with the saddle point method. We look for a translation-invariant saddle point:

$$\sigma_i = \sigma, \qquad \lambda_i = \lambda, \qquad \forall i. \tag{6.230}$$

The saddle-point equations are the following:

$$(2\lambda - K\omega(0))\sigma = h, \tag{6.231}$$

$$-\sigma^2 + n\lambda = \frac{n}{2N}\frac{\mathrm{d}}{\mathrm{d}\lambda}\Delta(\lambda). \tag{6.232}$$

One can easily see that

$$\frac{1}{2N}\frac{\mathrm{d}}{\mathrm{d}\lambda}\Delta(\lambda) = \frac{a_0^d}{(2\pi)^d}\int_{\text{Z.B.}} \frac{\mathrm{d}\boldsymbol{k}}{2\lambda - K\omega(\boldsymbol{k})}. \tag{6.233}$$

For $h = 0$, the first saddle point equation (6.231) implies that $\sigma = 0$ or $2\lambda = K\omega(0)$. Let us consider the first case, which, as we will see, holds above the critical temperature. The second equation (6.232) becomes a consistency equation for λ, of the form

$$\lambda = f(\lambda), \tag{6.234}$$

where

$$f(\lambda) = \frac{a_0^d}{2(2\pi)^d}\int_{\text{Z.B.}} \frac{\mathrm{d}\boldsymbol{k}}{2\lambda - K\omega(\boldsymbol{k})}. \tag{6.235}$$

Let us discuss this expression. In order for the integral over the ϕ that led us to this expression to be convergent, all the eigenvalues of the matrix $2\lambda\delta_{ij} - K_{ij}$ must be positive. This implies that $2\lambda - K\omega(0) > 0$—in other words, $2\lambda/K > \omega(0)$. Let us consider the behavior of the quantity $Kf(\lambda)$ as a function of λ/K. It increases as λ/K decreases, and therefore its maximum value is reached when $2\lambda/K=\omega(0)$.[1] Let us introduce the new variable $r=(2\lambda/K-\omega(0))/(2a_0^2)$, and calculate the value of $Kf(\lambda)$ for small values of r:

$$Kf(\lambda) \propto \int_{\text{Z.B.}} \mathrm{d}\boldsymbol{k}\, \frac{2a_0^2}{r+k^2} + \int_{\text{Z.B.}} \mathrm{d}\boldsymbol{k} \left[\frac{1}{2\lambda - K\omega(\boldsymbol{k})} - \frac{2a_0^2}{r+k^2}\right]. \tag{6.236}$$

The second term of this expression is also convergent and analytic for $r=0$. Let us discuss the first term's behavior. One has

$$I = \int_{\text{Z.B.}} \frac{\mathrm{d}\boldsymbol{k}}{r+k^2} = \text{const.} + \mathrm{O}\left(r^{(d-2)/2}\right). \tag{6.237}$$

We can distinguish two cases:

1. $d \le 2$: $\lim_{r\to 0} I = \infty$.
2. $d > 2$: $I = \text{const.} + \mathrm{O}\,(r^{(d-2)/2})$. The exponent $(d-2)/2$ is smaller than 1 for $2<d<4$; otherwise, it is larger than 1.

In the first case, for each temperature value, equation (6.232) admits a solution with a finite λ and $\sigma=0$. In other words, there is no spontaneous symmetry breaking. In the second case, it is possible to define a critical value for K_c by means of the equation

$$K_c\omega(0) = \frac{a_0^d}{(2\pi)^d} \int_{\text{Z.B.}} \frac{\mathrm{d}\boldsymbol{k}}{\omega(0) - \omega(\boldsymbol{k})}. \tag{6.238}$$

For $K > K_c$, the equation (6.232) no longer admits solutions with $\sigma=0$. We can therefore assume that $\sigma \neq 0$, which, because of (6.231), implies that for $h=0$, $2\lambda = K\omega(0)$. Equation (6.232) therefore becomes an equation for σ. We can easily see that $\sigma \propto n^{1/2}$, and we can therefore rescale it by defining $\sigma^2 = nm^2$. With a little algebra, one thus obtains

$$m^2 = 1 - K_c/K, \tag{6.239}$$

from which we can read the value of exponent β:

$$\beta = \frac{1}{2}. \tag{6.240}$$

This result notwithstanding, the exponents in this model are not the classical ones, at least for $2<d<4$. It is actually easy to see that for $K \to K_c$, one has $2\lambda - K_c \propto r \propto |K_c - K|^{2/(d-2)}$. Now from equation (6.231), we can obtain the expression per susceptibility per spin χ above the critical temperature (in other words, for $K<K_c$, $h=0$):

$$\chi = \left.\frac{\partial \phi}{\partial h}\right|_{h=0} = [2\lambda - K\omega(0)]^{-1}. \tag{6.241}$$

We thus obtain the critical behavior of the susceptibility and therefore the exponent γ:

[1] The analogy with the treatment of the Einstein condensation should be evident.

$$\chi \propto |K_c - K|^{-\gamma}; \qquad \gamma = \begin{cases} 2/(d-2), & \text{if } 2 < d < 4; \\ 1, & \text{if } d > 4. \end{cases} \tag{6.242}$$

(A more detailed analysis shows that for $d = 4$, logarithmic corrections are present.) From the expression of the correlation function, which is proportional to $1/(r + k^2)$, one also obtains the behavior of the coherence length ξ:

$$\xi = r^{-1/2} \propto |K_c - K|^{-\nu}; \qquad \nu = \begin{cases} 1/(d-2), & \text{if } 2 < d < 4; \\ 1/2, & \text{if } d > 4. \end{cases} \tag{6.243}$$

The advantage of this model is that it exhibits nonclassical behavior that can be calculated exactly, without resorting to renormalization methods. The $n \to \infty$ limit is a powerful tool for the study of complex systems.

Exercise 6.4 Discuss the behavior of the model for $h \neq 0$, and obtain the value of exponent δ directly. Compare this with the result obtained from the scaling laws:

$$\delta = \frac{d+2}{d-2}, \qquad 2 < d \leq 4.$$

6.11 Lower and Upper Critical Dimensions

One of the main lessons that we learned from the calculations we just reviewed is that it is useful to consider the system's dimensionality d as a variable. We have obtained some nonclassical critical exponents for $d < 4$. For $d > 4$, the Ginzburg criterion guarantees us that the exponents are the same as the classical ones. The same result can be obtained by observing that if $\epsilon < 0$, the recurrence formulas (6.148) and (6.149) tell us that the stable fixed point is Gaussian with $u = 0$. Let us note that this does not imply that the exponents are those from the Gaussian model—in fact, the model is not even defined for $r < 0$. Let us for instance consider the behavior of the order parameter. Since one has

$$m \propto \left(\frac{-r}{u}\right)^{1/2}, \qquad r \leq 0, \tag{6.244}$$

according to Landau theory, and since $u \to 0$ under renormalization, the singularity in the dependence from u changes the asymptotic behavior. The result is that the exponents are the *classical* ones, not the Gaussian ones, and more particularly that $\beta = 1/2$, instead of $(d-2)/4$, which would be what one would expect following the scaling laws.

The dimension d_c, below which exponents differ from the classical ones, is called the **upper critical dimension**. Since occasionally we may have to deal with problems that differ from the "usual" critical phenomena, it is useful to have a criterion at one's disposal to identify the upper critical dimension. This criterion was proposed by Toulouse [Toul75].

When the Kadanoff transformations admit of a nontrivial fixed point, the scaling laws apply, as we saw. They can be divided into two classes: those like $\alpha + 2\beta + \gamma = 2$, in which the system's dimensionality does not appear explicitly, and those like $2 - d\nu = \alpha$, in which it appears. The first are satisfied by the classical exponents; the second (generally) are not, since the classical exponents do not depend on dimensionality. The nonclassical exponents, on the other hand, satisfy the second group of scaling laws for all the values of d for which they are valid. At the upper critical dimension, the nontrivial fixed point is confused with the Gaussian (free of interaction), and the classical exponents therefore must satisfy the scaling laws: by setting $\alpha_c = 0$, $\nu_c = 1/2$, we obtain

$$d_c = \frac{2 - \alpha_c}{\nu_c} = 4. \tag{6.245}$$

For example, if a term proportional to ϕ^3 were present in the expansion of Landau free energy, one would obtain the classical exponents $\alpha = -1$, $\beta = 0$, $\gamma = 1$, and $d_c = 6$ would therefore follow. Since strictly speaking these terms are present in the van der Waals equation, we can ask ourselves why the Ising exponents that we obtained for $n = 1$ apply to the liquid–gas transition. The answer is that if the order parameter is a scalar, it is always possible to eliminate these terms by redefining ϕ by means of a translation: $\phi' = \phi - \phi_0$. When this transformation is forbidden because of the problem's symmetry, however, we are dealing with a new universality class and we obtain some new exponents, which must be calculated in an expansion in $d - 6$. This is the case of percolation, which we will discuss in chapter 10.

Another remarkable value of dimensionality is the one where symmetry breaking occurs at zero temperature. This is called the **lower critical dimension**. For $n = 1$ (Ising), it is equal to 1, while for $n \geq 2$, it is equal to 2. In this case, it is possible to expand the critical exponents into powers of $d - 2$. For $n = 2$ and $d = 2$, even though there is no symmetry breaking, there is a peculiar phase transition at finite temperature, which is called the **Kosterlitz-Thouless transition**.

Recommended Reading

Various recent texts contain reviews of the renormalization group at differing levels of detail. A "classic" text is D. J. Amit, *Field Theory, the Renormalization Group, and Critical Phenomena*, 2nd ed., Singapore: World Scientific, 1984. A recent discussion, focused and concise, can be found in J. Cardy, *Scaling and Renormalization in Statistical Physics*, Cambridge, UK: Cambridge University Press, 1996. One can also profitably consult J. J. Binney, N. J. Dowrick, A. J. Fisher, and M.E.J. Newman, *The Theory of Critical Phenomena*, Oxford, UK: Clarendon Press, 1992, and chapter 6 of M. Plischke and B. Birgersen's, *Equilibrium Statistical Physics*, 2nd ed., Singapore: World Scientific, 1994.

7 Classical Fluids

We shall treat the armor's steel as an ideal fluid.
—Lavrentiev and Shabat

The study of classical fluids by the methods of statistical mechanics was, historically speaking, the first field of application of these methods (see, for example, Boltzmann's *Gastheorie* [Bolt95]). In recent years, it has become the domain of a specialized and quite sophisticated community of researchers. In particular, there has been a lot of interest in the study of complex fluids, composed of mixtures of several chemical species, either anisotropic (liquid crystals) or nonhomogeneous (interfaces and microemulsions). Water is also a complex fluid, perhaps the most difficult of all, because of its special physical characteristics, which are determined by the hydrogen bond. There has also been a great deal of progress in understanding simple fluids, to the extent that by combining analytical and numerical methods, and on the basis of a microscopic model, we have arrived at a fairly satisfying description of the state diagram of systems like argon.

It is beyond the scope of this book to account for these developments. An introduction to the modern theory of classical fluids can be found in Hansen and McDonald [Hans86], which has frequently been my source in the pages that follow. My goal is to provide only an introduction to the methods that are used to tackle these problems and to show how they fit into the larger picture.

7.1 Partition Function for a Classical Fluid

Let us consider a model of a simple fluid, represented by means of a collection of N particles of mass m, placed in a container of volume V, and interacting by means of a pair potential $u(r)$ that depends only on the modulus of the distance r between particles. This model is already to a significant degree an idealization of a simple fluid. In effect:

- The particles' internal degrees of freedom are tacitly disregarded.
- Interactions between three or more bodies are excluded—these interactions are instead rather frequent; one need think only of the van der Waals interactions,

- The dependence of the interaction on the molecules' own orientation is also neglected, an effect that can lead to the formation of an anisotropic fluid.

This model, however, is already fairly useful for describing the behavior of fluids such as rare gases (excluding helium, whose properties are influenced by quantum effects) and certain molecular gases such as CO_2.

By introducing a single-particle external potential $v(\mathbf{r})$—which describes the interaction between the fluid and the container, for instance—the system's Hamiltonian can be written

$$\mathcal{H}(\{\mathbf{p},\mathbf{r}\}) = \sum_{i=1}^{N} \frac{p_i^2}{2m} + \sum_{i=1}^{N} v(\mathbf{r}_i) + U(\{\mathbf{r}\}), \tag{7.1}$$

where $(\mathbf{p}_i, \mathbf{r}_i)$ are, respectively, the momentum and position vectors for particle i, and where the potential interaction energy between particles is expressed as

$$U(\{\mathbf{r}\}) = \sum_{(ij)} u(|r_i - r_j|), \tag{7.2}$$

and in which the sum runs over all the distinct particle pairs. In this chapter, we will examine only three-dimensional systems.

The thermodynamic properties of this system can be obtained starting from the canonical partition function Z_N, which, taking into account the *correct Boltzmann counting* and the *volume of the phase space element* (see section 3.8), is expressed as follows:

$$Z_N = \frac{1}{N!} \int \prod_{i=1}^{N} \left(\frac{\mathrm{d}^3 \mathbf{p}_i \mathrm{d}^3 \mathbf{r}_i}{h^3} \right) \exp\left[-\frac{\mathcal{H}(\{\mathbf{p},\mathbf{r}\})}{k_B T} \right]. \tag{7.3}$$

The integral over momenta can be evaluated directly, so that one obtains

$$Z_N = \frac{1}{N!} \int \prod_{i=1}^{N} \left(\frac{\mathrm{d}^3 \mathbf{r}_i}{\lambda^3} \right) \exp\left\{ -\frac{1}{k_B T} \left[\sum_{i=1}^{N} v(\mathbf{r}_i) + U(\{\mathbf{r}\}) \right] \right\} = \frac{1}{N!} \lambda^{-3N} \Omega_N[v], \tag{7.4}$$

where λ is the thermal de Broglie wavelength:

$$\lambda = \sqrt{\frac{h^2}{2\pi m k_B T}}. \tag{7.5}$$

The integral

$$\Omega_N[v] = \int \prod_{i=1}^{N} \mathrm{d}^3 \mathbf{r}_i \exp\left\{ -\frac{1}{k_B T} \left[\sum_{i=1}^{N} v(\mathbf{r}_i) + U(\{\mathbf{r}\}) \right] \right\}. \tag{7.6}$$

is often called the **configuration integral**.

It is often useful to deal with the system in the grand canonical ensemble by introducing the chemical potential μ. The grand canonical partition function Z_{GC} is then given by

$$\begin{aligned} Z_{GC} &= \sum_{N=0}^{\infty} Z_N e^{\mu N / k_B T} \\ &= \sum_{N=0}^{\infty} \frac{z^N}{N!} \int \prod_{i=1}^{N} \left(\frac{\mathrm{d}^3 \mathbf{r}_i}{\lambda^3} \right) \exp\left\{ -\frac{1}{k_B T} \left[\sum_{i=1}^{N} v(\mathbf{r}_i) + U(\{\mathbf{r}\}) \right] \right\}, \end{aligned} \tag{7.7}$$

where we introduced the fugacity z, defined by

$$z = e^{\mu/k_B T}. \tag{7.8}$$

Let us note in particular that in the absence of interactions $[U(|\mathbf{r}|) = 0]$, the grand canonical partition function is equal to

$$Z_{GC} = Z^0_{GC} = \sum_{N=0}^{\infty} \frac{z^N}{N!} (Z_1[\nu])^N = \exp\{z Z_1[\nu]\}, \tag{7.9}$$

where $Z_1[\nu]$ is the canonical grand partition function of a single particle, expressed as a functional of the single-particle potential $\nu(\mathbf{r})$:

$$Z_1[\nu] = \int \frac{d^3\mathbf{r}}{\lambda^3} e^{-\nu(\mathbf{r})/k_B T}. \tag{7.10}$$

Therefore, if $\nu(\mathbf{r})$ is a confining potential in a box $\mathcal{B}$ of volume V—in other words, if

$$\nu(\mathbf{r}) = \nu_{\mathcal{B}}(\mathbf{r}) = \begin{cases} 0, & \text{if } \mathbf{r} \in \mathcal{B}, \\ +\infty, & \text{otherwise,} \end{cases} \tag{7.11}$$

one has $Z_1 = V/\lambda^3$, from which it follows that $Z_{GC} = \exp\{zV/\lambda^3\}$, what allows one to derive all the usual properties of ideal gases, as we saw in chapter 4.

In order to proceed, we need to specify some properties of the pair interaction's potential $u(r)$. In the systems we are considering, it exhibits a *hard core*—in other words, a strong repulsion over short distances, and an *attractive tail,* which rapidly decays over distances that are slightly larger than the *particle radius.* It is interesting that the interaction potential is basically determined by only one scale of lengths, i.e., the particle radius. This allows one to represent it fairly easily. One expression that has been particularly successful is the *12-6 potential,* also called the *Lennard-Jones potential,* in which only two parameters appear: the energy scale ε_0 and the length scale r_0. It is defined by

$$u(r) = 4\varepsilon_0 \left(\frac{r_0^{12}}{r^{12}} - \frac{r_0^6}{r^6} \right). \tag{7.12}$$

It is easy to see that ε_0 represents the greatest depth of the attractive potential, while r_0 satisfies $u(r_0) = 0$. This potential is shown in figure 7.1. Please note the rapid increase of the hard core and the almost equally rapid decrease of the attractive tail, such that already for $r = 2r_0$, one has $u(r_0) = -0.06\,\varepsilon_0$. More particularly, one has $\lim_{r\to\infty} r^3 u(r) = 0$; this relation is not valid for the Coulomb or dipolar potential, which are therefore considered *long range.* The potential's minimum is obtained at $r = 2^{1/6} r_0$.

Since the Lennard-Jones potential contains only two parameters, like the van der Waals equation of state, the equation of state of a Lennard-Jones fluid assumes a universal form in terms of the *reduced* physical parameters—in other words, when expressed as a function of critical temperature, pressure, and density. The validity of the Lennard-Jones potential can therefore be evaluated by comparing the equations of state for different fluids, and especially the values for their reduced critical temperature and density. In table 7.1, I report the values of the parameters of the Lennard-Jones potential estimated on the basis

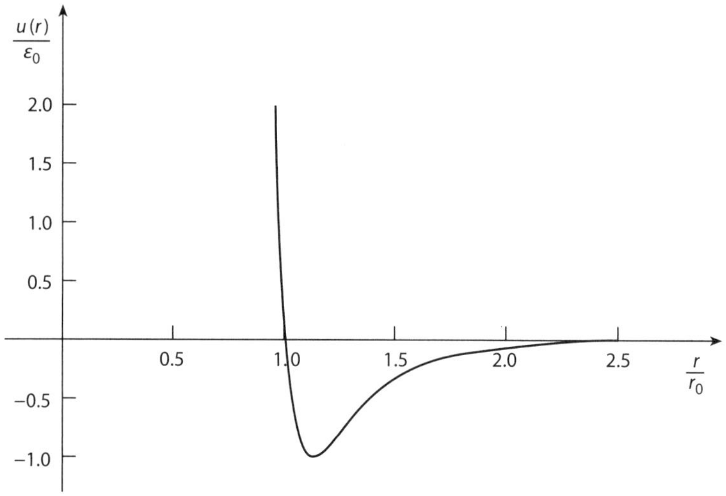

FIGURE 7.1. The Lennard-Jones potential.

Table 7.1 Parameter Values for the Lennard-Jones Potential for Some Simple Gases, Based on the Second Virial Coefficient, and the Corresponding Reduced Values for the Critical Temperature and the Critical Sensities

Gas	ε_0/k_B (K)	r_0 (Å)	$k_B T_c/\varepsilon_0$	$\rho_c r_0^3$
He	10.2	2.556	0.52	0.174
Ne	35.8	2.75	1.24	0.300
A	119.8	3.405	1.26	0.326
Kr	116.7	3.68	1.26	0.326
Xe	225.3	4.07	1.29	0.338
H_2	36.7	2.959	0.91	0.240
CH_4	148.2	3.817	1.29	0.338

Source: Hans86, p. 72.

of a measure of the second virial coefficient (in other words, on the basis of the first deviation of the equation of state from that of ideal gases; see section 7.3) for some simple gases. I have also reproduced the values of the critical temperature and density, expressed as a function of these parameters. If the Lennard-Jones potential were exact, the reduced critical temperature and density should be the same for all systems. One can see that the agreement is not very satisfactory. For lighter gases like H or He, a large part of the deviation is due to quantum effects. On the other hand, it is also clear that the model does not provide an exact description even of the heavier gases.

7.2 Reduced Densities

We have seen, however, that the interest of the methods of statistical mechanics does not lie mainly in the prediction of the equation of state of the thermodynamic systems, but rather in the possibility of giving a more detailed description of their behavior. It is possible in particular to evaluate the averages of a number of quasi-thermodynamic observables—extensive observables, which cannot be controlled from the exterior, and cannot therefore be counted as thermodynamic variables. In the case of fluids, a very broad class of these quantities can be expressed in terms of **reduced densities**—in other words, the correlation functions of the densities. In this section, we will define the reduced densities and see how they are connected to the thermodynamic properties of the fluid and how in certain cases they can be measured by means of scattering experiments. I refer the reader to the following sections to show how these quantities can be evaluated by means of a systematic power-series expansion of the densities (**virial expansion**) or, in a more subtle manner, of a function of the intensity of the pair potential's attractive tail, starting from the system of hard spheres taken as a reference.

The single-particle reduced density is simply the local average of the density:

$$\rho^{(1)}(\boldsymbol{x}) = \left\langle \sum_{i=1}^{N} \delta(\boldsymbol{x} - \boldsymbol{r}_i) \right\rangle. \tag{7.13}$$

Since all the particles are identical, it can be evaluated by calculating the relative weight of the configurations in which a single particle is found in $\boldsymbol{x}$:

$$\rho^{(1)}(\boldsymbol{x}) = (\Omega_N[v])^{-1} \int \prod_{i=1}^{N} \mathrm{d}^3\boldsymbol{r}_i \exp\left\{-\frac{1}{k_B T}\left[\sum_{i=1}^{N} v(\boldsymbol{r}_i) + U(\{\boldsymbol{r}\})\right]\right\}, \tag{7.14}$$

where we have assumed $\boldsymbol{r}_1 = \boldsymbol{x}$. Obviously, if $v(\boldsymbol{r})$ is equal to the box potential $v_B(\boldsymbol{r})$, $\rho^{(1)}(\boldsymbol{x}) = N/V = \rho$, at least not too close to the walls.

We can express the ℓ-particle reduced density in an analogous way:

$$\rho^{(\ell)}(\boldsymbol{x}_1, \dots, \boldsymbol{x}_\ell) = \left\langle \sum_{i_1,\dots,i_\ell} \prod_{\alpha=1}^{\ell} \delta(\boldsymbol{x}_\alpha - \boldsymbol{r}_{i_\alpha}) \right\rangle. \tag{7.15}$$

We obtain

$$\begin{aligned} \rho^{(\ell)}(\boldsymbol{x}_1, \dots, \boldsymbol{x}_\ell) &= (\Omega_N[v])^{-1} N(N-1)\cdots(N-\ell+1) \int \prod_{i=\ell+1}^{N} \mathrm{d}^3\boldsymbol{r}_i \\ &\quad \times \exp\left\{-\frac{1}{k_B T}\left[\sum_{i=1}^{N} v(\boldsymbol{r}_i) + U(\{\boldsymbol{r}\})\right]\right\}. \end{aligned} \tag{7.16}$$

in which one assumes that the positions of the first ℓ particles are given by $\boldsymbol{r}_\alpha = \boldsymbol{x}_\alpha$ $(\alpha = 1, \dots, \ell)$.

It is interesting to note that the $\rho^{(\ell)}$ can be expressed in terms of the functional derivatives of the integral over the configurations with respect to the single-particle potential $v(\boldsymbol{r})$:

$$\rho^{(\ell)}(x_1,\ldots,x_\ell) = (-k_BT)^\ell (\Omega_N[v])^{-1} \frac{\delta^\ell \Omega_N[v]}{\delta v(x_1)\cdots\delta v(x_\ell)}. \tag{7.17}$$

7.2.1 Pair Distribution

In addition to the single-particle density $\rho^{(1)} = N/V$, the reduced density that is most commonly of interest is the **pair density** $\rho^{(2)}(x_1, x_2)$. It can be interpreted as the joint probability density that two particles can be found in x_1 and x_2, and for $|x_1 - x_2| \to \infty$, it tends to the value $N(N-1)/V^2 \simeq \rho^2$. In a homogeneous and isotropic system (far from the walls), the pair density is a function only of $r = |x_1 - x_2|$. One usually introduces a function $g(r)$, called the **pair distribution**, by means of the relation

$$\rho^{(2)}(|x_1 - x_2|) = \rho^2 g(r). \tag{7.18}$$

One therefore has $\lim_{r\to\infty} g(r) = 1$.

It is easy to show that, if the interaction between particles is determined only by pair interactions, the system's interaction energy can be expressed as a function of the $g(r)$:

$$\langle U\rangle = \frac{V\rho^2}{2}\int_0^\infty \mathrm{d}^3 r\, u(r) g(r). \tag{7.19}$$

In effect, one has

$$\begin{aligned}\langle U\rangle &= \left\langle \sum_{(ij)} u(|\boldsymbol{r}_i - \boldsymbol{r}_j|)\right\rangle \\ &= \frac{1}{2}\sum_{i\neq j}{}' \int \mathrm{d}^3x_1 \mathrm{d}^3x_2 \langle u(|\boldsymbol{r}_i - \boldsymbol{r}_j|)\delta(x_1 - \boldsymbol{r}_1)\delta(x_1 - \boldsymbol{r}_1)\rangle \\ &= \frac{1}{2}\sum_{i\neq j}{}' \int \mathrm{d}^3x_1 \mathrm{d}^3x_2\, u(|\boldsymbol{r}_i - \boldsymbol{r}_j|)\langle \delta(x_1 - \boldsymbol{r}_1)\delta(x_1 - \boldsymbol{r}_1)\rangle \\ &= \frac{1}{2}\int \mathrm{d}^3x_1 \mathrm{d}^3x_2\, u(|\boldsymbol{r}_i - \boldsymbol{r}_j|)\rho^{(2)}(x_1, x_2) = \frac{V\rho^2}{2}\int \mathrm{d}^3 \boldsymbol{r} g(\boldsymbol{r}),\end{aligned} \tag{7.20}$$

where we have exploited the invariance of the $\rho^{(2)}(x_1, x_2)$ with respect to translations, the definition of $g(r)$, and the system's isotropy.

In the same hypotheses, it is possible to express the pressure in terms of the $g(r)$:

$$\frac{p}{k_BT} = \rho - \frac{1}{2dk_BT}\rho^2\int_0^\infty \mathrm{d}^3 r\, r\frac{\partial u}{\partial r} g(r). \tag{7.21}$$

In order to obtain this relation, let us suppose that the system is contained in a cubic box of side L, so that $V = L^3$. Let us note that the pressure is given by

$$\begin{aligned}\frac{p}{k_BT} &= \left.\frac{\partial \ln Z_N}{\partial V}\right)_{T,N} = \left.\frac{\partial \ln \Omega_N}{\partial V}\right)_{T,N} \\ &= \left.\frac{L}{dV}\frac{\partial \ln \Omega}{\partial L}\right)_{T,N}.\end{aligned} \tag{7.22}$$

On the other hand, if $v(\boldsymbol{r}) = v_B(\boldsymbol{r})$, the integral over the configurations can be written

$$\begin{aligned}\Omega_N &= \int_0^L \mathrm{d}r_{11} \int_0^L \mathrm{d}r_{12} \cdots \int_0^L \mathrm{d}r_{N3} \exp\left[-\frac{1}{k_B T} U(\{\mathbf{r}\})\right] \\ &= V^N \int_0^1 d\xi_{11} \cdots \int_0^1 d\xi_{N3}\, L \frac{\partial U(\{L\xi\})}{\partial L} \exp\left[-\frac{1}{k_B T} U(\{L\xi\})\right],\end{aligned} \tag{7.23}$$

where we have set $\mathbf{r}_i = (r_{i1}, \ldots, r_{i3}) = L(\xi_{i1}, \ldots, \xi_{i3})$. Therefore,

$$L\frac{\partial \Omega_N}{\partial L} = 3N - \frac{V^N}{k_B T \Omega_N} \int_0^1 d\xi_{11} \cdots \int_0^1 d\xi_{N3}\, L \frac{\partial U(\{L\xi\})}{\partial L} \exp\left[-\frac{1}{k_B T} U(\{L\xi\})\right], \tag{7.24}$$

Now,

$$\begin{aligned}L\frac{\partial U(\{L\xi\})}{\partial L} &= \frac{1}{2}\sum_{i\neq j} L \frac{\partial u(L|\xi_i - \xi_j|)}{\partial L} = \frac{1}{2}\sum_{i\neq j} L|\xi_i - \xi_j| \left.\frac{\partial u(r)}{\partial r}\right|_{r = L|\xi_i - \xi_j|} \\ &= \frac{1}{2}\sum_{i\neq j} r_{ij} \left.\frac{\partial u(r)}{\partial r}\right|_{r = r_{ij}},\end{aligned} \tag{7.25}$$

where $r_{ij} = |\mathbf{r}_i - \mathbf{r}_j|$. Therefore, equation (7.24) can be rewritten as

$$\begin{aligned}L\frac{\partial \ln \Omega_N}{\partial L} &= 3N - \frac{1}{k_B T \Omega_N} \int \prod_{i=1}^{N} \mathrm{d}^3 r_i \frac{1}{2} \sum_{i\neq j}{}' r_{ij} \frac{\partial u(r_{ij})}{\partial r_{ij}} \exp\left[-\frac{1}{k_B T} U(\{\mathbf{r}\})\right] \\ &= 3N - \frac{1}{2k_B T} \int \mathrm{d}^3 x_1 \int \mathrm{d}^3 x_2 |x_1 - x_2| \left.\frac{\partial u(r)}{\partial r}\right|_{r = |x_1 - x_2|} \rho^{(2)}(x_1, x_2) \\ &= 3N - \frac{V}{2k_B T} \rho^2 \int_0^\infty \mathrm{d}r\, 4\pi r^2 r \frac{\partial u}{\partial r} g(r).\end{aligned} \tag{7.26}$$

In the last step, we used the definition of $g(r)$ and the system's homogeneity and isotropy. By substituting in (7.22), we obtain (7.21). This expression is called the **pressure equation**, and we will obtain it again, in a different garb, in chapter 8.

We have obtained these properties of $g(r)$ in the canonical ensemble, in which N is fixed. The $g(r)$, however, is essentially a thermodynamic density—in other words, the ratio between two extensive variables. In effect, we have

$$\frac{N^2}{V} g(x) = \int \mathrm{d}^3 x_0 \left\langle \sum_{i=1}^{N} \sum_{j=1}^{N} \delta(x_0 - \mathbf{r}_i)\, \delta(x_0 + x - \mathbf{r}_j) \right\rangle. \tag{7.27}$$

The quantity on the right-hand side is an extensive thermodynamic variable. (It seems to be more than extensive, because it contains N^2 terms, but you can check and see that its value is actually proportional to the system's size, since for each particle i there will be no more than one particle j that will contribute to the sum.) By dividing it by the volume, we obtain a density—in other words, a quantity that is independent of the system's size. Thermodynamics guarantees us that all densities have the same value in a state of thermodynamic equilibrium, independently of the ensemble chosen to represent that state—therefore, the $g(r)$ has the same value in both the canonical and grand canonical ensembles. We can use this relation to relate the $g(r)$ to the fluctuations of the number of particles in the grand canonical ensemble and, by means of the relation between fluctuations and susceptibility, to the isothermal compressibility $K_T = -V^{-1}\partial V/\partial p)_T$, as in the exercise that follows.

Exercise 7.1 Derive the **compressibility equation of state**:

$$\rho \int \mathrm{d}^3 r \, [g(r) - 1] + 1 = \rho k_{\mathrm{B}} T K_T.$$

We must, however, not fall into the error of supposing that *all* thermodynamic properties can be obtained from $g(r)$, at least for systems with only pair interactions and isotropic potentials. To know $g(r)$ is a little bit like knowing the internal energy E as a function of the temperature T and V—it is useful information, but incomplete. More specifically, there is no way of evaluating the system's absolute entropy starting from the $g(r)$. This situation is unfortunate because entropy is crucial to an understanding of the behavior of liquids. The configurational entropy of a gas is complete in the sense that the particles can be essentially anywhere. That of a solid is minimal, since the particles perform only small oscillations around their equilibrium positions. A liquid, instead, has a sort of restricted entropy, and $g(r)$ provides only a partial description of this situation. By knowing $g(r)$, we are not able to predict the stability of the liquid phase as compared to that of the other possible phases. Structural glasses, which can be considered metastable phases, have $g(r)$'s are that are very similar to those of liquids, and it is not possible to distinguish them from liquids just by looking at them.

7.2.2 Reversible Work Theorem

Let us now examine another one of $g(r)$'s properties: it can be expressed as a function of the work $W_{\mathrm{rev}}(r)$ necessary to bring two particles reversibly from infinity to a distance r from each other. One has

$$g(r) = \exp\left[-\frac{W(r)}{k_{\mathrm{B}} T}\right]. \tag{7.28}$$

This relation is known as the **reversible work theorem**.

Let us consider the definition of $\rho^{(2)}$:

$$\rho^{(2)}(\boldsymbol{x}_1, \boldsymbol{x}_2) \propto \int \mathrm{d}^3 \boldsymbol{r}_3 \cdots \int \mathrm{d}^3 \boldsymbol{r}_N \exp\left[-\frac{U(\{\boldsymbol{r}\})}{k_{\mathrm{B}} T}\right], \tag{7.29}$$

where $\boldsymbol{r}_1 = \boldsymbol{x}_1$ and $\boldsymbol{r}_2 = \boldsymbol{x}_2$. Let us calculate $\partial \ln \rho^{(2)} / \partial \boldsymbol{x}_2$:

$$\begin{aligned} \frac{\partial \ln \rho^{(2)}}{\partial \boldsymbol{x}_2} &= -\frac{1}{k_{\mathrm{B}} T} \left[\rho^{(2)}(\boldsymbol{x}_1, \boldsymbol{x}_2)\right]^{-1} \int \mathrm{d}^3 \boldsymbol{r}_3 \cdots \int \mathrm{d}^3 \boldsymbol{r}_N \frac{\partial U(\{\boldsymbol{r}\})}{\partial \boldsymbol{x}_2} \exp\left[-\frac{U(\{\boldsymbol{r}\})}{k_{\mathrm{B}} T}\right] \\ &= \frac{1}{k_{\mathrm{B}} T} \langle \boldsymbol{F}_2(\boldsymbol{x}_1, \boldsymbol{x}_2) \rangle, \end{aligned} \tag{7.30}$$

where $\langle \boldsymbol{F}_2(\boldsymbol{x}_1, \boldsymbol{x}_2) \rangle$ is the average of the force that acts on the particle placed in $\boldsymbol{x}_2$, when another particle is placed in $\boldsymbol{x}_1$. The force we must apply to the particle in order to keep it in $\boldsymbol{x}_2$ is the opposite of this force. By integrating this relation between ∞ and $\boldsymbol{x}_2$, and taking into account the fact that $\rho^{(2)}(\boldsymbol{x}_1, \boldsymbol{x}_2) \to \rho^2$ for $|\boldsymbol{x}_1 - \boldsymbol{x}_2| \to \infty$, we obtain

$$\ln \rho^{(2)}(\boldsymbol{x}_1, \boldsymbol{x}_2) = \frac{1}{k_{\mathrm{B}} T} \int_{\infty}^{\boldsymbol{x}_2} \langle \boldsymbol{F}_2(\boldsymbol{x}_1, \boldsymbol{x}_2) \rangle \cdot \boldsymbol{x}_2 + \ln \rho^2. \tag{7.31}$$

Now,

$$W_{\text{rev}}(r) = -\int_{\infty}^{|x_2|=r} \langle \boldsymbol{F}_2(0,\boldsymbol{x}_2)\rangle \cdot \boldsymbol{x}_2, \tag{7.32}$$

and $\rho^2 g(r) = \rho^{(2)}(0, |\boldsymbol{x}_2| = r)$. Therefore,

$$\ln g(r) = -\frac{1}{k_B T} W_{\text{rev}}(r). \tag{7.33}$$

7.2.3 Measure of g(r)

The interest in g(r) is enhanced by the fact that its Fourier transform, the so-called **structure factor** $S(\boldsymbol{k})$, is directly measurable by means of neutron scattering experiments.

Neutrons interact only with the particles' nuclei, and the nuclei are so small that the interaction potential can be well represented by a delta function. In a scattering experiment, one prepares a monochromatic beam of neutrons, characterized by momentum $\boldsymbol{p}$, connected to the wave vector $\boldsymbol{q}$ by the de Broglie relation:

$$q = \frac{p}{\hbar}. \tag{7.34}$$

The neutrons are produced in a nuclear reactor and are selected by subjecting them to Bragg elastic scattering, which selects the neutrons with the correct value of $\boldsymbol{q}$.

The beam of monochromatic neutrons is then directed onto the sample, and one measures the flow of neutrons that have been scattered by a certain angle θ. The experiment's geometry is shown in figure 7.2. After the scattering, the neutrons are characterized by a wave vector $\boldsymbol{q}'$, whose modulus is equal to that of $\boldsymbol{q}$, because the scattering is elastic. The result of the experiment therefore, is the ratio between the outgoing flux, with a certain wave vector $\boldsymbol{q}'$, and the incoming flux, as a function of the difference wave vector $\boldsymbol{k}$. For an isotropic system, only $\boldsymbol{k}$'s modulus k counts and, as one can easily derive from the geometry shown in the illustration, is equal to

$$k = 2q \sin\frac{\theta}{2}, \tag{7.35}$$

where q is the modulus of $\boldsymbol{q}$. We see that in this manner, one can obtain a direct measurement of the structure function $S(\boldsymbol{k})$.

A wave function $\psi_q(r) = \psi_0 e^{-iq\cdot r}$ that interacts with a nucleus placed in $\boldsymbol{r}_i$ produces, in each direction determined by the wave vector $\boldsymbol{q}'$, a diffused wave:

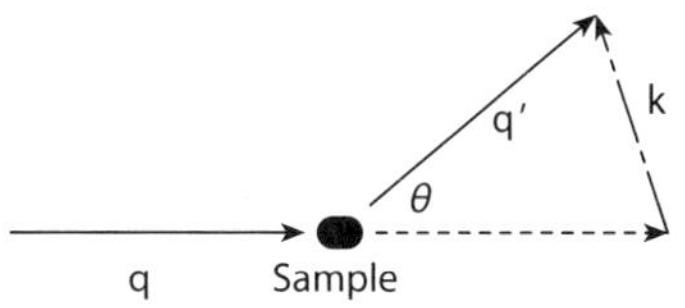

FIGURE 7.2. Scheme of a neutron scattering experiment.

$$\psi'_k(\boldsymbol{r}) = \psi_0 \phi(\boldsymbol{k}) \mathrm{e}^{-\mathrm{i}\boldsymbol{q}\cdot\boldsymbol{r}_i} \frac{\mathrm{e}^{-\mathrm{i}\boldsymbol{q}'\cdot(\boldsymbol{r}-\boldsymbol{r}_i)}}{|\boldsymbol{r}-\boldsymbol{r}_i|}, \tag{7.36}$$

where the coefficient $\phi(\boldsymbol{k})$ is the scattering amplitude expressed by the matrix element of the interaction potential between neutron and nucleus $u_{\mathrm{I}}(\boldsymbol{r})$:

$$\phi(\boldsymbol{k}) = \langle \boldsymbol{q}' | u_{\mathrm{I}} | \boldsymbol{q} \rangle \propto \int \mathrm{d}^3\boldsymbol{r}\, \mathrm{e}^{-\mathrm{i}\boldsymbol{q}'\cdot\boldsymbol{r}+\mathrm{i}\boldsymbol{q}\cdot\boldsymbol{r}} u_{\mathrm{I}}(\boldsymbol{r}) \propto \tilde{u}_{\mathrm{I}}(\boldsymbol{k}). \tag{7.37}$$

In the case we are considering, in which this potential is well approximated by a delta, it can be represented by a constant ϕ_0. In general, $\phi(\boldsymbol{k})$ will have a nontrivial dependence on $\boldsymbol{k}$ for wave vectors on the order of the inverse of the characteristic lengths of the scattering center. Since we are interested in much greater distances (and consequently in much smaller values of $|\boldsymbol{k}|$), we can disregard this dependence.

The wave function measured by the detector at a point $\boldsymbol{R}$ is the superposition of the wave functions scattered by the individual scattering centers:

$$\Psi(\boldsymbol{R}) = \psi_0 \phi_0 \sum_i \mathrm{e}^{-\mathrm{i}\boldsymbol{q}\cdot\boldsymbol{r}_i} \frac{\mathrm{e}^{-\mathrm{i}\boldsymbol{q}'\cdot(\boldsymbol{R}-\boldsymbol{r}_i)}}{|\boldsymbol{R}-\boldsymbol{r}_i|}. \tag{7.38}$$

If we assume that the sample is placed in the origin, one has $|\boldsymbol{R}-\boldsymbol{r}_i| \simeq |\boldsymbol{R}|$, and we can disregard the dependence on $\mathbf{r}_i$ in the denominator. The flux of neutrons in $\boldsymbol{R}$ is proportional to the square modulus of $\Psi(\boldsymbol{R})$:

$$|\Psi(\boldsymbol{R})|^2 = |\psi_0|^2 \frac{|\phi_0|^2}{R^2} \sum_{ij} \mathrm{e}^{-\mathrm{i}\boldsymbol{q}\cdot(\boldsymbol{r}_i-\boldsymbol{r}_j)} \mathrm{e}^{-\mathrm{i}\boldsymbol{q}'\cdot(\boldsymbol{r}_i-\boldsymbol{r}_j)}. \tag{7.39}$$

The dependence on $\boldsymbol{R}$ in the exponent cancels out, and one remains with

$$|\Psi(\boldsymbol{R})|^2 = |\psi_0|^2 \frac{|\phi_0|^2}{R^2} \sum_{ij} \mathrm{e}^{\mathrm{i}(\boldsymbol{q}-\boldsymbol{q}')\cdot(\boldsymbol{r}_i-\boldsymbol{r}_j)}. \tag{7.40}$$

From this quantity, one subtracts the term in which $i = j$. We obtain

$$|\Psi(\boldsymbol{R})|^2 = |\psi_0|^2 \frac{|\phi_0|^2}{R^2} \left[N + \sum_{(i\neq j)} \mathrm{e}^{\mathrm{i}(\boldsymbol{q}-\boldsymbol{q}')\cdot(\boldsymbol{r}_i-\boldsymbol{r}_j)} \right]. \tag{7.41}$$

By taking this expression's average with respect to thermal fluctuations, and taking into account the fact that $\boldsymbol{k} = \boldsymbol{q}' - \boldsymbol{q}$, we obtain

$$\begin{aligned} \langle |\Psi(R)|^2 \rangle &= |\psi_0|^2 \frac{|\phi_0|^2}{R^2} \left[N + \int \mathrm{d}^3\boldsymbol{x} \int \mathrm{d}^3\boldsymbol{x}' \mathrm{e}^{-\mathrm{i}\boldsymbol{k}\cdot(\boldsymbol{x}-\boldsymbol{x}')} \left\langle \sum_{(i\neq j)} \delta(\boldsymbol{x}-\boldsymbol{r}_i)\delta(\boldsymbol{x}'-\boldsymbol{r}_j) \right\rangle \right] \\ &= |\psi_0|^2 \frac{|\phi_0|^2}{R^2} \left[N + \int \mathrm{d}^3\boldsymbol{x} \int \mathrm{d}^3\boldsymbol{x}' \mathrm{e}^{-\mathrm{i}\boldsymbol{k}\cdot(\boldsymbol{x}-\boldsymbol{x}')} \rho^{(2)}(\boldsymbol{x},\boldsymbol{x}') \right] \\ &= |\psi_0|^2 \frac{|\phi_0|^2}{R^2} N \left[1 + \frac{N}{V} \int \mathrm{d}^3\boldsymbol{x}\, \mathrm{e}^{-\mathrm{i}\boldsymbol{k}\cdot\boldsymbol{x}} g(\boldsymbol{x}) \right]. \end{aligned} \tag{7.42}$$

The ratio between exit and entry flux I_0 allows one to get rid of the $|\psi_0|^2$ factor. On the other hand, by evaluating the exit flux in a solid angle element $\mathrm{d}\Omega = \sin\theta\,\mathrm{d}\theta\,\mathrm{d}\phi$, the area over which we integrate the flux becomes proportional to $|\boldsymbol{R}|^2$. This cancels out the dependence on $|\boldsymbol{R}|^2$ in the denominator. Since for $\boldsymbol{k}=0$, moreover, there is no way of distinguishing between forward-scattered particles and nonscattered particles, there is no reason to also use this formula for $\boldsymbol{k}=0$. It is usual to subtract a delta function at $\boldsymbol{k}=0$ from this expression, thus obtaining the following expression of the ratio between entrance and exit flux in the direction defined by $\boldsymbol{q}'$:

$$I(\theta) = I_0|\phi_0|^2 N\left[1 + \rho\int \mathrm{d}^3\boldsymbol{x}\,\mathrm{e}^{-\mathrm{i}\boldsymbol{k}\cdot\boldsymbol{x}}[g(|\boldsymbol{x}|) - 1]\right]. \tag{7.43}$$

We can now define the **factor structure** $S(\boldsymbol{k})$:

$$S(\boldsymbol{k}) = 1 + \rho\int \mathrm{d}^3\boldsymbol{x}[g(|\boldsymbol{x}|) - 1]\mathrm{e}^{-\mathrm{i}\boldsymbol{k}\cdot\boldsymbol{x}}. \tag{7.44}$$

Therefore, for $\boldsymbol{k} \neq 0$ one has

$$I(\theta) = I_0|\phi_0|^2 S(\boldsymbol{k}). \tag{7.45}$$

The measure of the flux therefore provides a measure of $S(\boldsymbol{k})$, which in turn allows us to determine the pair distribution $g(r)$ by means of an inverse Fourier transformation.

7.2.4 BBGKY Hierarchy

The various reduced densities are connected by a chain of equations, so that each $\rho^{(\ell)}$ can be expressed as a function of $\rho^{(\ell+1)}$ and the interaction potential. We are therefore dealing with an equation hierarchy, which has been given the name of the many authors who discovered it (actually, in the time-dependent version): Born, Bogolyubov, Green, Kirkwood, Yvon. It is therefore referred to as the **BBGKY hierarchy**.

Let us consider the derivative of $\rho^{(2)}(\boldsymbol{x}_1, \boldsymbol{x}_2)$ with respect to $\boldsymbol{x}_1$:

$$\begin{aligned}\partial_{x_1}\rho^{(2)}(\boldsymbol{x}_1,\boldsymbol{x}_2) &= \frac{N(N-1)}{\Omega_N}\partial_{x_1}\int \mathrm{d}^3\boldsymbol{r}_1\cdots\mathrm{d}^3\boldsymbol{r}_N \exp\left\{-\frac{1}{k_\mathrm{B}T}U(\{\boldsymbol{r}\})\right\}\\ &= -\frac{1}{k_\mathrm{B}T}(\partial_{x_1}u(\boldsymbol{x}_1 - \boldsymbol{x}_2))\rho^{(2)}(\boldsymbol{x}_1,\boldsymbol{x}_2)\\ &= -\frac{1}{k_\mathrm{B}T}\int \mathrm{d}^3\boldsymbol{x}_3(\partial_{x_1}u(\boldsymbol{x}_1 - \boldsymbol{x}_3))\rho^{(3)}(\boldsymbol{x}_1,\boldsymbol{x}_2,\boldsymbol{x}_3).\end{aligned} \tag{7.46}$$

Let us introduce the distribution functions [generalizations of the $g(r)$'s]:

$$\rho^{(2)}(\boldsymbol{x}_1,\boldsymbol{x}_2) = \rho^{(2)}g(\boldsymbol{x}_1,\boldsymbol{x}_2), \tag{7.47}$$

$$\rho^{(3)}(\boldsymbol{x}_1,\boldsymbol{x}_2,\boldsymbol{x}_3) = \rho^3 g^{(3)}(\boldsymbol{x}_1,\boldsymbol{x}_2,\boldsymbol{x}_3), \tag{7.48}$$

where $\rho = N/V$ is the particle density. We thus obtain

$$\begin{aligned}\partial_{x_1}g(\boldsymbol{x}_1,\boldsymbol{x}_2) &= -\frac{1}{k_\mathrm{B}T}(\partial_{x_1}u(\boldsymbol{x}_1 - \boldsymbol{x}_2))g(\boldsymbol{x}_1,\boldsymbol{x}_2)\\ &\quad -\frac{1}{k_\mathrm{B}T}\rho\int \mathrm{d}^3\boldsymbol{x}_3(\partial_{x_1}u(\boldsymbol{x}_1 - \boldsymbol{x}_3))g^{(3)}(\boldsymbol{x}_1,\boldsymbol{x}_2,\boldsymbol{x}_3).\end{aligned} \tag{7.49}$$

The other equations in the hierarchy can be derived in the same manner. Obviously, there is no way to solve directly these equations, since the equation for $\rho^{(\ell)}$, for example, involves

$\rho^{(\ell+1)}$, and so on. It is possible to use this hierarchy, however, as a starting point for some approximations, the most familiar being Kirkwood's **superposition approximation**. It consists of the representation of $g^{(3)}$ as a superposition of the pair distributions:

$$g^{(3)}(\boldsymbol{x}_1,\boldsymbol{x}_2,\boldsymbol{x}_3) = g(\boldsymbol{x}_1,\boldsymbol{x}_2)\,g(\boldsymbol{x}_2,\boldsymbol{x}_3)\,g(\boldsymbol{x}_3,\boldsymbol{x}_1). \tag{7.50}$$

This approximation allows one to convert equation (7.49) into a closed nonlinear equation for the $g(r)$, which is known as the Born-Green-Yvon (BGY) equation. This equation is fairly useful at low densities.

By substituting this approximation in (7.49), we obtain

$$\begin{aligned}\partial_{x_1} g(\boldsymbol{x}_1,\boldsymbol{x}_2) &= -\frac{1}{k_BT}[\partial_{x_1} u(\boldsymbol{x}_1-\boldsymbol{x}_2)]g(\boldsymbol{x}_1,\boldsymbol{x}_2)\\ &\quad -\frac{1}{k_BT}\rho g(\boldsymbol{x}_1-\boldsymbol{x}_2)\int \mathrm{d}^3\boldsymbol{x}_3[\partial_{x_1} u(\boldsymbol{x}_1-\boldsymbol{x}_3)]g(\boldsymbol{x}_2,\boldsymbol{x}_3)g(\boldsymbol{x}_3,\boldsymbol{x}_1).\end{aligned} \tag{7.51}$$

We can lighten the notation by setting $r_{ij} = |\boldsymbol{x}_i - \boldsymbol{x}_j|$. We can rearrange this expression by writing

$$\partial_{x_1}\left[\ln g(r_{12}) + \frac{u(r_{12})}{k_BT}\right] = -\frac{\rho}{k_BT}\int \mathrm{d}^3\boldsymbol{x}_3\left[\partial_{x_1} u(r_{13})\right]g(r_{23})\,g(r_{31}). \tag{7.52}$$

This is the BGY equation, and it can also be expressed in another form, which was derived by Kirkwood. Let us introduce the auxiliary function

$$E(r) = -\int_r^{\infty} \mathrm{d}r'\,\frac{\mathrm{d}u(r')}{\mathrm{d}r'}\,g(r'). \tag{7.53}$$

Then,

$$\frac{\partial E(r_{13})}{\partial \boldsymbol{x}_1} = \frac{\partial u(r_{13})}{\partial \boldsymbol{x}_1}\,g(r_{13}). \tag{7.54}$$

By substituting this expression in (7.52), we obtain

$$\partial_{x_1}\left[\ln g(r_{12}) + \frac{u(r_{12})}{k_BT}\right] = -\frac{\rho}{k_BT}\int \mathrm{d}^3\boldsymbol{x}_3\,\frac{\partial E(r_{13})}{\partial \boldsymbol{x}_1}\,g(r_{23}). \tag{7.55}$$

By integrating with respect to $\boldsymbol{x}_1$, we obtain

$$\ln g(r_{12}) + \frac{u(r_{12})}{k_BT} = -\frac{\rho}{k_BT}\int \mathrm{d}^3\boldsymbol{x}_3\,E(r_{13})\,g(r_{23}) + C, \tag{7.56}$$

where C is an integration constant. We can evaluate C by passing to the limit $r_{12}\to\infty$, while maintaining r_{13} finite. The left-hand side goes to zero, and $g(r_{23})\to 1$. Therefore,

$$C = \frac{\rho}{k_BT}\int \mathrm{d}^3\boldsymbol{x}\,E(|\boldsymbol{x}|). \tag{7.57}$$

By substituting this result in (7.56), we obtain

$$\ln g(r_{12}) + \frac{u(r_{12})}{k_BT} = -\frac{\rho}{k_BT}\int \mathrm{d}^3\boldsymbol{x}_3\,E(r_{13})\,[g(r_{23}) - 1]. \tag{7.58}$$

This equation is known as the **Kirkwood equation**.

7.2.5 *Direct Correlation Function*

The $g(r)$ is connected to the isothermal compressibility by the compressibility equation of state, which we obtained in exercise 7.1. We know that the compressibility diverges near the critical point—this is due to the fact that the correlation length diverges at the critical point. Ornstein and Zernicke had the idea of separating the effects that lead to the divergence of the correlation length in the $g(r)$, by introducing the **direct correlation function** $c(r)$, which is always short-range, and which one expects to be regular around the critical point. Actually, we know from the theory of critical phenomena that this is not exactly true, but can be so only in the framework of a mean-field theory.

It is useful to introduce the function $h(r) = g(r) - 1$, which goes to zero for $r \to \infty$. The direct correlation function $c(r)$ is then defined by

$$h(r) = c(r) + \rho \int \mathrm{d}^3 r' c(|\mathbf{r}'|)\, h(|\mathbf{r} - \mathbf{r}'|). \tag{7.59}$$

The correlations expressed by $h(r)$ are then divided into two contributions: the *direct* correlations between particles placed at a distance r, represented by $c(r)$, and the *indirect* correlations, in which the particle in the origin directly influences a particle placed in $\mathbf{r}'$, which in its turn (directly or indirectly) influences the particle placed in $\mathbf{r}$.

This integral equation can be resolved by Fourier transforming it:

$$\tilde{h}(\mathbf{k}) = \tilde{c}(\mathbf{k}) + \rho \tilde{c}(\mathbf{k})\, \tilde{h}(\mathbf{k}). \tag{7.60}$$

We thus obtain

$$\tilde{h}(\mathbf{k}) = \frac{c(\mathbf{k})}{1 - \rho \tilde{c}(\mathbf{k})}. \tag{7.61}$$

The divergence of isothermal compressibility is exhibited by the divergence of $h(\mathbf{k})$ for $k \to 0$. From this equation, we see that this occurs if $c(\mathbf{k} = 0) \to 1/\rho$ for $T \to T_c$.

These equations obviously do not provide us with any new information, if we are not able to evaluate the h or the c independently. We will see in the following sections that various approximation schemes can be formulated in terms of an additional relation between the h [and therefore the $g(r)$] and the c.

7.3 Virial Expansion

The interaction potential between particles exhibits a hard core, which makes a systematic expansion in which the interaction is treated as a perturbation impractical. If the particle density is low, however, it can play the role of a *small parameter*; one thus obtains a systematic expansion of the thermodynamic properties as a function of density, which is known as the **virial expansion**.

The virial expansion of the equation of state has the form

$$\frac{p}{k_B T} = \sum_{\ell=1}^{\infty} B_\ell(T)\, \rho^\ell, \tag{7.62}$$

where the coefficients $B_\ell(T)$ are called **virial coefficients**. Since the equation of state for ideal gases holds in the limit $\rho \to 0$, one obviously has

$$B_1(T) = 1, \qquad \forall T. \tag{7.63}$$

The expression of the virial coefficients as a function of T and the pair potential $u(r)$ is one of the first successes of "modern" statistical mechanics, and was derived by J. E. Mayer [Maye40]. In this section, I will attempt to show, as a general outline, how it can be obtained, and I will refer to, among others, the volume by Hansen and McDonald [Hans86] for further details.

Let us consider the canonical partition function Z_N:

$$Z_N = \frac{1}{N!}\frac{\Omega_N[\nu]}{\lambda^{3N}}. \tag{7.64}$$

If $\nu = \nu_{\mathcal{B}}$, given the expression of $\Omega_N[\nu]$, it can be written

$$Z_N = \frac{1}{N!}\int_{\mathcal{B}} \prod_{i=1}^{N} \left(\frac{\mathrm{d}^3 \boldsymbol{r}_i}{\lambda^3}\right) \prod_{(jk)} \mathrm{e}^{-u(|\boldsymbol{r}_j - \boldsymbol{r}_k|)/k_{\mathrm{B}}T}, \tag{7.65}$$

where the integral over each $\boldsymbol{r}_i$ is limited to the box $\mathcal{B}$. The last product runs over all different pairs of particles (jk). Now, each factor $\mathrm{e}^{-u(r)/k_{\mathrm{B}}T}$ should be approximately equal to 1 as soon as $u(r) \ll k_{\mathrm{B}}T$, which occurs as soon as $r \gg r_0$. For each pair of particles (jk), we can therefore introduce the **Mayer function** f_{jk}, defined by

$$f_{jk} = \mathrm{e}^{-u(|\boldsymbol{r}_j - \boldsymbol{r}_k|)/k_{\mathrm{B}}T} - 1. \tag{7.66}$$

It is easy to see that this function is bounded, equal to -1 in the origin, and goes to zero very rapidly when $r \gg r_0$. A plot of the Mayer function for the Lennard-Jones potential (and for $k_{\mathrm{B}}T = \varepsilon_0$) is shown in figure 7.3. The product in equation (7.65) can therefore be written

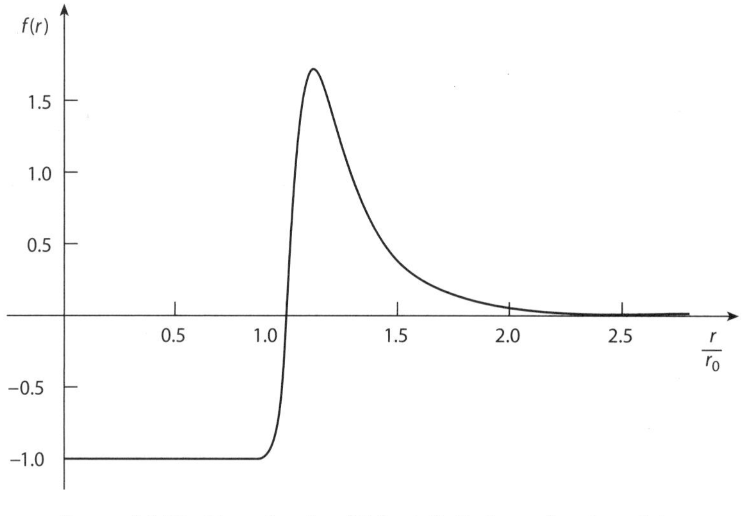

FIGURE 7.3. The Mayer function $f(r)$ for $\epsilon_0/k_{\mathrm{B}}T = 1$, as a function of r/r_0.

$$\prod_{(jk)}(1+f_{jk}) = 1+\sum_{(jk)} f_{jk} + \sum_{(jk)}\sum_{(\ell m)} f_{jk}f_{\ell m} + \cdots \tag{7.67}$$

The second term of this expression vanishes as soon as $r_{jk} = |\mathbf{r}_j - \mathbf{r}_k| \gg r_0$. Analogously, the third term vanishes if $r_{jk} \gg r_0$ or $r_{\ell m} \gg r_0$. Let us note that we can distinguish two types of contributions in the third term: those that concern four distinct particles, and those in which, for example, $\ell = j$, and which therefore concern only three particles.

If we stop at the first contribution, we obtain the following approximation for Z_N:

$$Z_N = \frac{1}{N!}\left(\frac{V}{\lambda^3}\right)^N\left[1+\frac{N(N-1)}{2V}\int \mathrm{d}^3\mathbf{r}\, f(r)\right]. \tag{7.68}$$

The Helmholtz free energy is given by $F = -k_{\mathrm{B}}T\ln Z_N$. We thus obtain

$$\begin{aligned} F &= F = -k_{\mathrm{B}}T\ln Z_N = F_0 - k_{\mathrm{B}}T\ln\left[1+\frac{N(N-1)}{2V}\int \mathrm{d}^3\mathbf{r}\, f(r)+\cdots\right] \\ &\simeq F_0 - k_{\mathrm{B}}T\frac{N^2}{2V}\int \mathrm{d}^3\mathbf{r}\, f(r), \end{aligned} \tag{7.69}$$

where $F_0 = -Nk_{\mathrm{B}}T\ln(\mathrm{e}V/N\lambda^3)$ is the free energy of the ideal gas. By taking the derivative with respect to V, we obtain the pressure:

$$p = -\left.\frac{\partial F}{\partial V}\right)_T = k_{\mathrm{B}}T\left[\rho - \frac{\rho^2}{2}\int \mathrm{d}^3\mathbf{r}\, f(r)\right]. \tag{7.70}$$

In this fashion, we obtain the expression of the second virial coefficient:

$$B_2(T) = -\frac{1}{2}\int \mathrm{d}^3\mathbf{r}\, f(r). \tag{7.71}$$

Exercise 7.2 By comparing (7.70) with (7.21), show that in this approximation the g(r) is expressed by

$$g(r) = 1 + f(r).$$

We now examine $B_2(T)$'s behavior. At high temperatures, in other words for $\varepsilon_0 \ll k_{\mathrm{B}}T$, we can approximate the Mayer function as follows:

$$f(r) \simeq \begin{cases} -1, & \text{if } r < r_0, \\ -u(r)/k_{\mathrm{B}}T, & \text{if } r > r_0. \end{cases} \tag{7.72}$$

Therefore,

$$B_2(T) = \frac{1}{2}\int \mathrm{d}^3\mathbf{r}\, f(r) \simeq b - \frac{a}{k_{\mathrm{B}}T}, \tag{7.73}$$

where

$$b = \frac{1}{2}\frac{4\pi}{3}r_0^3, \tag{7.74}$$

$$a = -\frac{1}{2}\int_0^\infty \mathrm{d}^3\mathbf{r}\, u(r) = -2\pi\int_0^\infty \mathrm{d}r\, r^2 u(r). \tag{7.75}$$

These constants are positive and can be compared to those that appear in the van der Waals equation. More particularly, we see that, since r_0 is equal to twice the volume of the hard core of a gas particle, b is equal to four times this volume.

We also see that $B_2(T)$ is positive at high temperatures but decreases as T decreases. There will therefore be a temperature T^* (called the **inversion temperature**) in which $B_2(T)$ changes sign. One exploits this phenomenon in a gas cooling system, based on the **Joule-Thompson process.**

Exercise 7.3 (Joule-Thompson Process) Consider a gas that expands through a throttle from a region at pressure p_1 to a region at pressure p_2, as shown in figure 7.4.

1. Show that the process occurs at constant enthalpy H.

2. Evaluate the temperature variation at constant enthalpy

$$\left.\frac{\partial T}{\partial p}\right)_H,$$

as a function of both the specific heat at constant pressure C_p and of the expansion coefficient

$$\left.\frac{\partial V}{\partial T}\right)_p,$$

3. Express this quantity as a function of the second virial coefficient $B_2(T)$, and evaluate the inversion temperature T^* in which $\partial T/\partial p)_H$ changes sign.

4. Estimate the order of magnitude of T^* as a function of the critical temperature T_c for a van der Waals gas.

7.3.1 *Higher Virial Coefficients*

In order to evaluate the further virial coefficients, it is convenient to use the grand canonical ensemble. Let us write the grand canonical partition function in the form (7.7):

$$Z_{GC} = \sum_{N=0}^{\infty} z^N Z_N = \sum_{N=0}^{\infty} \frac{z^N}{N!} \Omega_N, \tag{7.76}$$

where $z = e^{\mu/k_BT}$ is the *fugacity*. By using expression (7.65), we can express the various contributions that appear in this expression in terms of diagrams. A diagram with N particles contains N vertices, labeled $i = 1, \ldots, N$, and a certain number of lines ij that connect vertex i with vertex j. There can be no more than one line that directly connects two given vertices. The contribution of a diagram of this type, with N vertices, is equal to

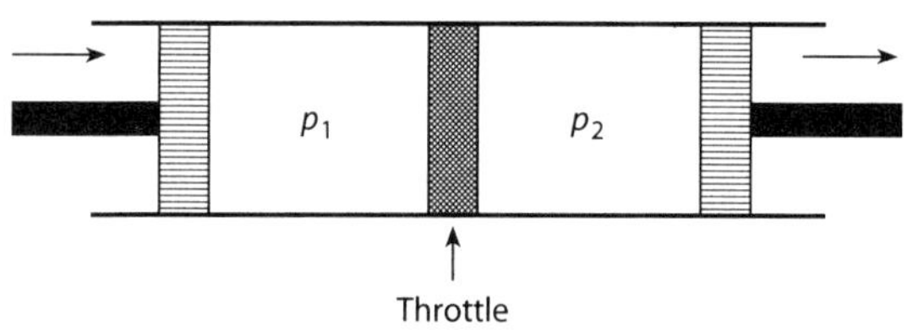

FIGURE 7.4. Joule-Thompson process.

$$\left(\frac{z^N}{N!}\right)\int_{\mathcal{B}}\left(\frac{\mathrm{d}^3\boldsymbol{r}_1}{\lambda^3}\right)\cdots\int_{\mathcal{B}}\left(\frac{\mathrm{d}^3\boldsymbol{r}_N}{\lambda^3}\right)\prod f_{ij},$$

where the product runs over all the lines ij that are present in the diagram, and the integrals are limited to the box $\mathcal{B}$. Evidently, each vertex j that is not connected by any line contributes a factor V/λ^3. If we therefore consider the contributions of the diagrams that contain a total of N vertices, but only k of which are connected by one or more lines, the corresponding contribution is equal to

$$\left(\frac{z^N}{N!}\right)\frac{N!}{k!(N-k)!}\left(\frac{V}{\lambda^3}\right)^{N-k}\int_{\mathcal{B}}\left(\frac{\mathrm{d}^3\boldsymbol{r}_1}{\lambda^3}\right)\cdots\int_{\mathcal{B}}\left(\frac{\mathrm{d}^3\boldsymbol{r}_k}{\lambda^3}\right)\prod f_{ij}$$
$$=\left(\frac{z^{N-k}}{(N-k)!}\right)\left(\frac{V}{\lambda^3}\right)^{N-k}\left(\frac{z^k}{k!}\right)\int_{\mathcal{B}}\left(\frac{\mathrm{d}^3\boldsymbol{r}_1}{\lambda^3}\right)\cdots\int_{\mathcal{B}}\left(\frac{\mathrm{d}^3\boldsymbol{r}_k}{\lambda^3}\right)\prod f_{ij}.$$

The combinatorial factor $N!/(k!(N-k)!$ is a result of the different ways in which we can choose the k particles that are connected among the N that are available. By summing over N, for a predetermined diagram, the first factor reproduces the Z^0_{GC}. We thus obtain

$$\begin{aligned} Z_{\mathrm{GC}} &= Z^0_{\mathrm{GC}}\left[1+\sum_k\left(\frac{z^k}{k!}\right)\sum_{\mathcal{G}(k)}\int_{\mathcal{B}}\prod_{i\in\mathcal{G}(k)}\left(\frac{\mathrm{d}^3\boldsymbol{r}_i}{\lambda^3}\right)\prod_{i\in\mathcal{G}(k)} f_{ij}\right] \\ &= Z^0_{\mathrm{GC}}Z'_{\mathrm{GC}}. \end{aligned} \tag{7.77}$$

where $\mathcal{G}(k)$ denotes all the diagrams with k vertices, the product over i runs over all the vertices, and that over ij runs through all the lines that appear in the diagram. Let us note that the factor $1/k!$ is removed by summing over all the ways in which the k vertices can be labeled, and we can therefore remove it if we only consider diagrams in which the vertices are not labeled. In certain cases, even if the labels of certain particles are exchanged, the connections that appear in the diagram remain the same. The simplest case is that in which only the line (ij) appears, which clearly remains invariant with respect to the exchange $i\leftrightarrow j$. In these cases, in order to avoid multiple counts, it is necessary to divide the diagram's contribution by the number of exchanges S that leave the diagram unchanged. In order to evaluate S, it is necessary to label the diagram once. Once S has been obtained, one can forget about the labeling.

We can now consider the fraction $\partial \ln Z_{\mathrm{GC}}/\partial z$. On the one hand, since $z=\mathrm{e}^{\mu/k_BT}$, we have

$$\left.\frac{\partial \ln Z_{\mathrm{GC}}}{\partial z}\right)_T=\frac{1}{z}\left.\frac{\partial \ln Z_{\mathrm{GC}}}{\partial \ln z}\right)_T=\frac{1}{z}N. \tag{7.78}$$

On the other hand, by taking the derivative of both sides of equation (7.77), we obtain

$$\left.\frac{\partial \ln Z_{\mathrm{GC}}}{\partial z}\right)_T=\frac{1}{Z'_{\mathrm{GC}}}\left[\frac{V}{\lambda^3}+\sum_k\left(\frac{z^k}{k!}\right)\sum_{\mathcal{G}'(k)}\int_{\mathcal{B}}\prod_{i\in\mathcal{G}'(k)}\left(\frac{\mathrm{d}^3\boldsymbol{r}_i}{\lambda^3}\right)\prod_{ij\in\mathcal{G}'(k)} f_{ij}\right], \tag{7.79}$$

where the sum runs over the diagrams $\mathcal{G}'(k)$ with $k+1$ vertices, of which one—let us refer to it as 0—is marked; it corresponds to the vertex with respect to which we took the derivative. We will call this vertex the **root** and denote it by means of a small, open circle. The term V/λ^3 comes from the derivative of ln Z^0_{GC}. Each one of the diagrams we are considering includes a certain number of vertices that are directly or indirectly connected to 0 by means of lines and possibly some other nonconnected subdiagrams. By focusing our attention on a given subdiagram connected to 0, with k vertices, and by summing over all the others, we obtain the product of that subdiagram for Z'_{GC}. which cancels out with the first factor. We thus obtain

$$\left.\frac{\partial \ln Z_{GC}}{\partial z}\right)_T = \frac{V}{\lambda^3} + \sum_k \left(\frac{z^k}{k!}\right) \sum_{\mathcal{G}^*(k)} \int_{\mathcal{B}} \prod_{i \in \mathcal{G}^*(k)} \left(\frac{d^3 \mathbf{r}_i}{\lambda^3}\right) \prod_{ij \in \mathcal{G}^*(k)} f_{ij}, \tag{7.80}$$

where the sum runs over all the diagrams *connected* to the root 0.

By introducing the notation

$$\frac{V}{\lambda^3} b_k = \sum_{\mathcal{G}^*(k)} \int_{\mathcal{B}} \prod_{i \in \mathcal{G}^*(k)} \left(\frac{d^3 \mathbf{r}_i}{\lambda^3}\right) \prod_{ij \in \mathcal{G}^*(k)} f_{ij}, \tag{7.81}$$

and taking (7.78) into account, we obtain

$$\frac{N}{z} = \frac{V}{\lambda^3}\left[1 + \sum_k \left(\frac{z^k}{k!}\right) b_k\right]. \tag{7.82}$$

Therefore,

$$\rho = \frac{N}{V} = \frac{1}{\lambda^3}\left[z + \sum_k \left(\frac{z^{k+1}}{k!}\right) b_k\right]. \tag{7.83}$$

We have thus obtained the expansion of density in powers of the fugacity. In order to reach our goal—the expression of pressure as a function of density—we must invert this relation and obtain the expression of pressure. As we will see, this operation has a clear meaning in terms of diagrams.

Let us define a vertex such that, were it removed, the remaining subdiagrams would be disconnected (see figure 7.5) as a **node** (or **articulation point**) of a diagram. The relative contributions to subdiagrams that are connected by a node factorize. Let us, for example, consider two subdiagrams $\mathcal{G}_1$ and $\mathcal{G}_2$, connected by a vertex in j_0. In the diagram's contribution, the lines $i_1 j_1$ that connect the vertices belonging to $\mathcal{G}_1$ to one another and to j_0, and analogous lines $i_2 j_2$ that connect the vertices belonging to $\mathcal{G}_2$ appear, but the lines that connect the subdiagrams to each other do not. By setting $\mathbf{r}^0_{i_1} = \mathbf{r}_{i_1} - \mathbf{r}_{j_0}$ for $i_1 \in \mathcal{G}_1$ (and analogously for $\mathcal{G}_2$), we see that we can freely integrate over the $\mathbf{r}^0_{i_1}, \mathbf{r}^0_{i_2}$. The contribution of subdiagrams $\mathcal{G}_1$ and $\mathcal{G}_2$ factorizes.

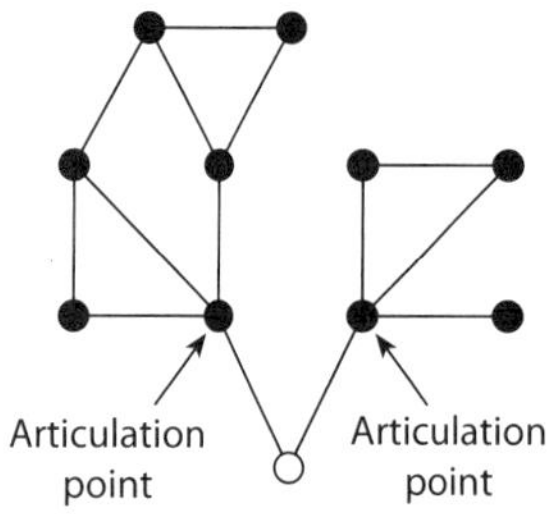

FIGURE 7.5. A diagram connected by two articulation points. The root is represented by a small open circle.

Let us now consider a diagram with $k+1$ vertices in expansion (7.80), in which there are no articulation points, except possibly the root. We can then obtain a set of diagrams with articulation points from this diagram, by "decorating" each of the k vertices with sub-diagrams, except for the marked point. By summing with respect to all the "decorations," we obtain the diagram's contribution multiplied by $(\rho\lambda^3)^k$. In fact, the contribution of the decorations (with their respective z factors) reproduces N/z, up to the factor V/λ^3 relative to the integration over the root's coordinates, and the z factor is provided by the vertex to which the decorations are attached. We have therefore arrived at

$$\frac{N}{z} = \frac{V}{\lambda^3} + \sum_k \frac{1}{k!}\beta'_k (\rho\lambda^3)^k, \tag{7.84}$$

where

$$\beta'_k = \sum_{\tilde{\mathcal{G}}(k)} \int_{\mathcal{B}} \prod_{i\in\tilde{\mathcal{G}}(k)} \left(\frac{\mathrm{d}^3\boldsymbol{r}_i}{\lambda^3}\right) \prod_{ij\in\tilde{\mathcal{G}}(k)} f_{ij}, \tag{7.85}$$

is the contribution of the diagrams of $k+1$ particles in which possibly only the root is an articulation point.

The structure of these diagrams is, generally, that of ℓ **irreducible** diagrams (in other words, diagrams free of articulation points), which join at the marked point. We will refer to the irreducible diagrams with $k+1$ vertices (of which one is a root) as $\mathcal{G}'(k)$,and to diagrams composed of ℓ irreducible diagrams, containing $k+1$ particles, that join at the root, as $\tilde{\mathcal{G}}^{(\ell)}(k)$. Then,

$$\begin{aligned} &\sum_k \frac{1}{k!} \sum_{\tilde{\mathcal{G}}^{(\ell)}(k)} (\rho\lambda^3)^k \int_{\mathcal{B}} \prod_{i\in\tilde{\mathcal{G}}(k)} \left(\frac{\mathrm{d}^3\boldsymbol{r}_i}{\lambda^3}\right) \prod_{ij\in\tilde{\mathcal{G}}(k)} f_{ij}, \\ &\quad = \frac{1}{\ell!}\frac{V}{\lambda^3}\left[\sum_{\tilde{\mathcal{G}}'(k)} (\rho\lambda^3)^k \int_{\mathcal{B}} \prod_{0<i\in\tilde{\mathcal{G}}'(k)} \left(\frac{\mathrm{d}^3\boldsymbol{r}_i}{\lambda^3}\right) \prod_{ij\in\tilde{\mathcal{G}}'(k)} f_{ij}\right]^\ell. \end{aligned} \tag{7.86}$$

The factor $1/\ell!$ is a result of the fact that one must not count diagrams that are obtained by exchanging irreducible diagrams with each other multiple times. Let us note that we have integrated explicitly over $\boldsymbol{r}_0$, thus obtaining the V/λ^3 factor. By defining β_k by means of the

$$\beta_k = \sum_{\tilde{\mathcal{G}}'(k)} \frac{1}{k!} \int_{\mathcal{B}} \prod_{i\in\tilde{\mathcal{G}}'(k)} \left(\frac{\mathrm{d}^3\boldsymbol{r}_i}{\lambda^3}\right) \prod_{ij\in\tilde{\mathcal{G}}'(k)} f_{ij}, \tag{7.87}$$

and taking the nontrivial term into account, we finally obtain

$$\frac{N}{z} = \frac{V}{\lambda^3}\exp\left[\sum_k \beta_k (\rho\lambda^3)^k\right]. \tag{7.88}$$

We have thus expressed the fugacity as a function of density. Let us take the logarithm of this expression:

$$\ln(\rho\lambda^3) - \frac{\mu}{k_{\mathrm{B}}T} = \sum_k \beta_k (\rho\lambda^3)^k. \tag{7.89}$$

From the Gibbs-Duhem equation

$$d\mu = -s dT + \frac{1}{\rho}\, dp, \tag{7.90}$$

we obtain

$$\left.\frac{\partial \mu}{\partial \rho}\right)_T = \frac{1}{\rho}\left.\frac{\partial p}{\partial \rho}\right)_T \tag{7.91}$$

Therefore,

$$p(T,\rho) = \int_0^\rho d\rho' \left.\frac{\partial p}{\partial \rho'}\right)_T = \int_0^\rho d\rho' \rho' \left.\frac{\partial \mu}{\partial \rho'}\right)_T . \tag{7.92}$$

But from equation (7.89), we obtain

$$\left.\frac{\partial \mu}{\partial \rho}\right)_T = \frac{k_B T}{\rho} - \lambda^3 k_B T \sum_k k\beta_k (\rho\lambda^3)^{k-1}. \tag{7.93}$$

By integrating, we finally obtain

$$p(T,\rho) = k_B T\rho - \lambda^3 k_B T \sum_k \frac{k}{k+1} \beta_k (\rho\lambda^3)^k. \tag{7.94}$$

This is the virial expansion of the equation of state. This expansion is fairly useful for describing the behavior of gases (at least in the inverse direction—in other words, to determine the interaction potential starting from the equation of state's measurement), but it has scarce predictive capacity for liquids. In this case, it is important to find extrapolation schemes that somehow account for all the orders in the expansion of the density function. On the other hand, one would like to be able to evaluate more microscopic properties like $g(r)$, and not only the equation of state, in the liquid phase.

In table 7.2, I reproduce the first virial coefficients for a hard sphere gas. It is interesting to note that it is possible to account almost exactly for the results obtained for the first virial coefficients by using an extrapolation formula due to Carnahan and Starling [Carn69]. One introduces the ratio of the volume occupied by the particles to the total volume of the system, called the **packing ratio**:

Table 7.2 Virial Coefficient for a Hard Sphere gas, with Particles of Diameter *d*

$B_2 = b$	$2/3\pi d^3$
B_3/b^2	$5/8$
B_4/b^3	0. 28695
B_5/b^4	0.1103 ± 0.0003
B_6/b^5	0.0386 ± 0.0004
B_7/b^6	0.0138 ± 0.0004

Source: Hans86, p. 94.

$$\eta = \frac{1}{6}\pi d^3 \rho, \tag{7.95}$$

where d is the particle's diameter. Then the Carnahan-Starling equation expresses pressure as follows:

$$\frac{p}{pk_{\mathrm{B}}T} = \frac{1+\eta+\eta^2-\eta^3}{(1-\eta)^3}. \tag{7.96}$$

This equation's predictions are almost indistinguishable from the Monte Carlo results.

7.3.2 *Integral Equations for the* g(r)

In this section, we derive some approximate integral equations that provide relations between the pair distribution $g(r)$ and the pair potential $u(r)$.

Let us consider the definition of the $\rho^{(2)}$:

$$\rho^{(2)}(\boldsymbol{x}_1, \boldsymbol{x}_2) \propto \int \mathrm{d}^3\boldsymbol{r}_3 \cdots \int \mathrm{d}^3\boldsymbol{r}_N \exp\left[-\frac{U(\{\boldsymbol{r}\})}{k_{\mathrm{B}}T}\right].$$

We can isolate the contribution of the direct interaction between particles 1 and 2 at $U(\{\boldsymbol{r}\})$:

$$\rho^{(2)}(\boldsymbol{x}_1, \boldsymbol{x}_2) \propto \exp\left(-\frac{u(r_{12})}{k_{\mathrm{B}}T}\right) \int \mathrm{d}^3\boldsymbol{r}_3 \cdots \int \mathrm{d}^3\boldsymbol{r}_N \exp\left[-\frac{U'(\{\boldsymbol{r}\})}{k_{\mathrm{B}}T}\right], \tag{7.97}$$

where $U'(\{\boldsymbol{r}\})$ contains the sum of $u(r_{ij})$ over all the pairs of particles, except for 1 and 2. We can therefore represent $\rho^{(2)}$ by means of a sum over diagrams, analogous to those we used for the virial expansion. By some additional considerations, one easily arrives at the following result:

$$\rho^2 g(r)\, \mathrm{e}^{u(r_{12})/k_{\mathrm{B}}T} = \left(\frac{V}{\lambda^3}\right)^2 \left[1 + \sum_{N=1}^{\infty} \frac{z}{N!} \int \prod_{i=3}^{N+2} \left(\frac{\mathrm{d}\boldsymbol{r}_3}{\lambda^3}\right) \prod_{(ij)}{}' (1+f_{ij})\right], \tag{7.98}$$

where $r = |\mathrm{r}_{12}| = |\boldsymbol{x}_1 - \boldsymbol{x}_2|$, and where the product runs over all the pairs of particles, except for 12. We can associate the usual diagrams with this expression, in which there are, however, now two marked points—that is to say, 1 and 2—over which we do not integrate. A direct line between 1 and 2, moreover, is not allowed. By summing all the decorations of vertices 1 and 2, and keeping everything else as it is, we obtain the ρ^2 factor, which is removed by passing from the $\rho^{(2)}$ to the $g(r)$. Therefore, $g(r)\mathrm{e}^{u(r)/k_{\mathrm{B}}T}$ is represented in terms of the diagrams containing the two marked points 1 and 2, without the decorations of vertices 1 and 2, and without the direct line between these points. In addition to the trivial diagram composed of points 1 and 2 in isolation, all the other diagrams connect these points. Let us attempt to classify these diagrams.

We will call **parallel connections** diagrams that can be separated into subdiagrams that have the following properties:

1. Each subdiagram connects 1 and 2.
2. Different subdiagrams are not connected to one another, except in 1 or 2.

A diagram that does not contain parallel connections is called an **irreducible channel**. So more generally, we will be able to write

$$g(r)\,e^{u(r)/k_BT} = 1 + \sum_k \frac{1}{k!}\left(\sum \text{irreducible channels}\right)^k. \tag{7.99}$$

The $1/k!$ comes, as usual, from the necessary correction to avoid multiple counting. Therefore,

$$\ln g(r) + \frac{u(r)}{k_BT} = \sum \text{irreducible channels}. \tag{7.100}$$

We can identify the articulation points in the irreducible channels, as we did earlier. If a specific vertex is an articulation point, we distinguish the vertex's decorations—in other words, the subdiagrams that are connected to the rest of the diagram only through that vertex and the subdiagrams that connect it to 1 and 2. By eliminating the decorations, we shift from the expansion in powers of z to that in powers of ρ.

Let us now consider the irreducible channels, free of decorations, that connect 1 and 2. We can distinguish the diagrams into two classes: the diagrams that contain articulation points, called **series** diagrams, and those that do not contain them, called **bridge** diagrams. We refer to the sum of all series diagrams as $\mathcal{S}$ and to the sum of all bridge diagrams as $\mathcal{B}$. We therefore have

$$\ln g(r) + \frac{u(r)}{k_BT} = \mathcal{S} + \mathcal{B}. \tag{7.101}$$

The first diagrams that contribute to $\mathcal{S}$ and to $\mathcal{B}$ are shown in figure 7.6. Our strategy is to identify classes of diagrams that can be resummed so as to express $\mathcal{S}$ and $\mathcal{B}$ as a function of the $g(r)$ itself. In this fashion, one obtains some equations for the $g(r)$ that one will be able to solve numerically if necessary.

Let us consider the expansion of $h(r) = g(r) - 1$. It is easy to see that one has

$$h(r_{12}) = \sum \text{diagrams that connect 1 and 2, without decoration.} \tag{7.102}$$

Among these diagrams, we can identify diagrams with articulation points and those without articulation points. We refer to the sum of diagrams without articulation points as $c(r_{12})$. One then has

$$h(r_{12}) = c(r_{12}) + \rho \int d^3r_i\, c(r_{1i})\, h(r_{i2}), \tag{7.103}$$

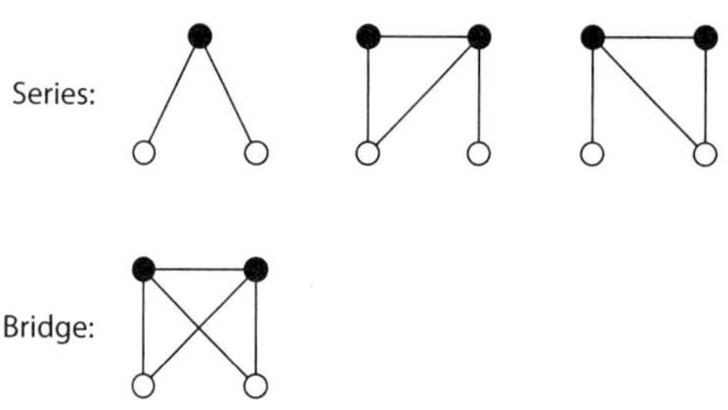

FIGURE 7.6. The first series and bridge diagrams.

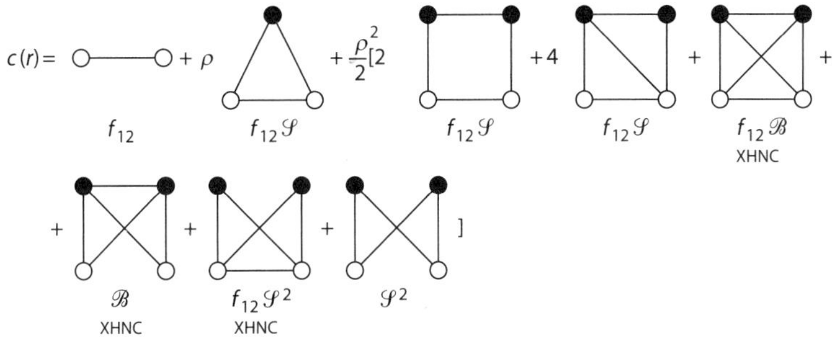

FIGURE 7.7. The first terms of the expansion of $c(r)$ in powers of ρ. The diagrams labeled XHNC are neglected in the HNC approximation.

where i is the first articulation point one finds coming from 1. If we compare this relation with equation (7.59), we see that the $c(r)$ that is so defined is the direct correlation function.

If we now assume that we can disregard the *bridge* diagrams, we see that the *series* diagrams reproduce only the second term of equation (7.103)—in other words, the $h(r) - c(r)$ term. We thus obtain the equation

$$\mathcal{S} \simeq h(r) - c(r), \tag{7.104}$$

in other words,

$$\ln g(r) + \frac{u(r)}{kt} = h(r) - c(r). \tag{7.105}$$

Since $h(r)$ and $c(r)$ can be expressed as a function of $g(r)$, this is a closed integral equation in $g(r)$. It is known as the **hypernetted chain equation** (HNC).

In order to understand the meaning of this approximation, let us consider the first terms of the expansion of $c(r)$ in figure 7.7. We can write

$$g_{\text{HNC}}(r) = e^{-u(r)/k_B T} e^{\mathcal{S}} = (1 + f_{12})\left(1 + \mathcal{S} + \frac{1}{2!}\mathcal{S}^2 + \cdots\right), \tag{7.106}$$

and therefore,

$$h_{\text{HNC}}(r) = f_{12}\left(1 + \mathcal{S} + \frac{1}{2!}\mathcal{S}^2 + \cdots\right) + \mathcal{S} + \frac{1}{2!}\mathcal{S}^2 + \cdots, \tag{7.107}$$

from which we obtain

$$c_{\text{HNC}}(r) = f_{12}\left(1 + \mathcal{S} + \frac{1}{2!}\mathcal{S}^2 + \cdots\right) + \frac{1}{2!}\mathcal{S}^2 + \cdots. \tag{7.108}$$

A more drastic approximation (but, strangely, a more effective one) is obtained by disregarding the terms that contain powers of $\mathcal{S}$ greater than 1 in this last equation. One thus obtains

$$c_{\text{PY}}(r) = f_{12}(1 + \mathcal{S}). \tag{7.109}$$

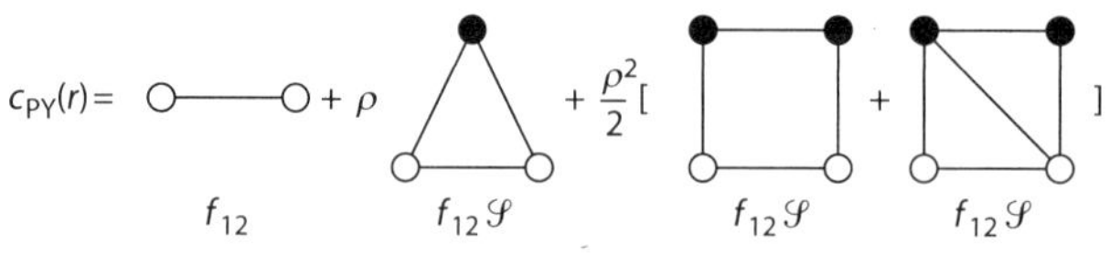

FIGURE 7.8. The first diagrams that contribute to the Percus-Yevick equation in an expansion in powers of ρ.

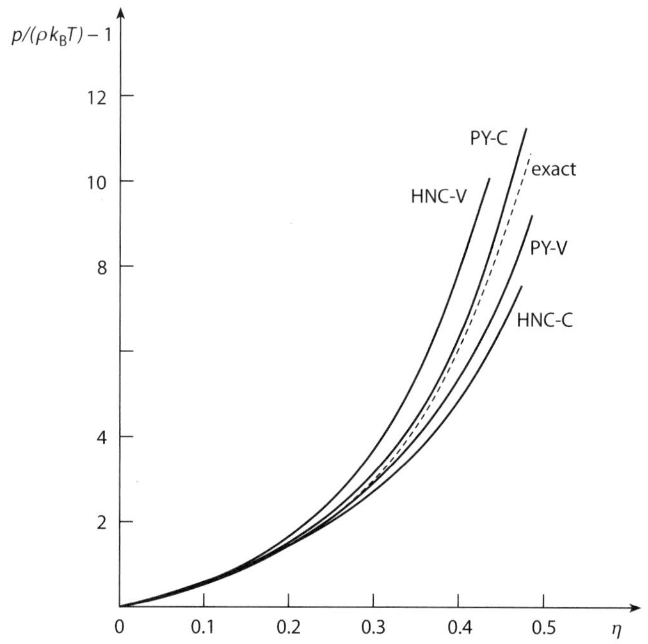

FIGURE 7.9. Equation of state of hard sphere gas, obtained by means of the hypernetted chain (HNC) approximation and the Percus-Yervick (PY) equation. The discontinuous line, with the "exact" designation, actually corresponds to the Carnahan-Starling extrapolation formula (7.96). The letters C and V refer to the two methods that allow one to obtain the equation of state from the $g(r)$: V refers to the pressure equation (7.21), while C refers to the compressibility equation (see exercise 7.1). On the x axis, the packing ratio $\eta = \pi \rho d^3/6$, is shown, where d is the sphere's diameter. On the y axis, the deviation with respect to the perfect gas equation—in other words, $p/\rho k_BT - 1$—is shown. Adapted from [Hans86], with permission.

From this, it follows that

$$h_{PY}(r) = c_{PY}(r) + \mathcal{S} = f_{12} + \mathcal{S} + f_{12}\mathcal{S}, \tag{7.110}$$

and therefore,

$$g_{PY}(r) = 1 + f_{12} + \mathcal{S} + f_{12}\mathcal{S} = (1 + f_{12})(1 + \mathcal{S}) = e^{-u(r)/k_BT}[g_{PY}(r) - c_{PY}(r)]. \tag{7.111}$$

This equation is known as the **Percus-Yevick equation**—surprisingly, it is analytically solvable in the case of the hard spheres. In figure 7.8, we show the first diagrams that contribute to the Percus-Yevick equation. In figure 7.9, I reproduce the results of the various approximations for the hard sphere equation of state.

In the next two subsections, we show that if the interaction potential satisfies certain stability conditions, the virial expansion (7.62) converges for $\rho < R_C$, where R_C depends on the properties of the potential and on the temperature. This analysis is at a somewhat higher mathematical level than the rest of this chapter, and these sections can be omitted in a first reading. They do provide, however, a good example of how rigorous results in statistical mechanics have been derived.

7.3.3 *Integral Equations for the Reduced Densities*

The convergence of the virial expansion (7.62) and the fugacity expansion (7.76) of the equation of state can be analyzed by showing that the infinite sequence of reduced densities $\rho = (\rho^{(\ell)}(x_1, \dots, x_\ell))$ satisfies a system of linear nonhomogeneous integral equations whose kernel is small if the fugacity (and therefore the density ρ) is small. This allows for an explicit estimate of the convergence radius of the expansions. In this section, we will derive the system of equations, while in the next, we will discuss the analyticity of its solution, and hence the convergence of the expansion. The following considerations require some concepts of functional analysis. Although I have tried to explain them as we need them, the discussion remains at a somewhat higher level than the rest of this chapter. These results are reported in [Ruel69, chapter 4] , where one can find directions to the original literature.

It is more convenient to carry out the analysis in the grand canonical ensemble. The reduced density $\rho^{(\ell)}(x_1, \dots, x_\ell)$ is defined as the joint probability distribution function of obtaining a configuration in which there is one particle in x_1, one in $x_2, \dots,$ and one in x_ℓ. Then, $\rho^{(\ell)}$ is expressed as a function of the fugacity z by

$$\rho^{(\ell)}(x_1, \dots, x_\ell) = Z_{\mathrm{GC}}^{-1}\left(\frac{z}{\lambda^d}\right)^{\ell} \sum_{n=0}^{\infty} \frac{z^n}{n!} \int \prod_{j=\ell+1}^{\ell+n} \left(\frac{\mathrm{d}x_j}{\lambda^d}\right) \exp\left[-\frac{U^{(\ell+n)}(x)}{k_B T}\right]. \tag{7.112}$$

Here, Z_{GC} is the grand canonical partition function given by equation (7.7), and $U^{(\ell+n)}(x)$ is the potential energy of a configuration in which there are $\ell + n$ particles, one at each of the points denoted by $x_1 (i = 1, \dots, \ell + n)$. We will assume until further notice that all coordinates are restricted in a region of volume V, and we will dispense with the one-particle potential $v(r)$. We will also introduce the more compact notations

$$(x)_\ell = (x_1, \dots, x_\ell), \tag{7.113}$$

$$(x)'_{\ell-1} = (x_2, \dots, x_\ell). \tag{7.114}$$

We can then write, e.g.,

$$\rho^{(\ell)}(x_1, \dots, x_\ell) = \rho(x)_\ell, \qquad U^{(\ell)}(x_1, \dots, x_\ell) = U(x)_\ell. \tag{7.115}$$

Let us first consider $\rho(x)_1$. We have

$$\rho(x)_1 = Z_{\mathrm{GC}}^{-1}\left(\frac{z}{\lambda^d}\right) \sum_{n=0}^{\infty} \frac{z^n}{n!} \int \prod_{j=2}^{1+n} \left(\frac{\mathrm{d}x_j}{\lambda^d}\right) \exp\left[-\frac{U(x)_{1+n}}{k_B T}\right]. \tag{7.116}$$

Let us isolate, in the interactions $U(x)_{1+n}$ that appear in equation (7.116), those that involve the particle located at x_1 from the interactions among the other particles:

$$\begin{aligned}\exp\left[-\frac{U(x)_{1+n}}{k_\mathrm{B}T}\right] &= \prod_{j=2}^{1+n} \mathrm{e}^{-u(x_1-x_j)/k_\mathrm{B}T}\exp\left[-\frac{U(x)'_n}{k_\mathrm{B}T}\right]\\ &= \prod_{j=2}^{1+n}(1+f_{1j})\exp\left[-\frac{U(x)'_n}{k_\mathrm{B}T}\right],\end{aligned} \tag{7.117}$$

where we have introduced the Mayer function f_{ij} defined in (7.66). Substituting this equation in (7.116) and collecting terms that contain the same f_{ij} factors, we obtain

$$\begin{aligned}\rho(x)_1 &= Z_{\mathrm{GC}}^{-1}\left(\frac{z}{\lambda^d}\right)\sum_{n=0}^{\infty}\sum_{s=0}^{n}\frac{z^n}{s!\,(n-s)!}\int\prod_{j=1}^{s}\left(\frac{\mathrm{d}r_j}{\lambda^d}\right)K(x_1,r_s)\\ &\quad\times\int\prod_{k=s+1}^{n}\left(\frac{\mathrm{d}r_k}{\lambda^d}\right)\exp\left[-\frac{U(r)_n}{k_\mathrm{B}T}\right].\end{aligned} \tag{7.118}$$

where we have set $r_j = x_{1+j}$, and we have defined

$$K(x_1,(r)_s) = \prod_{j=1}^{s}\left[\mathrm{e}^{-u(x_1-r_j)/k_\mathrm{B}T}-1\right]. \tag{7.119}$$

Taking into account equation (7.112), we obtain

$$\rho(x)_1 = \left(\frac{z}{\lambda^d}\right)\left[1+\sum_{s=1}^{\infty}\frac{1}{s!}\int\prod_{j=1}^{s}\mathrm{d}r_j\,K(x_1,(r)_s)\,\rho(r)_s\right]. \tag{7.120}$$

In the same way, one obtains a similar equation for the higher order reduced densities $\rho(x)_\ell$ with $\ell > 1$. We single out one particle and take into account the interaction potential between this particle and the other $\ell - 1$ particles. Defining

$$W^1(x)_\ell = \sum_{j=2}^{\ell} u(x_1-x_j), \tag{7.121}$$

we obtain

$$\begin{aligned}\rho(x)_\ell &= \left(\frac{z}{\lambda^d}\right)\exp\left[-W^1(x)_\ell/k_\mathrm{B}T\right]\\ &\quad\times\left[\rho(x)'_{\ell-1}+\sum_{s=1}^{\infty}\frac{1}{s!}\int\prod_{j=1}^{s}\mathrm{d}r_j\,K(x_1,(r)_s)\,\rho((x)'_{\ell-1},(r)_s)\right].\end{aligned} \tag{7.122}$$

The integral equations (7.120) and (7.122) are the **Kirkwood-Salzburg equations**. They form a linear inhomogeneous system for the sequence $\rho = (\rho(x)_\ell)$ of reduced densities, with $\ell \geq 1$. They may be written in the form

$$\rho = z(\alpha + \mathsf{K}\rho), \tag{7.123}$$

where the sequence $\alpha = (\alpha(x)_\ell)$ is defined by

$$\alpha(x)_1 = \lambda^{-d}, \tag{7.124}$$

$$\alpha(x)_\ell = 0, \qquad \text{for } \ell > 1, \tag{7.125}$$

and the operator K acting on a sequence of functions $\phi = (\phi(x)_\ell)$ is defined by

$$(\mathsf{K}\phi)(x)_1 = \lambda^{-d}\sum_{s=1}^{\infty}\frac{1}{s!}\int\prod_{j=1}^{s}\mathrm{d}r_j\,K(x_1,(r)_s)\,\phi(r)_s; \tag{7.126}$$

$$(\mathsf{K}\phi)(x)_\ell = \lambda^{-d}\exp\left[-W^1(x)_\ell/k_BT\right] \times\left[\phi(x)'_{\ell-1} + \sum_{s=1}^{\infty}\frac{1}{s!}\int\prod_{j=1}^{s}\mathrm{d}r_j\,K(x_1,(r)_s)\phi((x)'_{\ell-1}(r)_s)\right], \tag{7.127}$$

where the last equation holds for $\ell > 1$.

7.3.4 Convergence of the Fugacity and Virial Expansion

We will see that a sufficient condition for the existence of a finite convergence region of the expansions is that the pair potential $u(x)$ is regular. A pair potential $u(x)$ is regular if it is **stable**—i.e., for any given set $(x)_\ell$ of ℓ points, one has

$$U(x)_\ell = \sum_{1\le i<j\le \ell} u(x_i - x_j) \ge -\ell B \tag{7.128}$$

for some negative constant $-B$, and the integral of its Mayer function over all space is finite at all temperatures T:

$$C(T) = \int \mathrm{d}x\left(\mathrm{e}^{-u(x)/k_BT} - 1\right) < +\infty. \tag{7.129}$$

In a stable potential, it is impossible for particles to cluster in such a way as to make the potential energy per particle arbitrarily large and negative, which could happen, e.g., if a soft core is surrounded by an attractive potential region. A good recipe for stable potentials is that of combining a repulsive hard core with a bounded attractive interaction that decreases sufficiently fast at infinity.

Exercise 7.4 Show that if $C(T)$ is finite for *any* temperature T, it is also finite for *all* temperatures $0 < T < \infty$.

We need to determine which conditions the system of equations (7.123) has a solution, analytic in z, in a suitable functional space of sequences. We expect the reduced density $\rho(x)_\ell$ to be of the order of magnitude of ρ^ℓ, at least when the points $(x)_\ell$ are far apart from one another. Thus, we will define its magnitude by a set of norms depending on a factor ξ that will be adjusted when needed, which defines the order of magnitude of $\rho(x)_\ell$ as a function of ℓ. Given a sequence $\phi = (\phi(x)_\ell, (\ell \ge 1)$, and a positive number ξ, we can define the norm $\|\phi\|_\xi$ by[1]

$$\|\phi\|_\xi = \sup_\ell\left[\xi^{-\ell}\sup_{(x)\ell}|\phi(x)_\ell|\right]. \tag{7.130}$$

If this norm is finite for a value ξ^* of ξ, then it will be finite for all $\xi > \xi^*$. We will now show that if u is regular, it is possible to redefine the operator K in such a way that its norm

$$\|\mathsf{K}\|_\xi = \sup_{\phi:\|\phi\|_\xi=1}\|\mathsf{K}\phi\|_\xi \tag{7.131}$$

is finite for a given ξ.

[1] Note that the sup within parentheses in this expression should be understood as an *essential supremum*, i.e., as the smallest number larger than any value reached by $\phi(x)_l$, except possibly for a set of measure zero.

Given a sequence ϕ of norm $\|\phi\|_\xi$, we have

$$|(\mathsf{K}\phi)(x)_\ell| \leq \|\phi\|_\xi \exp[\xi C(T)] (e^{B/k_BT}\xi)^{\ell-1}. \tag{7.132}$$

Here, we have exploited (7.129) and the fact that, under our hypotheses, one has

$$W^1(x)_\ell > -(\ell-1)B. \tag{7.133}$$

The dependence on ℓ of (7.132) makes it awkward to use. This dependence arises from the fact that we cannot rule out a situation in which our singled-out point x_1 interacts attractively with a large number of points r_s, in equation (7.128). However, there is nothing special about the point x_1, we could choose any other point i instead, and a factor $W^i(x)_\ell$ would appear, instead of $W^1(x)_\ell$, in this equation. Now, it is easy to see that if u is regular, one has

$$\sum_{i=1}^{\ell} W^i(x)_\ell = \sum_{i\neq j}{}' u(x_i - x_j) = 2U(x)_\ell \geq -2\ell B.$$

Thus it is always possible to choose i such that $W^i(x)_\ell \geq -2B$. Therefore, we can redefine K in such a way that this inequality holds, and we obtain

$$|(\mathsf{K}\phi)(x)_\ell| \leq \|\phi\|_\xi \exp[\xi C(T)] e^{2B/k_BT} \xi^{\ell-1}. \tag{7.134}$$

which implies that

$$\|\mathsf{K}\|_\xi \leq \exp[\xi C(T)] e^{2B/k_BT} \xi^{-1}. \tag{7.135}$$

If z (considered now as a complex number) is such that $|z|\,\|\mathsf{K}\|_\xi < 1$, it is possible to define the operator $\mathsf{R}(z) = (1 - z\mathsf{K})^{-1}$ as a power series in z, which converges according to the norm (7.130):

$$(1 - z\mathsf{K})^{-1} = 1 + z\mathsf{K} + z^2\mathsf{K}\circ\mathsf{K} + \cdots \tag{7.136}$$

Here, 1 is the identity operator, and $\mathsf{A}\circ\mathsf{B}$ represents the composition of operators A and B.

Indeed, given the sequence ϕ of norm $\|\phi\|_\xi$, we have

$$\begin{aligned} |[\mathsf{R}(z)\phi](x)_\ell| &\leq \|\phi\|_\xi \xi^\ell \left(1 + |z|\|\mathsf{K}\|_\xi + |z|^2\|\mathsf{K}\|_\xi^2 + \cdots\right) \\ &= \frac{\|\phi\|_\xi \xi^\ell}{1 - |z|\|\mathsf{K}\|_\xi}. \end{aligned} \tag{7.137}$$

Thus the series (7.136) applied to any sequence ϕ of finite norm, converges absolutely and this convergence is uniform in any region such that $|z|\,\|\mathsf{K}\|_\xi = M < 1$. In view of equation (7.135), this corresponds to

$$|z| < \xi \exp[-\xi C(T)] e^{-2B/k_BT}. \tag{7.138}$$

The right-hand side is largest for $\xi = 1/C(T)$. We have derived, therefore, a bound on the radius of convergence z_C of the series (7.136):

$$z_C \geq \frac{e^{-2B/k_BT-1}}{C(T)}. \tag{7.139}$$

We can then express the sequence ρ of reduced densities by solving equation (7.123) via

$$\rho = z\mathsf{R}(z)\alpha. \tag{7.140}$$

In this equation, we can safely take the infinite-volume limit by simply extending the integrals appearing in the expression of K to the whole space; convergence is assured by the regularity condition (7.129). Since $\rho = \rho(\boldsymbol{x})_1$ in the infinite-volume system, the bound (7.139) holds for the expansion (7.83) of the density ρ in powers of the fugacity z. On the other hand, we have from (7.78), dividing both sides by V/z:

$$\frac{z}{V}\frac{\partial \ln Z_{\mathrm{GC}}}{\partial z} = \rho. \tag{7.141}$$

Integrating this equation, taking into account that $\rho(z=0)=0$, we have

$$\int_0^z \mathrm{d}z'\rho(z') = \lim_{V\to\infty}\frac{\ln Z_{\mathrm{GC}}}{V} = \frac{p}{k_{\mathrm{B}}T}. \tag{7.142}$$

Therefore, the same bound holds for the expansion of p in powers of z.

It is now possible to convert the bound (7.139) to a bound on the density for the virial expansion of the equation of state (7.62) Taking the derivative of (7.62)with respect to z, we obtain

$$\frac{1}{k_{\mathrm{B}}T}\frac{\mathrm{d}p}{\mathrm{d}z} = \frac{\mathrm{d}\rho}{\mathrm{d}z}\sum_{\ell=1}^{\infty}\ell B_\ell(T)\rho^{\ell-1}. \tag{7.143}$$

The theory of residues tells us that if $f(z)$ is a function of the complex variable z, analytic in a circle of radius z_{C} around the origin, then

$$\frac{1}{\ell!}\left.\frac{\mathrm{d}^\ell f(z)}{\mathrm{d}z^\ell}\right|_{z=0} = \oint_C \frac{\mathrm{d}z}{2\pi\mathrm{i}}\frac{f(z)}{z^{\ell+1}}, \tag{7.144}$$

where C is a circular path of radius smaller than z_{C}, which circles counterclockwise around the origin. Applying this result to (7.143) and taking into account (7.142), we obtain

$$\begin{aligned} B_\ell(T) &= \frac{1}{k_{\mathrm{B}}T}\oint_C \frac{\mathrm{d}\rho}{2\pi\mathrm{i}}\frac{\mathrm{d}z}{\mathrm{d}\rho}\frac{\mathrm{d}p}{\mathrm{d}z}\frac{1}{\ell\rho^\ell} \\ &= \frac{1}{k_{\mathrm{B}}T}\oint_{C'}\frac{\mathrm{d}z}{2\pi\mathrm{i}}\frac{\mathrm{d}p}{\mathrm{d}z}\frac{1}{\ell\rho^\ell} = \oint_{C'}\frac{\mathrm{d}z}{2\pi\mathrm{i}}\frac{1}{\ell\rho^{\ell-1}}, \end{aligned} \tag{7.145}$$

where C' is a circle of radius smaller than z_{C}. On the other hand, we have

$$\begin{aligned} \left|\rho - \frac{z}{\lambda^d}\right| &= |\rho(z)_1 - z\alpha(x)_1| \le \xi|z|\,\|\mathsf{R}(x)-1\|_\xi\|\alpha\|_\xi \\ &\le \xi|z|\,\|\alpha\|_\xi\sum_{\ell=1}^{\infty}|z|^\ell\|\mathsf{K}\|_\xi = \xi\lambda^{-d}\frac{|z|^2\|\mathsf{K}\|_\xi}{1-|z|\,\|\mathsf{K}\|_\xi} \\ &= \frac{\xi\lambda^{-d}|z|^2}{\|\mathsf{K}\|_\xi^{-1}-|z|}. \end{aligned} \tag{7.146}$$

Therefore, if $|z| \le \|\mathsf{K}\|_\xi^{-1}$, one has

$$|\rho| \geq \frac{|z|}{\lambda^d} - \frac{\xi \lambda^{-d} |z|^2}{\|\mathsf{K}\|_{\xi}^{-1} - |z|} \leq \lambda^{-d}(3 - 2\sqrt{2})\mathrm{e}^{-2B/k_BT-1}, \tag{7.147}$$

corresponding to $|z| = (2 - 2^{-1/2})\, z_C$, which maximizes the right-hand side. It follows that the virial expansion (7.62) converges for

$$|\rho| \leq \lambda^{-d}(3 - 2\sqrt{2})\mathrm{e}^{-2B/k_BT-1}, \tag{7.148}$$

and the virial coefficients $B_\ell(T)$ are bounded by

$$|B_\ell(T)| \leq \frac{1}{\ell}\left(\lambda^d \frac{\mathrm{e}^{-2B/k_BT-1}}{3 - 2^{-3/2}}\right)^{\ell-1}. \tag{7.149}$$

These estimates can be slightly improved—the convergence radius is at least R_C and $|B_\ell(T)| \leq R_C^{-\ell+1}/\ell$, where

$$R_C = \lambda^{-d}\theta(\mathrm{e}^{2B/k_BT} + 1)C(T), \tag{7.150}$$

with $\theta = 0.28952$.

7.4 Perturbation Theory

Another method to evaluate the behavior of liquids is based on an observation by Zwanzig [Zwan54]—namely, that the properties of a liquid are, as a first approximation, determined by the hard core, and only slightly modified by the potential's attractive tail. One can therefore attempt to express a fluid's properties by dividing the interaction potential u into a *known* part u_0 and a *perturbation* u_1, which we assume is small, and then calculate the deviations from the behavior with u_0 alone in powers of the perturbation. It is clear that the results we obtained suggest that we take a rigid sphere potential as our u_0, so as to be able to exploit the exceptional efficacy of the Carnahan-Starling formula (7.96). Given that the rigid sphere potential is infinite, however, this means that the u_1 perturbation is not, generally, small.

Barker and Henderson [Bark76] have shown that one can solve this problem by choosing a hard sphere diameter d that is dependent on temperature. Their prescription is

$$d = \int_0^{r_0} \mathrm{d}r \left[1 - \mathrm{e}^{-u_0(r)/k_BT}\right], \tag{7.151}$$

where this time $u_0(r)$ is the kernel of the Lennard-Jones potential—in other words,

$$u_0(r) = \theta(r_0 - r)\, u(r), \tag{7.152}$$

in which $\theta(x)$ is the Heaviside step function. Practically speaking, a system of *soft spheres* with potential $u_0(r)$ behaves like a system of hard spheres with a diameter d given by equation (7.151).

With this premise, we see how we can begin the calculation in perturbation theory. Let us consider the canonical partition function

$$Z_N = \frac{1}{N!}\int \prod_{i=1}^{N}\left(\frac{\mathrm{d}r_i}{\lambda_B^3}\right) \exp\left\{-\frac{1}{k_BT}\sum_{(ij)}[u_0(r_{ij}) + \lambda u_1(r)]\right\}. \tag{7.153}$$

We have introduced a perturbation parameter λ (which we will be set at 1 at the end of the calculations) and renamed the thermal de Broglie wavelength λ_B to avoid confusion.

We expand $\ln Z_N$ in powers of λ:

$$\ln Z_N = \ln Z_N(\lambda = 0) + \lambda \frac{\partial}{\partial \lambda} \ln Z_N + O(\lambda^2), \tag{7.154}$$

where

$$\frac{\partial \ln Z_N}{\partial \lambda} = -\frac{N(N-1)}{2V^2 k_B T} \left(\lambda_B^3\right)^2 \langle u_1(r_{12}) \rangle_0, \tag{7.155}$$

and where $\langle \cdots \rangle_0$ is the average over the unperturbed system ($\lambda = 0$). But one has

$$\langle u_1(r_{12}) \rangle_0 = V \int d^3 r\, u_1(r) g^{(0)}(r), \tag{7.156}$$

where $g^{(0)}(r)$ is the $g(r)$ of the unperturbed system, which can be expressed in terms of the $g(r)$ of hard spheres gas. From these expressions, we can obtain an expression for the Helmholtz free energy, for example,

$$\begin{aligned} F(T, V, N) &= -k_B T \ln Z_N = -k_B T \ln Z_N(\lambda = 0) - k_B T\, \lambda \frac{\partial \ln Z_N}{\partial \lambda}\bigg|_{\lambda=1} + \cdots \\ &= F_0(T, V, N) + N \frac{\rho}{2} \int d^3 r\, u_1(r) g^{(0)}(r) + \cdots, \end{aligned} \tag{7.157}$$

where $F_0(T, V, N)$ is the free energy of the unperturbed system. In figure 7.10, I show the $g(r)$ of a Lennard-Jones fluid calculated for $\rho = 0.85\ r_0^{-3}$ and $k_B T = 0.72\,\varepsilon$ (close to the triple

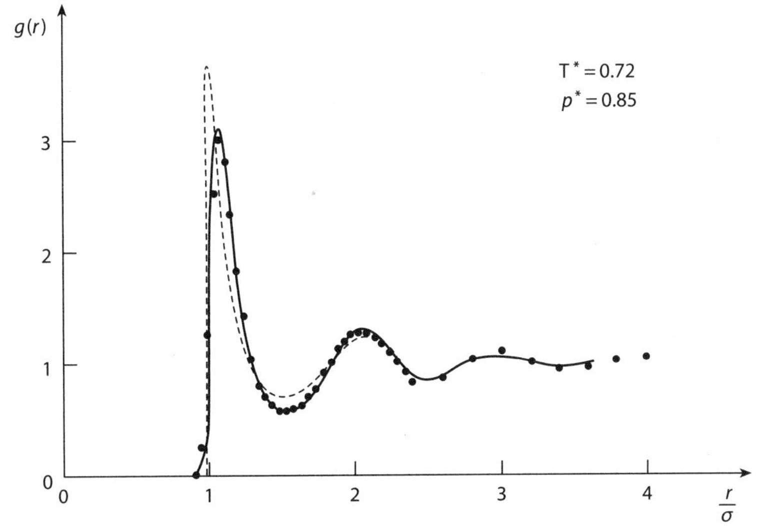

FIGURE 7.10. Pair distribution for a Lennard-Jones fluid for $\rho = 0.85\ r_0^{-3}$ and $k_B T = 0.72\,\varepsilon$ (close to the triple point), at zero (dashed line) and first orders (solid line) in perturbation theory, compared with Verlet's numerical simulations. Adapted from [Bark76], with permission.

point) at the first order in perturbation theory, compared to the results of numerical simulations. The agreement is fairly good and shows that the ideas behind the approximations are not too much off the mark.

7.5 Liquid Solutions

Let us now consider a fluid composed of a prevalent chemical species (the **solvent**) in which a small quantity of another chemical species is dissolved (the **solute**). If the solute's concentration is fairly small, we will be able to disregard the direct interactions between the solute's molecules (but obviously not those between solute and solvent). Let us denote the solute's number of particles and chemical potential by n and μ, respectively, and those of the solvent by N and μ_s. The Gibbs free energy of the solution $G(T, p, N, n)$ will contain a contribution proportional to n, as well as a contribution due to the entropy change, which derives from the fact that of the $(N+n)$ particles present in the solution, N are of one type and n of the other. This in entropy change (called **entropy of mixing**) is equal to

$$\Delta S = k_B \ln \frac{(N+n)!}{N!n!} \simeq -k_B \left(N \ln \frac{N}{N+n} + n \ln \frac{n}{N+n} \right) \simeq -k_B \left(-n + n \ln \frac{n}{N} \right). \quad (7.158)$$

Since G contains a term $-TS$, we obtain

$$G(T,p,N,n) = G(T,p,N,n=0) + n\psi(T,p,N) + nk_B T \ln \frac{n}{eN}. \quad (7.159)$$

By taking the derivative of equation (7.159) with respect to N and n, we obtain the corresponding chemical potentials:

$$\mu_s = \left.\frac{\partial G}{\partial N}\right)_{T,p,n} = \mu_s^0 - xk_B T, \quad (7.160)$$

$$\mu = \left.\frac{\partial G}{\partial n}\right)_{T,p,N} = \psi + k_B T \ln x, \quad (7.161)$$

where $x = n/N$ is the molar concentration of the solute, and μ_s^0 is the chemical potential of the pure solvent.

Let us now consider the equilibrium of two regions of the solvent separated by a semipermeable membrane, which will allow the solvent, but not the solute, to pass. In this situation, the chemical potential of the solvent (but not of the solute) must be equal on both sides of the membrane. We thus obtain

$$\mu_s(T,p_1,x_1) = \mu_s^0(T,p_1) - k_B T\, x_1 = \mu_s(T,p_2,x_2) = \mu_s^0(T,p_2) - k_B T\, x_2. \quad (7.162)$$

Expanding to the first order in the difference $\delta p = p_2 - p_1$, we obtain

$$\mu_s^0(T,p_2) = \mu_s^0(T,p_1) + \Delta p \left.\frac{\partial \mu_s^0}{\partial p}\right)_T = \mu_s^0(T,p_1) + \Delta p v, \quad (7.163)$$

where $v = V/N$ is the molecular volume of the pure solvent. (We have applied the Gibbs-Duhem equation.) Therefore,

$$\Delta p = k_{\mathrm{B}}T\,\frac{x_2 - x_1}{v}. \tag{7.164}$$

More particularly, if on one of the two sides we have pure solvent, we obtain

$$\Delta p = k_{\mathrm{B}}T\,\frac{x_1}{v} = k_{\mathrm{B}}T\,\frac{n}{Nv} = k_{\mathrm{B}}T\,\frac{n}{V}. \tag{7.165}$$

The osmotic pressure for a diluted solution therefore follows the same law as perfect gases. This result is known as **van 't Hoff's law**.

7.5.1 *Ionic Solutions*

Van 't Hoff's law is not valid if the interactions between the solute's molecules are long range. In this case, the free energy exhibits singular terms in the concentration of the solute. In practice, the most important case is that of ionic solutions. In order to clarify our ideas, let us assume that our salt is formed by a single cation and a single anion, and each one is monovalent, and that we denote the concentration of the cations (anions) by ν_+ (ν_-), both being equal to the molar concentration of the salt. We now show that the effect of these free charges is to *screen* the Coulomb interactions in the solution beyond a characteristic length called the **Debye length.**

Let us now consider a charge q placed in the origin: it produces an electric field, represented by an electrostatic potential $\phi(\mathbf{r})$, which polarizes the solution. In order to evaluate this effect, let us consider that if the local charge density in a generic point $\mathbf{r}$ is equal to $\rho(\mathbf{r})$, then the potential $\phi(\mathbf{r})$ is a solution of the Poisson equation:

$$-\nabla^2\phi = \frac{\rho}{\epsilon_0\epsilon_{\mathrm{r}}}. \tag{7.166}$$

On the other hand, if the electrostatic potential is equal to $\phi(\mathbf{r})$, the local density of ions with a $\pm e$ charge will be proportional to the Boltzmann factor:

$$\mathrm{e}^{\mp e\phi(r)/k_{\mathrm{B}}T}.$$

Since both the Poisson equation and the Boltzmann factor are present, this theory (which is due to Debye and Hückel [Deby23]) is also known as the **Poisson-Boltzmann theory**.

If the concentration ν is sufficiently small, to the extent that the $\phi(r)/k_{\mathrm{B}}T$ factor can be considered small throughout the solution (possibly leaving a microscopic volume around the point charges), we can expand the Boltzmann factor to the first order in $\phi(r)/k_{\mathrm{B}}T$, and obtain

$$\nu_\pm = \nu_\pm^0\left[1 \mp \frac{e\phi(\mathbf{r})}{k_{\mathrm{B}}T}\right]. \tag{7.167}$$

But then the local charge density is equal to

$$\rho(r) = e[\nu_+ - \nu_-] = 2\nu_0 e^2\,\frac{\phi(r)}{k_{\mathrm{B}}T}. \tag{7.168}$$

We thus obtain the equation for ϕ:

$$\nabla^2 \phi = \frac{\phi}{\ell_\mathrm{D}^2}, \tag{7.169}$$

where ℓ_D is the **Debye screen length**:

$$\ell_\mathrm{D}^2 = \frac{\epsilon_0 \epsilon_\mathrm{r} k_\mathrm{B} T}{2\nu_0 e^2}. \tag{7.170}$$

The solution of this equation, with a charge q at the origin, is given by

$$\phi_q(r) = \frac{q}{4\pi\epsilon_0\epsilon_\mathrm{r}} \frac{\mathrm{e}^{-r/\ell_\mathrm{D}}}{r}. \tag{7.171}$$

We therefore see that Coulomb's law is valid only for $r\ell_\mathrm{D} \ll 1$, after which the Coulomb interactions are screened. By substituting this expression in equation (7.168), we obtain the charge density $\rho(r)$ at a distance r from the origin:

$$\rho(r) = -2\nu_0 e^2 \frac{\phi(r)}{k_\mathrm{B}T} = -\frac{q\nu_0 e^2}{2\pi\epsilon_0\epsilon_\mathrm{r}} \frac{\mathrm{e}^{-r/\ell_\mathrm{D}}}{r} = -\frac{q}{4\pi\ell_\mathrm{D}^2} \frac{\mathrm{e}^{-r/\ell_\mathrm{D}}}{r}. \tag{7.172}$$

By expanding the expression of the potential around an ion of charge q for small values of the r/ℓ_D, we obtain

$$\phi(r) \simeq \frac{q}{4\pi\epsilon_0\epsilon_\mathrm{r}} \left(\frac{1}{r} - \frac{1}{\ell_\mathrm{D}} \right). \tag{7.173}$$

The first term represents the potential produced by the ion in question, while the second term represents the effect of the interactions with all the other ions in solution. The ion's potential energy is therefore given by $-q^2/(4\pi\epsilon_0\epsilon_\mathrm{r}\ell_\mathrm{D})$. The total electrostatic energy U_el will be obtained by summing this expression over all the ions present in the solution. One thus obtains

$$U = -\frac{2V\nu_0 e^2}{4\pi\epsilon_0\epsilon_\mathrm{r}\ell_\mathrm{D}} = -\frac{e^3 V}{4\pi} \left(\frac{2\nu_0}{\epsilon_0\epsilon_\mathrm{r}} \right)^{3/2} (k_\mathrm{B}T)^{-1/2}. \tag{7.174}$$

By integrating the thermodynamic relation

$$\frac{E}{T^2} = -\frac{\partial}{\partial T} \frac{F}{T},$$

we obtain the correction that must be added to the Helmholtz free energy:

$$F_\mathrm{el} = -\frac{e^3 V}{3\pi} \left(\frac{2\nu_0}{\epsilon_0\epsilon_\mathrm{r}} \right)^{3/2} (k_\mathrm{B}T)^{-1/2}. \tag{7.175}$$

We can now substitute ν_0 with its expression $\nu_0 = N_\mathrm{salt}/V$. We thus obtain

$$F_\mathrm{el}(T, V, N_\mathrm{salt}) = -\frac{e^3}{3\pi} \left(\frac{2N_\mathrm{salt}}{\epsilon_0\epsilon_\mathrm{r}} \right)^{3/2} (Vk_\mathrm{B}T)^{-1/2} \tag{7.176}$$

By taking the derivative F with respect to V, we obtain the electrostatic pressure of the ions, which is therefore proportional to $V^{-3/2}$ and to $T^{-1/2}$.

Exercise 7.5 Generalize this result to the case of a salt with composition $\sum_i n_i X_i$, where the ion X_i has a charge z_i. (One obviously has $\sum_i n_i z_i = 0$ due to neutrality.)

Let us note that the results we obtained also exhibit some paradoxical aspects if compared with the Ornstein-Zernike theory of $g(r)$. Because of the screening effect of the Coulombian potential, the correlation function $h(r)$ actually becomes short range (it vanishes over distances on the order of the Debye length), while the pair potential and the direct correlation are long range.

Recommended Reading

The "classic" reference is J. P. Hansen and I. R. MacDonald, *Theory of Simple Liquids*, 2nd ed., London: Academic Press, 1986. I have also followed D. L. Goodstein, *States of Matter*, New York: Dover, 1985.

8 Numerical Simulation

Meditate before thinking!

—Stanislaw J. Lec

8.1 Introduction

The invention of the computer has had a very significant impact on statistical mechanics. One the one hand, it has allowed us to tackle problems that required complicated calculations, and on the other, it has provided an important contribution to the understanding of fundamental concepts. In this chapter, we will briefly discuss some methods and problems related to the use of the computer in statistical mechanics.

We must distinguish between two ways of using the computer in theoretical physics: **calculation** and **simulation**. In calculation, the computer is used to make the contents of a theory explicit, undertaking calculations that would be too long or tedious to be reliably performed manually. When faced with a method of perturbation theory, for instance, such as high-temperature expansions, it is possible to entrust the evaluation of higher order contributions to a computer. A classic example of the use of a computer for these purposes is Haken and Appell's proof of the four-color theorem [Appe77]. In this case, the theory predicted that the theorem would be proven if a certain class of configurations of regions in the plane, which possessed a certain set of properties, were to be found. If a specific class of configurations did not possess some of these properties, the theory provided a method to modify it so as to make its satisfiability more probable. Both the generation of classes of configurations and the verification of the validity of the properties being sought, however, rapidly became too complicated to be performed manually. These operations were therefore assigned to the computer, which after a very long series of attempts and calculations, did identify the required class of configurations.

The spread of software for algebraic calculus, and also for integrated algebraic and numeric calculus, is an additional aspect of the development of the calculus function in theoretical physics. In all of these activities, the computer is an additional tool for the development and the verification of theory. It is beyond the scope of my reflections here to discuss the advantages and the problems that result from this use of the computer.

On the other hand, the development of computers has led to the appearance of a new form of scientific research: **simulation**, which is sometimes called the *third branch* of science [Pool92], since it is placed at the same level as theory and experiment. In a simulation, the evolution equations of a system are defined to such a detail that a computer can make its behavior explicit, and one looks to see what takes place. From a certain point of view, it is an experiment (one does in fact talk about *numerical experiments*). On the other (and this is where the most subtle dangers are hidden), the simulated system is defined exclusively by means of its evolution equations, and it remains to be seen whether and to what extent the simulated system is related to the real system.

In certain situations, the border between calculation and simulation is fairly difficult to define. Let us suppose that we want to calculate the behavior of the planets in our solar system for a certain period of time. As is well known, it is not possible to find an analytical solution to the problem of three or more gravitating bodies—however, many refined techniques for the approximate solution of the equations of motion in celestial mechanics exist. The power of computers allows for the problem's solution, at least for not too long time intervals, by solving the Newton equations of motion for the planets and the sun by brute force. In this case, we can say that we are "simulating" the evolution of the planets or, just as easily, that we are simply calculating the solution of the Newton equations.

These ambiguities do not occur for the systems of statistical mechanic—in this case, the distinction between simulation and calculation activities is usually fairly clear. So *theory* (and experiment) actually relates to macroscopic systems, which are well described by the thermodynamic limit, while *simulation* necessarily concerns systems of limited size—usually very small. The most serious technical problem one has to face, therefore, when interpreting simulation data consists of the extrapolation from finite systems to "infinite" systems in the thermodynamic limit.

Numerical experiments, on the other hand, also exhibit a number of advantages. Once the model has been defined, the calculation does not contain approximations and is perfectly explicit. This is an advantage compared to the analytic treatment, which usually requires intermediate hypotheses, approximations, and so on. The results can therefore be directly compared with real experiments so as to determine the model's validity, and with analytic solutions in order to evaluate the validity of the various approximations. It is also possible to define the model for situations that are not experimentally accessible, so as to check the validity of theoretical predictions—this is typically the case of the predictions, made on the basis of the renormalization group, that the critical exponents will assume classical values for spatial dimensions higher than 4. One should add that numerical experiments give one access to microscopic details (such as each particle's position and motion at each instant) that are not available in real experiments. Last but not least, one has complete control over a numerical experiment: one does not have to worry about the sample's purity, thermostatization, and so on.

Simulation activity is obviously *experimental*, and it is difficult to understand it unless one actually carries it out. I cannot therefore undertake an abstract discussion of the methods and the problems connected with numerical simulation here, but I refer the reader to Binder and Heermann [Bind88] as one possible practical guide to numerical simulation in statistical mechanics—one that can also be used as a source for simulation activities to be performed on one's PC at home.

I will focus on the simulation of thermodynamic systems at equilibrium. This activity can be performed in at least two different ways: **molecular dynamics** and the **Monte Carlo method**. The specific aim of molecular dynamics is the simulation of mechanical systems (such as fluids composed of interacting particles), while the Monte Carlo method is more flexible. In molecular dynamics, one sets out to simulate the evolution of a mechanical system composed of a large number of particles, solving its equations of motion numerically. One obvious advantage is that in this fashion, one gains direct information not only about the system's static properties, but also about its dynamic properties. In the Monte Carlo method, one instead solves a *fictitious* evolution process, chosen to replicate the equilibrium distribution predicted by statistical mechanics. This has a series of advantages: first, it is possible to simulate systems whose dynamics are not defined (like the Ising model), and second, it is possible to choose the fictitious dynamics so as to reach equilibrium as fast as possible, in order to optimize the quality of the data one gathers. On the other hand, while molecular dynamics can constitute an *experimental proof* of statistical mechanics' foundations—in the sense that it allows one to verify whether the predictions of statistical mechanics follow from the solution of the equations of motion, starting from "reasonable" but arbitrary initial conditions—the Monte Carlo method *presupposes* its validity, as a constraint on the choice of the fictitious dynamics.

In any case, in both situations the result of the simulation is a trajectory in phase space, starting from which one calculates the observables' averages, the correlation functions such as the $g(r)$, and so on. The problems related to the interpretation of this data are similar for both techniques.

8.2 Molecular Dynamics

In this section, we will show how one can perform the numerical simulation of a classical mechanical system by means of the **molecular dynamics** method. This consists of the explicit solution of Newton's equations of motion for the system of particles that makes up the system. One of its most remarkable advantages is that it allows one to evaluate equilibrium thermodynamic properties (such as the equation of state) and dynamic properties like viscosity, thermal conductivity, and so on at the same time. Its main disadvantage lies in the complexity of the calculations, which require significant resources even for very simple and small systems.

Let us specifically consider a model of a simple fluid, composed of N identical particles, of mass m, which evolve while subject to mutual interactions, described by a pair potential $U(r)$, where r is the distance between the two particles. Let us denote the momentum vector by $\boldsymbol{p}_i$ and the position vector of particle i by $\boldsymbol{r}_i$. This system's Hamiltonian is then given by

$$H = \sum_i \frac{p_i^2}{2m} + \frac{1}{2} \sum_{(i \neq j)} u(|\boldsymbol{r}_i - \boldsymbol{r}_j|). \tag{8.1}$$

The Newton equations of motion for the $3N$ particles derive from this Hamiltonian:

$$m\ddot{\boldsymbol{r}}_i = \sum_{j(\neq i)} f_{ij}, \tag{8.2}$$

where f_{ij} is the force with which particle j acts on particle i:

$$f_{ij} = -\frac{\partial u(|\boldsymbol{r}_i - \boldsymbol{r}_j|)}{\partial \boldsymbol{r}_i}. \tag{8.3}$$

It is possible to solve this set of differential equations numerically by introducing a small temporal increment τ and then calculating the positions $r(t+\tau) = (\boldsymbol{r}_i(t+\tau))$ and $p(t+\tau) = (m\dot{\boldsymbol{r}}_i(t+\tau))$, once their values at instant t are known. We thus obtain a discrete approximation for one phase space trajectory:

$$((r(t_0), p(t_0)), (r(t_0+\tau), p(t_0+\tau)), \ldots, (r(t_0+n\tau), p(t_0+n\tau)), \ldots).$$

Given the initial conditions (r_0, p_0) and a sequence of temporal intervals $\mathcal{N}$, we will have univocally defined a sequence of states in phase space. Now, if the hypotheses we made about statistical mechanics are valid, then if the number of particles N is sufficiently large, and we follow the system evolution for a time $\mathcal{N}\tau$ of the order of magnitude of the observation time that interests us, the average of the observables along this evolution must tend toward the ensemble's averages. We can informally set

$$\langle A(r,p) \rangle = \lim_{\mathcal{N} \to \infty} \frac{1}{\mathcal{N}} \sum_{k=1}^{\mathcal{N}} A(r(t_0+k\tau), p(t_0+k\tau)). \tag{8.4}$$

In the case we discussed, since the dynamics expressed by the Newton equations conserve energy, we must understand that the first average is performed over the microcanonical ensemble.

This conclusion should be taken with a great deal of caution. One of the first applications of computer simulation was performed by Fermi, Pasta, and Ulam on a one-dimensional model of coupled anharmonic oscillators [Ferm55]. To their great surprise, they saw that the system tended to return, after a long but not terribly complicated cycle, to a condition close to the one it had started from. Practically speaking, the system never "forgot" its initial state, and therefore the averages along the trajectory were limited to only one particular region of phase space. This phenomenon can appear in a variety of cases and can be difficult to identify. It usually becomes increasingly less probable as the system's size increases. It is, however, one of the simulator's tasks to verify that this phenomenon does not occur in the case being considered.

The quantities that can be obtained from a molecular dynamics trajectory are, as we saw, averages of thermodynamic observables—in other words, of functions $A(r, p)$ of the microstates $(r, p) = (\boldsymbol{r}_1, \boldsymbol{p}_1, \ldots, \boldsymbol{r}_N, \boldsymbol{p}_N)$. Many quantities that are thermodynamically interesting are *not* observables in this sense: entropy especially, which is a property of the *macrostate* (the logarithm of the accessible volume), and, consequently, temperature, pressure, chemical potentials, and so on, which are its derivatives. In order for simulations in molecular dynamics to be of some utility, it is therefore necessary to find a method to get around this difficulty. This can be accomplished, in the case of a classical mechanical system, by exploiting the equipartition theorem to calculate temperature and the virial theorem to calculate pressure.

8.2.1 *Temperature and Pressure in Molecular Dynamics*

For a classical mechanical system whose Hamiltonian is given by the sum of kinetic energy $K(p)$ and potential energy $U(r)$, where

$$K = \sum_{i=1}^{N} \sum_{\alpha=1}^{d} \frac{p_{i\alpha}^2}{2m_i}, \tag{8.5}$$

the equipartition theorem stipulates that $\langle K \rangle$ is equal to $1/2k_B T$ for each degree of freedom:

$$\langle K \rangle = \frac{1}{2} N d k_B T. \tag{8.6}$$

This observation suggests that we calculate T by evaluating $\langle K \rangle$, which is a mechanical observable. Let us note that in this manner, we proceed to calculate T as a function of internal energy E, which, in its turn, is determined by the initial conditions. In practice, there will be difficulties when approaching the lower temperatures, because in order to do so, it will be necessary to choose the initial conditions in such a way that the total energy is as small as possible. If one wants to simulate the system at low temperatures, one resorts to the subterfuge of artificially "cooling" the system, by simultaneously dividing the velocities of all the particles by a certain factor $\lambda > 1$ from time to time. By applying this transformation several times, one gradually leads the system to move closer to minimal internal energy configurations.

Knowledge of T as a function of internal energy E in principle allows one to calculate the entropy S by means of integration. The specific heat C can be obtained by taking the derivative of the temperature T with respect to the internal energy E:

$$\frac{1}{C} = \left.\frac{\partial T}{\partial E}\right)_N. \tag{8.7}$$

In this manner, it is also possible to obtain the entropy by integrating C/T numerically:

$$S(B) - S(A) = \int_A^B \frac{\mathrm{d}E}{T} = \int_A^B \mathrm{d}E \frac{1}{T(E)} \left(\frac{\partial T}{\partial E}\right)^{-1}, \tag{8.8}$$

where A and B are states of thermodynamic equilibrium. In order to be able to exploit this formula, however, it is essential to verify that the system has reached thermodynamic equilibrium in each of the intermediate states. Otherwise, the integral will come to depend on the path.

In order to calculate the pressure, one could evaluate the mean of the force exerted by the walls on the system, consistently with the mechanical definition of pressure. This requires that we explicitly introduce the effects of the walls into the simulation, however, whereas usually, in order to minimize finite size effects, one prefers to introduce periodic boundary conditions. For this reason, the pressure is usually calculated by resorting to the **virial theorem**. The quantity

$$\Pi = \sum_i \boldsymbol{p}_i \cdot \boldsymbol{r}_i, \tag{8.9}$$

is an observable called the **virial**. At equilibrium, its average is constant. Therefore,

$$\frac{\mathrm{d}}{\mathrm{d}t}\langle \Pi \rangle = 0 = \left\langle \sum_i \dot{\boldsymbol{p}}_i \cdot \boldsymbol{r}_i \right\rangle + \left\langle \sum_i \boldsymbol{p}_i \cdot \dot{\boldsymbol{r}}_i \right\rangle. \tag{8.10}$$

One obviously has

$$\left\langle \sum_i \boldsymbol{p}_i \cdot \dot{\boldsymbol{r}}_i \right\rangle = \frac{1}{m}\left\langle \sum_i p_i^2 \right\rangle = 2\langle K \rangle = Ndk_{\mathrm{B}}T, \tag{8.11}$$

because of the equipartition theorem.

On the other hand, due to Newton's law,

$$\dot{\boldsymbol{p}}_i = \boldsymbol{F}_i, \tag{8.12}$$

where $\boldsymbol{F}_i$ is the resulting force applied to particle i. If this force is due to external force $\boldsymbol{f}_i^{\mathrm{ext}}$ and interactions among particles, one has

$$\boldsymbol{F}_i = \boldsymbol{f}_i^{\mathrm{ext}} + \boldsymbol{f}_i^{\mathrm{int}} \tag{8.13}$$

where $\boldsymbol{f}_i^{\mathrm{int}} = -\nabla_i U(\{\boldsymbol{r}_i\})$ is the force that acts on particle i because of the interactions with all the other particles. We thus obtain

$$\left\langle \sum_i \dot{\boldsymbol{p}}_i \cdot \boldsymbol{r}_i \right\rangle = \sum_i \langle \boldsymbol{f}_i^{\mathrm{ext}} \cdot \boldsymbol{r}_i \rangle + \sum_i \langle \boldsymbol{f}_i^{\mathrm{int}} \cdot \boldsymbol{r}_i \rangle. \tag{8.14}$$

If the external force is applied by the walls, and the pressure is equal to p, one has

$$\sum_i \langle \boldsymbol{f}_i^{\mathrm{ext}} \cdot \boldsymbol{r}_i \rangle = -p \oint \mathrm{d}A\boldsymbol{n} \cdot \boldsymbol{r}, \tag{8.15}$$

where $\mathrm{d}A$ is the area element of the walls, $\boldsymbol{n}$ the corresponding normal versor (oriented toward the outside), and the integral is extended to the sample's external surface. Because of the Gauss theorem,

$$\oint \mathrm{d}A\boldsymbol{n} \cdot \boldsymbol{r} = \int \mathrm{d}\boldsymbol{r}(\nabla \cdot \boldsymbol{r}) = dV, \tag{8.16}$$

where V is the sample volume and d the space dimensionality. We thus obtain

$$\sum_i \langle \boldsymbol{f}_i^{\mathrm{ext}} \cdot \boldsymbol{r}_i \rangle = -dpV. \tag{8.17}$$

We can thus obtain an expression of pressure:

$$pV = Nk_{\mathrm{B}}T + \frac{1}{d}\left\langle \sum_i \boldsymbol{f}_i^{\mathrm{int}} \cdot \boldsymbol{r}_i \right\rangle, \tag{8.18}$$

Since the calculation of the dynamics requires the evaluation of the $\boldsymbol{r}_i$ and the $\boldsymbol{f}_i^{\mathrm{int}}$ at each step, it is possible to evaluate the pressure directly.

This expression is simplified if the internal forces are due only to pair interactions—in other words, if

$$f_i^{\text{int}} = \sum_{j(\neq i)}{}' f_{ij}, \tag{8.19}$$

where f_{ij} is the force that particle j applies on particle i and satisfies $f_{ij} = -f_{ji}$. We can then group the contribution of each pair of particles into the right-hand side of the equation (8.18), obtaining

$$\sum_i \sum_{j(\neq i)}{}' \langle f_{ij} \cdot r_i \rangle = \frac{1}{2} \sum_{i \neq j}{}' \langle f_{ij} \cdot r_{ij} \rangle, \tag{8.20}$$

where $r_{ij} = r_i - r_j$. By grouping all the terms, we thus obtain

$$\begin{aligned} pV &= Nk_B T + \frac{1}{d} \sum_{i \neq j}{}' \langle f_{ij} \cdot r_{ij} \rangle \\ &= Nk_B T - \frac{1}{d} \sum_{i \neq j}{}' \left\langle r_{ij} \cdot \frac{\partial u(r_{ij})}{\partial r_{ij}} \right\rangle. \end{aligned} \tag{8.21}$$

Exercise 8.1 (Ideal Gas in One Dimension) Consider a system composed of N particles of mass m in one dimension, contained in the half-line $r_i \geq 0$, $i = 1, 2, \ldots, N$, with an elastic wall at the origin. The particles are confined by a piston of mass M, whose instantaneous position is given by $L(t) > 0$, and which is acted on by a force p oriented toward the origin.

Resolve this problem in molecular dynamics, taking into account the fact that the particles interact only with the fixed wall at the origin (where they bounce elastically) and with the piston, by means of impacts in which energy and momentum are preserved. Check the enthalpy conservation $E + pL$. Show (numerically) that the velocity distribution tends to the Maxwell distribution and that the equation of state is given approximately by the law of ideal gases.

8.2.2 Verlet Algorithm

We owe the most frequently used algorithm for the resolution of Newton's equations to Verlet. Let us take the case of one degree of freedom that satisfies the equation of motion:

$$\ddot{x}(t) = f(x(t)). \tag{8.22}$$

By expanding $x(t + \tau)$ in series around t, we obtain

$$x(t + \tau) = x(t) + \tau \dot{x}(t) + \frac{1}{2} \tau^2 \ddot{x}(t) + \frac{1}{6} \tau^3 \dddot{x}(t) + O(\tau^4). \tag{8.23}$$

By summing with the corresponding expression for $x(t - \tau)$, we obtain

$$x(t + \tau) + x(t - \tau) = 2x(t) + \tau^2 \ddot{x}(t) + O(\tau^4), \tag{8.24}$$

and therefore,

$$x(t + \tau) = 2x(t) - x(t - \tau) + \tau^2 f(x(t)) + O(\tau^4). \tag{8.25}$$

This is a fourth-order algorithm (in other words, it disregards terms on the order of τ^4), like the most frequently utilized Runge-Kutta method, but it has the advantage of being symmetrical with respect to time reversal, thus maintaining (except for rounding errors) this important property. It is, moreover, very simple to put into practice. One needs to keep in mind that the method requires that one know $x(t)$ and the preceding value $x(t-\tau)$ at each instant. If as usually occurs, the initial positions and velocities are known, it is necessary to estimate the $x(t-\tau)$, and this introduces an error on the order of τ^3. This does not create difficulties in the case of statistical averages, but it can do so if one is looking for effective solutions of the differential equation for a given initial condition.

Exercise 8.2 Write the Verlet algorithm for the pendulum:

$$\ell\ddot{\theta} = -g\sin\theta.$$

Resolve the equations of motion for θ numerically, and check energy conservation.

In the case of simple fluid models, it is necessary to choose the initial condition with a certain amount of care. One usually prefers to initially arrange the particles on a regular lattice, so as to guarantee a certain value of the density, and assign a velocity, chosen randomly from a Gaussian distribution, to each of them. By changing the variance of the Gaussian, one can change the value of the temperature.

Exercise 8.3 Let us consider a one-dimensional fluid defined by the Hamiltonian

$$H = \sum_{i=1}^{N}\frac{p_i^2}{2m} + \sum_{i=1}^{N} u\,(r_{i+1} - r_i),$$

with periodic boundary conditions:

$$r_{i+N} = r_i, \qquad -L/2 \leq r_i \leq L/2, \qquad u\,(r_i + L) = u\,(r_i), \qquad \forall i.$$

Let us assume the Lennard-Jones form for $u(r)$:

$$u(r) = \epsilon_0\left[\left(\frac{r_0}{r}\right)^{12} - \left(\frac{r_0}{r}\right)^{6}\right].$$

1. By introducing r_0 as the length scale and $\tau = \sqrt{m/\epsilon_0}$ as a time scale, make the equations of motion adimensional. The explicit dependence on the adimensional density $\rho = (Nr_0)/L$ will remain.

2. Write the program for the solution of the equations of motion by means of the Verlet algorithm.

3. Write the expression of temperature and pressure as a function of the trajectory calculated by means of the Verlet algorithm.

4. Evaluate the order of magnitude (in real length and time) of the systems and durations you can simulate on your PC, by supposing that $m \sim 10^{-23}$ g, $\epsilon_0 \sim 1$ eV, and $r_0 \sim 1$ Å.

If you are courageous, launch the simulation! Choose the initial conditions so that the r_i are arranged on a regular lattice and the p_i follow a Gaussian distribution. Check the conservation of energy and monitor the approach to equilibrium. You can also compare your data to the exact solution. Be careful not to use excessively low values for energy, because the Fermi-Pasta-Ulam phenomenon mentioned earlier can make the approach to equilibrium in finite systems unreliable and slow.

The success of the method of molecular dynamics in the study of the statistical mechanics of fluids represents one of the most powerful arguments for the entire discipline's validity. In fact, it guarantees that the principal results of statistical mechanics descend directly from the properties of the equations of motion in the case of a fairly large system of particles, and it is reassuring to know that the required size of the system is not excessive. (Several thousand particles can be sufficient for simple systems.) The solution of the equations of motion is, however, often too complicated, or too slow to obtain, and one prefers to use a method that samples directly the probability distribution that defines the ensemble: the Monte Carlo method.

8.3 Random Sequences

The Monte Carlo method substitutes the "true" dynamics of the system being considered—*deterministic* dynamics, in which the system's further evolution is univocally defined by the present dynamic state—by a fictitious "stochastic" dynamics in which the state's present system determines only the probability that the system will be found in a specific successive state. The conditional probability of the next state x', given the present state x, is called **transition probability**. The actual realization of this dynamics requires that the computer be able to produce efficiently and rapidly very long sequences of random numbers, identically distributed and mutually independent:

> It may seem perverse to use a computer, the most precise and deterministic of all the machines conceived by the human mind, to produce "random" numbers. More than perverse, it can seem to be a conceptual impossibility. Any program, after all, will produce output that is entirely predictable, hence not truly "random." [Pres92, p. 265]

It is a common occurrence, however, to use a computer to generate random numbers: fastidious souls introduce the distinction between *truly* random numbers, when they are generated by a natural process that we describe only probabilistically (like the number of disintegrations of some radioactive element) and the *pseudo*-random (but reproducible) numbers that are produced by a computer. If one thinks about it carefully, however, the problem of why a deterministic system like a computer can produce sequences that are properly described as random is of the same order as the basic problem of statistical mechanics—in other words, of why a deterministic system (described by the equations of mechanics) can exhibit a random behavior. A discussion of these problems would take us very far.

Here we will adopt, following [Pres92], a pragmatic point of view: for our purposes, a program that produces sequences of random numbers is effective if, within the application that uses these numbers, it produces statistically equivalent results to those obtained by using the sequences produced by another random number generator. Good random number generators, in other words, can be exchanged with one another within applications. This definition appears to be circular, because it is based on a comparison between different generators. It is useful, however, because it is a fact that good random number generators exist and have extensive fields of application.

More pragmatically, randomness, like beauty, is in the eye of the beholder. If one looks too closely, no sequence is sufficiently random—but the existence of hidden correlations does not necessarily make your results unusable. In any case, it is always a good idea to verify that the program you are using passes a series of tests aimed at revealing secret correlations or, if it does not pass them, to check that the relevant effects are not too important:

> If all scientific papers whose results are in doubt because of bad rans [random number generators] were to disappear from library shelves, there would be a gap on each shelf about as big as your fist. [Pres92, p. 266]

We will now discuss the structure of the most widely used random number generators. They are based on the **linear congruence** mechanism, which generates a series of integers $I_1, I_2, I_3, \ldots,$ between 0 and $m-1$ (a large number) by means of the recurrence relation

$$I_{j+1} = aI_j + c \pmod{m}. \tag{8.26}$$

Remember that two integers, x and y, are congruent modulo m [which is written $x \equiv y \pmod{m}$], if the remainders of their division by m are equal. Equation (8.26) therefore imposes that to set I_{j+1} equal to the remainder of the division of aI_{j+c} by m.

Here m is called the **modulus**, and a and c are positive integers, called the **multiplier** and the **increment**, respectively. Sequence (8.26) will obviously repeat, at most after m steps, but if a, c, and m are properly chosen, the period will have the greatest value—in other words, m. In this case, all integers between 0 and $m-1$ will appear in the sequence once and once only, and therefore each "seed" I_0 is as good as any other. Once we have obtained I_j, we then obtain the real random number $x_j = I_j/m$, included between 0 (included) and 1 (excluded) and uniformly distributed in the interval.

A fairly effective linear congruence exploits the fact that in `C`, if an operation between two `long` operands causes overflow, only the lower order bits are kept. In a 32-bit machine, therefore, one can assume $m = 2^{32}$, and the mod operation is given us for free. At this point, a good choice of a and c (suggested by D. E. Knuth, one of the most influential thinkers in computer science [Knut81]) is the following:

$$a = 1664525, \qquad c = 1013904223. \tag{8.27}$$

Supposing therefore that `idum` is an `unsigned long int` in a 32-bit machine, we obtain a rapid and fairly efficient generator of random numbers, contained in a single line of `C` code:

```
idum =1664525L*idum+1013904223L
```

Be careful! This method is based on a property that, although it is guaranteed in C compilers, is not generally valid for FORTRAN compilers. This line, therefore, may well not have the intended effect if inserted in a FORTRAN program. This is only one of the (many) arguments that suggest we give FORTRAN a well-earned retirement.

The linear congruence method has the advantage of being very fast, because it requires only very few operations for each call—this explains its almost universal use. It does, however, produce correlations between successive calls. If we take k successive random numbers to represent a point in k-dimensional space, then the points will arrange themselves on $(k-1)$–dimensional planes, instead of tending to "fill" space. There will be at most $m^{1/k}$ planes, and usually there will be many fewer, unless a, c, and m are chosen very accurately. These correlations can (more or less) be reduced by "remixing" the numbers produced by the algorithm in the course of a certain number of successive calls. I invite you to consult Press and others [Pres92] for a discussion of the resulting algorithm.

8.4 Monte Carlo Method

Let us now suppose that we want to calculate an integral of the type

$$\langle A \rangle = \int_0^1 \mathrm{d}x\, A(x)\rho(x), \tag{8.28}$$

where $A(x)$ is a function of x, and $\rho(x) \geq 0$ can be interpreted as a probability distribution. The most direct way of numerically evaluating this integral is to evaluate $A(x)\rho(x)$ in $\mathcal{N}+1$ points uniformly arranged between 0 and 1, $x_i = i/\mathcal{N}$, $i = 0, 1, \ldots, \mathcal{N}$, and to set

$$\langle A \rangle \simeq \overline{A} = \frac{1}{\mathcal{N}+1}\sum_{i=0}^{\mathcal{N}} A(x_i)\rho(x_i). \tag{8.29}$$

If necessary, we can refine this estimate by using Simpson's rule and so on.

This method is less effective, however, if the integrand, and $\rho(x)$ in particular, is much larger in certain regions of the interval than in others. In fact, the regions in which $\rho(x) \simeq 0$ give a negligible contribution to $\langle A \rangle$. If $A(x)$ is a "smooth" function, one will get a better estimate of $\langle A \rangle$ by distributing the x_i so that $\rho(x_i)\Delta(x_i) = \rho(x_i)(x_{i+1} - x_i)$ is approximately constant. It is easy to obtain this distribution in one dimension by introducing the cumulative distribution $\Phi(x)$:

$$\Phi(x) = \int_0^x \mathrm{d}x'\rho(x). \tag{8.30}$$

By then choosing the x_i so that

$$\Phi(x_{i+1}) - \Phi(x_i) = \frac{1}{\mathcal{N}}, \tag{8.31}$$

we obtain the following expression of $\langle A \rangle$:

$$\langle A \rangle = \frac{1}{\mathcal{N}+1} \sum_{i=0}^{\mathcal{N}} A(x_i). \tag{8.32}$$

In this expression, all points give approximately the same contribution, at least if $A(x)$ is a slowly varying function.

This procedure is quite effective in one dimension, although it requires one to solve equation (8.31), which can require considerable effort. It has the shortcoming, however, that it cannot be adapted with efficiency to the case in which x varies in a space possessing more that a few dimensions.

Let us assume that we are dealing with a d-dimensional integral:

$$\langle A \rangle = \int \mathrm{d}^d x A(\boldsymbol{x}) \rho(\boldsymbol{x}), \tag{8.33}$$

where $\boldsymbol{x} = (x^\alpha)$, $\alpha = 1, \ldots, d$, $0 \le x^\alpha \le 1$ $(\forall \alpha)$. In order to obtain a uniform distribution of the $\boldsymbol{x}_i$, we will need a number of points that grows exponentially with d. This number can rapidly become too large even for moderate values of d. On the other hand, if we consider the expression

$$\overline{A} = \frac{1}{\mathcal{N}} \sum_{i=1}^{\mathcal{N}} A(\boldsymbol{x}_i) \rho(\boldsymbol{x}_i), \tag{8.34}$$

where the $\boldsymbol{x}_i$ are random, independent, and identically distributed vectors (uniformly in the hypercube unit in d dimensions, for example), as a result of the central limit theorem, we see that $\overline{A}$ is a Gaussian variable, whose average is equal to $\langle A \rangle$, and whose variance is proportional to $\mathcal{N}^{-1/2}$. We can therefore use $\overline{A}$ as an estimate of $\langle A \rangle$, and expect a random deviation ("statistical error") of the order of $\mathcal{N}^{-1/2}$. This approach to the asymptotic value is very slow (in order to improve $\langle A \rangle$'s estimate by one significant digit, one needs 100 times more points), but its behavior is independent of d.

In order to improve the convergence, it is obviously worth our while to avoid having many points $\boldsymbol{x}_i$ falling into regions in which $\rho(x)$ is small. As we saw earlier, the best situation is that in which the points' density is proportional to $\rho(x)$, so that the number of points contained in a small volume area, centered in $\boldsymbol{x}$ and of volume $\mathrm{d}^d\boldsymbol{x}$, is given by $\mathcal{N}$ $\rho(\boldsymbol{x})\mathrm{d}^d\boldsymbol{x}$. If we can obtain this situation, we will have

$$\langle A \rangle \simeq \overline{A} = \frac{1}{\mathcal{N}} \sum_{i=1}^{\mathcal{N}} A(\boldsymbol{x}_i), \tag{8.35}$$

with a statistical error on the order of $\mathcal{N}^{-1/2}$.

The idea of the Monte Carlo method is to evaluate the average (8.33) of a quantity $A(\boldsymbol{x})$ over a distribution $\rho(\boldsymbol{x})$ using expression (8.35), where the $\boldsymbol{x}_i$ are therefore random, independent, and distributed with a probability density $\rho(\boldsymbol{x})$.

One should immediately add that this "ideal" program cannot be realized. Except for trivial cases, it is not possible to find a procedure that allows one to extract some $\boldsymbol{x}$'s with a probability distribution $\rho(\boldsymbol{x})$, independently from one another. The central limit theorem, however, guarantees us that $\overline{A} = \langle A \rangle + O(\mathcal{N}^{-1/2})$ even if the $\boldsymbol{x}_i$ are *not* mutually independent. It is sufficient that the $\boldsymbol{x}_i$ sequence is such that:

1. The distribution of each of the x_i is given by $\rho(x)$.

2. The correlations between x_i and $x_{i+\ell}$ go to zero fairly rapidly as $|\ell|$ grows.

The problem is therefore reduced to finding an algorithm that will allow us to extract the x_i with probability distribution $\rho(x)$, and in such a way that correlations among the x_i along the sequence go to zero fairly rapidly. This problem can be resolved by introducing a suitable **Markov chain**.

8.4.1 Markov Chains

In order to clarify our ideas, let us consider a system that can assume only discrete states, which belong to a certain set Q. The **Markov chain** X is the sequence of random variables X_t, $t = 0, 1, 2, \ldots$ such that:

1. $X_t \in Q$, $\forall t$.

2. The conditional probability distribution of x_{t+1} depends only on the value of X_i, and not on the values of the other $X_{t'}$, with $t' \neq t, t+1$.

In other words, in a Markov chain, the probability of finding oneself in a state at instant $t+1$ depends only on the state in which the system finds itself at state t and not on the preceding (or following!) ones.

In appendix D, I provide a concise summary of the properties of Markov chains. In this section, I show how, under fairly stringent hypotheses, one can build a Markov process that allows one to sample an arbitrary distribution $p_\alpha\,(a \in Q)$. This proof is taken from [Youn10].

From property 2 of the Markov chains, it follows that the statistical properties of process X are determined by the conditional probabilities $W_{ab}(t) = P(X_{t+1} = a \mid X_t = b)$, $a, b \in Q$, $\forall t$, called **transition probabilities**, and by the initial condition $p_a^0 = P(X_0 = a)$. If the conditional probabilities $W_{ab}(t)$ do not depend on t, the process's dynamics are stationary.

Some important properties follow from the interpretation of the W_{ab} as conditional probabilities:

1. $W_{ab} \geq 0$, $\forall a, b \in Q$;

2. $\sum_{a \in Q} W_{ab} = 1, \forall b \in Q$.

Matrices W that satisfy these conditions are called **stochastic matrices**.

The transition probabilities W_{ab} determine the evolution of the probability $p_a(t) = P(X_t = a)$ according to the law

$$p_a(t+1) = \sum_{b \in Q} W_{ab} p_b(t). \tag{8.36}$$

By exploiting property 2 of stochastic matrices, this equation can be put in the form known as a **master equation**:

$$\Delta p_a(t) = p_a(t+1) - p_a(t) = \sum_{b(\neq a)}{}' [W_{ab} p_b(t) - W_{ba} p_a(t)]. \tag{8.37}$$

We want to look for the **stationary distribution** p_a^{eq} of the X_t—in other words, a distribution such that if $P(X_t = a) = p_a^{\text{eq}}$ at instant t, it remains equal to p_a^{eq} for all successive instants.

From equation (8.37), it follows that the p_a^{eq} must satisfy

$$\sum_{b(\neq a)}{}' [W_{ab} p_b^{\text{eq}} - W_{ba} p_a^{\text{eq}}] = 0, \qquad \forall a. \tag{8.38}$$

An important subclass of Markov chains with stationary dynamics is that in which the transition probabilities have the **detailed balance** property. One says that the stochastic matrix W has the detailed balance property if, given any three different states $a, b, c \in Q$, one has

$$W_{ab}\, W_{bc}\, W_{ca} = W_{ac}\, W_{cb}\, W_{ba}. \tag{8.39}$$

If the W matrix has the detailed balance property, the stationary distribution p_a^{eq} satisfies, in addition to equation (8.38), the stronger condition

$$W_{ab} p_b^{\text{eq}} - W_{ba} p_a^{\text{eq}} = 0, \qquad \forall a, b. \tag{8.40}$$

In this case, it is actually possible to construct the p_a^{eq} explicitly. Choose any reference state a_0, and assign it a (nonvanishing) probability p_{a_0}. Assign to any state a_1 such that $W_{a_1 a_0} \neq 0$ the probability $p_{a_1} = (W_{a_1 a_0} p_{a_0})/W_{a_0 a_1}$. Now look for all states a_2 such that there exists at least one a_1 for which $W_{a_2 a_1} \neq 0$, and assign the probability $p_{a_2} = (W_{a_2 a_1} p_{a_1})/W_{a_1 a_2}$ to each of them. The detailed balance property will make it so that if the same state b is reached from a_0, but following two different paths, the resulting probabilities will in any case be the same.

Exercise 8.4 Prove the previous assertion.

In this way one can assign a probability p_a to each state a in Q, up to an overall factor (p_0) that will be determined by the normalization condition.

If the matrix W has the detailed balance property, it is possible to prove, in an elementary fashion, that the distribution $p_a(t)$ converges to the p_a^{eq} defined in this manner, by means of an argument reported by A. P. Young.

Let us consider the quantity

$$\mathcal{I}[p] = \sum_{a \in Q} \frac{1}{p_a^{\text{eq}}} (p_a - p_a^{\text{eq}})^2 = \sum_{a \in Q} \frac{p_a^2}{p_a^{\text{eq}}} - 1. \tag{8.41}$$

It is easy to see that, if p is a normalized distribution,

$$\mathcal{I}[p^{\text{eq}}] = 0; \qquad \mathcal{I}[p] > 0, \qquad p \neq p^{\text{eq}}. \tag{8.42}$$

Let us assume that $p(t) = p$, and denote the corresponding distribution at instant $t + 1$ by p'. We will denote the value of $\mathcal{I}'$ at instant $t + 1$ by $\mathcal{I}$. We now show that

$$\Delta \mathcal{I} = \mathcal{I}' - \mathcal{I} \leq 0,$$

where the equation is valid only if $p = p^{\text{eq}}$.

We can write $\Delta\mathcal{I}$ in the form

$$\Delta\mathcal{I} = \sum_{abc} W_{ab} W_{ac} \frac{p_b p_c}{p_a^{\text{eq}}} - \sum_a \frac{p_a^2}{p_a^{\text{eq}}}. \tag{8.43}$$

As a result of the detailed balance condition, W_{ac} in the first term can be replaced by $W_{ca} p_a^{\text{eq}} / p_c^{\text{eq}}$. On the other hand, as a result of the normalization condition of the W_{ab}, we can introduce a sum $\sum_b W_{ab}$ in the second term and exchange the a and b indices. We thus obtain

$$\Delta\mathcal{I} = \sum_{abc} W_{ba} W_{ca} p_a^{\text{eq}} \frac{p_b p_c}{p_b^{\text{eq}} p_c^{\text{eq}}} - \sum_{a,b} W_{ab} \frac{p_b^2}{p_b^{\text{eq}}}. \tag{8.44}$$

By again applying the detailed balance property and introducing a $\sum_c W_{ca}$ factor, the second term of this expression can be written

$$-\sum_{abc} W_{ba} W_{ca} p_a^{\text{eq}} \left(\frac{p_b}{p_b^{eq}} \right)^2.$$

By taking the semisum of this expression as well as that of the one obtained by exchanging b and c, we finally obtain

$$\Delta\mathcal{I} = -\frac{1}{2} \sum_{abc} W_{ba} W_{ca} p_a^{\text{eq}} \left(\frac{p_b}{p_b^{eq}} - \frac{p_c}{p_c^{eq}} \right)^2. \tag{8.45}$$

This equation shows that $\Delta\mathcal{I} < 0$ unless, for each state a, all the states that can be reached from a in only one move have probabilities proportional to the stationary distribution. The most natural situation is that *all* the states satisfy this condition with the same proportionality constant (which must then be equal to 1). $\Delta\mathcal{I}$, however, vanishes even if $p_a = 0$ for certain "inaccessible" states and $p_b \propto p_b^{\text{eq}}$ for the remaining "accessible" states, as long as there are no transitions between accessible and inaccessible states. In order for the process to reach equilibrium, it is therefore sufficient for it to be **ergodic**—in other words, that transitions between any two arbitrary states can take place, as long as one waits for a sufficient amount of time.

At this point, we can consider the p_a^{eq} as the given distribution (to be sampled) and interpret our results in the following manner. The Markov chain defined by the transition probability matrix $\mathsf{W} = (W_{ab})$ allows us to sample the p_a^{eq} if the following conditions are satisfied:

1. W is ergodic. (In other words, W^n has elements that are all nonvanishing for a fairly large n.)

2. W has the detailed balance property, and more particularly, given any two states a and b, one has

$$W_{ab} p_b^{\text{eq}} = W_{ba} p_a^{\text{eq}}. \tag{8.46}$$

We note that this relation implies that if the direct transition $a \to b$ is forbidden (in other words, $W_{ba} = 0$), then the inverse transition $b \to a$ is forbidden as well.

These conditions, in any case, leave ample room for the choice of the transition probabilities, as the discussion in the next section makes clear.

8.4.2 Monte Carlo Algorithms in Statistical Mechanics

We can now see how one can construct some Markov chains that allow for the sampling of thermodynamic ensembles. In order to clarify our ideas, let us consider an Ising model in two dimensions, defined by the Hamiltonian

$$H(\sigma) = -\sum_{<ij>} J\sigma_i\sigma_j - \sum_i h\sigma_i, \tag{8.47}$$

where $\sigma_i = \pm 1$, $i = 1, 2, \ldots, N$, and the first sum is extended to nearest-neighbor pairs over a square lattice. The collection $\sigma = (\sigma_i)$ of the spin values of all the lattice's sites defines the system's microstate. The thermodynamic observables (in the canonical ensemble) are obtained as averages of functions of σ with respect to the Boltzmann distribution

$$P_\sigma^B = \frac{\mathrm{e}^{-H(\sigma)/k_BT}}{Z}, \tag{8.48}$$

where Z is the partition function:

$$Z = \sum_\sigma \mathrm{e}^{-H(\sigma)/k_BT}. \tag{8.49}$$

We thus obtain the expression of the internal energy:

$$E = \langle H \rangle = \sum_\sigma H(\sigma) P_\sigma^B, \tag{8.50}$$

and that of the magnetization M:

$$M = \left\langle \sum_i \sigma_i \right\rangle = \sum_\sigma \left(\sum_i \sigma_i \right) P_\sigma^B. \tag{8.51}$$

We want to define a Markov chain whose states correspond to the microstates of system σ, and for which the P_σ^B is the stationary distribution. If we impose that the probability distributions $W_{\sigma'\sigma}$ possess the detailed balance property, this result will be obtained if the matrix W (a $2^N \times 2^N$ matrix) is ergodic, and if for each pair of microstates (σ', σ), one has

$$W_{\sigma'\sigma} P_\sigma^B = W_{\sigma\sigma'} P_{\sigma'}^B. \tag{8.52}$$

This condition implies that either the direct transition $\sigma \to \sigma'$ is "forbidden" (but it *has* to take place sooner or later, following a finite number of steps; otherwise the matrix W is not ergodic) and therefore the inverse transition $\sigma' \to \sigma$ is forbidden as well, or the transition probabilities $W_{\sigma'\sigma}$ and $W_{\sigma\sigma'}$ can be expressed one as the function of the other:

$$W_{\sigma\sigma'} = W_{\sigma'\sigma} \frac{P_\sigma^B}{P_{\sigma'}^B} = W_{\sigma'\sigma} \exp\left[-\frac{H(\sigma) - H(\sigma')}{k_BT} \right]. \tag{8.53}$$

Let us remark that the partition function Z disappears from this relation, which is lucky for us, because if we knew how to calculate the Z, we would have no need to resort to the simulation!

One therefore still has to choose the class of "allowed" transitions (for which $W_{\sigma'\sigma} \neq 0$) and define the $W_{\sigma'\sigma}$ so as to satisfy the relation (8.53) Researchers' inventiveness has been unleashed, especially on solutions to the first problem, while the second has some standard solutions, the most common being the **Metropolis algorithm**.

Let us assume that $\sigma \to \sigma'$ is an allowed transition. Then, the relation (8.53) is satisfied if

$$W_{\sigma'\sigma} = \begin{cases} \kappa, & \text{if } H(\sigma') < H(\sigma), \\ \kappa \exp\{-[H(\sigma') - H(\sigma)]/k_B T\}, & \text{otherwise,} \end{cases} \quad (8.54)$$

where κ is a constant that guarantees the normalization of $\sum_{\sigma'} W_{\sigma'\sigma}$. Practically speaking, this can be realized as follows: Given a microstate σ, choose a microstate σ' so that the transition $\sigma \to \sigma'$ is allowed. (This "choice" is a delicate matter, because one needs to be sure that the probability of choosing σ starting from σ' is equal to that of choosing σ' starting from σ, in order to maintain the detailed balance property.) Then, if $H(\sigma') < H(\sigma)$, the transition takes place, and σ' is the system's new state; otherwise, one extracts a random number x, uniformly distributed between 0 and 1—if $x < \exp\{-[H(\sigma') - H(\sigma)]/k_B T\}$, the transition takes place; otherwise, the system's state remains equal to σ.

The advantage of this algorithm is that it requires us to compare the energy values of only the two states connected by the transition, which is useful if the number of states accessible in one move is large.

While the algorithm's most delicate step is the choice of allowed transitions, the slowest is without a doubt the evaluation of $\exp\{-[H(\sigma') - H(\sigma)]/k_B T\}$. One usually uses allowed transitions so that the difference $H(\sigma') - H(\sigma)$ contains only a few terms—if, for example, σ and σ' only differ by state σ_{i_0} of a single spin (and therefore $\sigma'_{i_0} = -\sigma_{i_0}$), one has

$$\Delta H = H(\sigma') - H(\sigma) = 2\sigma_{i_0}\left(\sum_{j \in \text{nn}(i_0)} J\sigma_j + h\right), \quad (8.55)$$

where the sum over j is extended to the nearest neighbors of i_0. This expression requires the evaluation of a small number of terms, and above all, it can assume only a small number of different values. In order to save time, it is therefore useful to create a table with all possible values of $\exp(-\Delta H/k_B T)$ and use them when necessary.

In the case of the Ising model over a square lattice, with interactions between nearest neighbors, the algorithm's effectiveness can be improved by observing that the lattice can be divided into two sublattices: that in which the sum of the coordinates is an even multiple of the lattice step a_0, and that in which it is an odd multiple. Now the nearest neighbors of a spin that belongs to the *even* sublattice all belong to the odd sublattice, and vice versa. It is therefore possible to *update* (decide whether to invert) all of a sublattice's spins without changing the value of ΔH for the other spins in the sublattice. This operation can be carried out in parallel (if the machine allows for parallel computing), thus saving a great deal of time.

We would like to comment on a point that is often neglected. If we update the two sublattices sequentially—the even one (0) and the odd one (1), and so forth—the transition matrix of the Markov chain is then given by

$$\mathrm{W} = \mathrm{W}^{(1)}\mathrm{W}^{(0)}, \quad (8.56)$$

where $\mathsf{W}^{(\alpha)}$ is the transition matrix for updating sublattice α $(\alpha = 0, 1)$. Now, the W matrix does *not* have the detailed balance property, even if $\mathsf{W}^{(0)}$ and $\mathsf{W}^{(1)}$ have it.

Exercise 8.5 Prove this assertion, by finding two configurations σ and σ' such that $W_{\sigma'\sigma} \neq 0$ while $W_{\sigma\sigma'} = 0$, for example.

This however does not prevent the distribution from reaching equilibrium, since the $\mathsf{W}^{(\alpha)}$ separately admit P_σ^B as a stationary distribution, and the unicity of the stationary distribution is guaranteed by the ergodicity of the W. But in more complex situations, problems of this kind can compromise convergence to equilibrium.

Exercise 8.6 (Microcanonical Ensemble) Describe a Markov chain (defined by the matrix W of the transition probabilities) that can sample the microcanonical ensemble:

$$P_\sigma^{\mathrm{M}} = \begin{cases} 1/|\Gamma|, & \text{if } E < H(\sigma) < E + \Delta E, \\ 0, & \text{otherwise}, \end{cases}$$

where $|\Gamma|$ is the number of accessible configurations.

Exercise 8.7 (One-Dimensional Ising Model) Consider a one-dimensional Ising model of spin N with free boundary conditions. Simulate the system with the Metropolis algorithm, alternately updating the even and odd spins. Evaluate the internal energy and the specific heat, using both the fluctuation–dissipation theorem and the calculation of energy variation. Compare with the exact result.

8.4.3 *Statistical Errors*

Let us suppose that you have chosen your algorithm with great care, so that you have a W that ensures convergence toward the P_σ^B. You will choose an initial configuration $\sigma(0)$, and by applying your algorithm, you will obtain a succession of configurations $[\sigma(0), \sigma(1), \ldots, \sigma(t), \ldots]$. Starting from this succession, you want to evaluate the observables, of type $\langle A \rangle$, where $A(\sigma)$ is some function of the configurations, like energy or instantaneous magnetization.

First, you will *discard* all configurations for a (long) initial stretch, $t < T_0$. Your initial condition on the probability distribution is $P_\sigma^0 = \delta_{\sigma\sigma(0)}$, and a certain amount of time will be needed in order for the sampled distribution of your Markov chain to approach P_σ^B.

The time T_0 must be chosen by monitoring the values of the observables (averaged over relatively brief time periods) until it becomes clear that they are varying only because of stochastic fluctuations, and no longer systematically.

But wait a minute: You can object that statistical mechanics is based on the fact that *almost all* the microstates that belong to the accessible region Γ have almost identical values of the thermodynamic observables, and therefore, for any $\sigma \in \Gamma$,

$$A(\sigma) = \langle A \rangle \left[1 + O(N^{-1/2})\right]. \tag{8.57}$$

Therefore, basically, all I would need to do is choose a random point in Γ and the entire lengthy simulation would become useless . . .

Actually, it is not so easy to choose a random point. The accessible region is very large, on the order of e^{S/k_B}, but it is still a negligible portion of the entire configuration space. In an Ising model in which $S = 0.5\ N k_B \ln 2$, for example, one has

$$\frac{|\Gamma|}{2^N} \sim 2^{-0.5\,N}, \tag{8.58}$$

which, for $N = 100$, gives us 10^{-15}. It is therefore practically impossible that a point chosen at random will fall in Γ. The only way to move closer to Γ is to allow the system to do so spontaneously, in the course of its evolution.

Since we have therefore discarded the simulation's *initial stretch*, we can set

$$\langle A \rangle \simeq [A]_T = \frac{1}{T} \sum_{t=T_0}^{T_0+T} A_{\sigma(t)}, \tag{8.59}$$

where we have denoted A's *current average* along a stretch of the trajectory of length T with $[A]_T$. This expression yields $\langle A \rangle$ up to a **statistical** error ΔA_T, of the order of $T^{-1/2}$. If the values of $A_{\sigma(t)}$ were independent for different values of t, one would have

$$\langle \Delta A_T^2 \rangle \simeq \frac{1}{T} \sum_{t=T_0}^{T_0+T} \left(A_{\sigma(t)} - [A]_T \right)^2. \tag{8.60}$$

The configurations obtained from the Markov chain, however, are not mutually independent: the configurations σ_t and $\sigma_{t'}$ will only be independent if $|t - t'|$ is larger than a certain characteristic correlation time τ_0. We can therefore expect

$$\langle \Delta A_T^2 \rangle \simeq \frac{\tau_0}{T} \sum_{t=T_0}^{T_0+T} \left(A_{\sigma(t)} - [A]_T \right)^2, \tag{8.61}$$

which can be much larger than equation (8.60). In order to be able to estimate the statistical error, it is necessary to have an idea of how large τ_0 is. One possibility is that of monitoring the configurations' **mutual overlap**:

$$C(t+\tau, t) = \frac{1}{N} \sum_i \sigma_i(t+\tau)\,\sigma_i(t). \tag{8.62}$$

If the configurations are independent, $C(t + \tau, t) \simeq m^2$, where m is the magnetization per spin (at least in the ferromagnetic Ising model). Therefore, if we value $C(t + \tau, t)$ as a function of τ, we can arrive at an estimate of τ_0 and hence also of the statistical error ΔA_T.

It is clearly very useful to define the algorithm in such a manner as to make τ_0 as small as possible. One can achieve this outcome by attempting to enlarge the configuration space σ' accessible by a given configuration σ, making it so that the overlap between σ and σ' is as small as possible. On the other hand, the transition $\sigma \to \sigma'$ is in danger of not being accepted if $H(\sigma)$ is very different from $H(\sigma')$. Looking for very different configurations that have approximately the same energy is an interesting problem that has been solved using sophisticated methods in the case of the Ising model.

Let us note that $C(t + \tau, t)$ can reach a plateau as τ grows even if the system remains trapped in a metastable region of configuration space—in this case, all the quantities we will calculate will be relative to this metastable region and will have nothing to do with equilibrium. This problem is important, because simulation times are always enormously shorter than the "real" times, and therefore a metastable state whose lifespan is much longer than the simulation time can have a very short life span in the real world.

8.4.4 Extrapolation to the Thermodynamic Limit

Let us now suppose that you have correctly evaluated correlation time τ_0 and that you have obtained some trajectories long enough to produce $\langle A \rangle$ averages with a negligible statistical error. The problem is now to interpret the data we have obtained in this fashion, to obtain some information about the thermodynamics of the Ising model, for instance. We might, for example, be interested in localizing the transition temperature T_c, estimating the critical exponents β and ν, and so on. At this point (setting aside the problem of boundary conditions and other idealizations), you will have correctly simulated the behavior of a sample whose real-world size is equal to a few nanometers, for a duration of some picoseconds. You will therefore want to extrapolate the data to obtain the behavior of a system in the thermodynamic limit.

The problem is that obviously the behavior of such a small system is qualitatively different from that of a macroscopic system. Let us consider the Ising model under T_c in a vanishing field. The infinite system exhibits a nonvanishing spontaneous magnetization (which is equal to $\pm m_0$ per spin), while we know that there cannot be spontaneous symmetry breaking in a finite system. In fact, if we calculate the instantaneous magnetization $M(t)$ in the finite system

$$M(t) = \sum_i \sigma_i(t), \tag{8.63}$$

we will see that it normally fluctuates around $\pm m_0 N + O(N^{1/2})$, but its sign changes from time to time. If we limited ourselves to calculating $\langle M \rangle$, averaging $M(t)$ over a time interval in which these inversions do not appear, we could not reduce our statistical errors sufficiently. It is therefore necessary to calculate some quantities that will not be influenced by these inversions. In our case, it is convenient to introduce the quantity m_{RMS}, defined by

$$m_{\text{RMS}}^2 = \frac{1}{N^2}\langle M^2(t) \rangle = \frac{1}{N^2}\left\langle \left[\sum_i \sigma_i(t)\right]^2 \right\rangle. \tag{8.64}$$

This quantity is well defined but has the drawback that it also remains nonvanishing for $T > T_c$. Therefore, we cannot immediately determine T_c. This property of m_{RMS} is, however, actually an advantage, because it allows us to have some fairly long runs so that we can reduce the systematic error over m_{RMS} to values smaller than m_{RMS} itself (something that would be rather difficult to do for quantities that vanish), and therefore allows us to produce reliable m_{RMS} curves as a function of T.

In order to utilize these data, one makes use of the theory of **finite-size scaling**, which I will attempt to summarize.

Let us consider a sample of linear dimensions L and discuss its spin susceptibility $\chi(L, T)$, for example. On the one hand, we expect that $\chi(L, T)$ will have the following scaling expression:

$$\chi(L, T) = L^{y}\hat{\chi}(L/\xi(T)). \tag{8.65}$$

On the other hand, if the coherence length ξ is much smaller than L, we expect the behavior of the sample to be substantially identical to that of an infinite system. For $L \gg \xi$, $\hat{\chi}(L/\xi) \sim \xi^{\gamma/\nu}$, therefore, and in order for χ to be substantially independent of L, one must have $y = \gamma/\nu$. Vice versa, for $L \ll \xi$, we expect that $\chi(L, T)$ (and therefore $\hat{\chi}[L/\xi(T)]$) be independent of ξ, and therefore $\chi(L/T) \sim L^{\gamma/\nu}$.

We can now observe that if $T > T_c$,

$$m_{\mathrm{RMS}}^2 = \frac{1}{N^2}\sum_{ij}\langle\sigma_i\sigma_j\rangle = \frac{k_B T}{N}\chi(L, T). \tag{8.66}$$

By substituting in this expression the behavior of $\chi(L, T)$, which we have found previously, we obtain (for $T \simeq T_c$)

$$m_{\mathrm{RMS}}^2 = L^{-d+\gamma/\nu}\hat{\chi}\left(\frac{L}{\xi}\right) = L^{-2\beta/\nu}\hat{\chi}\left(L|T - T_c|^{\nu}\right). \tag{8.67}$$

where we have exploited the scaling law

$$\beta = \frac{1}{2}(d\nu - \gamma). \tag{8.68}$$

Therefore, if one knows T_c and the exponents β and ν, the data relative to $m_{\mathrm{RMS}}^2 L^{2\beta/\nu}$, plotted as a function of $x = L|T - T_c|^{\nu}$ for various values of L, should fall on a single line.

It is possible to determine these values iteratively by exploiting the fact that equation (8.67) implies that, for each L, the temperature $T_0(L)$ where $\mathrm{d}^2 m_{\mathrm{RMS}}/\mathrm{d}T^2 = 0$ scales as

$$T_0(L) = T_c + O(L^{-1/\nu}), \tag{8.69}$$

while $m_{\mathrm{RMS}}[T_0(L)]$ goes as $L^{-\beta/\nu}$. By making use of the data relative to $T_0(L)$—in other words, those in which $m_{\mathrm{RMS}}(T)$ exhibits an inflection point one can try to determine T_c, ν and β/ν. The quality of the result can then be checked by replotting all the data in the form dictated by (8.67). An example of this *data collapse* is presented in figure 8.1.

Another technique exploits the properties of the distribution $p_L(m)$ of the instantaneous magnetization per spin $m = M/N$. For $T > T_c$, this distribution is a Gaussian with a width determined by the susceptibility $\chi(L, T)$:

$$p_L(m) \propto \exp\left[-\frac{m^2}{2k_B T\chi(L, T)}\right]. \tag{8.70}$$

For $T < T_c$, this distribution exhibits two peaks, and at low temperature, it becomes the sum of two narrow Gaussians. For $T \simeq T_c$, we can assume a scaling form for $p_L(m)$:

$$p_L(m) = L^{y}\hat{P}(mL^{y}, L/\xi(T)). \tag{8.71}$$

Let us evaluate the average of m's modulus:

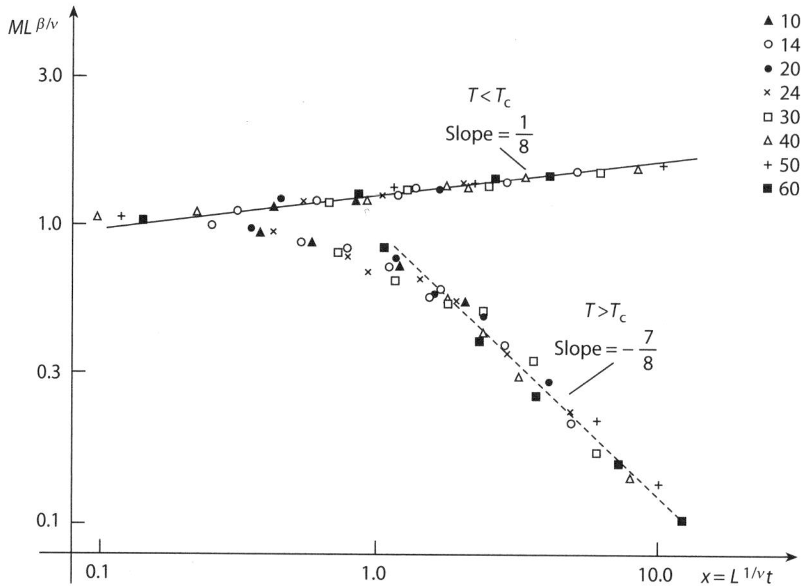

FIGURE 8.1. Rescaled magnetization $y = ML^{\beta/\nu}$ for the two-dimensional Ising model, as a function of the rescaled distance from the critical temperature $y = L^{1/\nu}t$, with $t = |T - T_c|$, and for both $T > T_c$ and $T < T_c$. Adapted from [Bark76], with permission.

$$\begin{aligned} \langle |m| \rangle &= L^y \int_{-\infty}^{\infty} \mathrm{d}m |m| \hat{P}(mL^y, L/\xi(T)) \\ &= L^{-y} \int_{-\infty}^{\infty} \mathrm{d}z |z| P(z, L/\xi(T)) = L^{-y} \Phi(L/\xi(T)), \end{aligned} \tag{8.72}$$

which is the usual scaling form for the order parameter and therefore implies $y = \beta/\nu$. Let us now consider the moments $\langle m^2 \rangle$ and $\langle m^4 \rangle$. Clearly

$$\langle m^2 \rangle = L^{-2/y} \int_{-\infty}^{\infty} \mathrm{d}z\, z^2 \hat{P}(z, L/\xi(T)), \tag{8.73}$$

$$\langle m^4 \rangle = L^{-4/y} \int_{-\infty}^{\infty} \mathrm{d}z\, z^4 \hat{P}(z, L/\xi(T)). \tag{8.74}$$

Therefore, the ratio

$$R = \frac{\langle m^4 \rangle}{\langle m^2 \rangle^2} = \Psi(L/\xi(T)) \tag{8.75}$$

tends to a well-defined limit (independent of L) for $\xi(T) \to \infty$. On the other hand, for $T \gg T_c$, $R = 3$, while we would expect $R \to 1$ for $T \to 0$. Therefore, the $R(T)$ curves, evaluated for different L's, intersect at a point at R which assumes a value between 1 and 3, in correspondence of the critical temperature for the infinite system.

8.5 Umbrella Sampling

Computing time is not free. If you have finally obtained a long Monte Carlo run for your model, at the temperature (let us say) T_0 and magnetic field h, it would be good not to throw your precious data away before having extracted as much relevant information as possible. For example, you could exploit it so as to obtain data for different temperature and magnetic field values.

In fact, if the trajectory $(\sigma(0), \sigma(1), \ldots)$ samples the canonical distribution of the Hamiltonian H at the temperature T_0, the A_σ observable's average at temperature T can be calculated using the relation

$$\begin{aligned}
\langle A \rangle_T &= \frac{1}{Z(T)} \sum_\sigma A(\sigma)\, E^{-H(\sigma)/k_B T} \\
&= \frac{1}{Z(T_0)} \frac{Z(T)}{Z(T_0)} \sum_\sigma A(\sigma) \exp\left[-H(\sigma)\left(\frac{1}{T} - \frac{1}{T_0}\right)\right] \mathrm{e}^{-H(\sigma)/k_B T_0} \\
&= \left\langle A \exp\left[-H\left(\frac{1}{T} - \frac{1}{T_0}\right)\right]\right\rangle_{T_0} \left\langle \exp\left[-H\left(\frac{1}{T} - \frac{1}{T_0}\right)\right]\right\rangle_{T_0}^{-1},
\end{aligned} \tag{8.76}$$

where $\langle \ldots \rangle_T$ denotes the canonical average at the temperature T.

Thus the average at temperature T can be obtained by evaluating suitable means along trajectories at the different temperature T_0. One can calculate the averages relative to the different values of the magnetic field and so on. in a similar fashion. One would do well, however, to keep in mind that this method works only if the perturbation (the argument of the exponential in the preceding expression) is very small. If the exponential factor becomes very large or very small, the results will no longer be trustworthy. Practically speaking, it is convenient to *always* use this trick to evaluate the effect of a small variation in the parameters, and therefore the respective susceptibilities (or, in the case of temperature, the specific heat). One will then be able to exploit the fluctuation–dissipation theorem to check that the fluctuations that have been estimated (and by means of which one calculates the statistical errors) agree with the susceptibilities. If this does not occur, one can infer that something is not working in the simulation—for example, that equilibrium has not actually been reached, or that the correlation time has been underestimated.

In a similar fashion, we can sample the canonical distribution with a Hamiltonian H_0, which is not identical to the one (let us call it H) that we are interested in, if this is to our advantage, as long as we are careful to take the correction into account. By denoting the average over the Hamiltonian H by $\langle \ldots \rangle_H$, in general, we will have

$$\langle A \rangle_H = \frac{\left\langle A \mathrm{e}^{-(H-H_0)/k_B T} \right\rangle_{H_0}}{\left\langle \mathrm{e}^{-(H-H_0)/k_B T} \right\rangle_{H_0}}. \tag{8.77}$$

This technique is called **umbrella sampling**—in addition to being used for the study of perturbations with respect to a given Hamiltonian, it can be used for the study of rare events. We can, in fact, be interested in the properties of a rare class of configurations in which certain properties are satisfied, and we can introduce H_0 in order to facilitate their appearance.

A fairly frequent case is that in which one wants to study the probability distribution of the order parameter in finite samples. This is the same as studying the $p_L(m)$ distribution we discussed in the preceding section using a histogram. The shape of the $p_L(m)$ actually allows us, for example, to distinguish between continuous and discontinuous transitions. In the vicinity of a discontinuous transition, the $p_L(m)$ exhibits peaks close to the different equilibrium values of m, while in the case of a discontinuous transition, the distribution is much broader. On the other hand, during a run at temperature $T < T_c$, only those configurations in which m is not far from the equilibrium value m_0 will be "spontaneously" sampled. So we can introduce a perturbation V_σ by defining $H_0 = H + V$, chosen so that H_0 favors values of m that differ from equilibrium values. One example is to set $V_\sigma = 0$ if m belongs to a *window* $m^* \leq m \leq m^* + \delta m$, and $V_\sigma = +\infty$ otherwise. We can then evaluate the histogram $p^*(m)$ in the window. By collating the windows, and imposing $p^*(m)$'s continuity [or rather that of $\ln p^*(m)$, which is connected to availability and varies much more slowly], we can reconstruct the behavior of the entire $p_L(m)$.

Exercise 8.8 Evaluate the specific heat in the one-dimensional Ising model using umbrella sampling and check the validity of the fluctuation–dissipation relation.

8.6 Discussion

One of the most interesting aspects of numerical simulation is the way in which it allows one to clarify (not resolve!) the fundamental conceptual problems of statistical mechanics. We have actually seen how a suitably defined Markov chain allows us to sample the canonical ensemble, for example. The runs one can actually perform, however, will allow us to evaluate the observables' mean only over a ridiculously small number of configurations. For $N = 100$, for instance, we have about 10^{30} possible configurations—in other words, ten million trillion trillions. How many *truly independent* configurations will you manage to sample? If you are *very* good and *very* patient, you will reach several hundred thousand (for 100 spins!). You are therefore not sampling the Boltzmann distribution in the same sense in which, for example, by throwing a die 10,000 times you are checking whether it is "honest," since the frequencies of the various results are equal to $1/6$ within 1%. The point is that we are not aiming to sample the distribution in phase space, but only to evaluate the observables' averages. Let us consider the correlation function $\langle \sigma_i \sigma_j \rangle$. It depends on the four possible configurations of σ_i and σ_j. We are therefore sampling the distribution of these four configurations over a *significant sample* of phase space: if we have 100,000 independent configurations, we can be fairly confident about our results, since the statistical errors will be small. This means, for example, that we will have difficulty calculating the correlations for ℓ spins, where ℓ is only slightly larger (5 or 6), because the number of configurations they can exhibit begins to become excessive. More generally, we will have to be careful not to squeeze our data too much—in other words, not to base our conclusions on events that, in our run, occurred a small number of times. If we really have to, we can use umbrella sampling to obtain a statistically significant sample.

Last, let us remark that we were able to obtain only the *observables'* average. There are important thermodynamic quantities that are *not* observable: first entropy, then free energy, and so on. Strictly speaking, one could express the partition function (and therefore free energy) as an average. Since

$$\left\langle e^{H/k_BT} \right\rangle = \frac{1}{Z} \sum_{\sigma} e^{H(\sigma)/k_BT} \, e^{-H(\sigma)/k_BT}, \tag{8.78}$$

we can write

$$Z = \frac{2^N}{\left\langle e^{H/k_BT} \right\rangle}. \tag{8.79}$$

Since H is extensive, however, the e^{H/k_BT} will be too large in a too extended region of configuration space for our results to be reliable. We had therefore better look for some methods to directly calculate quantities that behave slightly better, for instance the entropy S. One of the most frequently used methods (which works if your algorithm reaches equilibrium very fast) is to integrate the specific heat directly (evaluated by means of the umbrella sampling technique), according to the formula

$$\Delta S = \int_{\mathrm{A}}^{\mathrm{B}} dT \frac{C}{T}, \tag{8.80}$$

where A and B are states of thermodynamic equilibrium. Techniques have also been introduced that allow one to evaluate S directly, like the *coincidence method*, for instance, due to S.-K. Ma [Ma85, chapter 25].

Recommended Reading

Molecular dynamics is discussed by, among others, M. P. Allen and D. J. Tildesley, *Computer Simulations of Liquids*, Oxford, UK: Oxford University Press, 1987. An introductory text to the Monte Carlo method is K. Binder and D. W. Heermann, *Monte Carlo Simulation in Statistical Physics*, Berlin: Springer, 1988. More advanced contributions are found in K. Binder, *Monte Carlo Methods in Statistical Mechanics*, 2nd ed., Berlin: Springer, 1986. Nontraditional applications of numerical simulation are discussed in S. Moss de Oliveira, P.M.C. de Oliveira, and D. Stauffer, *Evolution, Money, War, and Computers*, Stuttgart: Teubner, 1999. The chapter on numerical simulation in D. Chandler, *Introduction to Modern Statistical Mechanics*, Oxford, UK: Oxford University Press, 1987, is excellent.

9 Dynamics

> Theories are made to seduce aesthetes, irritate philistines, and make everyone else laugh.
>
> —Amélie Nothomb

9.1 Brownian Motion

Brownian motion is the paradigm for dynamic phenomena in classical statistical mechanics. J. Perrin, in his classic book *Atoms*, describes it in the following manner [Perr16, pp. 83–85]:

> We have only to examine under the microscope a collection of small particles suspended in water to notice at once that each of them, instead of sinking steadily, is quickened by an extremely lively and *wholly haphazard* movement. Each particle spins hither and thither, rises, sinks, rises again, without ever tending to come to rest. . . . Of course, the phenomenon is not confined to suspensions in water, but takes place in all fluids, though more actively the less viscous the fluid. Thus it is just perceptible in glycerin, and extremely active, on the other hand, in gases. . . . In a given fluid the size of the grains is of great importance, the agitation being the more active the smaller the grains. This property was pointed out by Brown at the time of his original discovery. The nature of the grains appears to exert little influence, if any at all.

Although the connection between Brownian motion and molecular theory had already been seen previously,[1] Einstein [Eins05] was the first to provide a quantitative explanation of Brownian motion, in one his fundamental works of 1905. Subsequently taken up again by Einstein himself, by P. Langevin, and by L. S. Ornstein, and experimentally verified by J. Perrin himself, by T. Svedberg, and others at the beginning of the last century, it represented one the fundamental proofs of the corpuscular nature of matter. The most important papers by Einstein on this subject are collected in [Eins56]. Other important papers are [Lang08], [Orns20], [Perr08], and [Sved06] cited in the bibliography.

[1] Specifically, by O. Wiener in 1863 and by G. Gouy in 1888.

Let us consider a particle much larger than a molecule, of mass M, immersed in a fluid at thermal equilibrium, at a temperature T. If it moves with a velocity $\boldsymbol{v}$, because of the fluid's resistance, it will be subject to a force $\boldsymbol{F}_r = -\lambda \boldsymbol{v}$ where λ is a positive coefficient that depends on the particle's geometry and the fluid's viscosity. For a spherical particle of radius R, immersed in a fluid of viscosity η, Stokes's law gives

$$\lambda = 6\pi\eta R. \tag{9.1}$$

On the other hand, because of the ceaseless thermal agitation animating the liquid's molecules, the particle will be subjected to a random force $\boldsymbol{f}(t) = (f_1(t), f_2(t), f_3(t))$. We can reasonably assume that this random force has the following properties:

Isotropy: The force does not favor any particular direction. (For simplicity's sake, let us momentarily suppose that we disregard the effects of external forces, such as weight.)

Noncorrelation: The force fluctuates at each instant, and its value at an instant t is independent from that at instant t', as soon as $|t - t'|$ is larger than a microscopic time τ_{micr}.

Gaussian distribution: The force $\boldsymbol{f}(t)$ is the result of the independent actions performed at every instant by a large number of molecules. As a result of the central limit theorem, we can therefore expect that its distribution will be Gaussian.

Since it is Gaussian, the force distribution is determined by its average and its correlation, which can be determined by the other two properties. We thus obtain

$$\langle f_i(t) \rangle = 0, \qquad i = 1,2,3, \tag{9.2}$$

$$\langle f_i(t)\, f_j(t') \rangle = \Lambda \delta_{ij} \Delta(|t - t'|). \tag{9.3}$$

In equation (9.3), $\Delta(|t - t'|)$ is a function that rapidly tends to zero when its argument is larger than a very short time τ_{micr}, and it is arbitrarily normalized so that $\int_{-\infty}^{\infty} \mathrm{d}t \Delta(t) = 1$. It can therefore by represented by a Dirac delta.

By applying Newton's law to our particle, we obtain a **Langevin equation**:

$$M \frac{\mathrm{d}\boldsymbol{v}}{\mathrm{d}t} = -\lambda \boldsymbol{v} + \boldsymbol{f}(t). \tag{9.4}$$

This is an example of a **stochastic differential equation**—in other words, of a differential equation in whose definition random quantities appear, in our case, the random force $\boldsymbol{f}(t)$. We have derived it (and we will treat it) in a heuristic manner; it is however, possible to give it a rigorous definition and treatment, which is due to N. Wiener and K. Itō ([Wien30] and [Ito44]). An accessible introduction to the theory and literature of stochastic differential equations is available in [Gard83]. To simplify the discussion it is convenient to consider the problem in one dimension for the time being. By denoting the only components of $\boldsymbol{v}$ and $\boldsymbol{f}$ as v and f, we obtain

$$M \frac{\mathrm{d}v}{\mathrm{d}t} = -\lambda v + f(t). \tag{9.5}$$

This is a linear equation in $v(t)$, in which $f(t)$ appears as an additive term. It can be formally integrated, and one obtains the solution

$$v(t) = \exp\left(-\frac{\lambda}{M}t\right)v_0 + \int_0^t \mathrm{d}t' \exp\left[-\frac{\lambda}{M}(t-t')\right]\frac{f(t')}{M}. \tag{9.6}$$

In this equation, v_0 is $v(t)$'s value at instant $t = 0$. We see that $v(t)$ is a linear functional of $f(t)$. Since $f(t)$ is Gaussian, and since linear functions of random Gaussian variables are Gaussian, $v(t)$ is also a random Gaussian process. By taking the average of both sides of equation (9.6), we obtain [keeping in mind the equation (9.2)]

$$\langle v(t)\rangle = \exp\left(-\frac{\lambda}{M}t\right)\langle v_0\rangle. \tag{9.7}$$

If the Brownian particle is initially at thermodynamic equilibrium, one needs to have $\langle v_0\rangle = 0$, and therefore $v(t) = 0$, $\forall t$. Let us now evaluate $\langle v^2(t)\rangle$, taking the square of both sides of equation (9.6). We obtain

$$\begin{aligned}\langle v^2(t)\rangle &= \exp\left(-\frac{2\lambda}{M}t\right)\langle v_0^2\rangle + 2\int_0^t \mathrm{d}t' \exp\left[-\frac{\lambda}{M}(t-t')\right]\langle v_0 f(t)\rangle \\ &\quad + \frac{1}{M^2}\int_0^t \mathrm{d}t' \int_0^t \mathrm{d}t_1' \exp\left\{-\frac{\lambda}{M}\left[(t-t') + (t-t_1')\right]\right\}\langle f(t)\, f(t')\rangle .\end{aligned} \tag{9.8}$$

The second term vanishes because v_0 and $f(t)$ are independent, and each has a vanishing average. The third term contributes only for $t' = t_1'$ because of equation (9.3). We thus obtain

$$\begin{aligned}\langle v^2(t)\rangle &= \exp\left(-\frac{2\lambda}{M}t\right)\langle v_0^2\rangle \\ &\quad + \frac{\Lambda}{M^2}\int_0^t \mathrm{d}t' \exp\left[-\frac{2\lambda}{M}(t-t')\right].\end{aligned} \tag{9.9}$$

By evaluating the integral on the right hand side, we finally have

$$\langle v^2(t)\rangle = \langle v_0^2\rangle\, \mathrm{e}^{-(2\lambda/M)t} + \frac{\Lambda}{2M\lambda}\left(1 - \mathrm{e}^{-(2\lambda/M)t}\right). \tag{9.10}$$

We see that after a short time (on the order of $M/2\lambda$), one has $\langle v^2(t)\rangle \to \Lambda/2M\lambda$. The central idea of Einstein's 1905 article is that a Brownian particle differs from the solvent's molecules *only by its size*. The particle is therefore subject to the equipartition theorem, which stipulates that

$$\frac{1}{2}M\langle v^2\rangle = \frac{1}{2}k_{\mathrm{B}}T. \tag{9.11}$$

Comparing this relation with equation (9.10), we see that λ must satisfy

$$\Lambda = 2\lambda k_{\mathrm{B}}T. \tag{9.12}$$

One should note that if equation (9.11) is initially satisfied, then equation (9.10) implies that it remains identically satisfied for each successive value of t. The equation provides a relation between a quantity (λ) connected to the Brownian particle's mobility and a quantity (Λ) that expresses the fluctuations of the random force. The drawback of this relation is that Λ is not directly measurable.

Einstein himself, however, obtained a relation between λ and a directly observable quantity, connected to the amplitude of the Brownian particle's excursions. Following Langevin, let us multiply both sides of equation (9.4) by x and define

$$\Delta x(t) = x(t) - x(0), \tag{9.13}$$

$$z(t) = \frac{\mathrm{d}\Delta x^2}{\mathrm{d}t}. \tag{9.14}$$

We obtain

$$\frac{1}{2}M\frac{\mathrm{d}z}{\mathrm{d}t} - Mv^2 = -\frac{\lambda}{2}z + xf(t). \tag{9.15}$$

Let us take the average of this relation, and keep in mind the equipartition theorem, as well as the fact that $f(t)$ is independent of x and has a vanishing average. We obtain

$$\frac{1}{2}M\frac{\mathrm{d}\langle z\rangle}{\mathrm{d}t} + \frac{\lambda}{2}\langle z\rangle = k_{\mathrm{B}}T. \tag{9.16}$$

Therefore,

$$\langle z(t)\rangle = \frac{2k_{\mathrm{B}}T}{\lambda} + Ce^{-(\lambda/M)t}, \tag{9.17}$$

where C is a constant that depends on the initial conditions. After some time τ_{r} of the order of M/λ, the Brownian particle's displacements will tend to follow a law according to which $\Delta x^2(t)$ grows linearly with t:

$$\lim_{t\to\infty}\frac{\mathrm{d}\langle\Delta x^2\rangle}{\mathrm{d}t} = 2D = \frac{2k_{\mathrm{B}}T}{\lambda}. \tag{9.18}$$

The constant D that appears in this relation is called the **diffusion coefficient**. The relation between D and λ is also known as the **Einstein relation**. From what has preceded, one can see that the Einstein relation allows one to evaluate $k_{\mathrm{B}}T$ if one follows the displacements of a Brownian particle of known radius R. This is the most surprising result of Einstein's 1905 article. In particular, let us note that D is independent of the Brownian particle's mass M, and depends only on its radius R (via equation [9.1]) and on the solvent's viscosity η (as well as on temperature), agreeing with Perrin's qualitative observations. We will see how Einstein's derivation allows one to understand the deeper reasons for this relation.

For a particle with a radius of one micrometer, one has $D \simeq 2.6 \cdot 10^{-13}$ m^{2}s^{-1}, which corresponds to a displacement of about 5.6 μm per minute. The particles used by Perrin in his experiments had a density equal to about 1.2 times that of water, and a radius of this order of magnitude. For such particles, the relaxation time τ_{r} is of the order of $2.6 \cdot 10^{-10}$ s.

An equivalent expression of the diffusion coefficient D can be obtained by considering the velocity correlation function. By multiplying both sides of equation (9.4) by $v(0)$ and taking the average, we obtain

$$M\frac{\mathrm{d}}{\mathrm{d}t}\langle v(t)\,v(0)\rangle = -\lambda\langle v(t)\,v(0)\rangle + \langle f(t)\,v(0)\rangle. \tag{9.19}$$

The last term vanishes for $t > 0$. On the other hand,

$$\langle v^2(0)\rangle = \frac{k_B T}{M}, \tag{9.20}$$

because of the equipartition theorem, which provides us with the initial condition. We thus obtain

$$\langle v(t)\,v(0)\rangle = \frac{k_B T}{M}\,\mathrm{e}^{-(\lambda/M)t}, \tag{9.21}$$

Let us note that, since we are at equilibrium, one must have

$$\langle v(t+t')\,v(t')\rangle = \langle v(t)\,v(0)\rangle, \tag{9.22}$$

for any value of t'. This property is known as **time translation invariance**. More particularly, by setting $t' = -t$, we obtain

$$\langle v(t)\,v(0)\rangle = \langle v(0)\,v(-t)\rangle, \tag{9.23}$$

which allows us to extend the expression of the correlation function to include negative times. We thus obtain

$$\langle v(t)\,v(0)\rangle = \frac{k_B T}{M}\,\mathrm{e}^{-(\lambda/M)|t|}, \qquad \forall t. \tag{9.24}$$

Let us now consider the definition of the diffusion coefficient that follows from equation (9.18):

$$D = \lim_{t\to\infty}\frac{1}{2}\frac{\mathrm{d}\langle\Delta x^2\rangle}{\mathrm{d}t}.$$

Since $\Delta x = \int_0^t \mathrm{d}t'\,v(t')$, we can write the relation

$$\langle\Delta x^2(t)\rangle = \int_0^t \mathrm{d}t_1 \int_0^t \mathrm{d}t_2\,\langle v(t_1)\,v(t_2)\rangle. \tag{9.25}$$

By taking the derivative of this expression with respect to t, we obtain

$$\frac{\mathrm{d}}{\mathrm{d}t}\langle\Delta x^2(t)\rangle = 2\int_0^t \mathrm{d}t_1\,\langle v(t)\,v(t_1)\rangle,$$

and therefore

$$D = \lim_{t\to\infty}\frac{1}{2}\frac{\mathrm{d}\langle\Delta x^2\rangle}{\mathrm{d}t} = \lim_{t\to\infty}\int_0^t \mathrm{d}t_1\,\langle v(t)\,v(t_1)\rangle.$$

By exploiting time translation invariance, one can rewrite this relation in the form

$$D = \int_0^\infty \mathrm{d}t\,\langle v(t)\,v(0)\rangle. \tag{9.26}$$

This expression, which connects a kinetic coefficient (D) to a dynamic correlation function (in our case, of v) is a first example of a class of relations known as **Green-Kubo formulas.**

It is now easy to describe the Brownian particle's motion in d dimensions. It is useful to define the diffusion coefficient D as follows:

$$D = \lim_{t \to \infty} \frac{1}{2d} \frac{\mathrm{d}\langle \Delta r^2 \rangle}{\mathrm{d}t} = \lim_{t \to \infty} \frac{1}{2d} \frac{\mathrm{d}}{\mathrm{d}t} \sum_{i=1}^{d} \langle \Delta r_i^2 \rangle, \tag{9.27}$$

where $\Delta r = (\Delta r_i)$, $(i = 1, \ldots, d)$. The Einstein relation between D and λ thus remains unchanged, and the same applies for that between Λ and λ, and accounting for equation (9.3) and the equipartition theorem. The d components of the displacement are mutually independent. Relation (9.26) is modified as follows:

$$D = \frac{1}{d} \int_0^{\infty} \mathrm{d}t \, \langle \boldsymbol{v}(t) \cdot \boldsymbol{v}(0) \rangle. \tag{9.28}$$

Proving these conclusions becomes an easy exercise for the reader.

9.2 Fractal Properties of Brownian Trajectories

In figure 9.1, one can see the extremely irregular appearance of the trajectories traveled by a Brownian particle. When describing the trajectories of the Brownian particles he observed, Perrin [Perr16, pp. 115–116] notes that

> As a matter of fact diagrams of this sort, and even the next figure, in which a large number of displacements are traced on an arbitrary scale, give only a very meager idea of the extraordinary

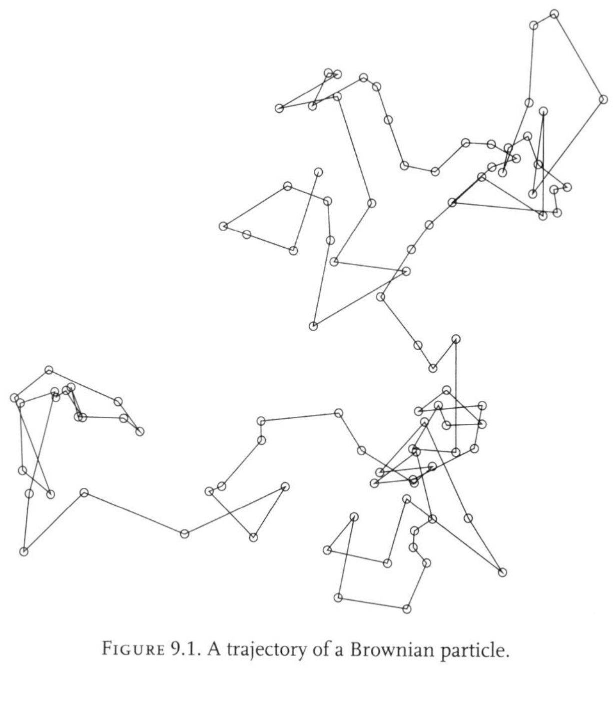

FIGURE 9.1. A trajectory of a Brownian particle.

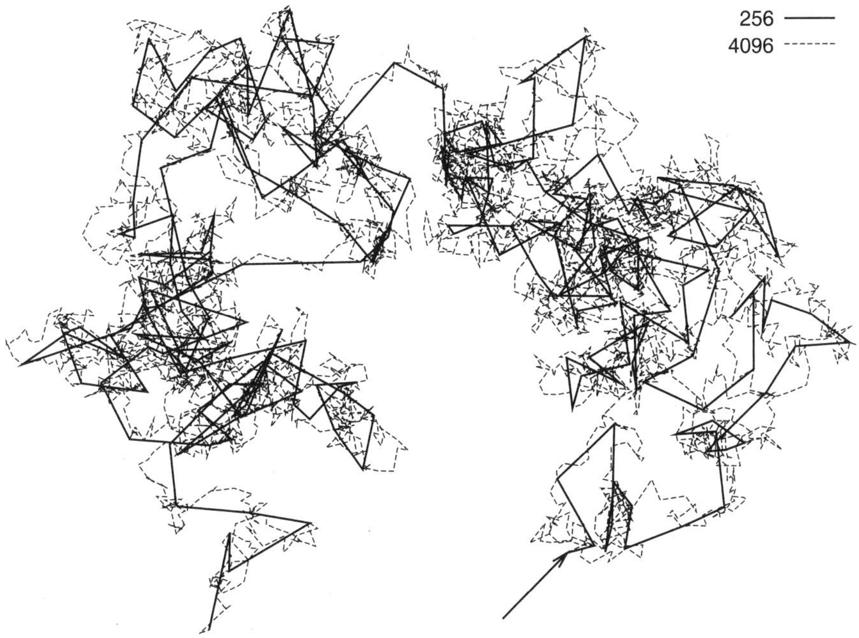

FIGURE 9.2. A Brownian trajectory described with increasing resolution. The particle's initial position is indicated by the arrow. Solid line: position at time intervals τ. Dashed line: position at time intervals $\tau/16$.

irregularity of the actual trajectory. For if the positions were to be marked at intervals of time 100 times shorter, each segment would be replaced by a polygonal contour relatively just as complicated as the whole figure, and so on. Obviously it becomes meaningless to speak of a tangent to a trajectory of this kind.

The fact that the Brownian trajectory becomes all the more complex the more densely the particle's positions are plotted is illustrated in figure 9.2. In that figure, we reproduce the particle's position over 256 successive time intervals equal to τ (the continuous line). In the same figure, the dashed line shows positions measured over 4096 time intervals, each one lasting $\tau/16$. In this manner, one can observe the trajectory's progressive increase in complexity. Only if we were able to observe the particle's displacements over times on the order of τ_r, the velocities' correlation time, would the Brownian trajectory obviously begin to look smooth. As long as the interval τ between successive observations remains much larger than τ_r, the velocity estimate obtained from the $\Delta r/\tau$ ratio will tend to grow as $\tau^{-1/2}$ with decreases of τ. This paradoxical behavior had also been noticed by Gouy, and it represented one of the obstacles to the understanding of Brownian motion that Einstein so brilliantly overcame.

The Brownian trajectory (at least as long as it is not measured over time intervals on the order of τ_r) is therefore a **fractal**, because it possesses scale-invariance properties: one of its segments, suitably enlarged, has the same statistical properties as the entire trajectory. The fractal concept was introduced by Mandelbrot [Mand77] to designate objects with extremely irregular shapes, which however still possess the property of scale-invariance. Historically, the first example of a fractal object (well before the introduction of the term) was precisely Brownian trajectories.

Let us suppose that we want to measure the length of the path traveled by a Brownian particle during a certain time interval T. We subdivide this interval into subintervals of $\tau = T/n$, and we measure the particle's displacement $\Delta \boldsymbol{r}_i$ in each of these intervals. The total length L_n of the path measured with this degree of resolution will be given by

$$L_n = \sum_{i=1}^{n} \sqrt{\Delta r_i^2}. \tag{9.29}$$

The length of the path L_n is obviously a random quantity, but we can easily assess its typical value—actually, $\sqrt{\Delta r_i^2}$ will be on the order of $\sqrt{2dD\tau} \simeq \sqrt{2D\mathcal{T}}/n^{1/2}$. Therefore,

$$L_n \propto n \cdot n^{-1/2} = n^{1/2}. \tag{9.30}$$

We see that as the resolution with which we describe the trajectory increases, the run's length increases! This differs from what happens when we are dealing with *smooth* curves—in this case, the total length tends to a limit, for $n \to \infty$, which is the *length* of the curve.

The manner in which length grows as resolution increases is described by a number that can be considered the trajectory's **fractal dimension**. An intuitive notion of an object's dimensions can be obtained by counting how the number of objects of size ℓ necessary to cover the object in question varies when ℓ changes. If we consider a smooth and finite curve in a plane with circles of radius ℓ, the number n_ℓ of circles of radius ℓ necessary to cover it will vary as L/ℓ, and will therefore grow as ℓ^{-D_C}, where $D_C = 1$. On the other hand, if we want to cover a regular plane figure (a polygon, for example) with circles of radius ℓ, the number of circles necessary will vary as ℓ^{-D_C}, with $D_C = 2$. We can therefore identify D_C with the dimension $D = 1$ for the line and $D = 2$ for the plane surface. We will call D_C the **covering dimension** of the figure in question.

Let us now assume that we have an infinitely detailed picture of the Brownian trajectory, and we ask ourselves how many circles of radius ℓ are necessary to cover it. Since the typical displacement over a time interval τ is proportional to $\sqrt{\tau}$, we can associate a time interval $\tau \propto \ell^2$ with each ℓ, during which the displacement will be on the order of ℓ. In that case, it is clear that the number of circles necessary to cover the trajectory will vary as τ^{-1}, and therefore as ℓ^{-D_C} with $D_C = 2$. We thus arrive at the result that *the coverage dimension of the Brownian trajectory is equal to 2.* This result is independent of the dimension of the space in which the Brownian motion occurs, as long as it is at least equal to 2. The Brownian trajectory is therefore a line with fractal dimension equal to 2.[2]

[2]There are various ways to generalize the concept of dimension of fractal objects, and *coverage dimension* is only one of them. These ways are not equivalent, and for this reason, it is best to use the concept of *fractal dimension* in an informal sense, and each time specify what particular concept of dimension we are referring to.

9.3 Smoluchowski Equation

Let us now consider the evolution of the probability density $P(\boldsymbol{r}, t)$ that the Brownian particle can be found around point $\boldsymbol{r}$ at instant t. In order to simplify our reasoning, let us once again limit ourselves to considering motion in one dimension, and let us denote $\boldsymbol{r}$'s only coordinate with x.

As a consequence of the law of total probabilities, we can evaluate the probability density $P(x, t+\Delta t)$ that the particle is found close to x at time $t+\Delta t$ by summing over all its possible positions x' at time t. Then $P(x, t+\Delta t)$ is given by the integral over x' of the probability density $P(x', t)$ that the particle was found at x' at time t, times the conditional probability $P(x, t+\Delta t \mid x', t)$, that starting from x' at time t, it had reached x at time $t+\Delta t$.

$$P(x, t+\Delta t) = \int \mathrm{d}x'\, P(x, t+\Delta t \mid x', t)\, P(x', t). \tag{9.31}$$

In a time interval of duration $\Delta t \gg \tau_{\mathrm{r}}$, the Brownian particle undergoes a displacement $\Delta x = x - x'$ whose average vanishes and whose variance is equal to $2D\Delta t$. We will denote the distribution of this displacement by $\phi(\Delta x)$. If Δt is small enough, we can expand $P(x', t)$ on the right-hand side of equation (9.31) in a Taylor series in powers of Δx around x. Changing the integration variable to Δx, we obtain

$$P(x, t+\Delta t) = \int \mathrm{d}\Delta x\, \phi(\Delta x) \left[P(x, t) - \Delta x \left.\frac{\partial P}{\partial x'}\right|_x + \frac{1}{2}\Delta x^2 \left.\frac{\partial^2 P}{(\partial x')^2}\right|_x + \cdots \right]. \tag{9.32}$$

The second term in square brackets vanishes because Δx's average vanishes, and by taking into account what we previously discussed, we obtain

$$P(x, t+\Delta t) = P(x, t) + D\Delta t \frac{\partial^2 P}{\partial x^2} + \mathrm{o}(\Delta t). \tag{9.33}$$

Passing to the limit for $\Delta t \to 0$, we see that the $P(x, t)$ satisfies the **diffusion equation** (also called the **Fick equation**):

$$\frac{\partial P}{\partial t} = D \frac{\partial^2 P}{\partial x^2}. \tag{9.34}$$

It is interesting to consider the specific solution of this equation that satisfies the initial condition $P(x, t=0) = \delta(x)$, where $\delta(x)$ is a Dirac delta. It is obviously none other than the particle's displacement distribution Δx during a time interval of duration t. In order to arrive at this solution, let us transform both sides of equation (9.34) according to Fourier and obtain

$$\frac{\partial P_k}{\partial t} = -Dk^2 P_k. \tag{9.35}$$

Since $P_k(t=0) = \int \mathrm{d}x\, \mathrm{e}^{ikx} \delta(x) = 1$, we obtain

$$P_k(t) = \mathrm{e}^{-Dk^2 t}, \qquad t > 0. \tag{9.36}$$

By antitransforming according to Fourier, we finally arrive at

$$P(x,t) = (4\pi Dt)^{-1/2} \exp\left(-\frac{x^2}{4Dt}\right). \tag{9.37}$$

The displacements' distribution is therefore Gaussian, as was to be expected on the basis of the central limit theorem, since the displacement is the result of the sum of a large number of independent displacements.

Let us now suppose that we have a suspension of a very large number (N) of identical Brownian particles, and denote the local density of Brownian articles by $\rho(x,t)$. If the suspension is sufficiently diluted, to the extent that the Brownian particles can be considered independent, the $\rho(x,t)$ will obey the same equation (9.34). If at a certain instant the density $\rho(x,t)$ is not uniform, therefore, a flow of particles proportional to the derivative of ρ will be produced. This flow is called the **diffusion current** and is expressed as follows:

$$j_{\mathrm{D}} = -D\frac{\partial \rho}{\partial x}. \tag{9.38}$$

Einstein, in his 1905 article, remarked that the existence of this current was a necessary consequence of van 't Hoff's theory of the thermodynamics of diluted solutions, since there was no reason, from the point of view of the molecular theory of heat, to consider a suspension of Brownian particles as substantially different from a diluted solution.

At this point, it is possible to obtain the Einstein relation between the resistance coefficient λ and the diffusion constant D directly, by supposing that the particles are subject to a uniform and direct force F—for example, in the direction of the growing x's. As a consequence of this force, the particles will be driven, on average, by a velocity v_{lim} equal to F/λ. There will therefore by a systematic current j_F equal to

$$j_{\mathrm{F}} = \rho v_{\mathrm{lim}} = \rho\frac{F}{\lambda}. \tag{9.39}$$

At equilibrium, the two currents should compensate one another on average, and therefore p_{eq} must satisfy the relation

$$\rho\frac{F}{\lambda} = -D\frac{\partial \rho}{\partial x}. \tag{9.40}$$

This equation admits

$$\rho_{\mathrm{eq}}(x) \propto \exp\left(\frac{Fx}{\lambda D}\right). \tag{9.41}$$

as a solution. On the other hand, we know that $p_{\mathrm{eq}}(x)$ is given by the Maxwell-Boltzmann distribution, expressed as a function of the potential $U(x)$ associated with force F by

$$\rho_{\mathrm{eq}}(x) \propto \exp\left[-\frac{U(x)}{k_{\mathrm{B}}T}\right]. \tag{9.42}$$

Since for a uniform force F, one has $U(x) = -Fx + \text{const.}$, by comparing equations (9.41) and (9.42), we obtain the Einstein relation in the form

$$\lambda D = k_B T, \tag{9.43}$$

independently of the force F.

Let us now try to obtain the equation satisfied by the $\rho(x,t)$ [or by the single-particle distribution $P(x,t)$] in the presence of a nonuniform force-field $F(x)$, which derives from a potential $U(x)$, and for which therefore $F(x) = -U'(x)$. In order to accomplish this, let us generalize the diffusion equation's derivation in a suitable fashion, starting from equation (9.31), which we rewrite as follows:

$$P(x,t+\Delta t) = \int \mathrm{d}x' \phi(\Delta x, x')\, P(x',t).$$

Let us remark that in this case, the distribution $\phi(\Delta x, x')$ of the displacement $\Delta x = x - x'$ also depends on the starting point x', since the system is not uniform. This makes it complicated to continue on this path, and for this reason, one prefers to introduce a trick.

Let us therefore introduce an arbitrary function $f(x)$ and evaluate its average with respect to the distribution $P(x,t+\Delta t)$:

$$\langle f \rangle_{t+\Delta t} = \int \mathrm{d}x\, f(x)\, P(x,t+\Delta t) = \int \mathrm{d}x \int \mathrm{d}x' f(x)\, \phi(\Delta x, x')\, P(x',t). \tag{9.44}$$

For each value of x', we expand $f(x)$ in a series of Δx around x', and we change the integration variable, setting $x = x' + \Delta x$. We thus obtain

$$\langle f \rangle_{t+\Delta t} = \int \mathrm{d}\Delta x \int \mathrm{d}x' \left[f(x') + \Delta x\, f'(x') + \frac{1}{2} \Delta x^2 f''(x') + \cdots \right] \phi(\Delta x, x')\, P(x',t). \tag{9.45}$$

At this point, we can evaluate the integral over Δx, by exploiting the following relations that follow immediately from the considerations made earlier:

$$\langle \Delta x \rangle = \int \mathrm{d}\Delta x\, \Delta x\, \phi(\Delta x, x') = \frac{F(x')}{\lambda} \Delta t, \tag{9.46}$$

$$\langle \Delta x^2 \rangle = \int \mathrm{d}\Delta x\, \Delta x^2 \phi(\Delta x, x') = 2D\, \Delta t. \tag{9.47}$$

In this manner, by changing the integration variable, we obtain

$$\langle f \rangle_{t+\Delta t} \simeq \int \mathrm{d}x \left[f(x) + \frac{F(x)}{\lambda} \Delta t\, f'(x) + D \Delta t\, f''(x) \right] P(x,t).$$

Integrating by parts, we finally obtain

$$\langle f \rangle_{t+\Delta t} = \int \mathrm{d}x\, f(x) \left(P(x,t) + \Delta t \left\{ \frac{\partial}{\partial x} \left[-\frac{F(x)}{\lambda} P(x,t) \right] + \frac{\partial^2}{\partial x^2} [D\, P(x,t)] \right\} \right).$$

Therefore,

$$\frac{\mathrm{d}}{\mathrm{d}t} \langle f \rangle_t = \int \mathrm{d}x\, f(x) \left\{ \frac{\partial}{\partial x} \left[-\frac{F(x)}{\lambda} P(x,t) \right] + \frac{\partial^2}{\partial x^2} [D\, P(x,t)] \right\}.$$

Since one obviously has

$$\frac{\mathrm{d}}{\mathrm{d}t}\langle f\rangle_t = \int \mathrm{d}x\, f(x)\,\frac{\partial P}{\partial t},$$

and the $f(x)$ is arbitrary, we obtain the equation satisfied by the $P(x, t)$:

$$\frac{\partial P}{\partial t} = \frac{\partial}{\partial x}\left[-\frac{F(x)}{\lambda}P + \frac{\partial}{\partial x}(DP)\right]. \tag{9.48}$$

This equation is known as the **Smoluchowski equation.** It is easy to see that the time-independent Maxwell-Boltzmann distribution is its identical solution, if λ and D are connected by the Einstein relation (9.43).

At this point, it is easy to see that the corresponding equation for Brownian motion in d dimensions, subject to the force $\boldsymbol{F}$, and under the hypothesis that the fluctuating force $\boldsymbol{f}$ satisfies equations (9.2) and (9.3), where Λ is expressed by equation (9.12), has the following expression:

$$\frac{\partial P}{\partial t} = \frac{1}{\lambda}\nabla\cdot[-\boldsymbol{F}(\boldsymbol{r})P + k_{\mathrm{B}}T\nabla P]. \tag{9.49}$$

9.4 Diffusion Processes and the Fokker-Planck Equation

The considerations reported in the preceding section can be easily extended to more general systems. Let us assume that we are dealing with a physical system whose behavior is described by a collection of random functions of time $\boldsymbol{X}(t)$, where $\boldsymbol{X} = (X_1, \ldots, X_n)$, in which the following hypotheses are satisfied:

1. The process is **Markovian**—in other words, the conditional probability

   $$P(\boldsymbol{x}, t \mid \boldsymbol{x}_k, t_k; \boldsymbol{x}_{k-1}, t_{k-1}; \ldots; \boldsymbol{x}_0, t_0),$$

   that $\boldsymbol{X}(t) = \boldsymbol{x}$, given that $\boldsymbol{X}(t_k) = x_k, \ldots, \boldsymbol{X}(t_0) = \boldsymbol{x}_0$, with $t_0 < t_1 < \ldots < t_k$, depends only on (x_k, t_k).

2. The process's increment $\Delta\boldsymbol{X}$ during the $|t, t+\Delta t|$ time interval satisfies the following conditions:

   - The expected value of $\Delta\boldsymbol{X} = \boldsymbol{X}(t+\Delta t) - \boldsymbol{X}(t)$ is proportional to Δt:

     $$\langle \Delta\boldsymbol{X}\rangle = \boldsymbol{b}(\boldsymbol{x})\Delta t + \mathrm{o}(\Delta t), \tag{9.50}$$

     where $\boldsymbol{x} = \boldsymbol{X}(t)$, and where $\boldsymbol{b}(\boldsymbol{x}) = [b_1(\boldsymbol{x}), \ldots, b_n(\boldsymbol{x})]$ is a collection of n functions of the n variables x_i eventually dependent on time t.

   - The expected value of the products $\Delta X_i \Delta X_j$, $(i, j \in \{1, \ldots, n\})$ is also proportional to Δt:

     $$\langle \Delta X_i \Delta X_j\rangle = 2\Lambda_{ij}(\boldsymbol{x})\Delta t + \mathrm{o}(\Delta t), \tag{9.51}$$

     where $\Lambda_{ij}(\boldsymbol{x})$ is a symmetric positive semidefinite matrix that depends on $\boldsymbol{x}$ and eventually on time t.

If these conditions are satisfied, the random function $\boldsymbol{X}(t)$ is called a **diffusion process.** By applying the method that we used to derive the Smoluchowski equation, we can see from the probability density $P(x, t)$ that $\boldsymbol{X}(t) = \boldsymbol{x}$ satisfies the following equation:

$$\frac{\partial P}{\partial t} = \sum_{i=1}^{n} \frac{\partial}{\partial x_i} \left\{ -b_i(x)\, P(x,t) + \sum_j \frac{\partial}{\partial x_j} \left[\Lambda_{ij}(x)\, P(x,t) \right] \right\}. \tag{9.52}$$

This equation is known as the **Fokker-Planck equation**, or also as the **forward Kolmogorov equation.**

To see how the Fokker-Planck equation is useful, let us analyze the motion of a Brownian particle in one dimension, subject to an applied force, with a time resolution comparable to the velocity relaxation time τ_r. It is then necessary to take both the particle's position x and its momentum p into account. By applying the reasoning given earlier, we see that

$$\langle \Delta x \rangle = \frac{p}{M} \Delta t + o(\Delta t), \tag{9.53}$$

$$\langle \Delta p \rangle = \left[F(x) - \frac{\lambda p}{M} \right] \Delta t + o(\Delta t), \tag{9.54}$$

$$\langle \Delta p^2 \rangle = 2\Lambda \Delta t + o(\Delta t), \tag{9.55}$$

and also that $\langle \Delta x^2 \rangle$ and $\langle \Delta x \Delta p \rangle$ vanish at order Δt. We thus obtain a diffusion process that satisfies the Fokker-Planck equation

$$\frac{\partial P}{\partial t} = \frac{\partial}{\partial x}\left(-\frac{p}{M} P \right) + \frac{\partial}{\partial p} \left\{ \left[-F(x) + \frac{\lambda p}{M} \right] P \right\} + \lambda k_B T \frac{\partial^2 P}{\partial p^2}, \tag{9.56}$$

where we have exploited the Einstein relation. This equation (which can be almost immediately generalized to d dimensions) is called the **Klein-Kramers equation**.

9.5 Correlation Functions

Let us observe that the preceding results can also be read differently. Let us assume that we are not dealing with a single Brownian particle, but with a distribution of *noninteracting* Brownian particles, distributed in space according to a distribution $\rho(\mathbf{r},t)$. Equation (9.52) can be rewritten as follows:

$$\frac{\partial \rho}{\partial t} = -\nabla \cdot J, \tag{9.57}$$

where the particle current J is given by

$$J = -\frac{\nabla U}{\lambda} - D\nabla\rho. \tag{9.58}$$

The proportionality coefficient between particle density ρ and the probability density P is obviously the total number of particles N. Let us return to considering the problem in the absence of an external potential and write the diffusion equation in Fourier space. By defining

$$\tilde{\rho}(\mathbf{k},t) = \int d\mathbf{r}\, e^{i\mathbf{k}\cdot\mathbf{r}} \rho(\mathbf{r},t), \tag{9.59}$$

we obtain

$$\frac{\partial}{\partial t}\tilde{\rho}(\boldsymbol{k},t) = -Dk^2\tilde{\rho}(\boldsymbol{k},t). \tag{9.60}$$

A possible perturbation with respect to the (uniform) equilibrium state therefore disappears in a characteristic time $\Omega(\boldsymbol{k})^{-1}$, where

$$\Omega(\boldsymbol{k}) = Dk^2. \tag{9.61}$$

Obviously, the distribution $\rho(\boldsymbol{r},t)$ can also now be regarded as a random quantity—since each Brownian particle is subjected to an independent random force, the corresponding density will fluctuate over time. In this case, the Fokker-Planck equation will be valid *on average*, for the same reason that the frequency of a random result is, on average, equal to its probability. A moment of reflection, however, shows us that while the continuity equation (9.57) must be valid for the fluctuating density ρ instant by instant, so as to maintain the total number of particles constant, the relation (9.58) between the current $\boldsymbol{J}$ and the density ρ is valid only on average. In fact, this relation is a statistical constraint between $\boldsymbol{J}$ and ρ, which we obtained by taking the average of the displacements $\boldsymbol{r}$. It is therefore a **constitutive relation** between density and current, containing the phenomenological coefficient λ, related to viscosity.

We now calculate the correlation function $C(\boldsymbol{k},t) = \langle\tilde{\rho}(\boldsymbol{k},t)\tilde{\rho}(-\boldsymbol{k},0)\rangle$:

$$\frac{\partial}{\partial t}\langle\tilde{\rho}(\boldsymbol{k},t)\tilde{\rho}(-\boldsymbol{k},0)\rangle = -Dk^2\langle\tilde{\rho}(\boldsymbol{k},t)\tilde{\rho}(-\boldsymbol{k},0)\rangle, \tag{9.62}$$

for $t \geq 0$. Therefore,

$$C(\boldsymbol{k},t) = \exp(-Dk^2 t)\, C(\boldsymbol{k},0), \qquad t \geq 0. \tag{9.63}$$

In order to extend this result to negative times, let us assume that the system is at equilibrium and that $\langle\tilde{\rho}(\boldsymbol{k},t)\tilde{\rho}(-\boldsymbol{k},t_0)\rangle$ satisfies therefore time translation invariance. We thus obtain

$$\langle\tilde{\rho}(\boldsymbol{k},-t)\tilde{\rho}(-\boldsymbol{k},0)\rangle = \langle\tilde{\rho}(\boldsymbol{k},0)\tilde{\rho}(-\boldsymbol{k},t)\rangle. \tag{9.64}$$

From equation (9.63), one has

$$C(\boldsymbol{k},t) = \exp(-Dk^2|t|)\, C(\boldsymbol{k},0), \tag{9.65}$$

and transforming according to Fourier with respect to t,

$$\begin{aligned} \tilde{C}(\boldsymbol{k},\omega) &= \int_{-\infty}^{\infty} \mathrm{d}t\, \mathrm{e}^{-i\omega t} C(\boldsymbol{k},t) \\ &= C(\boldsymbol{k},0)\left[\int_{-\infty}^{0} \mathrm{e}^{-i\omega t + Dk^2 t} + \int_{0}^{+\infty} \mathrm{d}t\, \mathrm{e}^{-i\omega t - Dk^2 t}\right] \\ &= C(\boldsymbol{k},0)\,\frac{2Dk^2}{\omega^2 + (Dk^2)^2}. \end{aligned} \tag{9.66}$$

The correlation function $\tilde{C}(\boldsymbol{k},\omega)$ is therefore a *Lorentzian*, centered in $\omega = 0$, and with a width equal to $\Omega(\boldsymbol{k}) = Dk^2$.

The coefficient $C(\mathbf{k},0)$ can be related to the system's equilibrium properties. For each value of t, because of invariance by temporal translations, one has in fact

$$C(\mathbf{k},0) = \int \mathrm{d}\mathbf{r}\, \mathrm{e}^{\mathrm{i}\mathbf{k}\cdot\mathbf{r}} \langle \rho(\mathbf{r},t)\rho(-\mathbf{k},t)\rangle. \tag{9.67}$$

Our system of Brownian particles behaves like a diluted gas with average density equal to $\bar{\rho} = \langle \rho(\mathbf{r},t)\rangle$. At equilibrium, since under our hypothesis the Brownian particles do not interact, we can assume that the density fluctuations at different points in space are independent. More specifically, let us consider an element of volume $\mathrm{d}V$ such that, on average, it contains a very small number of particles $\bar{\nu} = \bar{\rho}\,\mathrm{d}V$. The number of particles actually contained in this volume obeys therefore a Poisson distribution, and one has $\langle(\nu - \bar{\nu})^2\rangle = \bar{\nu}$. From this, it follows that the variance in the number of particles N contained in a small but macroscopic volume V is equal to its average N.

We can therefore express the instantaneous local density $\rho(\mathbf{r},t)$ measured over fairly large volumes (yet still small compared to k^{-1}) as follows:

$$\rho(\mathbf{r},t) = \bar{\rho} + \delta\rho(\mathbf{r},t), \tag{9.68}$$

where $\bar{\rho}$ is the average density, and $\delta\rho(\mathbf{r},t)$ is a random Gaussian function, whose spatial and temporal correlations decay very rapidly, and whose average vanishes. The fact that one can approximate $\delta\rho(\mathbf{r}, t)$ with a Gaussian function derives once again from the central limit theorem. We can therefore set

$$\langle \delta\rho(\mathbf{r},t)\,\delta\rho(\mathbf{r}',t)\rangle \propto \delta(\mathbf{r}-\mathbf{r}'). \tag{9.69}$$

On the other hand,

$$\int \mathrm{d}\mathbf{r} \int \mathrm{d}\mathbf{r}' \langle \delta\rho(\mathbf{r},t)\,\delta\rho(\mathbf{r}',t)\rangle = \langle \Delta N^2\rangle. \tag{9.70}$$

However, $\langle \Delta N^2\rangle = N$, and because of translational invariance, one of the two integrals gives us the sample's volume V. Therefore,

$$\langle \delta\rho(\mathbf{r},t)\,\delta\rho(\mathbf{r}',t)\rangle = \frac{N}{V}\delta(\mathbf{r}-\mathbf{r}') = \bar{\rho}\,\delta(\mathbf{r}-\mathbf{r}'). \tag{9.71}$$

By transforming this relation according to Fourier, we obtain

$$C(\mathbf{k},0) = \int \mathrm{d}\mathbf{r}\, \mathrm{e}^{\mathrm{i}\mathbf{k}\cdot\mathbf{r}} \langle \delta\rho(\mathbf{r},t)\,\delta\rho(\mathbf{r}',t)\rangle = \bar{\rho}. \tag{9.72}$$

The final expression is therefore

$$\tilde{C}(\mathbf{k},\omega) = \frac{2\bar{\rho}Dk^2}{\omega^2 + (Dk^2)^2}. \tag{9.73}$$

Let us observe that for $\mathbf{k} \neq 0$ one has $\lim_{\omega\to 0}\tilde{C}(\mathbf{k},\omega) = 2\bar{\rho}/Dk^2$, while for $\omega \neq 0$ one has $\lim_{\mathbf{k}\to 0}\tilde{C}(\mathbf{k},\omega) = 0$. Therefore $\lim_{\mathbf{k}\to 0}\lim_{\omega\to 0}\tilde{C}(\mathbf{k},\omega) = \infty$, while $\lim_{\omega\to 0}\lim_{\mathbf{k}\to 0}\tilde{C}(\mathbf{k},\omega) = 0$. This noncommutativity of limits appears quite generally in dynamic correlation functions.

9.6 Kubo Formula and Sum Rules

One can relate the diffusion coefficient D to the current's correlation function. By applying the continuity equation

$$\frac{\partial}{\partial t}\hat{\rho} + \nabla \cdot J = 0, \tag{9.74}$$

to the correlation function $\langle \hat{\rho}(\mathbf{r},t)\,\hat{\rho}(\mathbf{r}',t')\rangle$, one obtains

$$\frac{\partial}{\partial t}\frac{\partial}{\partial t'}\langle \hat{\rho}(\mathbf{r},t)\,\hat{\rho}(\mathbf{r}',t')\rangle = \sum_{ij}\frac{\partial}{\partial r_i}\frac{\partial}{\partial r_j}\langle J_i(\mathbf{r},t)\,J_j(\mathbf{r}',t')\rangle. \tag{9.75}$$

By transforming according to Fourier, and taking the space and time translation invariance into account, we obtain

$$\omega^2\langle \tilde{\rho}(\mathbf{k},\omega)\,\tilde{\rho}(-\mathbf{k},-\omega)\rangle = \sum_{ij} k_i k_j \left\langle \tilde{J}_i(\mathbf{k},\omega)\,\tilde{J}_j(-\mathbf{k},-\omega)\right\rangle. \tag{9.76}$$

Due to the problem's isotropy, the correlation function $\left\langle \tilde{J}_i(\mathbf{k},\omega)\,\tilde{J}_j(-\mathbf{k},-\omega)\right\rangle$ can be expanded into a scalar term (independent of the direction of $\mathbf{k}$), a tensorial quantity (which transforms like the collection of products $k_i k_j$), and higher order terms:

$$\left\langle \tilde{J}_i(\mathbf{k},\omega)\,\tilde{J}_j(-\mathbf{k},-\omega)\right\rangle = a_0\delta_{ij} + a_1 k^2 + \cdots + b_0 k_i k_j + b_1 k^2 k_i k_j + \cdots \tag{9.77}$$

Therefore,

$$\sum_{ij} k_i k_j \left\langle \tilde{J}_i(\mathbf{k},\omega)\,\tilde{J}_j(-\mathbf{k},-\omega)\right\rangle = a_0 k^2 + O(k^4). \tag{9.78}$$

On the other hand, following in the path of equation (9.77), one has

$$a_0 = \frac{1}{3}\lim_{k\to 0}\langle \tilde{J}(\mathbf{k},\omega)\cdot\tilde{J}(-\mathbf{k},-\omega)\rangle. \tag{9.79}$$

Taking equation (9.73), into account, we obtain:

$$D\bar{\rho} = \frac{1}{2}\lim_{\omega\to 0}\lim_{k\to 0}\frac{\omega^2}{k^2}\langle \tilde{\rho}(\mathbf{k},\omega)\,\tilde{\rho}(-\mathbf{k},-\omega)\rangle = \frac{1}{6}\lim_{\omega\to 0}\lim_{k\to 0}\langle \tilde{J}(\mathbf{k},\omega)\cdot\tilde{J}(-\mathbf{k},-\omega)\rangle. \tag{9.80}$$

This equation, which relates the diffusion coefficient to a correlation function, is a new example of the **Kubo formula**. The importance of these relations is that they allow us to calculate dynamic quantities (like the diffusion coefficient) in terms of correlation functions at thermodynamic equilibrium.

Another relation, apparently trivial but extremely important in the applications, comes from the observation that the correlation function $C(\mathbf{k}, 0)$ for equal times is obtained as a Fourier antitransform of the dynamic correlation $\tilde{C}(\mathbf{k},\omega)$:

$$C(\mathbf{k},0) = \int \frac{d\omega}{2\pi}\,\tilde{C}(\mathbf{k},\omega). \tag{9.81}$$

The reader is invited to check that this relation is actually valid for our equation (9.66). The interest in this **sum rule** is connected especially to those cases in which the dynamic correlation function is not perfectly known. In these cases, it imposes an important constraint on the approximations.

9.7 Generalized Brownian Motion

It is interesting to note that we had to wait almost 50 years before it became obvious that one could apply the theory of Brownian motion to *any* observable in a macroscopic system. It was only in the 1950s, with some works by S. Nakajima, H. Mori, and R. Zwanzig ([Naka58], [Zwan60], [Mori65]), that the theory of generalized Brownian motion was developed, one that aimed to provide a general description of the observables' evolution. In this section, we will see how the Brownian motion scheme naturally arises when one wants to describe the evolution of a system not in terms of its microscopic variables (whose number is on the order of Avogadro's constant), but in terms of a limited number of variables (which can however be *fields* defined in each point of space, even if they vary slowly). We will additionally derive a certain number of useful general relations that are the natural outcome of this approach. On the other hand, we will not attempt to formally justify the description in terms of generalized Brownian motion, on the basis of the projection scheme (see [Kubo85, section 2.5]). In fact, it is relatively easy to distinguish a systematic and a random part in the equation of motion for an observable. On the other hand, it is very difficult to justify, in the general case, all the simplifying assumptions that hold for the random force in the Brownian motion, like Gaussian distribution, noncorrelation, and the like.

Let us suppose that we have a thermodynamic system, of which X is an observable. We will denote the value of X by x. As a consequence of Einstein's theory of fluctuations, the probability that X can assume a specific value x is given, at thermodynamic equilibrium, for a certain value of temperature T, by

$$\rho_{\mathrm{eq}}(x) \propto \exp\left[-\frac{\mathcal{H}(x)}{k_{\mathrm{B}}T}\right], \tag{9.82}$$

where $\mathcal{H}(x)$ is the availability expressed as a function of x:

$$\mathcal{H}(x) = -k_{\mathrm{B}}T \ln \int d\nu\, \delta(X(\nu) - x) \exp\left[-\frac{\mathcal{H}(\nu)}{k_{\mathrm{B}}T}\right]. \tag{9.83}$$

In this expression, $\nu = (p, q)$ is the point in the system's phase space, and $H(\nu)$ is the Hamiltonian. The minimum value of $\mathcal{H}(x)$ is reached for the equilibrium value of X, x^* and is equal to the Helmholtz free energy:

$$\mathcal{H}(x) \geq \mathcal{H}(x^*) = F(T) = -k_{\mathrm{B}}T \ln \int d\nu \exp\left[-\frac{\mathcal{H}(\nu)}{k_{\mathrm{B}}T}\right]. \tag{9.84}$$

The apparent contradiction between equations (9.83) and (9.84) is resolved by considering that almost all the available phase space corresponds to the value of x^* in the thermodynamic limit.

In a finite system, X is a random quantity, which varies according to the system's evolution. Using the analogy with the theory of Brownian motion, we will suppose that it is governed by an evolution equation that contains both a systematic and a random part:

$$\frac{dX}{dt} = v(X(t)) + f_x(t). \tag{9.85}$$

Let us therefore make the following hypotheses, which can be valid for a suitable choice of X:

- The quantity X varies over time scales that are long when compared to those in which the random force $f_x(t)$ varies—this allows us to describe $f_x(t)$ as a random function that is correlated only over very brief time periods.
- The random force $f_x(t)$ is a result of the effects of a great number of independent processes that allows X to interact with the system's other degrees of freedom—this allows us to describe it as a random Gaussian function, with a vanishing average [since its average is taken into account by the systematic velocity $v(x)$].
- We will additionally suppose (for simplicity's sake, but also because we expect to be describing *small* fluctuations), that the correlation of $f_x(t)$ does not depend on the instantaneous value of X.
- The process described by equation (9.85) must be compatible with the equilibrium distribution (9.82).

Given these hypotheses, the random force $f_x(t)$ satisfies the relations

$$\langle f_x(t) \rangle = 0, \qquad \langle f_x(t)\, f_x(t') \rangle = \Lambda \delta(t - t'), \tag{9.86}$$

in which Λ satisfies a relation analogous to Einstein's, which we will determine further on. By following the previous section's line of reasoning, it is easy to see that the probability distribution $\rho(x, t)$ satisfies the Fokker-Planck equation:

$$\frac{\partial}{\partial t} \rho(x,t) = -\frac{\partial}{\partial x}\left[v(x)\rho(x,t) - \frac{1}{2}\Lambda \frac{\partial}{\partial x}\rho(x,t)\right]. \tag{9.87}$$

By imposing that the equilibrium distribution $p_{\text{eq}}(x)$ be stationary, one obtains the expression of the systematic velocity $v(x)$:

$$v(x) = -\frac{\Lambda}{2k_B T}\frac{\partial \mathcal{H}(x)}{\partial x}. \tag{9.88}$$

This has the form of a phenomenological equation of the type $v = \Gamma F$, where F is the force, and Γ is a kinetic coefficient (analogous to the resistance for motion in fluids). One usually assumes that Γ depends less strongly than Λ on temperature, and one therefore writes the Langevin equation in the form

$$\frac{dx}{dt} = -\Gamma \frac{\partial \mathcal{H}(x)}{\partial x} + f_x(t), \tag{9.89}$$

$$\langle f_x(t) \rangle = 0; \qquad \langle f_x(t)\, f_x(t') \rangle = 2k_B T \Gamma \delta(t - t'). \tag{9.90}$$

By means of these essentially phenomenological considerations, we have obtained a nonlinear Langevin equation satisfied by X's fluctuations. Recall that the availability $\mathcal{H}(x)$ reaches its minimum for X's equilibrium value x^*, and that both X and $\mathcal{H}$ are extensive quantities with respect to the system's size (which we will denote by N). Therefore,

$$\mathcal{H}(x) = \mathcal{H}(x^*) + \frac{1}{2}\left.\frac{\partial^2 \mathcal{H}}{\partial x^2}\right|_{x=x^*}(x-x^*)^2 + \cdots \tag{9.91}$$

The characteristic values of $|x - x^*|$ are on the order of $N^{1/2}$, and further terms of this expansion are usually negligible in the thermodynamic limit.

In this approximation, the equation satisfied by X is linear. By introducing the deviation $\delta x = x - x^*$, we obtain

$$\frac{\mathrm{d}\delta x}{\mathrm{d}t} = -\Gamma \chi_x^{-1}\delta x + f_x(t), \tag{9.92}$$

where we have introduced the susceptibility χ_x, defined by

$$\chi_x^{-1} = \left.\frac{\partial^2 \mathcal{H}}{\partial x^2}\right|_{x=x^*}. \tag{9.93}$$

It is easy to see that χ_x is actually a susceptibility. Let us imagine that we introduce a perturbation of the form $-\lambda X(\nu)$ into the Hamiltonian $H(\nu)$ and calculate the variation in X's equilibrium value. We will then actually obtain

$$\left.\frac{\partial \langle X\rangle}{\partial \lambda}\right|_{\lambda=0} = \chi_x. \tag{9.94}$$

Near the critical point, the order parameter's susceptibility χ_x diverges. In this case, it is necessary to take the further terms in the expansion (9.91) into account.

From this equation, retracing the preceding chapter's calculations once again, we can obtain the behavior of X's correlation function:

$$C_x(t) = \langle X(t)\,X(0)\rangle = k_{\mathrm{B}} T \chi_x \exp\left(-\Gamma \chi_x^{-1}|t|\right), \tag{9.95}$$

and the corresponding Fourier transform

$$\tilde{C}_x(\omega) = \int \mathrm{d}\omega\, \mathrm{e}^{-i\omega t}\, C_x(t) = \frac{2\Gamma k_{\mathrm{B}} T}{\omega^2 + (\Gamma \chi_x^{-1})^2}. \tag{9.96}$$

The characteristic relaxation time τ_{relax} is given by χ_x/Γ. It tends to diverge in the proximity of the critical point, if X is the order parameter. On the one hand, this guarantees us that the approximations made over the correlations of the random forces become more acceptable, while on the other, it makes validity of the expansion (9.91) more doubtful.

If, we continue however to describe the system in the context of the linearized Langevin equation, we obtain that $\tau_{\mathrm{relax}} \propto \chi_a \propto |T - T_c|^{\gamma}$. This is the *classic* result, which is usually attributed to van Hove. The divergence of the relaxation times in proximity of the critical point is a general phenomenon and is known as **critical slowing down**.

9.8 Time Reversal

In order to proceed further, we must assume that we have chosen the quantities X_i with a defined parity with respect to time reversal. Let us recall that the X_i are ultimately some functions $X_i(\nu)$ defined over the phase space to which ν belongs. The time-reversal operator $\mathcal{I}$ is defined in this space. Given $\nu = (p, q)$, one has

$$\nu' = \mathcal{I}\nu = (-p, q). \tag{9.97}$$

The Hamiltonian is invariant due to time reversal, in the absence of magnetic fields,[3] and if it does not explicitly depend on time. Consequently, if $\nu(t)$ is a solution of the canonical equations of motion that satisfies the initial condition $\nu(0, \nu_0) = \nu_0$, then $\nu_I(t) = \mathcal{I}\nu(-t)$ is a solution of the equations of motion that satisfies the initial condition $\nu_I(0) = \mathcal{I}\nu_0$.

We can define our observables in such a way that they are eigenfunctions of $\mathcal{I}$:

$$\mathcal{I}X_i(\nu) = X_i(\mathcal{I}\nu) = X_i(-p, q) = \epsilon_i X_i(p, q), \tag{9.98}$$

where $\epsilon_i = \pm 1$ is X_i's parity.

Let us now calculate the effect of time reversal on the correlation function $\langle X_i(t) X_j(t') \rangle$. Let us assume that we fix a state ν_0 at $t' = 0$. We shift to the inverse $\nu_0' = \mathcal{I}\nu_0$, and we allow the system to evolve. After a time t, we will have obtained the state $\mathcal{I}\nu(-t)$. The corresponding value of X_i is equal to $\mathcal{I}X_i(-t) = \epsilon_i X_i(-t)$. On the other hand, by definition we have $X_j(\nu_0') = \epsilon_j X_j(\nu_0)$, and, since the Hamiltonian is invariant, ν_0's and ν_0''s probability are equal at equilibrium.

This line of reasoning shows that, at equilibrium,

$$\langle X_i(t) X_j(0) \rangle = \epsilon_i \epsilon_j \langle X_i(-t) X_j(0) \rangle. \tag{9.99}$$

Therefore, the correlation functions of quantities that have the same parity with respect to time reversal are *even* functions of time; the others are odd.

9.9 Response Functions

The previous argument can be generalized to the case in which, instead of a single variable X, we are dealing with a collection $X = (X_1, \ldots, X_p)$ of dynamic variables. The equilibrium probability distribution is expressed as a function of the availability $\mathcal{H}(x) = \mathcal{H}(x_1, \ldots, x_p)$. The Langevin equation can be written

$$\frac{dx_i(t)}{dt} = v_i[x(t)] + f_i(t). \tag{9.100}$$

If we assume (as we did earlier) that the correlation of random forces does not depend on the values of the x, we can choose the X_i so that it is diagonal:

[3] And we also need to disregard the violation of time reversal characteristic of weak interactions.

$$\langle f_i(t) f_j(t') \rangle = 2k_B T \Gamma_i \delta_{ij} \delta(t - t'). \tag{9.101}$$

By imposing the stationariness of the equilibrium distribution, we obtain the expression of the velocities $v_i(x)$:

$$v_i(x) = -\Gamma_i \frac{\partial \mathcal{H}(x)}{\partial x_i}. \tag{9.102}$$

With this definition, the Langevin equation assumes the form

$$\frac{dx_i}{dt} = -\Gamma_i \frac{\partial \mathcal{H}(x(t))}{\partial x_i} + f_i(t), \tag{9.103}$$

and the Fokker-Planck equation assumes the following one:

$$\frac{\partial}{\partial t} \rho(x,t) = \sum_i \frac{\partial}{\partial x_i} \left[\Gamma_i \frac{\partial \mathcal{H}(x(t))}{\partial x_i} \rho(x,t) + k_B T \Gamma_i \frac{\partial \rho(x,t)}{\partial x_i} \right]. \tag{9.104}$$

Let us now suppose that we perturb our system, initially (at time $t = 0$) at thermodynamic equilibrium, by adding an independent term of the form $-\sum_i h_i(t) X_i(\nu)$ to the Hamiltonian. This perturbation of the Hamiltonian reverberates in an analogous perturbation of the availability $\mathcal{H}$, which comes to depend on time in the form

$$\mathcal{H}(x,t) = \mathcal{H}_0(x) - \sum_i h_i(t) x_i, \tag{9.105}$$

where $\mathcal{H}_0(x)$ is the "unperturbed" expression we have considered thus far. In order to persuade ourselves of this result, it is sufficient to imagine performing the calculation of the integral (9.83) in the presence of the external field h, but keeping the value of the X_i variables constrained. One can then take the contribution due to the field out of the integral, and one obtains expression (9.105).

Because of the perturbation, the average values of the observables X_i will come to depend on time. Let us redefine the observables in such a fashion that their average equilibrium values vanish. We want to calculate their average values at instant $t > 0$, to the first order in the perturbation $h(t)$. In general, we will have

$$\langle X_i \rangle_t = \int_0^t \mathrm{d}t' \sum_j \chi_{ij}(t,t') h_j(t'). \tag{9.106}$$

This equation defines the matrix $x_{ij}(t, t')$ of the response functions. The integral is limited to the time interval $[0, t]$ because the perturbation $h(t)$ vanishes at negative times, and because the average value $\langle X_i(\mathrm{t}) \rangle$ does not depend on the perturbation's values for times successive to t (**causality**). Moreover, since in the absence of perturbation the system is at equilibrium, the response function must satisfy time translation invariance:

$$\chi(t,t') = \chi(t+\tau, t'+\tau), \qquad \forall \tau. \tag{9.107}$$

We can therefore write $\chi(t, t') = \chi(t - t', 0)$, and we can consider the second argument put to zero to be understood. Due to causality, one also has

$$\chi(t) = 0, \qquad \text{for } t < 0. \tag{9.108}$$

We now want to show that the response function $\chi(t)$ can be expressed in terms of a correlation function at equilibrium. To help clarify our ideas, let us suppose that we consider a single variable X that satisfies the Langevin equation

$$\frac{\mathrm{d}x}{\mathrm{d}t} = -\Gamma\frac{\partial \mathcal{H}}{\partial x} + f(t), \tag{9.109}$$

where $f(t)$ is the usual stochastic Gaussian force, with a vanishing average and a correlation given by $\langle f(t)\, f(t')\rangle = 2\Gamma k_{\mathrm{B}}T\,\delta(t-t')$. We will denote its correlation $\langle X(t)X(t')\rangle$ by $C(t-t')$ and the response function $\delta\langle X(t)\rangle/\delta h(t')$ by $\chi(t-t')$. We will additionally assume that X's average vanishes at equilibrium. Let us introduce the conditional probability $\mathcal{P}(x,t\,|\,x',t')$ that $X(t)=x$ if $X(t')=x'$. We will always assume $t'<t$.

Let us consider the effect of the perturbation

$$\mathcal{H} \longrightarrow \mathcal{H} - h(t)\,X, \tag{9.110}$$

where $h(t) = h\delta(t-t')$. Then,

$$\langle X(t)\rangle = h\chi(t,t'). \tag{9.111}$$

In a small time interval around t', X's evolution is determined by the perturbation, which is much larger than the systematic part. It is therefore easy to see that

$$\lim_{t\to t'^{+}} X(t) = X(t') + \Gamma h. \tag{9.112}$$

Let us calculate $\langle X(t)\rangle$. One has

$$\langle X(t)\rangle = \int \mathrm{d}x \int \mathrm{d}x'\, x\,\mathcal{P}(x,t\,|\,x'+\Gamma h, t')\,\rho_0(x'), \tag{9.113}$$

where $\rho_0(x) = \mathrm{e}^{-\mathcal{H}(x)}/Z$ is the equilibrium distribution.

By changing the integration variable, we obtain

$$\begin{aligned} \langle X(t)\rangle &= \int \mathrm{d}x \int \mathrm{d}x'\, x\,\mathcal{P}(x,t\,|\,x',t')\,\rho_0(x'-\Gamma h) \\ &\simeq \int \mathrm{d}x \int \mathrm{d}x'\, x\,\mathcal{P}(x,t\,|\,x',t')\,(-\Gamma h)\frac{\partial \rho_0}{\partial x'} \\ &= \frac{h}{k_{\mathrm{B}}T}\int \mathrm{d}x \int \mathrm{d}x'\, x\,\mathcal{P}(x,t\,|\,x',t')\left(\Gamma\frac{\partial \mathcal{H}}{\partial x'}\right)\rho_0(x'), \end{aligned} \tag{9.114}$$

where we have exploited the fact that the average X vanishes at equilibrium. Therefore,

$$\chi(t,t') = \frac{1}{k_{\mathrm{B}}T}\left\langle X(t)\left(\Gamma\frac{\partial \mathcal{H}}{\partial x'}(t')\right)\right\rangle. \tag{9.115}$$

On the other hand, by substituting equation (9.109), one has

$$\chi(t,t') = \frac{1}{k_{\mathrm{B}}T}\left[-\langle X(t)\,\dot{X}(t')\rangle + \langle X(t) f(t')\rangle\right]. \tag{9.116}$$

In order to evaluate the second term, we use the relation [valid for each collection of Gaussian variables $\xi = (\xi_i)$ whose averages vanish]

$$\langle F(\xi)\,\xi_i\rangle = \sum_j C_{ij}\left\langle \frac{\partial F}{\partial \xi_j}\right\rangle, \tag{9.117}$$

where C_{ij} is the correlation matrix of the ξ_i.

Exercise 9.1 Prove the preceding relation.

So (taking some liberties with the discretization of times) we obtain

$$\langle X(t) f(t')\rangle = 2\Gamma k_B T\left\langle \frac{\delta X(t)}{\delta f(t')}\right\rangle = 2k_B T\left\langle \frac{\delta X(t)}{\delta h(t')}\right\rangle = 2k_B T\chi(t,t'). \tag{9.118}$$

Therefore,

$$\chi(t,t') = \frac{1}{k_B T}\langle X(t)\dot{X}(t')\rangle. \tag{9.119}$$

By exploiting time translation invariance, and taking into account the fact that $\chi(t,t')=0$ for $t<t'$, we can extend this expression to all values of t and t'. We thus obtain the most widely known expression of the response function:

$$\chi(t) = -\frac{\theta(t)}{k_B T}\langle \dot{X}(t)X(0)\rangle. \tag{9.120}$$

More generally, one will have

$$\chi_{ij}(t) = -\frac{\theta(t)}{k_B T}\frac{\mathrm{d}}{\mathrm{d}t}\langle X_i(t)X_j(0)\rangle. \tag{9.121}$$

We have derived this equation in the context of the Langevin equation—it is possible, however, to see that the result is also valid in the case in which the X_i are mechanical quantities, and the system's evolution is supported by the canonical equations of motion (and therefore the probability distribution obeys the Liouville equation).

Time-reversal invariance has an important consequence on the symmetry of the matrix $x_{ij}(t)$. For $t>0$, in fact we obtain

$$\begin{aligned}\chi_{ij}(t) &= -\frac{1}{k_B T}\frac{\mathrm{d}}{\mathrm{d}t}\langle X_i(t)X_j(0)\rangle\\ &= -\frac{1}{k_B T}\epsilon_i\epsilon_j\frac{\mathrm{d}}{\mathrm{d}t}\langle X_i(-t)X_j(0)\rangle\\ &= -\frac{1}{k_B T}\epsilon_i\epsilon_j\frac{\mathrm{d}}{\mathrm{d}t}\langle X_i(0)X_j(t)\rangle = \epsilon_i\epsilon_j\chi_{ji}(t).\end{aligned} \tag{9.122}$$

This relation is at the origin of the **Onsager relations**, which express the symmetry of the kinetic coefficients. We will discuss these relations further on.

9.10 Fluctuation–Dissipation Theorem

Let us return to equation (9.121), which we can write in the form

$$\chi_{ij}(t) = \frac{\theta(t)}{k_B T}\left(-\frac{d}{dt}\right)\langle X_i(t) X_j(0)\rangle = \frac{\theta(t)}{k_B T}\left(-\frac{d}{dt}\right) C_{ij}(t). \tag{9.123}$$

where $\theta(t)$ is the Heaviside step function:

$$\theta(t) = \begin{cases} 1, & \text{if } t > 0, \\ 1/2, & \text{if } t = 0, \\ 0, & \text{if } t < 0. \end{cases} \tag{9.124}$$

By transforming according to Fourier, we obtain

$$\tilde{\chi}_{ij}(\omega) = \int_{-\infty}^{\infty} dt\, e^{-i\omega t} \chi_{ij}(t) = -\frac{1}{k_B T}\int_{-\infty}^{\infty} \frac{d\omega'}{2\pi}\, \tilde{\theta}(\omega - \omega')\,(i\omega')\,\tilde{C}_{ij}(\omega'), \tag{9.125}$$

where $\tilde{\theta}(\omega)$ is the Fourier transform of the step function.

The integral that defines this Fourier transform does not converge absolutely, unless one adds to it an exponentially decreasing factor. We can therefore set

$$\tilde{\theta}(\omega) = \lim_{\epsilon \to 0^+} \int_0^{\infty} dt\, e^{-i\omega t - \epsilon t} = \lim_{\epsilon \to 0^+} \frac{1}{i\omega + \epsilon}. \tag{9.126}$$

One has

$$\frac{1}{i\omega + \epsilon} = \frac{-i\omega + \epsilon}{\omega^2 + \epsilon^2}. \tag{9.127}$$

Let us consider the integral of this expression multiplied by any regular function $\phi(\omega)$:

$$\begin{aligned} \lim_{\epsilon \to 0^+} \int_{-\infty}^{\infty} d\omega \frac{1}{i\omega + \epsilon}\phi(\omega) &= \lim_{\epsilon \to 0^+} \Big[-i \int_{-\infty}^{-\epsilon} \frac{d\omega}{\omega}\phi(\omega) - i \int_{\epsilon}^{\infty} \frac{d\omega}{\omega}\phi(\omega) \\ &+ \int_{-\infty}^{\infty} d\omega \frac{\epsilon}{\omega^2 + \epsilon^2}\phi(\omega) \Big] = -i\mathcal{P}\int_{-\infty}^{\infty} \frac{d\omega}{\omega}\phi(\omega) + \pi\phi(0), \end{aligned} \tag{9.128}$$

where $\mathcal{P}$ denotes the integral's principal part:

$$\mathcal{P}\int_{-\infty}^{\infty} \frac{d\omega}{\omega}\phi(\omega) = \lim_{\epsilon \to 0^+}\left[\int_{-\infty}^{-\epsilon} \frac{d\omega}{\omega}\phi(\omega) + \int_{\epsilon}^{\infty} \frac{d\omega}{\omega}\phi(\omega)\right]. \tag{9.129}$$

This result can be formally written

$$\lim_{\epsilon \to 0^+} \frac{1}{i\omega + \epsilon} = -i\mathcal{P}\frac{1}{\omega} + \pi\delta(\omega). \tag{9.130}$$

From equation (9.125), we thus obtain:

$$\tilde{\chi}_{ij}(\omega) = \frac{1}{k_B T}\left[\mathcal{P}\int_{-\infty}^{\infty} \frac{d\omega'}{2\pi}\frac{\omega'}{\omega - \omega'}\tilde{C}_{ij}(\omega') - \frac{i\omega}{2}\tilde{C}_{ij}(\omega)\right]. \tag{9.131}$$

If the $C_{ij}(t)$ function is even, its Fourier transform is real. This expression therefore allows one to derive both the real χ' and the imaginary part χ'' of the response function, mutually related by the **Kramers-Kronig relation**, which expresses $\tilde{\chi}(\omega)$'s analyticity (with our conventions) in the lower semiplane of the complex plane:

$$\chi'_{ij}(\omega) = -\mathcal{P}\int_{-\infty}^{\infty}\frac{d\omega'}{\pi}\frac{1}{\omega-\omega'}\chi''_{ij}(\omega'). \tag{9.132}$$

(If the function is odd, its transform is purely imaginary, and one has an analogous relation in which the real and imaginary parts are switched.)

To close these considerations, let us obtain the relation between the correlation and the response in Fourier space:

$$\chi''_{ij}(\omega) = -\frac{\omega}{2k_B T}\tilde{C}_{ij}(\omega). \tag{9.133}$$

This is the most frequently used form of the relation between fluctuation and response, and is called the **fluctuation–dissipation theorem**. The imaginary part of the response function χ'' is connected to the dissipation of the external work performed by the perturbation.

Let us show this in a simple case in which we are dealing with a single observable. By applying the perturbation $-h(t)A$ to the system's Hamiltonian, the power $W(t)$ absorbed at instant t is given by

$$W(t) = -h(t)\left\langle\frac{dA(t)}{dt}\right\rangle_t. \tag{9.134}$$

By expressing $\langle A(t)\rangle_t$ by means of the response function, we obtain

$$W(t) = -h(t)\frac{d}{dt}\int_0^t dt'\chi(t-t')\,h(t'). \tag{9.135}$$

Let us calculate the average power absorbed:

$$\overline{W} = \lim_{t\to\infty}\frac{1}{t}\int_0^t dt'\,W(t') = -\lim_{t\to\infty}\frac{1}{t}\int_0^t dt'h(t)\frac{d}{dt}\int_0^t dt'\chi(t-t')\,h(t'). \tag{9.136}$$

The expression in the right-hand side can be expressed by means of Fourier transforms:

$$-\lim_{t\to\infty}\frac{1}{t}\int_0^t dt'h(t)\frac{d}{dt}\int_0^t dt'\chi(t-t')\,h(t') = \int_{-\infty}^{\infty}\frac{d\omega}{2\pi}\tilde{h}(-\omega)(-i\omega)\,\tilde{\chi}(\omega)\,\tilde{h}(\omega). \tag{9.137}$$

Since $h(t)$ is real, its Fourier transform satisfies the relation $\tilde{h}(-\omega)=\tilde{h}^*(\omega)$. The product $\tilde{h}(-\omega)\,\tilde{h}(\omega)=|\tilde{h}(\omega)|^2$ is therefore real and even. Since the $\chi(\omega)$ also satisfies the same relation, the part of the integral that survives is proportional to the even part of $\omega\tilde{\chi}(\omega)$, and therefore to $\omega\chi''(\omega)$. We thus obtain the expression of absorbed power:

$$\overline{W} = \int_{-\infty}^{\infty}\frac{d\omega}{2\pi}\,\omega\chi''(\omega)|\tilde{h}(\omega)|^2. \tag{9.138}$$

9.11 Onsager Reciprocity Relations

Let us assume that X_k, $k = 1, 2, \ldots, r$ are observables, defined in such a manner that their average value at equilibrium vanishes:

$$\langle X_k \rangle = 0, \qquad \forall k. \tag{9.139}$$

We will always denote the equilibrium average by $\langle \ldots \rangle$. From the fluctuation–dissipation theorem, we can evaluate the nonequilibrium average of the observable X_k at instant t, when we add a perturbation of type $-\sum_\ell h_\ell(t)\, X_\ell$ to the Hamiltonian. We have

$$\langle X_k \rangle_t = \int_{-\infty}^{t} \mathrm{d}t' \sum_\ell \chi_{k\ell}(t-t')\, h_\ell(t'), \tag{9.140}$$

where

$$\chi_{k\ell}(t-t') = -\frac{\theta(t-t')}{k_{\mathrm{B}}T} \frac{\mathrm{d}}{\mathrm{d}t} \langle X_k(t)\, X_\ell(t') \rangle. \tag{9.141}$$

Let us now consider a perturbation of the form

$$h_\ell(t) = \lambda \delta_{\ell j} \theta(-t)\, \mathrm{e}^{\epsilon t}, \tag{9.142}$$

where ϵ is a very small positive number, and $|\lambda| \ll 1$. This perturbation is proportional to the observable X_j, "switches on" very slowly for large and negative t's, is almost constant for small and negative t's, and then suddenly goes to 0 for $t = 0$. So for $t > 0$, one then has

$$\langle X_i \rangle_t = -\frac{\lambda}{k_{\mathrm{B}}T} \int_{-\infty}^{0} \mathrm{d}t' \frac{\mathrm{d}}{\mathrm{d}t} \langle X_i(t)\, X_j(t') \rangle\, \mathrm{e}^{\epsilon t'}. \tag{9.143}$$

Integrating by parts, we obtain

$$\begin{aligned} \langle X_i \rangle_t &= \frac{\lambda}{k_{\mathrm{B}}T} \int_{-\infty}^{0} \mathrm{d}t' \frac{\mathrm{d}}{\mathrm{d}t} \langle X_i(t)\, X_j(t') \rangle\, \mathrm{e}^{\epsilon t'} \\ &= \frac{\lambda}{k_{\mathrm{B}}T} \left\{ \left[\langle X_i(t)\, X_j(t') \rangle\, \mathrm{e}^{\epsilon t'} \right]_{-\infty}^{0} + \epsilon \int_{-\infty}^{0} \langle X_i(t)\, X_j(t') \rangle\, e^{\epsilon t'} \right\} \\ &= \frac{\lambda}{k_{\mathrm{B}}T} \langle X_i(t)\, X_j(0) \rangle + O(\epsilon). \end{aligned} \tag{9.144}$$

This result is called **Onsager's regression hypothesis**. Lars Onsager postulated it in 1931 [Onsa31], arguing that the system cannot "know" whether a nonvanishing value of the observable X_i is due to a spontaneous fluctuation or a perturbation that occurred in the past—the regression of X_i's average toward its equilibrium value must be proportional to the regression of the correlation function towards zero. The fluctuation–dissipation theorem therefore allows us to *prove* Onsager's regression hypothesis.

In a linear regime, it is reasonable to assume that the nonequilibrium averages $\langle X_i \rangle_t$ satisfy a linear differential equation of the form

$$\frac{\mathrm{d}}{\mathrm{d}t} \langle X_i \rangle_t = -\sum_k M_{ik} \langle X_k \rangle_t. \tag{9.145}$$

In reality, this form is valid only if one operates a *coarse graining* over time—in other words, if one gives up observing the dynamics over time intervals that are too small (on the order of noise's correlation time).

Onsager's regression hypothesis implies that the correlation function satisfies an analogous differential equation:

$$\frac{\mathrm{d}}{\mathrm{d}t}\langle X_i(t)\,X_j(0)\rangle = -\sum_k M_{ik}\langle X_k(t)\,X_j(0)\rangle, \qquad \forall k, t>0. \tag{9.146}$$

If X_i and X_j are both time-reversal invariant, and if the Hamiltonian is also invariant, one has

$$\langle X_i(t)\,X_j(0)\rangle = \langle X_j(t)\,X_i(0)\rangle. \tag{9.147}$$

By taking the derivative of both sides of this equation with respect to t and substituting equation (9.146), we obtain

$$\sum_k M_{ik}\langle X_k(t)\,X_j(0)\rangle = \sum_k M_{jk}\langle X_k(t)\,X_i(0)\rangle. \tag{9.148}$$

By setting $t=0$ in this relation, we obtain

$$\sum_k M_{ik}\langle X_k X_j\rangle = \sum_k M_{jk}\langle X_k X_i\rangle, \tag{9.149}$$

where

$$\langle X_i X_j\rangle = \langle X_i(0)\,X_j(0)\rangle \tag{9.150}$$

is the instantaneous correlation at equilibrium of the observables X_i and X_j.

By introducing the matrix $\mathsf{L} = (L_{ij})$, where

$$L_{ij} = \frac{1}{k_B}\sum_k M_{ik}\langle X_k X_j\rangle, \tag{9.151}$$

this result can be written

$$L_{ij} = L_{ji} \tag{9.152}$$

in other words $\mathsf{L}^T = \mathsf{L}$, which implies that the L matrix is *symmetrical*.

These relations are of great importance, as regards both fundamentals and applications, to the extent that they earned Onsager the 1968 Nobel Prize for chemistry. They are called the **Onsager reciprocity relations**.

9.12 Affinities and Fluxes

The importance of the Onsager reciprocity relations can be better understood by considering a system's response to a perturbation that takes it out of thermodynamic equilibrium. In order to clarify our ideas, let us consider a global system, divided into two subsystems: one is the actual **system**; the other we will call the **reservoir**.

In this section, we will denote a set of extensive thermodynamic variables with $X = (X_0, X_1, \ldots, X_r)$, and express thermodynamic entropy as their function. The system's global entropy (system + reservoir) will be largest at equilibrium. Out of equilibrium instead, in general one will have

$$F_i = \left.\frac{\partial S^{\text{tot}}}{\partial X_i}\right)_{X_i^{\text{tot}}} = \frac{\partial S}{\partial X_i} - \frac{\partial S^{(\mathrm{r})}}{\partial X_i^{(\mathrm{r})}} \neq 0. \tag{9.153}$$

In this equation S^{tot} is total entropy, S is the system's entropy, and $S^{(\mathrm{r})}$ is that of the reservoir, and analogously for the X_i variables. We assume that the equilibrium state is constrained by X_i's total value. The equilibrium is therefore characterized by the vanishing of F_i, which can be considered the *force* that takes the system out of equilibrium. These generalized forces are usually called **affinities**.

For example, for $i = 0$ ($X_0 = E$ is the internal energy), one has

$$F = \frac{1}{T} - \frac{1}{T^{(\mathrm{r})}}, \tag{9.154}$$

in other words, this affinity is equal to the difference between the inverses of the temperatures of the system and of the reservoir. If $X_i = V$ (the volume), one instead has

$$F_V = \frac{p}{T} - \frac{p^{(\mathrm{r})}}{T^{(\mathrm{r})}}. \tag{9.155}$$

An important case (above all in kinetic chemistry) is when one perturbs the number of moles of chemical species k. One then has

$$F_k = -\frac{\mu_k}{T} + \frac{\mu_k^{(\mathrm{r})}}{T^{(\mathrm{r})}}, \tag{9.156}$$

Because of the perturbation that takes the system out of equilibrium, the X_k quantities will start to vary over time. We can define the **fluxes** J_i in terms of the time derivatives of X_i:

$$J_i = \frac{\mathrm{d}X_i}{\mathrm{d}t} \tag{9.157}$$

Generally speaking, entropy will also come to depend on time. Since the system's entropy is a function of the extensive variables X_k, one has

$$\frac{\mathrm{d}S^{\mathrm{tot}}}{\mathrm{d}t} = \sum_i \left(\frac{\partial S}{\partial X_i} - \frac{\partial S^{(\mathrm{r})}}{\partial X_i} \right) \frac{\mathrm{d}X_i}{\mathrm{d}t} = \sum_i F_i J_i. \tag{9.158}$$

The production of entropy is therefore given by the sum of the products of the affinities times the corresponding fluxes. The second principle of thermodynamics guarantees that since the global system is isolated, total entropy cannot but increase. One therefore has

$$\frac{\mathrm{d}S^{\mathrm{tot}}}{\mathrm{d}t} = \sum_i F_i J_i \geq 0. \tag{9.159}$$

In a generic situation of nonequilibrium, it will generally be possible to express the fluxes J_i as a function of the affinities F_j and of the system's equilibrium parameters. One will therefore have $J_i = J_i(\{F_j\})$. The fluxes vanish at equilibrium, however, when the affinities vanish. As a first approximation, therefore, it is possible to write some linear **phenomenological relations** between fluxes and affinities:

$$J_i = \sum_j L_{ij} F_j, \tag{9.160}$$

where we have introduced the **matrix of the kinetic coefficients** $\mathsf{L} = (L_{ij})$, whose elements are given by

$$L_{ij} = \left.\frac{\partial J_i}{\partial F_j}\right|_{F=0}. \tag{9.161}$$

These will generally be a function of the parameters that identify the system's state of equilibrium, such as temperature, pressure, and so on. We will now see that the matrix L thus introduced can be identified with the matrix L defined earlier. Let us in fact assume that we can identify the A_i observables of the preceding section with the X_i. We then have the equation

$$\frac{\mathrm{d}X_i}{\mathrm{d}t} = \sum_j L_{ij} F_j. \tag{9.162}$$

Multiplying by $X_j(0)$ and taking the correlation at equilibrium, we obtain

$$\frac{\mathrm{d}}{\mathrm{d}t}\langle X_i(t)X_j(0)\rangle_{\mathrm{c}} = \sum_k L_{ik}\langle F_k(t)X_j(0)\rangle_{\mathrm{c}}. \tag{9.163}$$

We have denoted the correlation

$$\langle AB\rangle_{\mathrm{c}} = \langle AB\rangle - \langle A\rangle\langle B\rangle. \tag{9.164}$$

by $\langle \ldots \rangle_{\mathrm{c}}$. On the other hand, we had the equation

$$\frac{\mathrm{d}}{\mathrm{d}t}\langle X_i(t)X_j(0)\rangle_{\mathrm{c}} = -\sum_k M_{ik}\langle X_k(t)X_j(0)\rangle_{\mathrm{c}}, \tag{9.165}$$

and therefore,

$$-\sum_k M_{ik}\langle X_k(t)X_j(0)\rangle_{\mathrm{c}} = \sum_k L_{ik}\langle F_k(t)X_j(0)\rangle_{\mathrm{c}}. \tag{9.166}$$

By setting $t = 0$ in this equation, we obtain

$$-\sum_k M_{ik}\langle X_k X_j(0)\rangle_{\mathrm{c}} = \sum_k L_{ik}\langle F_k X_j\rangle_{\mathrm{c}}. \tag{9.167}$$

Let us recall that the $\langle X_i X_j\rangle_{\mathrm{c}}$ are the instantaneous correlations at equilibrium.

Let us calculate the correlations $\langle F_k X_j\rangle_{\mathrm{c}}$—they are given by

$$\left\langle \frac{\partial S}{\partial X_k} X_j \right\rangle_{\mathrm{c}} - \left\langle \frac{\partial S^{(\mathrm{r})}}{\partial X_k^{(\mathrm{r})}} X_j \right\rangle_{\mathrm{c}}. \tag{9.168}$$

Since one assumes that the reservoir is very large, $\partial S^{(\mathrm{r})}/\partial X_k^{(\mathrm{r})}$ does not fluctuate, and the second term therefore vanishes. We remain with

$$\left\langle \frac{\partial S}{\partial X_k} X_j \right\rangle_{\mathrm{c}} = \mathcal{N}\int \prod_i \mathrm{d}X_i \frac{\partial S}{\partial X_k} X_j \mathrm{e}^{S(X_0,\ldots,X_r)/k_{\mathrm{B}}} - \left\langle \frac{\partial S}{\partial X_k} \right\rangle \langle X_j\rangle, \tag{9.169}$$

where $\mathcal{N}$ is a renormalization constant and we have made use of the fundamental postulate. It is easy to see that $\langle \partial S/\partial X_k\rangle$ vanishes. Integrating the first terms by parts, we obtain

$$\langle F_k X_j\rangle = -k_{\mathrm{B}}\delta_{kj}. \tag{9.170}$$

By substituting, we finally obtain

$$L_{ij} = \frac{1}{k_B} \sum_k M_{ik} \langle X_k X_j \rangle. \tag{9.171}$$

Onsager's reciprocity relations therefore express the symmetry of the matrix L, which relates the affinities to the fluxes.

9.13 Variational Principle

In a linear regime, the production of entropy $\dot{S}$ becomes a quadratic form in the affinities F:

$$\dot{S} = \frac{\mathrm{d}S}{\mathrm{d}t} = \sum_i J_i F_i = \sum_{ij} L_{ij} F_i F_j. \tag{9.172}$$

Since we know that the production of entropy is positive out of equilibrium, this quadratic form must be positive definitive. More specifically, this implies that the diagonal elements of the matrix L are positive:

$$L_{ii} > 0, \qquad \forall i. \tag{9.173}$$

Let us now calculate the derivative of $\dot{S}$ with respect to time:

$$\frac{\mathrm{d}}{\mathrm{d}t}\dot{S} = 2\sum_{ij}\sum_k \frac{\partial F_i}{\partial X_k}\frac{\mathrm{d}X_k}{\mathrm{d}t} L_{ij} F_j = 2\sum_{ik} \frac{\partial F_i}{\partial X_k} J_k J_i. \tag{9.174}$$

The last relation follows from the fact that $J_k = \mathrm{d}X_k/\mathrm{d}t$ and from the relation $\sum_j L_{ij} F_j = J_i$.

On the other hand,

$$\frac{\partial F_i}{\partial X_k} = \frac{\partial}{\partial X_k}\left(\frac{\partial S}{\partial X_i} - \frac{\partial S^{(\mathrm{r})}}{\partial X_i^{(\mathrm{r})}}\right) = \frac{\partial^2 S}{\partial X_k \partial X_i}, \tag{9.175}$$

because the second term (the one relative to the reservoir) does not depend on X_i. Therefore,

$$\frac{\mathrm{d}\dot{S}}{\mathrm{d}t} = 2\sum_{ik} \frac{\partial^2 S}{\partial X_i \partial X_k} J_i J_k. \tag{9.176}$$

The thermodynamic condition of entropy's convexity implies that the quadratic form

$$\sum_{ij} \frac{\partial^2 S}{\partial X_i \partial X_j} \xi_i \xi_j \tag{9.177}$$

is negative definite—in other words, is negative for each nonvanishing vector (ξ_i). We thus obtain (positing the J_i flows instead of the generic vector ξ_i)

$$\frac{\mathrm{d}\dot{S}}{\mathrm{d}t} \leq 0. \tag{9.178}$$

Therefore, in the linear regime, the production of entropy *decreases* during the system's evolution until it reaches a minimum.

Let us now consider a system that is *kept* out of equilibrium. A first case is that in which one maintains an affinity F_i fixed and different from zero, while the other affinities are free to change. Then, the result we just obtained that the stationary state is reached when all the other fluxes J_k, with $k \neq 1$, vanish. In fact, since $\dot{S} = \sum_{ij} L_{ij} F_i F_j$, one has

$$\left.\frac{\partial \dot{S}}{\partial X_k}\right)_X = 2\sum_j L_{kj} F_j = J_k, \tag{9.179}$$

and we obtain $\partial \dot{S}/\partial X_k = 0$ for $J_k = 0$ $(k \neq 1)$.

A more interesting case is obtained if the system is in contact with *two* (or more) reservoirs, which we shall denote by (1) and (2). The production of entropy can then be written

$$\dot{S} = \sum_i \left(\frac{\partial S}{\partial X_i} \frac{dX_i}{dt} + \frac{\partial S^{(1)}}{\partial X_i^{(1)}} \frac{dX_i^{(1)}}{dt} + \frac{\partial S^{(2)}}{\partial X_i^{(2)}} \frac{dX_i^{(2)}}{dt} \right). \tag{9.180}$$

If the system's dimensions are small, the variation of the system's variable X_i is negligible with respect to those of the reservoirs, and one obtains

$$\dot{S} = \sum_i \left(\frac{\partial S^{(1)}}{\partial X_i^{(1)}} \frac{dX_i^{(1)}}{dt} + \frac{\partial S^{(2)}}{\partial X_i^{(2)}} \frac{dX_i^{(2)}}{dt} \right) = \sum_i \left(\frac{\partial S^{(2)}}{\partial X_i^{(2)}} - \frac{\partial S^{(1)}}{\partial X_i^{(1)}} \right) \frac{dX_i^{(2)}}{dt}, \tag{9.181}$$

where we have exploited the fact that if dX_i/dt is negligible,

$$\frac{dX_i^{(1)}}{dt} = -\frac{dX_i^{(2)}}{dt}. \tag{9.182}$$

In this situation, it is natural to define the flux $\mathcal{J}_i$ that goes through the system:

$$\mathcal{J}_i = -\frac{dX_i^{(1)}}{dt} = \frac{dX_i^{(2)}}{dt}. \tag{9.183}$$

The corresponding affinity is given by

$$\mathcal{F}_i = \frac{\partial S^{(2)}}{\partial X_i^{(2)}} - \frac{\partial S^{(1)}}{\partial X_i^{(1)}}, \tag{9.184}$$

and therefore corresponds to the *difference* between the affinities with respect to the two sources. The production of entropy therefore assumes the usual form:

$$\dot{S} = \sum_i \mathcal{F}_i \mathcal{J}_i. \tag{9.185}$$

It is possible to apply the preceding arguments to this situation, and one can thus arrive at the result that the production of entropy $\dot{S}$ decreases as the system evolves. In the present case, however, the system cannot reach equilibrium (at least as long as the affinities $\mathcal{F}_i$ are maintained different from zero), but it instead reaches a stationary state, characterized by nonvanishing values of the currents $\mathcal{J}_i$.

In this manner, we have obtained the variational principle:

The nonequilibrium stationary state reached by the system in a linear regime corresponds to the minimum value of the production of entropy $\dot{S}$ compatible with the constraints imposed on the system.

9.14 An Application

In this section, we show a simple application of Onsager's reciprocity relations. Let us consider two containers, each containing a gas, which are in contact by means of a wall that allows for the very slow exchange of gas from one container to the other by means of a throttle (or simply a small hole). The exchange of matter obviously also involves an exchange of internal energy. By focusing our attention on the first container (which we will consider the *system*), we will be able to write the expressions of the affinities F_E, associated with the internal energy, and F_N, associated with the number of particles:

$$F_E = \frac{\partial S}{\partial E} - \frac{\partial S^{(r)}}{\partial E^{(r)}} = \frac{1}{T} - \frac{1}{T^{(r)}}, \qquad F_N = \frac{\partial S}{\partial N} - \frac{\partial S^{(r)}}{\partial N^{(r)}} = \frac{\mu}{T} - \frac{\mu^{(r)}}{T^{(r)}}. \tag{9.186}$$

The corresponding fluxes are obviously

$$J_E = \frac{\mathrm{d}E}{\mathrm{d}t}, \qquad J_N = \frac{\mathrm{d}N}{\mathrm{d}t}. \tag{9.187}$$

We can write the phenomenological linear relations between affinities and fluxes in the form

$$J_E = L_{EE}F_E + L_{EN}F_N, \tag{9.188}$$

$$J_N = L_{NE}F_E + L_{NN}F_N. \tag{9.189}$$

Thermodynamics and the Onsager relations imply that

$$\begin{aligned} &L_{EE} > 0, \qquad L_{NN} > 0, \\ &L_{NE} = L_{EN}, \qquad L_{EE}L_{NN} - L_{NE}L_{EN} > 0. \end{aligned} \tag{9.190}$$

Let us now consider two situations:

- **Mechanocaloric effect:** The two containers are at the same temperature (but not at the same pressure), and therefore $F_E = 0$. The heat flow between the two containers is given by

$$J_E = L_{EN}F_N = -L_{EN}\Delta\left(\frac{\mu}{T}\right), \tag{9.191}$$

while the particle flow is given by

$$J_N = -L_{NN}F_N. \tag{9.192}$$

The ratio between the two flows is given by

$$J_E/J_N = L_{EN}/L_{NN}. \tag{9.193}$$

- **Thermomechanical effect:** The flow of matter between the two containers vanishes: $J_N = 0$. This implies the existence of a relation between the affinities in temperature and in number of particles:

$$L_{NE}\Delta\left(\frac{1}{T}\right) = L_{NN}\Delta\left(\frac{\mu}{T}\right). \tag{9.194}$$

The relation between the two affinities is therefore

$$\left[\Delta\left(\frac{\mu}{T}\right)\Big/\Delta\left(\frac{1}{T}\right)\right] = \frac{L_{NE}}{L_{NN}} = \frac{L_{EN}}{L_{NN}}, \tag{9.195}$$

where the last relation derives from the Onsager reciprocity relations.

The ratio between the flows in the first case is equal to the ratio between the affinities in the second. We therefore see how the Onsager reciprocity relations can relate apparently very different phenomena.

In order to make this relation more explicit, let us express the affinity F_N in terms of the difference in pressure. As a result of the Gibbs-Duhem equation, one has

$$d\mu = -s\mathrm{d}T + \frac{1}{\rho}\,\mathrm{d}p, \tag{9.196}$$

where s is the entropy per particle, and $\rho = N/V$ is the numerical density of the gas. In the case of not too large differences in temperature and pressure, we therefore have

$$F_E = \Delta\left(\frac{1}{T}\right) \simeq -\frac{\Delta T}{T^2}, \tag{9.197}$$

$$F_N = -\Delta\left(\frac{\mu}{T}\right) \simeq -\frac{\Delta \mu}{T} + \mu\frac{\Delta T}{T^2} = -\frac{h}{T^2}\Delta T + \frac{1}{\rho T}\Delta p, \tag{9.198}$$

where $h = \mu + T$ is the enthalpy per particle. The ratio between the temperature and the pressure difference in the mechanocaloric effect satisfies the relation

$$\frac{\Delta p}{\Delta T} = \left(h - \frac{L_{EN}}{L_{NN}}\right)\frac{\rho}{T}. \tag{9.199}$$

Let us now consider the case in which the contact between the two containers occurs by means of a little hole, much smaller than the mean free path of the gas. In this case, a particle passes from one container to the next conserving the same momentum it had in the first container, since there are no impacts that allow it to reequilibrate while passing. A setup of this kind is called **Knudsen gas**. If the hole's section is A, the number of particles $\mathrm{d}N$ driven by velocity $\boldsymbol{v}$ that exit the first container through the hole in the unit of time is given by

$$\mathrm{d}N(\boldsymbol{v}) = \rho A f(\boldsymbol{v})\theta(v_x)\, v_x d\boldsymbol{v}, \tag{9.200}$$

where we have assumed that the hole is parallel to the (yz) plane and that the particles, in order to exit, must move in the direction of positive x's. In this equation, $f(\boldsymbol{v})$ is the Maxwell distribution:

$$f(\boldsymbol{v}) = \left(\frac{m}{2\pi k_{\mathrm{B}}T}\right)^{3/2} \exp\left(-\frac{mv^2}{2k_{\mathrm{B}}T}\right). \tag{9.201}$$

Let us fix the modulus v of the velocity and integrate over its direction. We obtain the number $\mathrm{d}N(v)$ of particles driven by speed v:

$$\mathrm{d}N(v) = \rho A\pi\left(\frac{m}{2\pi k_{\mathrm{B}}T}\right)^{3/2} v^3 \exp\left(-\frac{mv^2}{2k_{\mathrm{B}}T}\right)\mathrm{d}v. \tag{9.202}$$

The average energy of a particle that reaches the hole is given by

$$\left\langle\frac{1}{2}\,mv^2\right\rangle = \left(\int \mathrm{d}N(v)\frac{1}{2}\,mv^2\right)\Big/\left(\int \mathrm{d}N(v)\right) = 2k_{\mathrm{B}}T. \tag{9.203}$$

The average transferred energy per particle (from left to right) is $2k_BT$. If the temperatures of the two containers do not differ by much, the ratio between the energy flow and the particles' flow is therefore equal to $2k_BT$. We thus obtain

$$\frac{L_{EN}}{L_{NN}} = 2k_BT. \tag{9.204}$$

Recommended Reading

This chapter owes much of its structure to the presentation that can be found in chapter 8 of D. Chandler, *Introduction to Modern Statistical Mechanics*, Oxford, UK: Oxford University Press, 1987. The theory of linear response and the connection with correlation functions are wonderfully presented in D. Forster, *Hydrodynamic Fluctuations, Broken Symmetry, and Correlation Functions*, Reading, MA: Benjamin, 1975. One can also usefully consult L. E. Reichl, *A Modern Course in Statistical Physics*, Austin: University of Texas Press, 1980, and R. Kubo, M. Toda, and N. Hashitsume, *Statistical Physics II*, Berlin: Springer, 1985. I invite everyone to read Einstein's original articles on the theory of Brownian motion, reprinted in A. Einstein, *Investigations on the Theory of Brownian Movement*, New York: Dover, 1956.

10 Complex Systems

Every rule has its exceptions, including this one.
—Anonymous

In this chapter, we examine systems in statistical mechanics that are characterized by disorder—not the usual thermal disorder, represented by the Boltzmann-Gibbs distribution, but a "frozen" disorder, in which the Hamiltonian itself is random.

The study of these systems is obviously of great importance as far as its applications are concerned—one need only to remember that all metal alloys are composed of mixtures of different metals, smelted together, and then cooled. In these systems, at room temperature, the arrangement of atoms with different chemical natures does not change even over very long time periods, and it is therefore set by the initial conditions. The semiconductors used in electronics also generally exhibit impurities (*doping*) distributed in a random fashion, and these remain blocked inside the solid matrix at room temperature.

Dealing with statistical mechanics systems with a random Hamiltonian has opened a new field of study. Among other applications, it is now possible to consider problems of a strictly geometric nature by means of a clever generalization of the statistical approach. An interesting case (and a very important one, also from the point of view of applications) is that of **percolation**. In the problem of percolation, one considers a random, but fixed, distribution of particles, e.g., on a lattice. One divides the particles present in the sample into clusters, such that neighboring particles belong to the same cluster. The problem is then to characterize the size and spatial extension of the clusters of connected particles as a function of the overall particle density. It is possible to formulate the problem of percolation in a way that highlights its analogy with a phase transition problem in statistical mechanics. This is achieved by introducing a clever limit procedure. Based on this analogy, it is possible to identify the universal properties of the percolation problem. It turns out that the knowledge obtained in this way sheds new light on the properties of the ordinary phase transitions that we have discussed in preceding chapters.

Ordinary and spin glasses (ferromagnets with random interactions) are a class of systems characterized by significant disorder, and they are at the edge of our understanding of statistical mechanics. In these systems, it is still not completely clear how we can apply the fundamental concepts of statistical mechanics, such as thermodynamic equilibrium, the Boltzmann-Gibbs distribution, and so on. Also these systems are characterized by frozen (or **quenched**) disorder. Some innovative techniques have been introduced to allow the representation of systems with quenched disorder in terms of the ordinary systems of statistical mechanics, where the configurations are not frozen but change over the observation time (one speaks of **annealed** disorder). The best known of these techniques is the **replica method**. It has produced surprising results, at least for those few models that can be treated exactly (even if not rigorously). The extent to which these results are valid for the real world is the object of intense debate. In this situation, numerical simulations perform an essential role because they allow one to obtain information about the behavior of realistic, but analytically intractable, systems—information that is not experimentally accessible.

We will start our discussion of disordered systems by dealing with the problem of the structure of linear polymers in a solvent, although, strictly speaking, quenched disorder is not present in this system. It is, however, a useful starting point, especially because it allows us to see how a strictly geometrical problem (what are the geometrical properties of random walks without intersections?) can be treated as the limit of an ordinary problem in statistical mechanics.

10.1 Linear Polymers in Solution

Polymers are molecules obtained by connecting a large number of simple units (the **monomers**) using covalent bonds. In **linear** polymers, the monomers are arranged in a sequence. Biologically important examples are **proteins**, in which the monomers are amino acids, connected to one another by means of the peptide bond –NH–CO–, and nucleic acids (DNA and RNA), whose units are nucleotides (molecules formed by a base—adenine, guanine, cytosine, and thymine [or uracil]—bonded with a sugar—deoxyribose or ribose—which in turn is bonded to a phosphate) that are connected by means of a phosphodiester bond –O–P–O–. Plastic materials are also usually made of linear polymers—polyethylene for instance, has the structure $\ldots\text{–}CH_2\text{– }CH_2\text{–}\ldots$, which is a result of the polymerization of ethylene $CH_2 = CH_2$.

We must therefore distinguish between **homogeneous** polymers such as polyethylene, in which the monomers all belong to one chemical species, or to several ones that succeed one another in regular fashion, and **heterogeneous** polymers such as proteins and nucleic acids, in which the monomers do not follow one another with regularity along the molecule. In the remainder of this section, we will deal only with homogeneous linear polymers.

In these structures, the bond between successive monomers has a certain flexibility and allows them to rotate fairly freely around the axis that connects them. In this manner, the polymer, as soon as it is composed of a large enough number of monomers, can be treated like a flexible chain.

Linear polymers are usually produced in a *good solvent*—a solvent whose affinity for the monomers is greater than the affinity the monomers have for one another. In this situation, the solvent interposes itself between monomers and tends to keep the monomers apart. In a bad solvent, on the other hand, the solvent is chased off by the polymer, which therefore folds in on itself and tends to form a compact ball and to precipitate from the solution.

10.1.1 The Gaussian Model

Let us therefore consider an isolated polymer in a good solvent, and ask ourselves what configurations it might assume. We will denote the distance between the center of one monomer and the center of the next by a_0, and the **degree of polymerization**—in other words, the number of monomers that constitute the molecule—by N. Let us assume that the bond's structure allows three successive monomers to assume any configuration, as long as the distance between their respective centers remains equal to a_0.

If we temporarily disregard the **excluded volume interaction**—in other words, the significant repulsion that various monomers that are close to one another in ambient space are subject to—we obtain a model of the linear polymer called the **Rayleigh chain**. This model can be formally defined as follows:

1. A polymer composed of $N+1$ monomers (and by N covalent bonds between one polymer and the next) is represented by $N+1$ variables $\boldsymbol{r}_i$, $i=0,\ldots,N$, where $\boldsymbol{r}_i \in \mathbb{R}^3$ is the vector that represents the $i+1$-th monomer center's position.
2. The covalent bond that connects successive monomers is represented by the constraint
$$|\boldsymbol{r}_{i+1} - \boldsymbol{r}_i| = a_0, \qquad i=0,\ldots,N-1. \tag{10.1}$$
3. All the configurations that satisfy the constraint (10.1) have the same energy (which can be considered vanishing).

The polymer will assume the shape of a noncompact ball, occupying a certain region of space. We want to calculate the typical linear dimensions of this region. We can first ask ourselves what the characteristic distance is between the first monomer (which we can assume placed in the origin: $\boldsymbol{r}_0 = 0$) and the last, placed in $\boldsymbol{r}_N$. We can easily answer this question by observing that, in our hypotheses, the vectors

$$\boldsymbol{y}_i = \boldsymbol{r}_i - \boldsymbol{r}_{i-1}, \qquad i=1,\ldots,N, \tag{10.2}$$

which represent the difference in position between the successive monomers, are independent, identically distributed random vectors. Their distribution is actually uniform over a sphere of radius a_0:

$$p(\boldsymbol{y}) \propto \delta\left(y^2 - a_0^2\right), \tag{10.3}$$

where the proportionality constant is determined by the normalization condition

$$\int \mathrm{d}^3 y\, p(\boldsymbol{y}) = 1. \tag{10.4}$$

It is easy to see that

$$\langle y \rangle = 0, \tag{10.5}$$

$$\langle y^2 \rangle = a_0^2, \tag{10.6}$$

Since $\boldsymbol{r}_0 = 0$, one has the relation

$$\boldsymbol{r}_N = \sum_{i=1}^{N} \boldsymbol{y}_i, \tag{10.7}$$

from which it follows that

$$\langle \boldsymbol{r}_N \rangle = 0, \tag{10.8}$$

$$\langle r_N^2 \rangle = N a_0^2. \tag{10.9}$$

The ball's radius therefore increases as $N^{1/2}$. In order to have a measure of this linear dimension that is more easily accessible to the experiment, one usually considers the **radius of gyration** R_G, which is defined by

$$R_G^2 = \frac{1}{N+1} \sum_i |\boldsymbol{r}_i - \boldsymbol{R}_{CM}|^2. \tag{10.10}$$

Here, $\boldsymbol{R}_{CM}$ is the position of the polymer's center of mass, defined by

$$\boldsymbol{R}_{CM} = \frac{1}{N+1} \sum_i \boldsymbol{r}_i \tag{10.11}$$

The ball's moment of inertia is equal to the ball's mass times the square of the radius of gyration.

It is easy to calculate $\langle R_G^2 \rangle$ by using the expression

$$R_G^2 = \frac{2}{N(N+1)} \sum_{i<j} |\boldsymbol{r}_i - \boldsymbol{r}_j|^2. \tag{10.12}$$

Actually, if $i<j$, one has

$$\langle |\boldsymbol{r}_i - \boldsymbol{r}_j|^2 \rangle = \left\langle \left| \sum_{k=i+1}^{j} \boldsymbol{y}_k \right|^2 \right\rangle = \sum_{k=i+1}^{j} \langle |\boldsymbol{y}_k|^2 \rangle = (j-i) a_0^2. \tag{10.13}$$

We thus obtain

$$\langle R_G^2 \rangle = \frac{2}{N(N+1)} \sum_{i<j} (j-i) a_0^2 = \frac{(N+2)\, a_0^2}{3}. \tag{10.14}$$

Therefore,

$$R_N = \sqrt{\langle R_G^2 \rangle} \simeq \frac{a_0}{\sqrt{3}} N^{1/2}. \tag{10.15}$$

Actually, by exploiting the central limit theorem, we can say a lot more. Since $\boldsymbol{r}_N = \sum_i \boldsymbol{y}_i$, where the $\boldsymbol{y}_i$ are independent, identically distributed random vectors, with a vanishing average and variance equal to a_0^2, we can conclude that $\boldsymbol{r}_N$ is, for a sufficiently large N, a random Gaussian vector, with a vanishing average and variance equal to Na_0^2:

$$P_N(\boldsymbol{r}_N) \propto \exp\left(-\frac{r^2}{2Na_0^2}\right). \tag{10.16}$$

Since this relation necessarily follows from our hypothesis that the excluded interaction volume is negligible (independently from the details of the y_i's distribution), the linear polymer model in which this hypothesis is made is called a **Gaussian model**.

We can now try to formulate the linear polymer's Gaussian model in a form that emphasizes its analogy with phase transitions. Let us introduce $\Gamma_N(\boldsymbol{r})$, which counts the number of configurations in which $\boldsymbol{r}_0 = 0$ and $\boldsymbol{r}_N = \boldsymbol{r}$:

$$\Gamma_N(\boldsymbol{r}) = \int \prod_{i=1}^{N} \left\{ \mathrm{d}^d \boldsymbol{r}_i \delta(|\boldsymbol{r}_i - \boldsymbol{r}_{i-1}|^2 - a_0^2) \right\} \delta(\boldsymbol{r}_N - \boldsymbol{r}). \tag{10.17}$$

One obviously has

$$\Gamma_{N+1}(\boldsymbol{r}) = \mathcal{S} \int \mathrm{d}^d \boldsymbol{r}_N p(\boldsymbol{r} - \boldsymbol{r}_N) \Gamma_N(\boldsymbol{r}_N), \tag{10.18}$$

where $p(\boldsymbol{y})$ is the probability distribution of the distance between successive monomers, and $\mathcal{S}$ is a constant, which depends on the details of the distribution of the $\boldsymbol{y}_i$. Let us note that here we have assumed, for generality's sake, that we are in a space with dimensionality d, not necessarily equal to 3.

This equation can be resolved by shifting to the Fourier transforms. We define, for some function $f(\boldsymbol{r})$, its Fourier transform $\tilde{f}(\boldsymbol{k})$ by means of

$$\tilde{f}(\boldsymbol{k}) = \int \mathrm{d}^d \boldsymbol{r} \mathrm{e}^{\mathrm{i}(\boldsymbol{k}\cdot\boldsymbol{r})} f(\boldsymbol{r}), \tag{10.19}$$

so that

$$f(\boldsymbol{r}) = \int \frac{\mathrm{d}^d \boldsymbol{k}}{(2\pi)^d} \mathrm{e}^{\mathrm{i}(\boldsymbol{k}\cdot\boldsymbol{r})} \tilde{f}(\boldsymbol{k}). \tag{10.20}$$

One then has

$$\tilde{\Gamma}_{N+1}(\boldsymbol{k}) = \frac{\mathcal{S}}{(2\pi)^d} \tilde{p}(\boldsymbol{k}) \tilde{\Gamma}_N(\boldsymbol{k}). \tag{10.21}$$

Let us set $\Gamma_0(\boldsymbol{r}) = \delta(\boldsymbol{r})$, and obtain

$$\tilde{\Gamma}_N(\boldsymbol{k}) = \left[\frac{\mathcal{S}}{(2\pi)^d}\right]^N \tilde{p}^N(\boldsymbol{k}). \tag{10.22}$$

If $p(\boldsymbol{r})$ is such that $\langle \boldsymbol{r} \rangle = 0$ and $\langle \boldsymbol{r}^2 \rangle = a_0^2$, then, for small values of $|\boldsymbol{k}|$, one has

$$\tilde{p}(\boldsymbol{k}) \simeq \mathrm{e}^{-a_0^2 k^2/2}. \tag{10.23}$$

Therefore,

$$\tilde{\Gamma}_N(\boldsymbol{k}) \simeq \left[\frac{\mathcal{S}}{(2\pi)^d}\right]^N \mathrm{e}^{-N a_0^2 k^2/2}. \tag{10.24}$$

In order to calculate $\boldsymbol{r}$'s distribution, we can calculate the Fourier antitransform of $\Gamma_N(\boldsymbol{k})$. One has

$$\Gamma_N(\boldsymbol{r}) = \int \frac{\mathrm{d}\boldsymbol{k}}{(2\pi)^d}\, \mathrm{e}^{-\mathrm{i}(\boldsymbol{k}\cdot\boldsymbol{r})}\tilde{\Gamma}_N(\boldsymbol{k}) \sim \mathrm{e}^{-r^2/(2Na_0^2)}, \tag{10.25}$$

which is obviously valid only for $r^2 \gg a_0^2$, so that the integral (10.25) will be dominated by small values of k.

Let us go over to the grand canonical ensemble, by introducing the fugacity z:

$$\tilde{\Gamma}_{\mathrm{GC}}(\boldsymbol{k}, z) = \sum_{N=0}^{\infty} z^N \tilde{\Gamma}_N(\boldsymbol{k}). \tag{10.26}$$

We obtain

$$\tilde{\Gamma}_{\mathrm{GC}}(\boldsymbol{k}, z) \simeq \left[1 - \frac{z\mathcal{S}}{(2\pi)^d}\,\mathrm{e}^{-a_0^2k^2/2}\right]^{-1} \propto \left(z_{\mathrm{c}} - z + ck^2\right)^{-1}, \tag{10.27}$$

where we have assumed that $ck^2 \ll 1$, and we have introduced

$$z_{\mathrm{c}} = \frac{(2\pi)^d}{\mathcal{S}}, \tag{10.28}$$

$$c = \frac{a_0^2}{2}. \tag{10.29}$$

Exercise 10.1 (Polymer over Lattice) Suppose that you arrange a polymer over a simple cubic lattice, so that the possible values of the $\boldsymbol{r}_i$ are of the form $\boldsymbol{r}_i = a_0(n_1, \dots, n_d)$, where the n_α, $\alpha = 1, \dots, d$ are integers. One imposes the constraint that $\boldsymbol{y}_i = \boldsymbol{r}_i + 1 - \boldsymbol{r}_i$ be equal to one of the lattice base vectors $\boldsymbol{y}_i = a_0\boldsymbol{e}_\alpha$, where $\boldsymbol{e}_\alpha = (\delta_{1\alpha}, \dots, \delta_{d\alpha})$. Show that for the $\tilde{\Gamma}_{\mathrm{GC}}(\boldsymbol{k}, z)$, one obtains an expression analogous to equation (10.27), and calculate the relative values of z_{c} and the constant c.

This expression is analogous to the Fourier transform of the correlation function of a system close to the critical point, in the Ornstein-Zernike approximation. The fugacity z has the role of the inverse of the temperature, β, and z_{c} that of the critical value. When z increases, getting closer to z_{c}, the average number of monomers in the linear polymer diverges. Susceptibility has as an analog $\tilde{\Gamma}_{\mathrm{GC}}(\boldsymbol{k} = 0, z)$, which is proportional to the total number of configurations of a linear polymer with fugacity z.

Exercise 10.2 Show that if $\tilde{\Gamma}_{\mathrm{GC}}(\boldsymbol{k} = 0, z) \propto |z_{\mathrm{c}} - z|^{-\gamma}$, one has

$$\Gamma_N = \int \mathrm{d}^d\boldsymbol{r}\,\Gamma_N(\boldsymbol{r}) \propto z_{\mathrm{c}}^{-N} N^{\gamma-1}.$$

In our case, therefore, $\Gamma_N \propto z_{\mathrm{c}}^{-N}$, while the factor $N^{\gamma-1}$ (which is called the **enhancement factor**) is equal to 1.

From these results, we can see that the radius of gyration R_{G} is analogous to the coherence length in critical phenomena. We thus have $R_{\mathrm{G}} \sim |z_{\mathrm{c}} - z|^{-\nu}$, with $\nu = 1/2$ (for the Gaussian model). This corresponds to $R_{\mathrm{G}} \sim N^{\nu}$.

A quantity analogous to the free energy can be obtained by considering the number of structures that return to the origin $\Gamma_N(0)$. This is obviously given by

$$\Gamma_N(0) = \int \frac{\mathrm{d}^d \boldsymbol{k}}{(2\pi)^d} \tilde{\Gamma}_N(\boldsymbol{k}). \tag{10.30}$$

This integral is dominated , for $2 < d \leq 4$, by small values of k.

Exercise 10.3 Show that, for $d < 4$, one has

$$\Gamma(0,z) = \int \frac{\mathrm{d}^d \boldsymbol{k}}{(2\pi)^d} \tilde{\Gamma}_{\mathrm{GC}}(\boldsymbol{k}) \simeq \text{const.} + \text{const.} \times |z_c - z|^{(d-2)/2}.$$

What occurs for $d = 4$? And what about for $d > 4$? Show that this result implies that, for $2 < d \leq 4$, one has

$$\Gamma_N(0) \sim N^{-d/2}.$$

We thus obtain $\Gamma_N(0) \sim N^{d\nu}$, with $\nu = 1/2$. As a result of the scaling laws, we know that $d\nu = 2 - \alpha$. In the Gaussian model, one therefore has $\alpha = (4 - d)/2$.

10.1.2 *The Flory Argument*

The results discussed thus far have been obtained by assuming that the excluded volume interactions are negligible. One can ask oneself how important they actually are. A first estimate can be obtained by calculating the number of encounters in a Gaussian polymer—in other words, the number of times in which monomers that are distant from each other on the chain come to be close to each other in space. We can estimate this quantity by supposing that the polymer is uniformly distributed within a region of space whose radius is proportional to R_G. The monomers' density ρ is therefore equal to $N/R_G^d \sim N^{1-d/2}$. The encounter probability per unit of volume is proportional to ρ^2, and multiplying by the volume R_G^d, we obtain the number of encounters:

$$\mathcal{N} \sim R_G^d \rho^2 \sim \frac{N^2}{R_G^d} \sim N^{2-d/2}. \tag{10.31}$$

If $d > 4$, we obtain $N \to 0$ for $N \to \infty$! This obviously means that the number of encounters remains limited when the polymer's length diverges. On the other hand, if $d < 4$, the number of encounters diverges, and the excluded volume interaction becomes relevant. (For $d = 4$, a more careful analysis shows that the number of superpositions increases logarithmically with N, and that therefore this is a marginal case.)

A slight modification in this line of reasoning allows us to obtain an excellent estimate of the exponent ν in the presence of excluded volume interactions. Let us assume that we consider R_G to be a variational parameter and look for the polymer's free energy minimum. The free energy is the difference between the interaction energy E and the entropy S multiplied by the temperature T, which we consider constant. From the calculation we just performed, we obtain an estimate of E:

$$E \sim \frac{N^2}{R_G^d}. \tag{10.32}$$

The entropy can be estimated by supposing that the polymer is still in a Gaussian ensemble, and as a result, the number of configurations of a width equal to R_G is proportional to exp(−const. × R_G^2/N). Entropy is this quantity's logarithm. By taking all the constants as understood, the free energy is therefore given by

$$F = \frac{N^2}{R_G^d} + \frac{R_G^2}{N}. \tag{10.33}$$

For $d > 4$, for a Gaussian polymer, the first term is negligible compared to the second one, and the effect of the excluded volume interaction is also negligible. For $d < 4$ instead, this expression's minimum is obtained by

$$R_G \sim N^{3/(d+2)}. \tag{10.34}$$

In this way, we obtain the **Flory estimate** of exponent ν, valid for $d < 4$:

$$\nu = \frac{3}{d+2}. \tag{10.35}$$

This expression is remarkable because it is exact for $d = 1$ ($\nu = 1$), $d = 2$ ($\nu = 3/4$], which can be obtained with other methods), and it obviously predicts $\nu = 1/2$ for $d = 4$. In $d = 3$, it predicts $\nu = 0.6$, compared with experimental results, which give $\nu = 0.56 \pm 0.01$. It is not clear, however, why these results are so good. We should add that up to now, no other argument of comparable simplicity for predicting the value of exponent γ, which appears in the enhancement factor, has been found.

10.1.3 Linear Polymers and Critical Phenomena

We can thus conclude that close analogy exists between the (essentially geometrical) problem of the conformation of a linear polymer with excluded volume interactions and ordinary critical phenomena. This analogy can be represented by the following list:

$$\begin{aligned} N &\leftrightarrow |T - T_c|^{-1} \\ R_G &\leftrightarrow \xi \\ \Gamma_N(\mathbf{r}) &\leftrightarrow G(\mathbf{r}, T) \\ \Gamma_N(0) &\leftrightarrow \Delta F \end{aligned}$$

It can also be put on a more formally explicit foundation. Actually, as de Gennes showed in 1972 (see [Genn79]), the properties of linear polymers with excluded volume interactions correspond to the critical properties of a Heisenberg model with an n component spin, in the limit $n \to 0$.

To show this relation, we wll use an argument put forward by G. Sarma [Sarm79]. We start by proving the following **geometrical lemma**:

Let $\mathbf{s} = (s_\alpha)$ be a classical n component spin, with norm $s^2 = \sum_{\alpha=1}^{n} s_\alpha^2 = n$. Let us consider the moments of **s***'s components $\langle s_{\alpha_1}^{p_1} \cdots s_{\alpha_k}^{p_k} \rangle$, where $\langle \ldots \rangle$ is the geometric mean over the sphere of radius $\sqrt{n}$ with n dimensions. Then these moments all cancel out in the limit $n \to 0$, except the trivial one (for which $p_i = 0\ \forall i$) and those of the type $\langle s_\alpha^2 \rangle$, which are all equal to 1.*

This result can be easily obtained considering the generating function of these moments:

$$f(\boldsymbol{k}) = \left\langle \exp\left(\mathrm{i}\sum_{\alpha} k_{\alpha} s_{\alpha}\right)\right\rangle. \tag{10.36}$$

This can depend on k^2 only, and it satisfies the differential equation

$$-\nabla^2 f = \left(\sum_{\alpha} s_{\alpha}^2\right) f = nf, \tag{10.37}$$

which in spherical coordinates is written

$$-\left(\frac{\mathrm{d}^2}{\mathrm{d}k^2} + \frac{n-1}{k}\frac{\mathrm{d}}{\mathrm{d}k}\right) f = nf, \tag{10.38}$$

The boundary conditions for this equation are easily obtained:

$$f(0) = 1, \tag{10.39}$$

$$\left.\frac{\partial^2 f}{\partial k_{\alpha}^2}\right|_{k=0} = \langle s_{\alpha}^2 \rangle = 1, \tag{10.40}$$

and therefore, for $k^2 \ll 1$, one has

$$f(\boldsymbol{k}) \simeq 1 - \frac{1}{2}k^2 + \cdots \tag{10.41}$$

Now, $f(\boldsymbol{k}) = 1 - 1/2\,k^2$ is the solution of equation (10.38) for $n = 0$ and satisfies the initial conditions—it follows that it is the generating function for $n = 0$. By taking the derivative of the generating function with respect to k_{α} at $k_{\alpha} = 0$, we can obtain the moments of s_{α}. We can thus see that the only nonvanishing such moment, apart from the trivial one, is $\langle s_{\alpha}^2 \rangle = 1$.

Let us now consider a simple cubic lattice in d dimensions. Each lattice site $\boldsymbol{r}$ is associated with a spin $\mathbf{s}_r$ with n dimensions. We will denote the interaction between nearest neighbors by J. One writes the Hamiltonian as follows:

$$H = -J \sum_{<rr'>} (\mathbf{s}_r \cdot \mathbf{s}_{r'}), \tag{10.42}$$

where the sum is extended to all nearest-neighbor pairs of sites in the lattice. The partition function Z takes the following expression:

$$Z = \prod_{r} \mathrm{d}\mathbf{s}_r \exp(-\beta H), \tag{10.43}$$

where the integral over $\boldsymbol{r}$ contains only the angular part, since the length of $\mathbf{s}_r$ is fixed.

One can expand Z in the following form:

$$Z = \left\langle \prod_{<rr'>,\alpha} \left(1 + K s_{\alpha r} s_{\alpha r'} + \tfrac{1}{2} K^2 s_{\alpha r}^2 s_{\alpha r'}^2 + \cdots\right)\right\rangle, \tag{10.44}$$

where $K = \beta J$. We will see that it is not necessary to consider further terms in the limit $n \to 0$. The different terms that contribute to Z can be represented by graphs. A typical term is a *cycle* on the lattice corresponding to a product of $K s_{\alpha r}\, s_{\alpha r'}$, with α fixed. Moreover, because of the geometrical lemma, all the averages except $\langle s_{\alpha}^2 \rangle$ vanish, and therefore one

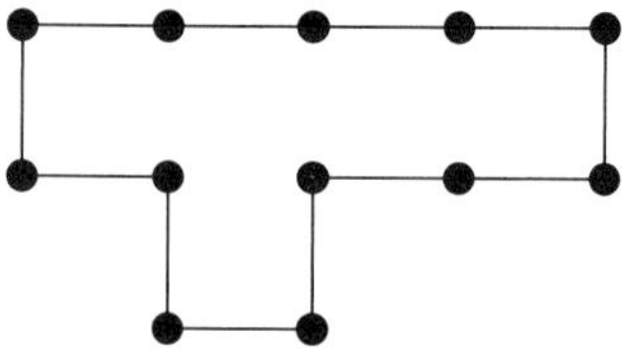

FIGURE 10.1. A typical term of the expansion of Z in equation (10.44).

cannot pass through each point on the lattice more than once, so the limit $n \to 0$ translates the excluded volume constraint. Diagrams cannot intersect themselves nor be overlapped by other diagrams (see figure 10.1). Each cycle's contribution is clearly K^N, where N is the number of bonds, but since this cycle can appear with any component index α whatsoever, one obtains nK^N, which tends to zero when n tends to zero. One thus obtains $\lim_{n\to 0} Z = 1$. From what we have stated, it is obvious that $\lim_{n\to 0} Z/n$ is equal to the number of configurations of a loop polymer over a lattice with excluded volume.

Let us now consider the correlation function between two spins s_{r_1} and s_{r_2}. Since $Z = 1$, the correlation function for component 1 has the following expression:

$$C_{11}(\mathbf{r}_1, \mathbf{r}_2) = \langle s_{1r_1} s_{1r_2} \rangle = \left\langle s_{1r_1} s_{1r_2} \left(1 + K s_{\alpha r} s_{\alpha r'} + \frac{1}{2} K^2 s_{\alpha r}^2 s_{\alpha r'}^2 + \cdots \right)\right\rangle. \tag{10.45}$$

The various contributions to $C_{11}(\mathbf{r}_1, \mathbf{r}_2)$ can be represented by graphs. Each cycle's contribution vanishes when one sums over α, and therefore, only those walks without intersection that connect $\mathbf{r}_1$ to $\mathbf{r}_2$ remain (see figure 10.2). For each bond $\langle \mathbf{r}\mathbf{r}' \rangle$ one has to consider the term $s_{1r}\ s_{1r'}$ in which one does not sum over α. Each graph of this type represents the configuration of an excluded volume polymer over the lattice. Each configuration is counted once and only once, and its weight is proportional to K^N, where N is the number of lattice steps that appear in the configuration. Consequently,

$$C_{11}(\mathbf{r}_1, \mathbf{r}_2) = \langle s_{1r_1} s_{1r_2} \rangle = \sum_N K^N \Gamma_N(\mathbf{r}_1 - \mathbf{r}_2), \tag{10.46}$$

where, as we saw earlier, $\Gamma_N(\mathbf{r})$ is the number of excluded volume configurations that connect the origin with point $\mathbf{r}$ on the lattice. We have thus established a correspondence between the problem of the excluded volume polymer and the zero component magnetic model. It is easy to obtain the behavior of the various quantities we are interested in in the limit $N \to \infty$ (which corresponds to the critical limit $T \to T_c$ for the magnetic model) from this correspondence.

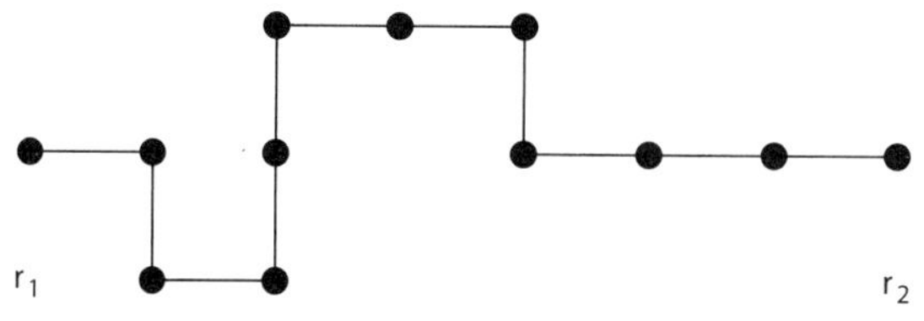

FIGURE 10.2. A term that contributes to $C_{11}(\mathbf{r}_1, \mathbf{r}_2)$.

Let us now consider the magnetic problem in the presence of an external magnetic field $\boldsymbol{h} = (h_1, 0, \dots, 0)$. In the limit $n \to 0$, this ensemble describes a solution that contains several excluded volume polymers.

Exercise 10.4 Show that the coefficient h_1^{2p} in $\lim_{n \to 0} Z/n$ can be represented by graphs that represent the configurations of p polymers over an excluded volume lattice. The contribution of each graph is proportional to K^N, where N is the total number of bonds present in the configuration.

In this manner, h_1^2 plays fugacity's role for polymers, while K plays that of fugacity for monomers. We note that in this approach, the length of each polymer is not separately fixed, but only the total length is (via K). Therefore, the ensemble described in this fashion contains polymers of all lengths (smaller than N). One says that this ensemble describes a **polydisperse** solution.

For further discussion, I refer the reader to the books by de Gennes [Genn79] and des Cloiseaux and Jannink [Cloi90] cited in the bibliography.

10.2 Percolation

The problem of percolation was introduced by Hammersley and his collaborators in the 1950s [Broa57]. In many situations that deal with random systems, it is necessary to evaluate the properties that concern the connectivity of impurities. A ceramic filter, for example, is composed of a solid matrix that is prepared with a certain number of holes, randomly arranged, which occupy a certain fraction of the volume. In order for the filter to work, it is obviously necessary for the holes to form part of a path that connects one side of the sample to the opposite one. In this case one says that the holes **percolate**. We would like to determine the minimal fraction of the holes' volume that allows us to be basically certain that they percolate. Analogously, if we consider a system composed of a casual assortment of conducting and insulated balls. We can ask what is the minimal fraction of conducting balls necessary for the sample to conduct current? And what will be the sample's resistance as a function of the number of conductive balls be?

We can formalize the percolation problem in the following manner. Let us consider a simple cubic lattice in d dimensions. Let us assume that the bonds between the lattice's nearest neighbors can be either present or absent. We define the relation *being connected* recursively between lattice sites, in the following manner:

1. Each lattice site $\boldsymbol{r}$ is connected to itself by definition.
2. If $\boldsymbol{r}_1$ is a nearest neighbor of a site $\boldsymbol{r}$ connected to $\boldsymbol{r}_2$, and if the bond $\langle \boldsymbol{r}_1 \boldsymbol{r} \rangle$ is present, then the $\boldsymbol{r}_1$ and $\boldsymbol{r}_2$ sites are connected.

Exercise 10.5 Show that the relation *being connected* is an equivalence relation, since it possesses the reflexivity, symmetry, and transitivity properties.

Given a configuration of the bonds over the lattice, and since the relation *being connected* is an equivalence relation, the lattice sites are divided into equivalence classes of sites that are connected to one another. These classes are called **clusters**. The **percolation** problem consists of determining the geometrical properties of these clusters, once the statistical properties of the bond distributions are known.

What we have defined is known as the **bond percolation** problem. A closely connected problem is that of **site percolation**, in which the lattice sites can be occupied or empty. In this case, two different sites are connected if they are both occupied and if one is the nearest neighbor of a site connected to the other. An occupied site is connected to itself.

The *standard* problem of bond percolation corresponds to the case in which each bond's presence is a random property, independent for different bonds, and with the same presence probability p for each bond. Analogously, in the case of site percolation, one can assume that each site is occupied with probability p independently from the others.

10.2.1 *Analogy with Magnetic Phenomena*

One can establish an analogy between magnetic phenomena—for example, in the Ising model—and percolation, by supposing that a spin variable is associated with each site, and that there is a ferromagnetic interaction of intensity J, where J is very large, between connected nearest-neighbor sites. In order to clarify our ideas, let us consider bond percolation. Then each site belongs to exactly one cluster. When p is small, only finite clusters are present, and therefore, in the absence of an external field, the system's total magnetization vanishes. On the other hand, when p is greater than the percolation threshold p_c, an infinite cluster will be present, and if the interaction is strong enough, all of this cluster's spins will be oriented in the same direction—the magnetization per spin will therefore be proportional to the probability $P(p)$ that a site belongs to the infinite cluster. In this manner, one can establish an analogy between $P(p)$ and spontaneous magnetization.

We will now denote the average number of clusters containing s sites per lattice site by $\nu_s(p)$. The probability that a specific site belongs to a cluster of size s is given by $s\nu_s(p)$. We can then express P as a function of $\nu_s(p)$ be means of the relation

$$\sum_s s\nu_s(p) + P(p) = 1. \tag{10.47}$$

Exercise 10.6 In similar fashion, show that the analog of magnetic susceptibility per spin with vanishing field χ is given by

$$\chi = \frac{J}{k_B T}\sum_s s^2 \nu_s(p).$$

As a result, χ can be expressed as a function of the average cluster size, defined by

$$S(p) = \frac{\sum_s s^2 \nu_s(p)}{\sum_s s\nu_s(p)}. \tag{10.48}$$

The magnetic system's partition function Z is obtained by summing the Boltzmann factor $\exp(-H/k_B T)$ over all the spin configurations. In our case, since the interaction is very strong, this factor is large if the connected neighbors point in the same direction; otherwise, it practically vanishes. Therefore,

$$Z \simeq \sum_{\mathcal{S}} \left[\exp\left(\frac{J}{k_B T} \right) \right]^{\mathcal{B}},$$

where the sum runs over all allowed spin configurations $\mathcal{S}$, and $\mathcal{B}$ denotes the number of bonds that are present. A configuration is allowed if all the spins in a connected cluster point in the same direction. The number of allowed configurations is therefore equal to 2^L, where L is the number of clusters. The free energy F is proportional to the logarithm of Z. The number of bonds present is proportional to p in the case of bond percolation, and to p^2 in the case of site percolation, because the occupation probabilities are mutually independent. One therefore obtains

$$F \propto \ln Z \simeq \ln \sum_{\mathcal{S}} \simeq \ln 2 \times \mathcal{C},$$

where $\mathcal{C}$ is the number of independent clusters which appear in the current site or bond configuration.

Exercise 10.7 Show that the analog of the magnetic correlation function between site i and site j is given by the average probability that i and j belong to the same finite cluster.

10.2.2 Percolation in One Dimension

Let us now consider the standard problem of site percolation in one dimension. This problem is slightly pathological, but it can be resolved exactly.

We therefore have a one-dimensional lattice composed of $N+1$ sites ($i = 0, 1, \ldots, N$). Each site can be present with probability p, independently from the other sites. Let us now consider two sites at distance r from each other. What is the probability $C(p, r)$ that they belong to the same cluster? It is easy to see that they will belong to the same cluster only if they are both occupied and if all the $r-1$ sites that separate them are also occupied. Therefore,

$$C(p,r) = p^{r+1}. \tag{10.51}$$

Let us note that for $r \to \infty$, one always has $C(p, r) \to 0$, except when p is exactly equal to 1. In one dimension, therefore, the infinite cluster is present only for $p = 1$.

Let us now calculate $\nu_s(p)$—in other words, the probability that a site belongs to a cluster that contains exactly s sites. The probability that it belongs to a cluster that contains a single site for example is given by

$$\nu_1(p) = p(1-p)^2, \tag{10.52}$$

since it must be occupied, while its two nearest neighbors must not be occupied. One can immediately see that, in general,

$$\nu_s(p) = sp^s(1-p)^2. \tag{10.53}$$

The factor s derives from the fact that the chosen site can be any one of the cluster's s sites. The clusters' average size is therefore given by

$$S(p) = \frac{\sum_s s^2 \nu_s(p)}{\sum_s \nu_s(p)}. \tag{10.54}$$

A simple calculation shows that in our case

$$S(p) = \frac{1+p}{1-p}, \tag{10.55}$$

and that therefore $S(p)$ diverges as $(1-p)^{-1}$.

10.2.3 *Percolation on the Bethe Lattice*

Percolation in one dimension is disappointing because the infinite cluster appears only when $p = 1$, where the system is no longer random. It is, however, possible to define a problem that is almost as simple for which a **percolation threshold** exists—in other words, there is a finite value p_c of the occupation probability p such that, for $p > p_c$, an infinite cluster arises. This problem is percolation on the **Bethe lattice** (also known as the **Cayley tree**).

The Bethe lattice (see figure 10.3) is an ideal lattice without cycles, which is defined recursively. One starts from a site 0, called the **root**, which is surrounded by ζ nearest neighbors. For each of the nearest neighbors (layer 1 of the lattice), one defines $\zeta - 1$ nearest neighbors (layer 2); and then for each site in layer 2, one defines $\zeta - 1$ nearest neighbors (layer 3); and so forth. Layer ℓ contains roughly $(\zeta - 1)^\ell$ sites. At each step, therefore, a finite fraction of sites belongs to the last layer. If we consider the points contained inside a sphere or a box in a lattice of finite size d, it is easy to convince oneself that the ratio between the points contained on the surface and those contained in the volume goes to zero as the volume to the power $-1/d$; since in the Bethe lattice this ratio tends to a finite

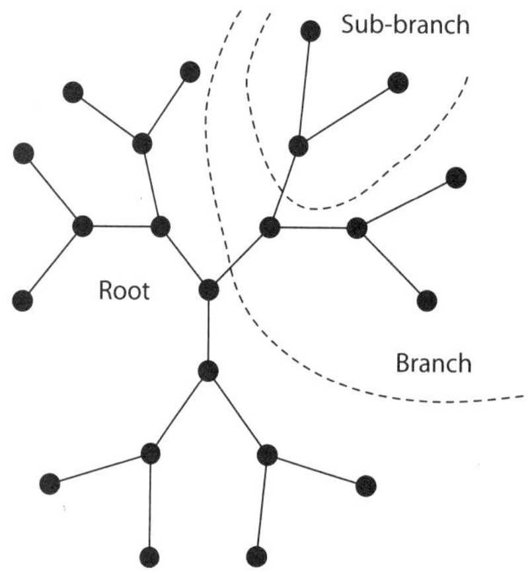

FIGURE 10.3. The origin's neighborhood in the Bethe lattice.

value, it can only be immersed in a space of infinite dimensionality. We will denote the probability that an occupied site belongs to an infinite cluster with $P(p)$. Let us now show that there exists a value p_c such that, if $p > p_c$, $P(p) > 0$.

Let us consider the origin and its ζ nearest neighbors. We will denote the probability that one of its nearest neighbors is *not* connected to infinity by $Q(p)$. This will occur if the nearest neighbor is not occupied, or if it is occupied but all its $\zeta - 1$ nearest neighbors are not connected to infinity. The point is that each of these nearest neighbors can eventually be connected to infinity by means of one of its sub-branches, independent from the others—therefore, the probability that all $\zeta - 1$ neighbors are not connected is simply given by $Q^{\zeta-1}$. We thus obtain

$$Q = (1 - p) + pQ^{\zeta-1}. \tag{10.56}$$

This equation always admits the solution $Q = 1$, but it is easy to see that if one assumes $Q \neq 1$, one obtains the equation

$$\frac{1}{p} = 1 + Q + \cdots + Q^{\zeta-2}. \tag{10.57}$$

This equation admits a solution with $0 < Q < 1$ if $p > (\zeta - 1)^{-1}$. The probability $P(p)$ that the origin, if occupied, belongs to the infinite cluster, is equal to the probability that at least one of its ζ nearest neighbors is connected to infinity. We therefore have

$$P(p) = 1 - Q^{\zeta}. \tag{10.58}$$

Obviously, $P(p) > 0$ if $Q < 1$, which can occur, as we saw, only if $p > p_c$. We have thus identified the **percolation threshold** p_c for the Bethe lattice with coordination number ζ:

$$p_c = \frac{1}{\zeta - 1}. \tag{10.59}$$

When p is slightly larger than p_c, $(1 - Q)$ [and therefore, $P(p)$] are roughly proportional to $(p - p_c)$. We can consider $P(p)$ as the order parameter of the percolation transition, and p as the analog of the inverse of temperature. We can therefore introduce the exponent β by means of the relation

$$P(p) \sim |p - p_c|^{\beta}. \tag{10.60}$$

We thus obtain $\beta = 1$ on the Bethe lattice.

Analogously, it is possible to calculate the average cluster size $S(p)$. If we denote the average size of clusters contained in one branch of the Bethe lattice with T, it is easy to see that on the one hand, one has

$$S(p) = p(1 + \zeta T), \tag{10.61}$$

while on the other,

$$T = p(1 + (\zeta - 1)T). \tag{10.62}$$

These relations are valid for $p < p_c$.

Exercise 10.8 Prove the previous two relations.

We thus obtain

$$S(p) = p\frac{1-(\zeta-2)p}{1-(\zeta-1)p} \qquad \text{for } p<p_c, \tag{10.63}$$

while $S=\infty$ for $p \geq p_c$.

The distribution $\nu_s(p)$ of cluster sizes can be obtained by observing that the probability that a set of connected Bethe lattice points s can form a cluster is equal to the probability p^s that they are all occupied, multiplied by the probability $(1-p)^{s'}$ that the s' points that make up its boundary are all empty. In the Bethe lattice, s' does not depend from the cluster's shape and is equal to

$$s' = (\zeta-2)\,s+2. \tag{10.64}$$

Exercise 10.9 Prove the previous relation.

The difficulty arises from the fact that the number of sets of Bethe lattice points that are connected to each other is not easy to estimate. If, however, we want to calculate the ratio between $\nu_s(p)$ and $\nu_s(p_c)$, this number cancels out between numerator and denominator, and we obtain

$$\frac{\nu_s(p)}{\nu_s(p_c)} = \left(\frac{p}{p_c}\right)^s \left(\frac{1-p}{1-p_c}\right)^{(\zeta-2)s+2} = \left(\frac{p(1-p)^{\zeta-2}}{p_c(1-p_c)^{\zeta-2}}\right)^s \left(\frac{1-p}{1-p_c}\right)^2. \tag{10.65}$$

For $p \neq p_c$, this ratio decays exponentially in s: $\nu_s(p)/\nu_s(p_c) = \exp[-c(p)s]$, where

$$c\,(p) = -\ln\left[\frac{p}{p_c}\left(\frac{1-p}{1-p_c}\right)^{\zeta-2}\right]. \tag{10.66}$$

For $p \simeq p_c$, one has $c(p) \sim (p-p_c)^2$. We have thus obtained

$$\nu_s(p) = \nu_s(p_c)\,\mathrm{e}^{-c(p)s}, \tag{10.67}$$

where $c(p)$ vanishes for $p \to p_c$ as a power law:

$$c\,(p) \sim |p-p_c|^{1/\sigma}. \tag{10.68}$$

The exponent σ defined in this manner is equal to 1/2 on the Bethe lattice. Let us suppose that $\nu_s(p_c)$ decays as a power of s:

$$\nu_s(p_c) \propto s^{-\tau}, \tag{10.69}$$

which defines the **Fisher exponent** τ. We thus obtain

$$\nu_s(p) \propto s^{-\tau}\mathrm{e}^{-c(p)s}. \tag{10.70}$$

We can now calculate the average cluster size $S(p)$ for $p < p_c$. Let me repeat that $\nu_s(p)$ is the average number of clusters with s sites per lattice site. Therefore, $\nu_s(p)s$ is the probability that an arbitrary site belongs to a cluster with s sites, and each of these clusters contributes s sites to the average. We thus obtain $S(p) = \sum_s s^2 \nu_s(p)$, while $\sum_s s\nu_s(p) = 1$, since $P(p) = 0$ for $p < p_c$. Therefore,

$$S(p) = \sum_s s^2 \nu_s(p) = \sum_s s^{2-\tau} e^{-c(p)s} \propto [c(p)]^{\tau-3}. \tag{10.71}$$

For $p \simeq p_c$ (but $p < p_c$), one therefore has $S \sim [c(p)]^{-3+\tau} \sim (p_c - p)^{(\tau-3)/\sigma}$. On the other hand, we had calculated that $\sigma = 1/2$ and $S \sim (p_c - p)^{-1}$, from which we obtain the value of τ:

$$\tau = \frac{5}{2}. \tag{10.72}$$

For $p > p_c$, one has $\sum_s s\nu_s(p) = 1 - P(p)$. We note that if S is defined as the average size of the finite clusters only, one has $S = \sum_s s^2 \nu_s(p)$ also for $p > p_c$, and therefore $S \sim |p_c - p|^{-1}$ also for $p < p_c$.

These relations can be summarized in a scaling scheme. The percolation transition's order parameter is P, and the exponent β is defined by the relation $P \sim |p_c - p|^{\beta}$ for $p \to p_c^+$. The clusters' average size is susceptibility's analog, and we can therefore set $S \sim |p_c - p|^{-\gamma}$. If we introduce the exponents σ and τ from the cluster distribution $\nu_s(p)$ (equations [10.68] and [10.70]), we obtain the relations

$$\beta = (\tau - 2)/\sigma; \qquad \gamma = (3 - \tau)/\sigma. \tag{10.73}$$

For the Bethe lattice, we thus have $\beta = 1$, $\gamma = 1$. More generally, we can consider all moments of the cluster distribution M_k:

$$M_k = \sum_s s^k \nu_s(p). \tag{10.74}$$

Exercise 10.10 Show that, if $k > \tau - 1$,

$$M_k \propto |p - p_c|^{\Delta_k},$$

where $\Delta_k = (\tau - 1 - k)/\sigma$.

If the sum in equation (10.74) does not diverge for $p \to p_c$, it contains both a regular and a singular part. Let us consider the case $k = 0$, for instance—in other words, the average number of clusters per site. Since

$$\frac{d^2}{dc^2} M_0 = M_2 \sim c^{\tau-3}, \tag{10.75}$$

we obtain

$$M_0 = M_0(p_c) + \text{const.}\ c + \text{const.}\ c^2 + \text{const.}\ c^{\tau-1}. \tag{10.76}$$

The first three terms are the *regular* part of M_0. The singular part of M_0 therefore behaves like $c^{\tau-1} \sim |p - p_c|^{(\tau-1)/\sigma}$. The corresponding exponent is normally defined as $2 - \alpha$. We thus obtain

$$2-\alpha=\frac{\tau-1}{\sigma}=2\beta+\gamma, \tag{10.77}$$

so that $\alpha=-1$ in the Bethe lattice.

10.2.4 Relation to the Potts Model

Just as a relation exists between the problem of polymers in a solution and the critical behavior of a magnetic system, in the limit $n\to 0$, it is also possible to relate the percolation problem to an analogous limit in a specific statistical system: the Potts model. This relation allows one to establish the analogy between percolation and critical phenomena that we alluded to previously on more rigorous foundations, but on the other hand, it is not very useful for the purposes of calculation. I will discuss these results by following Lubensky's exposition [Lube78].

The Potts model is a generalization of the Ising model. A variable σ_i, which can assume q values $\sigma_i\in\{1,\dots,q\}$, is assigned to each site of a lattice (for example, a simple cubic one) in d dimensions. Each time that the σ_i's pertaining to nearest-neighbor sites have the same value, the energy is decreased by a certain amount. We can also add an external field $h>0$, which favors the state $\sigma_i=1$. It is convenient to define the Hamiltonian in the following fashion:

$$H(\{\sigma\})=-\sum_{<ij>}J(q\delta_{\sigma_i\sigma_j}-1)-\sum_i h_0(q\delta_{\sigma_i 1}-1), \tag{10.78}$$

where the delta is a Kronecker delta, and the sum runs over all the nearest-neighbor pairs.

Let us now consider the contributions to the partition function Z due to the different configurations of a lattice containing N sites and $N_B=1/2\zeta N$ bonds (where ζ is the lattice's coordination number). It is useful to introduce a factor q^{-N} into the definition of Z, and we therefore obtain

$$\begin{aligned} Z &= q^{-N}\sum_{\{\sigma\}}\exp(-H/k_BT) \\ &= \exp\{(q-1)[KN_B+hN]\}Z'. \end{aligned} \tag{10.79}$$

where we have defined

$$K=\frac{J}{k_BT}, \qquad h=\frac{h_0}{k_BT}, \tag{10.80}$$

and where

$$\begin{aligned} Z' &= q^{-N}\sum_{\{\sigma\}}\exp\left[qK\sum_{<ij>}(\delta_{\sigma_i\sigma_j}-1)+qh\sum_i(\delta_{\sigma_i 1}-1)\right] \\ &= q^{-N}(1-p)^{N_B}\sum_{\{\sigma\}}\prod_{<ij>}\left(1+\frac{p}{1-p}\delta_{\sigma_i\sigma_j}\right) \\ &\quad\times\exp\left[qh\sum_i(\delta_{\sigma_i 1}-1)\right], \end{aligned} \tag{10.81}$$

where

$$p = 1 - e^{-qK}. \tag{10.82}$$

The expression of Z' can be represented by diagrams $\mathcal{G}$, which contain a certain number of lattice bonds N_b. Each $\langle i,j \rangle$ bond that is present carries with it a factor $[p/(1-p)]\delta_{\sigma_i\sigma_j}$, while the factor is equal to 1 for missing bonds. It is easy to see that the parameter p represents the probability that a bond will be present, because the weight of a $\mathcal{G}$ diagram with N_b bonds is proportional to $p^{N_b}(1-p)^{N_B - N_b}$. The Potts variables in each connected cluster that appears in the diagram $\mathcal{G}$ must be in the same state because of the Kronecker delta, so that once the trace over the σ's has been evaluated, a cluster with b bonds and s sites has the following weight:

$$\left(\frac{p}{1-p}\right)^b \left[1 + (q-1)\,e^{-qhs}\right]. \tag{10.83}$$

(Isolated sites are counted as clusters that contain a single site.)

We will now denote the number of clusters of the diagram $\mathcal{G}$ that have s sites and b bonds by $N\mathcal{K}(\mathcal{G}, b, s)$. We then have

$$\begin{aligned} Z' &= q^{-N}(1-p)^b \sum_{\mathcal{G}} \left[\left(\frac{p}{1-p}\right)^{N\mathcal{K}(\mathcal{G},b,s)\,b}\right] \\ &\times \prod_{bs} \left[1 + (q-1)\,e^{-qhs}\right]^{N\mathcal{K}(\mathcal{G},b,s)}. \end{aligned} \tag{10.84}$$

The total number of occupied bonds N_b is given by

$$N_b = \sum_{bs} b N\mathcal{K}(\mathcal{G}, b, s). \tag{10.85}$$

Therefore,

$$\begin{aligned} Z' &= q^{-N} \sum_{\mathcal{G}} p^{N_b}(1-p)^{N_B - N_b} \prod_{bs} \left[1 + (q-1)\,e^{-qhs}\right]^{N\mathcal{K}(\mathcal{G},b,s)} \\ &= q^{-N} \sum_{\mathcal{G}} P(\mathcal{G}) \prod_{bs} \left[1 + (q-1)\,e^{-qhs}\right]^{N\mathcal{K}(\mathcal{G},b,s)}, \end{aligned} \tag{10.86}$$

where the sum runs over all the diagrams $\mathcal{G}$, weighted with the relative probability

$$P(\mathcal{G}) = (1-p)^{N_B - N_b} p^{N_b}. \tag{10.87}$$

This fundamental result was derived in 1972 by Fortuin and Kasteleyn [Fort72]. We can exploit it to express the characteristic quantities of percolation in terms of the Z. We will now consider the limit $q \to 1$, which corresponds to percolation, and the limit $q \to 0$, which describes the statistics of trees (diagrams without cycles) on a lattice, in particular.

10.2.5 The $q \to 1$ Limit and Percolation

It is obvious from equation (10.86) that in this limit, $Z = 1$, and therefore $\ln Z = 0$. Let us consider the following expression of the free energy per site:

$$f_1 = \lim_{N\to\infty} \lim_{q\to 1} \frac{1}{N(q-1)} \ln Z = \lim_{N\to\infty} \left.\frac{\partial \ln Z}{\partial q}\right|_{q\to 1}. \tag{10.88}$$

From equations (10.79) and (10.86), we obtain

$$\begin{aligned} f_1 &= -1+\frac{\zeta}{2}K+h+\sum_{\mathcal{G}} P(\mathcal{G})\sum_{bs} \mathcal{K}(\mathcal{G};b,s)\,e^{-hs} \\ &= -1+\frac{\zeta}{2}K+h+\sum_{s} \mathcal{K}(s)\,e^{-hs}, \end{aligned} \tag{10.89}$$

where

$$\mathcal{K}(s)=\sum_{\mathcal{G}} P(\mathcal{G})\sum_{b} K(\mathcal{G};b,s), \tag{10.90}$$

is the average number of clusters containing s sites per site. One should note that only clusters containing a finite number of sites contribute to equation (10.89) in the limit $N\to\infty$, so that all the sums are restricted to finite values of s and b.

In the limit $h\to 0^+$, we obtain

$$f_1 = -1-\frac{\zeta}{2}\ln(1-p)+\langle N_c\rangle, \tag{10.91}$$

where $\langle N_c\rangle=\sum_s \mathcal{K}(s)$ is the average number of clusters that contain s sites per site.

The first derivative of f_1 with respect to h gives us probability $P(p)$ that a site belongs to the infinite cluster:

$$\begin{aligned} \frac{\partial f_1}{\partial h} &= 1-\sum_s s\mathcal{K}(s)\,e^{-hs} \\ &\xrightarrow[h\to 0]{} 1-\sum_s s\mathcal{K}(s) = P(p), \end{aligned} \tag{10.92}$$

since $s\mathcal{K}(s)$ is the probability that a site belongs to a cluster containing s sites and the sum runs over only finite clusters.

The second derivative of f_1 gives us the average cluster size per site $S(p)$:

$$\left.\frac{\partial^2 f_1}{\partial h^2}\right|_{h=0} = \sum_s s^2\mathcal{K}(s). \tag{10.93}$$

Let us now consider the correlation function between the sites located in $\boldsymbol{r}$ and $\boldsymbol{r}'$, defined by

$$(q-1)D(\boldsymbol{r},\boldsymbol{r}')=\langle q\delta_{\sigma_r\sigma_{r'}}-1\rangle. \tag{10.94}$$

The factor $(q-1)$ was introduced for convenience's sake, as we will see immediately. We note that equation (10.86), in the limit $h\to 0$, is reduced to

$$Z'=\sum_{\mathcal{G}} P(\mathcal{G})q^{N_c-N}. \tag{10.95}$$

Given two sites, placed in $\boldsymbol{r}$ and $\boldsymbol{r}'$, respectively, we can separate the graphs into two sets: the first $\mathcal{G}_1$, in which $\boldsymbol{r}$ and $\boldsymbol{r}'$ belong to the same cluster, and the second $\mathcal{G}_2$, in which they belong to different clusters. If $\mathcal{G}\in\mathcal{G}_1$, to multiply by $\delta_{\sigma_r\sigma_{r'}}$ does not change anything, while if $\mathcal{G}\in\mathcal{G}_2$, this multiplication reduces the number of clusters to one. Therefore,

$$
\begin{aligned}
(q-1)D(\mathbf{r},\mathbf{r}') &= \langle q\delta_{\sigma_r\sigma_{r'}} - 1\rangle \\
&\frac{1}{Z'}\left[\sum_{\mathcal{G}\in\mathcal{G}_1}(q-1)P(\mathcal{G})q^{N_c(\mathcal{G})-N}\right. \\
&\quad \left. + \sum_{\mathcal{G}\in\mathcal{G}_2}(qq^{-1}-1)P(\mathcal{G})q^{N_c(\mathcal{G})-N}\right] \\
&= \frac{q-1}{Z'}\sum_{\mathcal{G}\in\mathcal{G}_1} q^{N_c(\mathcal{G})-N}C(\mathbf{r},\mathbf{r}'),
\end{aligned}
\tag{10.96}
$$

where

$$
C(\mathbf{r},\mathbf{r}') = \begin{cases} 1, & \text{if } \mathbf{r} \text{ and } \mathbf{r}' \text{ belong to the same cluster;} \\ 0, & \text{otherwise} \end{cases}
\tag{10.97}
$$

In the limit $q \to 1$, we obtain

$$
\lim_{q\to 1} D_q(\mathbf{r},\mathbf{r}') = \sum_{\mathcal{G}} P(\mathcal{G})C(\mathbf{r},\mathbf{r}') = \langle C(\mathbf{r},\mathbf{r}')\rangle.
\tag{10.98}
$$

Let us remark that the average probability that the sites placed in $\mathbf{r}$ and $\mathbf{r}'$ belong to the same finite cluster is given by

$$
G(\mathbf{r},\mathbf{r}') = \langle C(\mathbf{r},\mathbf{r}')\rangle - P^2(p).
\tag{10.99}
$$

10.2.6 The Limit $q \to 0$ and the Statistics of Trees on a Lattice

In the limit $q \to 0$, $p = qK + O(q^2)$. Given the fundamental relation

$$
N_b + N_c - N_L = N,
\tag{10.100}
$$

where N_L is the number of closed loops, we obtain the following expression for Z' in the limit $q \to 0$:

$$
\begin{aligned}
\lim_{q\to 0} Z' &= \lim_{q\to 0}\sum_{\mathcal{G}} q^{N_L(\mathcal{G})}\prod_{bs}(1+hs)^{N\mathcal{K}(\mathcal{G};b,s)} \\
&= \sum_{\mathcal{G}_t}\prod_{s}(1+hs)^{N\mathcal{K}(\mathcal{G}_t;b,s)} = Z_0'(h).
\end{aligned}
\tag{10.101}
$$

In this expression, $\mathcal{G}_t$ refers to trees—in other words, to graphs without loops—and $N\mathcal{K}_t(\mathcal{G}_t; b, s)$ is the number of clusters in configuration $\mathcal{G}_t$ that contain s sites. The free energy per site is given by

$$
f_0 = \lim_{N\to\infty}\lim_{q\to 0}\frac{1}{N(q-1)}\ln Z = \frac{\zeta}{2}K + h - \lim_{N\to\infty}\ln Z_0'.
\tag{10.102}
$$

The first and second derivatives of f_0 with respect to H are

$$
\frac{\partial f_0}{\partial h} = 1 - \frac{1}{Z_0'}\sum_{\mathcal{G}_t} K^{N_b(\mathcal{G}_t)}\sum_s s\mathcal{K}(\mathcal{G}_t, s) = 1 - \langle s\rangle_t,
\tag{10.103}
$$

$$
\frac{\partial^2 f_0}{\partial h^2} = \frac{1}{Z_0'}\sum_{\mathcal{G}_t} K^{N_b(\mathcal{G}_t)}\sum_s s^2\mathcal{K}(\mathcal{G}_t, s) = \langle s^2\rangle_t,
\tag{10.104}
$$

respectively, where $\langle s \rangle_t$ and $\langle s^2 \rangle_t$ are the average and the second moment of the trees' size, respectively, mediated with respect to the distribution $Z_0'^{-1} K^{N_b(\mathcal{G}_t)}$.

Exercise 10.11 Show that

$$\lim_{q\to 0} D_q(\mathbf{r},\mathbf{r}') = \langle C(\mathbf{r},\mathbf{r}') \rangle_t,$$

where $\langle \ldots \rangle_t$ represents the average with respect to the distribution $Z_0'^{-1} K^{N_b(\mathcal{G}_t)}$. Show that in the limit $K \to \infty$, one obtains the average with respect to the spanning trees—in other words, to those trees that visit all the sites.

If a graph contains a *spanning tree*, it contains a single cluster and $N_b = N - 1$ bonds. We will denote the set of spanning trees with $\mathcal{G}_s$, and the set of graphics containing exactly two clusters, obtained by removing a single bond from each member of $\mathcal{G}_s$], we will denote by $\mathcal{G}_2$. We then have

$$\begin{aligned} \langle C(\mathbf{r},\mathbf{r}') \rangle_t &= \frac{N_s + K^{-1} \sum_{\mathcal{G}\in\mathcal{G}_2} C(\mathbf{r},\mathbf{r}') + \cdots}{N_s + K^{-1} \sum_{\mathcal{G}\in\mathcal{G}_2} 1 + \cdots} \\ &= 1 - K^{-1} N_s^{-1} \sum_{\mathcal{G}\in\mathcal{G}_2} [1 - C(\mathbf{r},\mathbf{r}')], \end{aligned} \tag{10.105}$$

where N_s is the number of spanning trees on the lattice. The factor $[1 - C(\mathbf{r},\mathbf{r}')]$ is equal to 0 if $\mathbf{r}$ and $\mathbf{r}'$ belong to the same cluster, and 1 otherwise. Therefore, $\sum_{\mathcal{G}\in\mathcal{G}_2} [1 - C(\mathbf{r},\mathbf{r}')]$ is equal to the number $N_s(\mathbf{r},\mathbf{r}')$ of diagrams such that $\mathbf{r}$ and $\mathbf{r}'$ belong to two different clusters, but once again form a spanning tree by relocating a single bond. If each bond has conductance σ, the resistance $R(\mathbf{r},\mathbf{r}')$ between the two sites is given by

$$R(\mathbf{r},\mathbf{r}') = \frac{1}{\sigma} \frac{N_s(\mathbf{r},\mathbf{r}')}{N_s}. \tag{10.106}$$

(This is a classic theorem derived by Kirchoff [Kirc45]). We therefore have

$$\langle C(\mathbf{r},\mathbf{r}'; H = \sigma J) \rangle_t = 1 - \frac{1}{J} R(\mathbf{r},\mathbf{r}') + O(J^{-2}). \tag{10.107}$$

This result can be used to calculate the behavior of the conductivity of a percolating net in an expansion in $d = 6 - \epsilon$ dimensions.

The Potts model can represent a starting point for the expansion of the critical exponents in $d = 6 - \epsilon$ dimensions, which is, however, not very useful from the point of view of actual calculation. For this purpose, the fixed dimension renormalization schemes of percolation, which we will discuss shortly, are more useful.

10.2.7 *Renormalization in Real Space*

Let us consider bond percolation over a square lattice in two dimensions. We want to renormalize the bonds on an elementary 2×2 cell (see figure 10.4). The renormalization scheme is presented in the illustration. It can be interpreted supposing that the

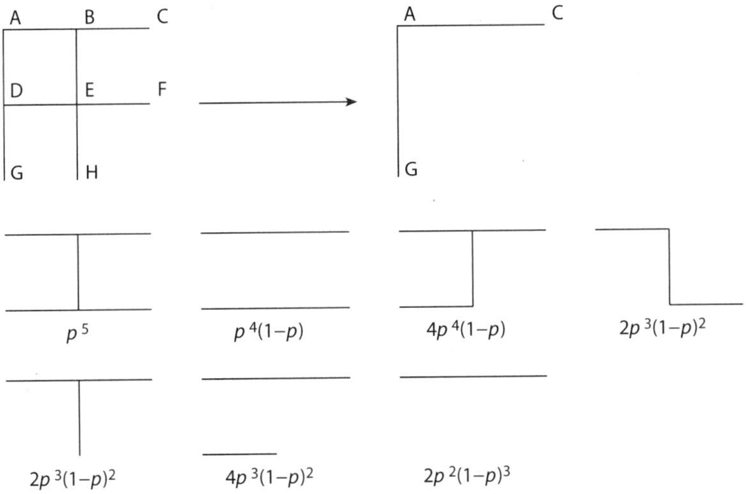

FIGURE 10.4. The configurations that contribute to the renormalization of bond percolation in the case of a 2×2 cell.

renormalized AC bond is considered present if one can go from A or from D to C or F in the original system. Analogously, the renormalized AG bond is considered present if one can go from A or B to G or H. Let us now consider the renormalized AC bond—the AB, DE, BE, BC, and EF bonds contribute to it in the original system. The renormalized AC bond will be present in the configurations shown in the illustration, to which the relative probabilities are associated. One thus obtains the recurrence relation that corresponds to a rescaling by a factor $b = 2$:

$$\begin{aligned} p' &= p^5 + 5p^4(1-p) + 8p^3(1-p)^2 + 2p^2(1-p)^3 \\ &= 2p^5 - 5p^4 + 2p^3 + 2p^2. \end{aligned} \tag{10.108}$$

In addition to the two trivial fixed points $p = 0$ and $p = 1$, this relation admits the fixed point $p = p^* = 1/2$, which is consistent with the exact value of p_c for the square lattice. One additionally has

$$\lambda = \left.\frac{dp'}{dp}\right|_{p=p^*} = \frac{13}{8}, \tag{10.109}$$

from which it follows that

$$\nu = \frac{\ln b}{\ln \lambda} \simeq 1.43, \tag{10.110}$$

which is to be compared with the value 4/3, which is presumably exact in two dimensions.

10.2.8 *Migdal-Kadanoff Recurrence Relations for Percolation*

Let us consider a simple cubic lattice in d dimensions. One can obtain a fairly simple recurrence relation by imagining that we are modifying the lattice's structure in a suitable

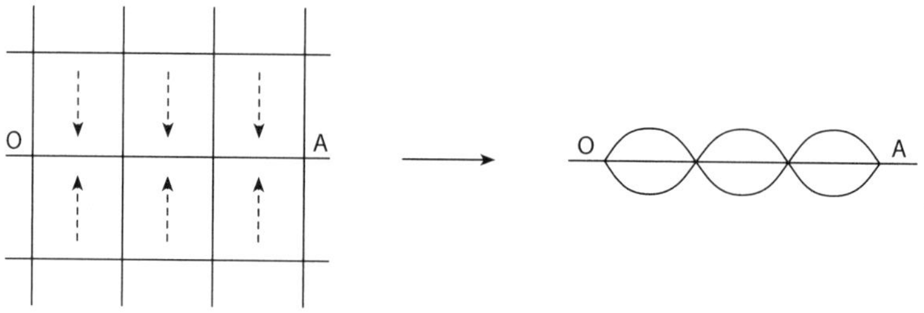

FIGURE 10.5. Scheme showing the shift in renormalization bonds for $b = 3$ and $d = 2$.

fashion before renormalization, so as to make the renormalization itself as close to a one-dimensional transformation as possible. Let us now focus our attention on the origin, and assume that the scaling factor is b. So the first vertex that remains after renormalization, in direction x, is located b lattice steps from the origin. We will denote this vertex with A. Let us imagine that we shift the bonds parallel to the x axis that connect vertices in the $(d-1)$ transversal directions, in such a manner that they connect vertices sequential to the origin along the x axis. In this way, the connection between the origin and A is composed of a sequence of b "bundles" of bonds, each one composed of $1 + (d-1)(b-1)$ bonds. This process is shown schematically in figure 10.5. This type of recurrence relations was introduced by A. A. Migdal [Migd75] and generalized by Leo P. Kadanoff [Kada76], and is therefore known as the Migdal-Kadanoff relation.

We can therefore apply the recurrence relation in a direct fashion: the bond between the origin and A in the renormalized system is present if, for *each* of the b bundles, *at least* one other bond is present. This leads to the following recurrence relation:

$$p \longrightarrow p' = \left\{1 - (1-p)^{[1+(d-1)(b-1)]}\right\}^{b}. \tag{10.111}$$

It is better to take the $b \to 1$ limit and calculate $\mathrm{d}p'/\mathrm{d}b$:

$$W(p) = \left.\frac{\mathrm{d}p'}{\mathrm{d}b}\right|_{b=1} = p\ln p - (d-1)(1-p)\ln(1-p). \tag{10.112}$$

For $d = 2$, we obtain the result shown in figure 10.6, from which one sees that there is an unstable fixed point for $p = p^* = 0.5$. For this point, one has

$$\lambda = \left.\frac{\mathrm{d}W}{\mathrm{d}p}\right|_{p=p^*} = 0.6137, \tag{10.113}$$

which corresponds to an exponent $\nu = 1/\lambda = 1.6294$, to be compared to the value $\nu = 4/3 = 1.3333$, which is presumably exact.

Exercise 10.12 Calculate the exponent ν for $d = 3$, and compare it with the numerical value $\nu \simeq 0.9$.

Exercise 10.13 Show that the recurrence relation provides the correct result for the exponent ν in the limit $d \to \infty$.

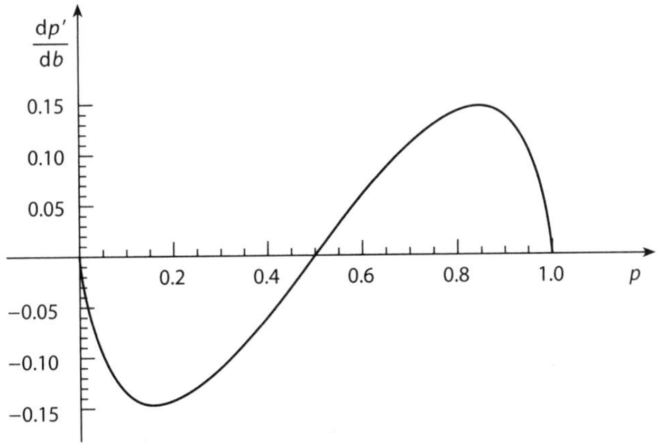

FIGURE 10.6. Renormalization flow for p according to equation (10.111) in $d = 2$.

Exercise 10.14 The Migdal-Kadanoff recurrence relation can also be derived for more "traditional" statistical systems. Once the bonds have been shifted, one obtains a one-dimensional system for which decimation is exact. Apply this method to the d-dimensional Ising model, and show that the recurrence relation for $K = J/k_BT$ in a d-dimensional Ising model, with a scaling factor b, is given by

$$e^{2K'} = \frac{\cosh^b(K_1) + \sinh^b(K_1)}{\cosh^b(K_1) - \sinh^b(K_1)},$$

where

$$K_1 = [1 + (d-1)(b-1)]\,K.$$

Calculate the equation for the infinitesimal transformation corresponding to $b \to 1^+$. Identify the fixed point for $d = 2$ and $d = 3$, and compare the value for the critical temperature and the exponent ν obtained in this fashion with the exact solution or available numerical values. Discuss the limits $d \to 1$ and $d \to \infty$.

10.2.9 *Fractal Structure of a Percolation Cluster*

Percolative systems provide a good example of the relevance of the concept of **fractals** introduced by B. Mandelbrot [Mand77], which we have already briefly seen in section 9.2. Fractal objects are geometrical objects whose dimension is smaller than the dimension of the space they are immersed in and that often assume noninteger values. A classic example of a fractal object immersed in one dimension is the **Cantor set**: it can be obtained by considering the closed interval [0, 1] and deleting its (open) middle third—in other words, the interval $(1/3, 2/3)$. One thus obtains two closed intervals $[0, 1/3]$ and $[2/3, 1]$, from which *their* middle intervals are deleted, and so forth. The points that remain at the end of this construction constitute the Cantor set: they can be characterized as the points included between 0 and 1 such that their base 3 coordinates do not contain the digit 1. The

FIGURE 10.7. Sierpiński gasket (the first five levels of construction are shown).

resulting set is a point *dust* that does not contain any interval. We can define its dimension d_H (called the **Hausdorff dimension**) as follows. Let us imagine that we cover the interval $[0, 1]$ with intervals of length ℓ. The number $\mathcal{N}(\ell)$ of intervals that contain points belonging to the set grows as a power of ℓ: $\mathcal{N}(\ell) \sim \ell^{-d_H}$. In our case, each time the length ℓ is divided by 3, the number $\mathcal{N}(\ell)$ is multiplied by 2. We thus obtain $d_H = \ln 2/\ln 3 = 0.630927\ldots$, a number between 0 and 1, as we anticipated.

Exercise 10.15 Calculate the Hausdorff dimension of the **Sierpiński gasket** (see figure 10.7), recursively defined as follows. Consider an equilateral triangle, and divide it into four equal triangles. The middle triangle is removed, and the construction is repeated for the remaining three triangles and so forth. Note that the dimension d_H obtained in this manner satisfies $1 < d_H < 2$.

Exercise 10.16 Calculate the Hausdorff dimension of the **von Koch curve** (see figure 10.8), recursively defined as follows. Consider a segment of unitary length in the plane. The segment's middle third is replaced with two segments of 1/3 each, as in the figure. The construction is repeated for each of the four segments thus obtained.

The introduction of the concept of fractal object has allowed us to quantitatively describe the properties of a great number of natural or mathematical objects that were previously considered too irregular to be treated mathematically. These objects are defined because they possess, at least approximately, a symmetry under scale transformations. Each of the three triangles that remain after the first step in the construction of the Sierpiński gasket, for instance, is a 1:2 scale copy of the gasket itself.

An alternative definition of the Hausdorff dimension, which works well in these simple cases, is obtained by evaluating the number of points in the set contained within an ℓ

FIGURE 10.8. The first steps in the construction of the von Koch curve.

radius from a point belonging to the set. This obviously makes sense if ℓ is sufficiently larger than the points' size, set, for example, by the lattice step a_0.

Let us consider, for example, a regular D-dimensional manifold immersed in a d-dimensional space. If we take a point that belongs to this manifold and trace a sphere of radius ℓ with a center in that point, the volume of the part of the manifold contained in this sphere will be proportional to ℓ^D and will contain a number of points proportional to ℓ^D/a_0^d. This confirms our intuitive idea that the manifold has D dimensions. Let us now suppose that we are dealing with a Cantor set, constructed by iterating the basic construction a finite (but large) number of times, so that the smallest intervals are of a_0 length. The number of points contained within a circle of ℓ radius then grows as $\ell^{\ln 2/\ln 3}$, because if ℓ is multiplied by 3, the number of points doubles. We can thus see that this definition of dimension (in this case) gives us the same result as the preceding definition.

Exercise 10.17 Derive the Hausdorff dimensions of the Sierpiński gasket and the von Koch curve according to this definition.

We saw that the critical fluctuations also possess an analogous property of symmetry under scale transformations. Similarly, the set of occupied points in the lattice, in a site percolation problem, at the percolation threshold p_c, is a system that admits a symmetry under scale transformations, at least as long as the scales being considered are large enough when compared to the lattice step. We thus come upon the idea of describing the infinite cluster at the percolation threshold (the so-called incipient percolation cluster) as a fractal characterized by a certain Hausdorff dimension.

The scale invariance is obviously only rough: On the one hand, it is no longer true if the lengths we are interested in are on the order of the lattice step a_0. On the other, if we are not *exactly* at the percolation threshold, the cluster is no longer fractal over distances on the order of the coherence length ξ. Between these two lengths, however, there can be some orders of magnitude of distance for which a fractal description is appropriate.

Let us now consider a site percolation model in d dimensions. Let us suppose that p is close to, but slightly larger than the percolation threshold p_c. We consider a point belonging to the infinite cluster and evaluate the number $\mathcal{M}(\ell)$ of points of the infinite cluster that are contained within a d-dimensional sphere of radius ℓ. In general, we expect $\mathcal{M}(\ell)$ to be described as a scaling expression, as a function of ℓ and of the coherence length ξ:

$$\mathcal{M}(\ell) = \ell^{\psi} \widehat{\mathcal{M}}(\ell/\xi). \tag{10.114}$$

If $\ell \gg \xi$, we obviously expect the irregularities of the infinite cluster, on average, to disappear, and therefore that

$$\mathcal{M}(\ell) \sim P\ell^{d}, \tag{10.115}$$

where P is the percolation order parameter. On the other hand, if $\ell \ll \xi$, the fractal irregularities of the infinite cluster will be dominant, and $\mathcal{M}(\ell) \sim \ell^{d_{\mathrm{H}}}$, while it will be practically independent of ξ. We can therefore define $\widehat{\mathcal{M}}(x) \simeq$ const., for $x \ll 1$, which implies that $\psi = d_H$. Since $P \sim |p - p_c|^{\beta} \sim \xi^{-\beta/\nu}$, for $x \gg 1$, we have

$$\widehat{\mathcal{M}}(x) \sim x^{\beta/\nu} \tag{10.116}$$

and therefore $\mathcal{M}(\ell) \sim \ell^{d-\beta/\nu}$. But this is compatible with expression (10.115) only if

$$d_{\mathrm{H}} = d - \beta/\nu. \tag{10.117}$$

In this manner, we have related the fractal dimension of the percolation cluster to the usual critical exponents.

10.3 Disordered Systems

The problems pertaining to polymers and percolation are peculiar because they concern purely geometrical properties of random systems. There is, moreover, a subtle, but important, physical distinction between the polymer and percolation problems. In fact, one can suppose that in the course of observation time, a linear polymer will explore the variety of physically accessible configurations like an ordinary system in statistical mechanics.[1] In this case, ordinary statistical mechanics and thermodynamics (the expression of entropy as a function of accessible phase space volume, for instance) maintain their validity. In the case of percolation, on the contrary, the systems are composed of units of much

[1] Actually certain configurations can be *kinetically* inaccessible, because they might require that the free ends of the polymer *knot* themselves a large number of times with the rest of the polymer. In our approach, we have disregarded these details.

larger than molecular scale, and once prepared, their percolative properties remain fixed. The space of percolative configurations is therefore not explored by a single system during an experiment—instead, each system is an **instance** of a random system extracted from a specific distribution. A physically interpretable entropy is not associated with this distribution's variety of *accessible configurations*. Even more important is the fact that the average properties of random systems of this sort can be evaluated experimentally only by performing measures over a large number of samples, each one corresponding to a different and independent instance of the random system. In practice, for *extensive* properties, this average is obtained by considering as independent instances the juxtaposed subsystems that a macroscopic sample can be considered to be composed of. Properties such as conductivity, however, are not extensive, and in this case, one can expect that there will be important fluctuations in macroscopic samples as well.

In a general disordered system, we can expect that having fixed the preparation procedure, the details of the system's structure will differ from sample to sample. We will be able to produce samples like alloys of certain proportions of two metals, for instance. The samples are prepared in the liquid state, at high temperature, and then rapidly cooled. In these conditions, the mutual positions of the atoms of different chemical natures do not vary at ambient temperature during the observation time—they can be considered *frozen*. If the preparation temperature is high enough, it is reasonable to assume that an atom of each chemical species with a probability proportional to its concentration is present at each site of the metallic lattice, and independently for different sites. Once the sample has been prepared, this structure remains blocked.

We can describe a system of this type by means of the random Hamiltonian concept. Let us consider, for instance, a d-dimensional Ising system, in which there is a spin variable $\sigma_i = \pm 1$ in each site i of the lattice. In addition to the σ_i variables, however, which change configuration during observation time (and are therefore called **fast** or **thermal** variables), there are some variables J_{ij} (associated with the bonds between sites i and j, for instance) that are fixed at the time the sample is prepared but that then do *not* change during observation time. These variables are called **quenched.** One supposes that each of these variables, at the moment the sample is prepared, assumes a random value distributed according to a certain law. In the simplest case, for each pair (i, j), the variables J_{ij} are independent and identically distributed. For each configuration of the quenched variables $J = (J_{ij})$, we thus obtain a Hamiltonian $H_J(\sigma)$, where $\sigma = \sigma_i$. For each instance (or, as one often says, **realization**) of disorder, we are able to evaluate the thermodynamic properties by means of the usual expressions of statistical mechanics—they will, however, depend on the particular configuration J of the quenched variables. The magnetization M for instance will be given by

$$M(J) = \left\langle \sum_i \sigma_i \right\rangle_J = \frac{1}{Z_J} \sum_\sigma \left(\sum_i \sigma_i \right) e^{-H_J(\sigma)/k_B T}, \tag{10.118}$$

where the partition function Z_J is given by

$$Z_J = \sum_\sigma e^{-H_J(\sigma)/k_B T}. \tag{10.119}$$

The average over the quenched variables J will be different from the one over the fast variables σ, due to the presence of the partition function Z_J in the denominator of the expression of M. Indeed, one has to normalize the Boltzmann factor $e^{-H_J(\sigma)/k_BT}$ separately for each realization of the disorder, evaluate the average of each thermodynamic observable over the distribution obtained in this way, and, eventually, evaluate the average over the distribution of the J of the thermal average obtained for each fixed J. In this manner, we will obtain the thermodynamic observable's **disorder average**:

$$\begin{aligned} \overline{M_J} &= \overline{\left\langle \sum_i \sigma_i \right\rangle_J} = \sum_J \mathcal{P}(J) \left\langle \sum_i \sigma_i \right\rangle_J \\ &= \sum_J \mathcal{P}(J) \frac{1}{Z_J} \sum_\sigma \left(\sum_i \sigma_i \right) e^{-H_J(\sigma)/k_BT}, \end{aligned} \tag{10.120}$$

where $\mathcal{P}(J)$ is the probability distribution of the J's. This average is also called the **quenched** average, while the average over the fast variables is, in contrast, called **annealed**. These expressions refer to the quench technique by which a metallic object is cooled rapidly in order to increase its hardness. The procedure by which a quenched metallic object is brought back to high temperature and then cooled slowly to make it more malleable is called *annealing*.

10.3.1 *Random Ferromagnets (Spin Glasses)*

The reflections in the preceding section can be made more concrete if one considers a specific physical system. A **random ferromagnet** (also called a **spin glass**) is a magnetic system in which the interaction among spins varies randomly from sample to sample. Experimentally, it is prepared by diluting the magnetic impurities in a nonmetallic magnetic matrix—for example, by diluting atoms of Mn or Fe in a matrix of Cu or Au, with a relative concentration on the order of a few percentage points. In this situation, the exchange interactions between the spins of the electrons located in the impurities and those of the conduction electrons give rise to a long-distance spin–spin interaction, whose intensity decays as the inverse of the distance cubed, but whose amplitude oscillates rapidly, changing sign over distances on the order of the lattice step.

Experimentally, one observes that at low temperatures, samples cooled in the absence of a field do not exhibit a measurable magnetization. However, the magnetic susceptibility to very small fields exhibits a cusp at temperatures on the order of a dozen degrees Kelvin. This cusp rounds out rapidly as the magnetic field increases.

A simple interpretation of this behavior (derived by Fischer [Fisc77]) can be obtained as follows. As we well know, the local susceptibility $\chi_{ij} = \partial \langle \sigma_i \rangle / \partial h_j$ of a spin placed in i with respect to a magnetic field applied in j is proportional to the spins' correlation function:

$$k_BT\chi_{ij} = \langle \sigma_i \sigma_j \rangle_c = \langle \sigma_i \sigma_j \rangle - \langle \sigma_i \rangle \langle \sigma_j \rangle. \tag{10.121}$$

If one considers a given sample in a vanishing field, its global susceptibility will be given by the average over the sample of the χ_{ij}. In a spin glass, we can expect both the $\langle \sigma_i \rangle$ and the $\langle \sigma_i \sigma_j \rangle$ to be random quantities, independent for different sites, and with a vanishing average. If this hypothesis is true, only the local terms contribute to the average susceptibility:

$$k_B T \left[\chi_{ij}\right]_{av} \simeq \left[\langle\sigma_i^2\rangle\right]_{av} - \left[\langle\sigma_i\rangle^2\right]_{av}, \tag{10.122}$$

where $[\ldots]_{av}$ represents the average over the sample. If we assume that the system can be found in a well-defined thermodynamic state, in which $\langle\sigma_i\rangle = m_i$ and that $\sigma_i = \pm 1$, this expression can be written

$$k_B T \left[\chi_{ij}\right]_{av} \simeq 1 - q_{EA}, \tag{10.123}$$

where

$$q_{EA} = \left[\langle\sigma_i\rangle^2\right]_{av} \tag{10.124}$$

is called the **Edwards-Anderson order parameter.**

This argument can account for the experimentally observed cusp, but it does not explain a whole series of "strange" behaviors, especially dynamic ones, exhibited by these systems. More particularly, one has realized that the hypothesis that the system is well described by an ensemble at equilibrium is not justified. The presence of the cusp, however, suggests that there is a temperature T_G below which the system's behavior changes radically. Understanding what the nature of the thermodynamic state below this temperature is (as well as whether one can even talk about a thermodynamic state!) is still an open problem in statistical mechanics.

The random ferromagnet can be described by generalizing the Ising model to a situation in which the exchange integral J_{ij} relative to the pair of nearest neighbors $\langle ij \rangle$ has a random character—it is, in other words, an independent variable for pairs of different spins, identically distributed according to some law (for example, $J_{ij} = \pm J_0$) with an equal probability for each sign, or a Gaussian distribution. The model we have just described is called the **Edwards-Anderson model**. Although it has been the subject of intense studies for more than thirty years by now, we are still quite far from understanding its behavior. In the sections that follow, I will try to deal with even less realistic models, whose behavior is simpler, so as to highlight the specific phenomena that are produced by this disorder in interactions.

10.3.2 *Frustration*

The presence of $\langle ij \rangle$ bonds with positive and negative J_{ij} interaction coefficients has an important consequence. Let us consider a two-dimensional Ising system, and focus our attention on an elementary cell like the one shown in figure 10.9. Obviously, the positive J_{ij} coefficients favor the parallel alignment of the spins σ_i and σ_j, while negative coefficients favor their antiparallel alignment. Now, it is easy to see that in the situation illustrated, it is not possible to find an arrangement of the four spins placed at the cell's vertices that will *satisfy* all four bonds. In effect, if we, for example, decide to arrange all the σ_i *up*, the negative bond (shown with a discontinuous line) will not be satisfied; if we then choose,

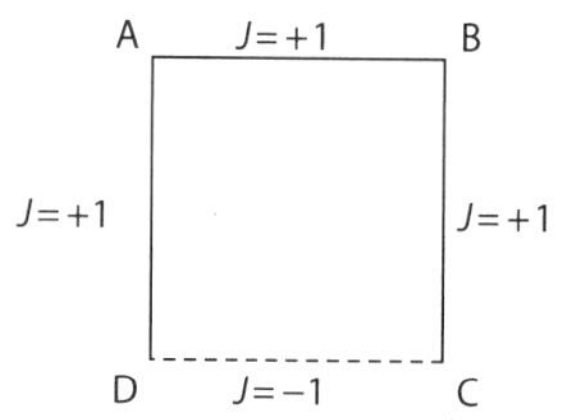

FIGURE 10.9. A frustrated plaquette in a two-dimensional Ising model with competing interactions.

for example, to arrange the spin located in C or in D *down*, so as to satisfy that bond, then either the AC or the BD bond, respectively, will not be satisfied. This phenomenon is called **frustration**, and the cell is called **frustrated**. That a cell is frustrated is made apparent by the fact that the product of the J_{ij}'s over all the cell bonds is negative:

$$J_{\mathrm{AB}}\, J_{\mathrm{BC}}\, J_{\mathrm{CD}}\, J_{\mathrm{DA}} < 0. \tag{10.125}$$

In the presence of frustration, the minimal energy states tend to be very numerous, because it is often possible to pass from one state to the next by simply changing the choice of which bonds not to satisfy. Frustration is therefore often associated with a large degeneration of the ground state.

Frustration does not only occur in random systems. For example, in an antiferromagnetic two-dimensional Ising model, in which the spins are arranged over a triangular lattice, all the cells are frustrated. In this case, it has been proven that the system does not exhibit phase transitions at a finite temperature with a vanishing magnetic field, while it exhibits particular forms of order if the magnetic field breaks the degeneration of the ground state caused by the frustration.

10.3.3 *Random Energy Model*

Let us consider a system composed of N Ising spins $\sigma_i = \pm 1$. The possible configurations are therefore 2^N. In the **random energy model** (REM), introduced by B. Derrida [Derr81] in 1981, one supposes that one associates an energy value $E(\sigma)$ with each configuration $\sigma = (\sigma_i)$, independently extracted from the same distribution $\mathcal{P}(E)$. In what follows, we shall assume that $\mathcal{P}(E)$ is a Gaussian distibution with variance equal to $NJ_0^2/2$:

$$\mathcal{P}(E) \propto \exp\left(-\frac{E^2}{NJ_0^2}\right). \tag{10.126}$$

Let us now see that there exists a temperature T_0, such that for $T < T_0$, the system will be essentially *frozen* in a single configuration. In fact, let us consider the number of states $\nu(E)$ whose energy lies between E and $E + \mathrm{d}E$. This number fluctuates from one sample to the next, but its average $\nu(E)$ can be easily calculated in terms of the $\mathcal{P}(E)$:

$$\overline{\nu(E)} = 2^N \mathcal{P}(E)\,\mathrm{d}E \simeq 2^N \exp\left(-\frac{E^2}{NJ_0^2}\right)\left(N\pi J_0^2\right)^{-1/2}\mathrm{d}E. \tag{10.127}$$

Let us choose the width dE of the energy interval to be proportional to N^α, with $0 < \alpha < 1$. Then it is easy to see that equation (10.127) implies the existence of a critical value E_0 of the energy, given by $E_0/N = J_0\sqrt{\ln 2}$, which separates two clearly different behaviors of $\overline{\nu(E)}$ in the limit $N \to \infty$. For $|E| < E_0$, the average number of states is much larger than 1, and increases exponentially as N increases. Since the energy levels are independent, the fluctuations of the level density $\nu(E)$ will be proportional to $\sqrt{\overline{\nu(E)}}$ and will be therefore much smaller than the average.

$$\nu(E) \simeq \overline{\nu(E)}. \tag{10.128}$$

If $|E| > E_0$, the average $\overline{\nu(E)}$ will be much smaller than 1. For almost all systems, therefore, one will have $\nu(E) = 0$, and only for a small number of them will one have $\nu(E) = 1$. The probability that one will have a state with energy in this interval vanishes exponentially with N.

It is then clear that in the thermodynamic limit, if $|E| < E_0$, the entropy $S(E)$ is given by

$$S(E) = Nk_B\left[\ln 2 - \left(\frac{E}{NJ_0}\right)^2\right]. \tag{10.129}$$

On the other hand, $S(E) = 0$ for $|E| > E_0$. The temperature T is obviously defined by

$$\frac{1}{T} = \frac{\partial S}{\partial E}. \tag{10.130}$$

Let us introduce the transition temperature T_0 by means of the condition

$$\frac{1}{T_0} = \left.\frac{\partial S}{\partial E}\right|_{E=-E_0} = \frac{2\sqrt{\ln 2}\, k_B}{J_0}. \tag{10.131}$$

By expressing E as a function of T by means of equation (10.129), we obtain the free energy $F = E - TS(E)$:

$$F = \begin{cases} -Nk_BT\ln 2 - NJ_0^2/4k_BT, & \text{for } T > T_0 \\ -E_0 = -NJ_0\sqrt{\ln 2}, & \text{for } T < T_0. \end{cases} \tag{10.132}$$

Therefore, as the temperature becomes smaller and smaller, the system's internal energy decreases until it reaches $-E_0$. For $T < T_0$, the system's energy remains close to $-E_0$ because there are no states with lower energy.

However, this does not mean that for $T < T_0$, the system can always be found in a *single* well-defined state. In reality, we are dealing with a phase coexistence, in which each of the coexisting phases has vanishing entropy, because the system occupies a number of states that grows with N more slowly than an exponential. We can discuss this coexistence using an approach developed by Bouchaud and Mézard [Bouc97].

The model's partition function has the following expression:

$$Z = \sum_{i=1}^{M} z_i, \tag{10.133}$$

where $z_i = \exp(-E_i/k_BT)$, $M = 2^N$, and the E_i are distributed according to equation (10.126). We are interested in the tail of distribution, when the E_i are large and negative. Let us discuss the distribution of the minimum energy E^*. Let us introduce the cumulative distribution function $\Phi(E)$:

$$\Phi(E) = \int_{-\infty}^{E} \mathrm{d}E' \, \mathcal{P}(E'). \tag{10.134}$$

The distribution $\mathcal{P}_M(E^*)$ of E^* has the expression

$$\mathcal{P}_M(E^*) = M\mathcal{P}(E^*)[1 - \Phi(E^*)]^{M-1} = -\frac{\mathrm{d}}{\mathrm{d}E^*}[1 - \Phi(E^*)]^M. \tag{10.135}$$

For large values of M, we can set

$$[1 - \Phi(E^*)]^M \simeq \exp[-M\Phi(E^*)]. \tag{10.136}$$

In correspondence with the critical value of the energy $E_c = -E_0 = -NJ_0\sqrt{\ln 2}$, the cumulative distribution function of E^* becomes very small:

$$M\Phi(E_c) \simeq 1. \tag{10.137}$$

Let us now define $E^* = E_c + \epsilon$, with $\epsilon \ll |E_c|$. One then has, at the first order

$$[1 - \Phi(E^*)]^M \simeq \left[1 - M\exp\left(\left.\frac{\mathrm{d}\ln\Phi(E)}{\mathrm{d}E}\right|_{E=E_c}\epsilon\right)\right]^M \simeq \exp\left[-\exp\left(-2\sqrt{\ln 2}\,\epsilon/J_0\right)\right]. \tag{10.138}$$

By introducing the rescaled variable $u = 2\sqrt{\ln 2}(\epsilon/J_0)$, we obtain u's distribution:

$$P^*(u) = \exp(u - \exp u). \tag{10.139}$$

This distribution of the extreme values of a random variable possesses a much more general validity, and is called the **Gumbel distribution**.

Exercise 10.18 Show the validity of the Gumbel distribution for the minimum value of a collection M of random variables E_i, identically distributed, whose distribution for large and negative E is given by

$$P(E) \sim \exp\left[-B|E|^{\delta}\right],$$

with $B, \delta > 0$.

The Gumbel distribution decreases exponentially for large and negative u (and much faster for positive u), and exhibits an absolute maximum for $u = 0$, which confirms that E_c is the most probable value of minimum energy.

We can now consider the tail of the distribution of the Boltzmann factors $z = \exp(-E/k_BT)$, for large values of z, when $E \sim E_c$. It decays like a power law:

$$P(z) = z^{-1-\mu}, \tag{10.140}$$

where

$$\mu = \frac{T}{T_0}, \tag{10.141}$$

in which T_0 is the critical temperature given by equation (10.131).

Exercise 10.19 Derive this relation.

The partition function Z has different behaviors for $\mu < 1$, where Z's average diverges and only a small number of terms contributes to Z, and for $\mu > 1$, where all the 2^N terms each give a small contribution to Z. In order to characterize this behavior, we introduce the relative weights

$$w_i = \frac{z_i}{Z}. \tag{10.142}$$

Let us assume that $T < T_0$. One has

$$w_i = w(z_i) = \frac{z_i}{z_i + Z'}, \tag{10.143}$$

where $Z' = \sum_{k(\neq i)} z_k$ is independent from z_i and on the order of $M^{1/\mu}$. Given a value w of the weight, let $\zeta(w)$ be the corresponding value of z. Then the distribution of the weights w at fixed Z' is given by

$$\begin{aligned} P_w(w) &= \int \mathrm{d}z\, \delta\left(w - \frac{z}{z + Z'}\right) P_z(z) = \left.\frac{\mathrm{d}z}{\mathrm{d}w}\right|_{w=w(z)} P_z(\zeta(w)) \\ &= \frac{Z'}{(1-w)^2}\, P_z\left(\frac{Z'w}{1-w}\right). \end{aligned} \tag{10.144}$$

In order to have a nonvanishing weight w in the limit of large M's, z_i must be large. We can therefore apply the asymptotic form for the distribution of z's, thus obtaining, for $w \gg M^{-1/\mu}$,

$$P_w(w) \propto \frac{1}{M}(1-\mu)^{\mu-1} w^{-\mu-1}. \tag{10.145}$$

In this manner, we can calculate the moments $Y_k = \overline{w^k}$: if all the weights are of the same order of magnitude, then $Y_k \sim M^{1-k} \to 0$ for $k > 1$, while only a finite number of states contributes to Z, the Y_k moments remain finite when $M \to \infty$. In our case, we obtain, for $\mu < 1$,

$$Y_k = M \int_0^1 \mathrm{d}w\, w^k P_w(w) = \frac{\Gamma(k-\mu)}{\Gamma(k)\Gamma(1-\mu)}, \tag{10.146}$$

where Γ is Euler's gamma function. Since $\mu = T/T_0$, we obtain that Y_2 goes linearly to zero for $T \to T_0$, while $Y_k = [\Gamma'_{\mathrm{E}}(k) - \Gamma'_{\mathrm{E}}(1)]/\Gamma(k)](T/T_0)$ for $T \to 0$. Last, the system's energy per degree of freedom is constant in the low-temperature phase and is given by

$$\frac{\bar{E}}{N} = \frac{E_c}{N} + \langle u \rangle \frac{J_0}{2N}\sqrt{\ln 2} \simeq -\frac{J_0\sqrt{\ln 2}}{N} + \mathrm{O}\left(\frac{1}{N}\right). \tag{10.147}$$

The average $\langle u \rangle$ is evaluated with the Gumbel distribution, and is equal to $\langle u \rangle = \Gamma'(1)$.

10.3.4 The Random Energy Model as Limit of a Spin Model

The definition of the random energy model is very abstract and can leave one perplexed. One can, however, obtain this model as a limit to a succession of models of mean-field random spins of a more familiar kind.

Let us consider some models with random interactions between p spins, defined by the Hamiltonian

$$H_J^{(p)}(\sigma) = -\sum_{\{i_1,\ldots,i_p\}} J_{i_1,\ldots,i_p}\sigma_{i_1}\cdots\sigma_{i_p}, \tag{10.148}$$

where the indices $i_1, \ldots, i_p$ are between 1 and N, and the sum runs over all the sets $\{i_1, \ldots, i_p\}$ of p distinct indices. The coefficients $J_{i_1,\ldots,i_p}$ are random variables, independent for different sets $\{i_1, \ldots, i_p\}$, identically distributed with the Gaussian distribution

$$P_J(J) = \sqrt{\frac{N^{p-1}}{\pi p!}}\exp\left(-\frac{J^2N^{p-1}}{J_0^2 p!}\right). \tag{10.149}$$

The power of N is chosen so as to guarantee a good thermodynamic limit, as we will see further on.

We now want to show that, in the limit $p \to \infty$ suitably defined, the system described by this Hamiltonian is a random energy system like the one we just described. First, let us consider a certain configuration σ and let us evaluate the probability density $\mathcal{P}(E)$ of the energy values for some configuration $\sigma = (\sigma_i)$:

$$\mathcal{P}(E) = \int \prod_{\{i_1,\ldots,i_p\}} J_{i_1,\ldots,i_p} P_J(J_{i_1,\ldots,i_p})\,\delta\left(E - H_J^{(p)}(\sigma)\right). \tag{10.150}$$

As usual, we can make use of the Fourier representation of the delta function and evaluate the resulting integral by the saddle point method. We thus have

$$\begin{aligned} P(E) &= \int \frac{\mathrm{d}\lambda}{2\pi\mathrm{i}} \int \prod_{\{i_1,\ldots,i_p\}} J_{i_1,\ldots,i_p} P_J(J_{i_1,\ldots,i_p}) \\ &\quad\times \exp\left(\mathrm{i}\lambda E + \mathrm{i}\lambda \sum_{\{i_1,\ldots,i_p\}} J_{i_1,\ldots,i_p}\sigma_{i_1}\cdots\sigma_{i_p}\right) \\ &= \int \frac{\mathrm{d}\lambda}{2\pi\mathrm{i}} \exp\left(\mathrm{i}\lambda E - \sum_{\{i_1,\ldots,i_p\}} \frac{\lambda^2 J_0^2 p!}{4N^{p-1}}\sigma_{i_1}^2\cdots\sigma_{i_p}^2\right) \end{aligned} \tag{10.151}$$

where we have calculated the Gaussian integral over the J's. Obviously, $\sigma_i^2 = 1$, and therefore, the terms under summation in this expression are all the same. It is easy to see that they are $N(N-1)\cdots(N-p+1)/p! \simeq N^p/p!$. We thus obtain

$$P(E) = \int \frac{\mathrm{d}\lambda}{2\pi\mathrm{i}} \exp\left(\mathrm{i}\lambda E - \frac{\lambda^2 N J_0^2}{4}\right) \tag{10.152}$$

and calculating the Gaussian integral one more time, we finally obtain

$$P(E) \propto \exp\left(-\frac{E^2}{NJ_0^2}\right). \tag{10.153}$$

We now calculate the joint distribution $P(E_1, E_2)$ of the energies of two configurations $\sigma^{(1)}$ and $\sigma^{(2)}$:

$$P(E_1, E_2) = \overline{\delta\left(E_1 - H_J\left(\sigma^{(1)}\right)\right)\delta\left(E_2 - H_J\left(\sigma^{(2)}\right)\right)}. \tag{10.154}$$

We obtain

$$\begin{aligned} P(E_1, E_2) &= \int \frac{\mathrm{d}\lambda_1}{2\pi \mathrm{i}} \int \frac{\mathrm{d}\lambda_2}{2\pi \mathrm{i}} \int \prod_{\{i_1, \dots, i_p\}} J_{i_1, \dots, i_p} P_J(J_{i_1, \dots, i_p}) \exp(\mathrm{i}\lambda_1 + \mathrm{i}\lambda_2 E_2) \\ &\times \exp\left[\sum_{\{i_1, \dots, i_p\}} J_{i_1, \dots, i_p}\left(\mathrm{i}\lambda_1 \sigma_{i_1}^{(1)} \cdots \sigma_{i_p}^{(1)} + \mathrm{i}\lambda_2 \sigma_{i_1}^{(2)} \cdots \sigma_{i_p}^{(2)}\right)\right]. \end{aligned} \tag{10.15}$$

By evaluating the Gaussian integral over the J's, we obtain

$$\begin{aligned} P(E_1, E_2) &= \int \frac{\mathrm{d}\lambda_1}{2\pi \mathrm{i}} \int \frac{\mathrm{d}\lambda_2}{2\pi \mathrm{i}} \exp(\mathrm{i}\lambda_1 E_1 + \mathrm{i}\lambda_2 E_2) \\ &\times \exp\left[-\sum_{\{i_1, \dots, i_p\}} \frac{J_0^2 p!}{4N^{p-1}}\left(\lambda_1 \sigma_{i_1}^{(1)} \cdots \sigma_{i_p}^{(1)} + \lambda_2 \sigma_{i_1}^{(2)} \cdots \sigma_{i_p}^{(2)}\right)^2\right]. \end{aligned} \tag{10.156}$$

The quantity under summation can be written

$$\frac{J_0^2 p!}{4N^{p-1}}\left[\lambda_1^2 + \lambda_2^2 + 2\lambda_1\lambda_2\left(\sigma_{i_1}^{(1)}\sigma_{i_1}^{(2)}\right)\cdots\left(\sigma_{i_1}^{(1)}\sigma_{i_1}^{(2)}\right)\right]. \tag{10.157}$$

Let us introduce the **overlap** $q(\sigma^{(1)},\sigma^{2)})$ between the configurations $\sigma^{(1)}$ and $\sigma^{(2)}$:

$$q\left(\sigma^{(1)},\sigma^{(2)}\right) = \frac{1}{N}\sum_i \sigma_i^{(1)}\sigma_i^{(2)}. \tag{10.158}$$

This quantity is equal to 1 if $\sigma^{(1)}$ and $\sigma^{(2)}$ differ by a finite number of spins and is otherwise smaller than 1. If $\sigma^{(1)}$ and $\sigma^{(2)}$ are opposites, $q(\sigma^{(1)}, \sigma^{(2)}) = -1$. Therefore, q is a good measure of the resemblance between spin configurations and has a fairly important role in the theory.

A little algebra shows that the sum over the indices in equation (10.157) is equal (up to terms negligible for a large N) to

$$\begin{aligned} &\sum_{\{i_1, \dots, i_p\}} \left(\sigma_{i_1}^{(1)}\sigma_{i_1}^{(2)}\right)\cdots\left(\sigma_{i_1}^{(1)}\sigma_{i_1}^{(2)}\right) \\ &= \frac{N^p}{p!}\left[q\left(\sigma^{(1)},\sigma^{(2)}\right)\right]^p. \end{aligned} \tag{10.159}$$

We thus obtain

$$\begin{aligned} P(E_1, E_2) &= \int \frac{\mathrm{d}\lambda_1}{2\pi \mathrm{i}} \int \frac{\mathrm{d}\lambda_2}{2\pi \mathrm{i}} \exp(\mathrm{i}\lambda_1 E_1 + \mathrm{i}\lambda_2 E_2) \\ &\times \exp\left[-\frac{NJ_0^2}{4}\left(\lambda_1^2 + \lambda_2^2 + 2\lambda_1\lambda_2 q^p\right)\right]. \end{aligned} \tag{10.160}$$

Let us consider two configurations that are neither identical nor opposite, such that $|q| < 1$. If we take the $p \to \infty$ limit of this expression, the term proportional to $\lambda_1\lambda_2$ vanishes, and therefore $P(E_1, E_2)$ factorizes into $P(E_1)$ and $P(E_2)$—in other words, in this limit, the energies for different configurations are independent random variables. Thus the random energy model is the limit, for $p \to \infty$, of a model with interactions for p spins like the one considered in this section.

10.3.5 The Replica Method

The greatest mathematical difficulty in calculating the properties of a disordered system arises from the fact that, in the quenched average, it is necessary to evaluate the average over the distribution of the slow variables of the *logarithm* of the partition function Z, rather than of the partition function itself. Actually, evaluating the average of the partition function is analogous to handling the quenched variables the same as the fast ones. Several methods have been introduced to get around this problem—among these, one of the most popular is the **replica method**.

Let us consider a random partition function Z_J, a function of a set of quenched variables $J = (J_i)$. Our goal is to calculate the average $\overline{\ln Z}$ of Z_J's logarithm over the probability distribution $P(J)$ of the quenched variables. The replica method is based on the observation that

$$\ln Z = \lim_{n\to 0} \frac{Z^n - 1}{n}, \tag{10.161}$$

where n is a real number. On the other hand, if n is an integer, Z_J^n can be obtained as the partition function of a system made of n identical replicas of the system (characterized by the *same* realization of disorder J). If we were allowed to first evaluate the average over the J's and then take the limit $n \to 0$, we could obtain the result we sought as follows:

$$\overline{\ln Z} = \lim_{n\to 0} \overline{\frac{Z^n - 1}{n}} = \lim_{n\to 0} \frac{\overline{Z^n} - 1}{n}. \tag{10.162}$$

It should be possible, in principle, to evaluate $\overline{Z^n}$ at least in mean-field models, for every integer value of n. The problem is to then extend the result analytically to real values of n so as to be able to take the limit $n \to 0$. It has been found that this program can be run through in the case of the spin glass problem, but at the cost of unexpected difficulties—and a framework in which the method can be justified with some rigor is not yet available. We will limit ourselves to applying it in the case we already discussed—namely, the random energy model.

Let us introduce the generating function of energy distribution:

$$\exp[g(\lambda)] = \int_{-\infty}^{+\infty} \mathrm{d}E\,\mathcal{P}(E)\,\mathrm{e}^{-\lambda E}. \tag{10.163}$$

This expression can be easily calculated with the saddle point method, which gives

$$g(\lambda) = N\left(\frac{\lambda J_0}{2}\right)^2. \tag{10.164}$$

In the replica method, we must evaluate Z's moments:

$$\overline{Z^n} = \overline{\sum_{i_1,\dots,i_n} z_{i_1}\cdots z_{i_n}} = \overline{\sum_{i_1,\dots,i_n} \exp\left(-\frac{1}{k_B T}\sum_i E_i \sum_{a=1}^{n} \delta_{i,i_a}\right)}. \tag{10.165}$$

By means of the generating function, it is easy to evaluate the average:

$$\overline{Z^n} = \sum_{i_1,\dots,i_n} \exp\left[\sum_i g\left(\frac{1}{k_B T}\sum_{a=1}^{n} \delta_{i,i_a}\right)\right]. \tag{10.166}$$

Now, we must understand which configurations of indices will dominate the sum we have obtained in this manner. The simplest hypothesis is that all indices i_a are different—this corresponds to the most numerous configurations. In this case,

$$\overline{Z^n} = M(M-1)\cdots(M-n+1)\exp\left[ng\left(\frac{1}{k_B T}\right)\right] \simeq \exp\left\{n\left[\ln M + g\left(\frac{1}{k_B T}\right)\right]\right\}. \tag{10.167}$$

By passing to the limit $n \to 0$, we obtain the free energy per degree of freedom $f = -(k_B T/N)\,\overline{\ln Z}$:

$$f = f_0 = k_B T \ln 2 - \left(\frac{J_0^2}{4k_B T}\right), \tag{10.168}$$

which corresponds to what we had obtained with the microcanonical method. This result, however, cannot be true for all temperatures. In fact, the entropy per particle is given by $s_0 = -\mathrm{d}f_0/\mathrm{d}T$, is equal to

$$s_0 = k_B\left(\ln 2 - \frac{J_0^2}{4k_B T^2}\right), \tag{10.169}$$

and becomes negative for $T < T_0$, where $k_B T_0 = J_0/(2\sqrt{\ln 2})$ is our usual transition temperature.

For $T < T_0$, one needs to consider configurations in which the n indices are divided into n/m groups of m equal indices, which can be denoted as follows:

$$\begin{aligned} i_1 = i_2 = \cdots = i_m &= k_1, \\ i_{m+1} = i_{m+2} = \cdots = i_{2m} &= k_2, \\ \cdots \quad \cdots &\quad \cdots \\ i_{n-m+1} = i_{n-m+2} = \cdots = i_n &= k_{n/m}, \end{aligned}$$

in which the indices $k_1, \dots, k_{n/m}$ are different one from the other.

These configurations' contribution to $\overline{Z^n}$ is equal to

$$\overline{Z^n} = M(M-1)\cdots(M-n/m+1)\exp\left[\frac{n}{m}g\left(\frac{m}{k_B T}\right)\right]\frac{n!}{(m!)^{n/m}}. \tag{10.170}$$

which gives

$$f(T) = f_0(T/m). \tag{10.171}$$

The minimum of this free energy is obtained when

$$\frac{\partial f}{\partial m} = 0 = s_0(T/m). \tag{10.172}$$

Therefore,

$$m = \frac{T}{T_0} = \mu. \tag{10.173}$$

We have therefore obtained the previous result—in other words, that the system remains frozen (with vanishing entropy)—at the temperature T_0. We can also evaluate the moments Y_k, obtaining

$$\begin{aligned} Y_k &= \sum_i \overline{\frac{z_i^k}{Z^k}} = \lim_{n\to 0} \overline{\sum_i z_i^k Z^{n-k}} \\ &= \lim_{n\to 0} \frac{1}{n(n-1)\cdots(n-k+1)} \sum_{a_1,\dots,a_k}{}' \sum_{i_1,\dots,i_n} \overline{z_{i_1}\cdots z_{i_n}} \prod_{j=1}^{k} \delta_{i_{a_1}, i_{a_j}}, \end{aligned} \tag{10.174}$$

where the sum over the a's runs over replica indices (between 1 and n) that are different from one another, but to which identical configurations correspond, because of the Kronecker delta's product. This implies that the k replica indices $a_1, \dots, a_k$ must be chosen within the same group of m indices. Once a_1 has been chosen, this can be performed in $(m-1)\cdots(m-k+1)$ different ways. Therefore,

$$Y_k = \lim_{n\to 0} \frac{n(m-1)\cdots(m-k+1)}{n(n-1)\cdots(n-k+1)} \overline{Z^n}. \tag{10.175}$$

Let us remark that

$$\lim_{n\to 0} (n-1)\cdots(n-k+1) = (-1)^{k-1}(k-1)! = (-1)^{k-1}\Gamma(k), \tag{10.176}$$

while

$$(m-1)\cdots(m-k+1) = (-1)^{k-1}(1-m)\cdots(k-1-m), \tag{10.177}$$

which can be written, exploiting the relation $z\Gamma(z) = \Gamma(z+1)$, in the form

$$(m-1)\cdots(m-k+1) = (-1)^{k-1}\frac{\Gamma(k-m)}{\Gamma(1-m)}. \tag{10.178}$$

By exploiting these relations and the fact that $m = \mu$, we obtain

$$Y_k = \frac{\Gamma(k-\mu)}{\Gamma(k)\Gamma(1-\mu)}, \tag{10.179}$$

which agrees with the direct calculation. In this calculation, we have assumed that the properties of the replicated system are not invariant with respect to any permutation of the replicas. One calls this **replica symmetry breaking**. In our case, the replicas were grouped into n/m groups. In other cases, replicas within the same first-level group can be grouped into second-level groups, and so forth. The number of levels that are obtained in this fashion is called the *number of steps* in replica symmetry breaking. We will not discuss these replica symmetry breaking steps here because doing so would take us too far afield.

10.3.6 The Hopfield Model

The **Hopfield model** was introduced by J. J. Hopfield in 1982 [Hopf82] to describe some neural networks. I will discuss this motivation at the end of this section. For the time being, it can be considered a spin glass model that is exactly solvable in mean-field theory.

The Hopfield model is defined by the Hamiltonian

$$H = -\frac{1}{N}\sum_{(ij)} J_{ij}\sigma_i\sigma_j, \tag{10.180}$$

where $\sigma_i = \pm 1$ for $i = 1, 2, \dots, N$, and the sum runs over all the (ij) pairs of indices. We are therefore dealing with a model defined in the mean field, and the factor $1/N$ is introduced in order to obtain a good thermodynamic limit.

The coupling constant J_{ij} for each random pair (ij) is given by the following expression:

$$J_{ij} = \sum_{a=1}^{p} \xi_i^a \xi_j^a. \tag{10.181}$$

In this expression, we have introduced p vectors $\xi^a (a = 1, \dots, p)$, where $\xi_i^a = \pm 1$ for $i = 1, 2, \dots, N$, and where the components ξ_i^a are independent random variables, identically distributed, with a $1/2$ probability of being equal to $+1$ or -1.

Following this definition, the coupling constants relative to different pairs are *not* independent. One effectively has

$$\begin{aligned} J_{ij}J_{jk} &= \frac{1}{N^2}\sum_{ab} \xi_i^a\xi_j^a\xi_j^b\xi_k^b = \frac{1}{N^2}\left(\sum_a \xi_i^a\xi_k^a + \sum_a \sum_{b(\neq a)} \xi_i^a\xi_j^a\xi_j^b\xi_k^b\right) \\ &= \frac{1}{N^2}\left[J_{ik} + \mathrm{O}\left(\sqrt{p(p-1)}\right)\right]. \end{aligned} \tag{10.182}$$

A systematic correlation therefore exists between bonds that share a common vertex.

In any case, the distribution of the J's is symmetrical, and one can expect that given three spins i, j and k, the product $J_{ij}J_{jk}J_{ki}$ will fairly often be negative. If p (as we shall assume) is finite, this model can be solved in the thermodynamic limit $N \to \infty$. This solution is due to Amit, Gutfreund, and Sompolinsky [Amit85].

Let us calculate the model's partition function for a given sample. We first note that

$$H_\xi(\sigma) = -\frac{1}{N}\sum_{(ij)} J_{ij}\sigma_i\sigma_j = -\frac{1}{2N}\sum_{ij}\sum_a \xi_i^a\sigma_i\xi_j^a\sigma_j + \frac{p}{2}. \tag{10.183}$$

We have made explicit the fact that H depends on the random vectors ξ. By disregarding the last term (which comes from the terms in which $i = j$) and by introducing the notation

$$(\sigma \cdot \xi) = \frac{1}{N}\sum_i \sigma_i\xi_i, \tag{10.184}$$

the Hamiltonian can be written

$$H_\xi(\sigma) = -\frac{N}{2}\sum_a (\sigma \cdot \xi^a)^2. \tag{10.185}$$

The canonical partition function (for a given sample) therefore assumes the form

$$Z = \sum_\sigma \exp\left[-\frac{H_\xi(\sigma)}{k_B T}\right] = \sum_\sigma \exp\left[\frac{N}{2k_B T}\sum_a (\sigma \cdot \xi^a)^2\right]. \tag{10.186}$$

Let us introduce the vector $\boldsymbol{m}$ with p components, defined by

$$\boldsymbol{m} = (m^a), \qquad m^a = (\xi^a \cdot \sigma). \tag{10.187}$$

The partition function can then be written as follows:

$$Z = \int \prod_a \mathrm{d}m^a \exp\left[\frac{N}{2k_{\mathrm{B}}T} \sum_a (m^a)^2\right] \sum_\sigma \delta(m^a - (\xi^a \cdot \sigma)). \tag{10.188}$$

This expression can be evaluated by using the integral representation of the delta:

$$\delta(x) = \int_{-\mathrm{i}\infty}^{+\mathrm{i}\infty} \frac{\mathrm{d}y}{2\pi\mathrm{i}} \, \mathrm{e}^{-xy}. \tag{10.189}$$

We thus obtain

$$Z = \int \prod_a \mathrm{d}m^a \int_{-\mathrm{i}\infty}^{+\mathrm{i}\infty} \prod_a \frac{N\mathrm{d}\lambda^a}{2\pi\mathrm{i}} \sum_\sigma \exp\left\{\sum_a \left[\frac{N}{2k_B T}(m^a)^2 - N\lambda^a m^a + \lambda^a \sum_i \xi_i^a \lambda^a \sigma_i\right]\right\}. \tag{10.190}$$

We have introduced the factors N in order to simplify some of the expressions that follow.

Let us now evaluate the expression

$$\mathrm{e}^{N\mathcal{F}} = \sum_\sigma \exp\left(\lambda^a \sum_i \xi_i^a \lambda^a \sigma_i\right). \tag{10.191}$$

One has

$$\begin{aligned} \mathrm{e}^{N\mathcal{F}} &= \prod_i \sum_{\sigma_i} \mathrm{e}^{\lambda^a \xi_i^a \sigma_i} \\ &= \prod_i 2\cosh\left(\sum_a \lambda^a \xi_i^a\right) \\ &= \exp\left\{\sum_i \ln\left[2\cosh\left(\sum_a \lambda^a \xi_i^a\right)\right]\right\}. \end{aligned} \tag{10.192}$$

We see that the argument of the exponential contains the sum of N random, independent, and identically distributed terms. As a result of the central limit theorem, this sum tends (in the thermodynamic limit) to N times the average of each term, evaluated with respect to the distribution of the ξ's. We thus obtain

$$\mathcal{F} = \overline{\ln\left[2\cosh\left(\sum_a \lambda^a \xi^a\right)\right]}, \tag{10.193}$$

where $\overline{\ldots}$ is the average over the distribution of the ξ's:

$$\overline{f(\xi^1, \ldots, \xi^p)} = \frac{1}{2^p} \sum_{\{\xi^a = \pm 1\}} f(\xi^1, \ldots, \xi^p). \tag{10.194}$$

In this fashion, we can observe that, in the Hopfield model, the thermodynamic properties are self-averaging.

The partition function effectively assumes the form

$$Z = \int \prod_a \mathrm{d}m^a \int_{-\mathrm{i}\infty}^{+\mathrm{i}\infty} \prod_a \frac{N\mathrm{d}\lambda^a}{2\pi\mathrm{i}} \exp\left\{\sum_a \left[\frac{N}{2k_B T}(m^a)^2 - N\lambda^a m^a + N\mathcal{F}(\lambda)\right]\right\}. \tag{10.195}$$

This expression can be evaluated by means of the saddle point method. The saddle point equations are

$$\lambda_c^a = \frac{m_c^a}{k_B T}, \tag{10.196}$$

$$m^a = \left.\frac{\partial F}{\partial \lambda^a}\right|_{\lambda_c} = \overline{\xi^a \tanh\left(\sum_b \lambda_c^b \xi^b\right)}. \tag{10.197}$$

Let us denote the number of nonvanishing components of $\boldsymbol{m}_c$ (and therefore, due to equation (10.196), also of λ_c) with $r (0 < r \leq p)$. The simplest and most important case is that in which $r = 1$. In this case, one has the following expression for the only nonvanishing component of $\boldsymbol{m}_c$:

$$m_c^a = \overline{\xi^a \tanh \xi^a \lambda_c^a} = \tanh \lambda_c^a, \tag{10.198}$$

and, by making use of equation (10.196), one finally has

$$m_c^a = \tanh \frac{m_c^a}{k_B T}. \tag{10.199}$$

This equation admits nonvanishing solutions for $k_B T < k_B T_c = 1$. Below this temperature, m_c's behavior is the same as that of the mean-field solution of the Ising model.

One can prove (and for this purpose, I refer the reader to Amit's original article [Amit85]) that the saddle points with $r > 1$ correspond to metastable states. The states of thermodynamic equilibrium (at least for finite p's) therefore correspond to $r = 1$—in other words, to states in which $(\langle \sigma \rangle \cdot \xi^a) \neq 0$ for only one of the ξ^a vectors. Since the Hopfield Hamiltonian admits the inversion symmetry $\sigma \to -\sigma$, these states come in pairs, with opposite values of m^a. We see, therefore, that the Hopfield Hamiltonian admits $2p$ pure phases of equilibrium at low temperature, which are generally not connected by symmetry transformations. The origin of this fact is clearly to be sought in the phenomenon of frustration.

10.3.7 *The Hopfield Model and Neural Networks*

The Hopfield model was introduced to describe a network of **formal neurons** that possess the property of associative memory. A formal neuron is a two-state automaton (introduced by McCulloch and Pitts in 1943 [McCu43]) that represents the dynamic behavior of a neural cell (**neuron**), in a very simplified manner.

In figure 10.10, I reproduce the fundamental structure of a neuron from the cerebral cortex. Two series of tree-like projections depart from the cell's body—on the one hand, the **dendrites**, on the other the **axon** (a more or less cylindrical projection that can reach respectable lengths), which in its turn branches out into a large number of **terminal branches**. These are in direct contact with the dendrites of other nerve cells. The contacts are called **synapses** and are endowed with a complex structure.

A neuron's activity is manifested in the emission of an **action potential**—an electrochemical excitation of invariable amplitude that is born in the cell's body and moves at constant speed along the axon. When the action potential reaches the terminal branches, it excites the synapses, which discharge special chemical substances (the **neurotransmitters**) in the intercellular medium.

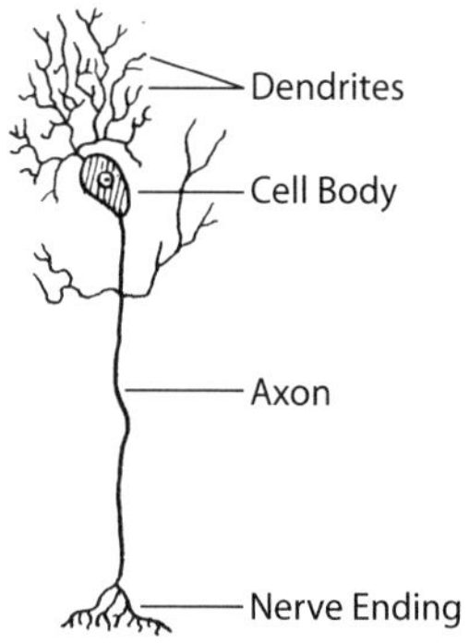

FIGURE 10.10. A pyramidal neuron in the cerebral cortex.

The neurotransmitters are recognized by particular receptors placed on the cell's membrane *downstream* from the synapse. The emission of the action potential is determined, in a complex and as yet not completely known fashion, by the activity of the receptors. McCulloch and Pitts suggested that one represent each neuron's state by means of a binary variable $\tau = 0, 1$—$\tau = 0$ if the neuron is inactive (does not emit action potential), $\tau = 1$ if it is active (is emitting action potential, or is in the *refractory* period that follows each emission, in which it cannot emit the action potential again).

The probability $p_0(t)$ that the neuron we are considering (and that we will denote by i) will be active at the next instant t is therefore monitored by an expression containing the activities τ_j of all neurons *upstream* from neuron i:

$$p_i(t) = f\left(\sum_j J_{ij}\tau_j - \theta_i\right). \tag{10.200}$$

In this expression, the J_{ij} coefficients are defined over each pair (i,j) of neurons and they represent the effectiveness of the synaptic connection between neuron j (above) and neuron i (below). A positive value of J shows that the synapse is an **exciter synapse**—in other words, it tends to excite the downstream neuron when the one upstream is excited. A negative value of J shows that the synapse is an **inhibitor synapse**—in other words, it tends to block the excitation of the downstream neuron when the upstream neuron is excited. The quantity θ_i (which varies from neuron to neuron) is a threshold that determines the dendrite excitation level beyond which it is probable that the downstream neuron will become active.

The $f(x)$ function is not determined—for simplicity's sake, one usually assumes that it is the same for all neurons. The simplest case is obviously that in which it is a simple step function:

$$f(x) = \begin{cases} 0, & \text{if } x \le 0, \\ 1, & \text{if } x = 0. \end{cases} \tag{10.201}$$

Obviously, however, functions of the Fermi type

$$f(x) = \frac{1}{e^{-\beta x} + 1}, \tag{10.202}$$

are also acceptable. In this case, the parameter β plays to some extent the role of the inverse of temperature.

One usually assumes that a synchronization mechanism between neurons does not exist, and the system's evolution can therefore be described as follows: during each time interval, a neuron i is chosen (at random) and the expression $h_i = \sum_j J_{ij}\tau_j - \theta_i$, is evaluated; one then sets $\tau_i = 1$ with a probability equal to $f(h_i)$. This mechanism is known as **asynchronous updating.**

If we assume that J_{ij} is a symmetrical matrix and that $f(x)$ is given by equation (10.201), it is easy to see that the dynamical process we have defined coincide with the Monte Carlo dynamics relative to an Ising system, whose variables are represented by the τ_i's, and whose Hamiltonian is equal to

$$H = -\frac{1}{2}\sum_{ij}{}' J_{ij}\tau_i\tau_j + \sum_i \theta_i\tau_i. \tag{10.203}$$

The transformation

$$\tau_i = \frac{1+\sigma_i}{2} \tag{10.204}$$

transforms this Hamiltonian into that characteristic of the usual Ising model, in the variables $\sigma_i = \pm 1$. In this particular case, therefore, we can expect the dynamically stable states of the neural network's evolution to be the local minima of the Hamiltonian. We can then convince ourselves that by taking expression (10.202) instead, one obtains the Monte Carlo dynamics of the same system at a temperature given by $k_B T = \beta^{-1}$. In this case, we expect that the long-term behavior will be described by the system's statistical mechanics—more particularly, that the system's behavior can be well described by the thermodynamic equilibrium phases corresponding to the Hamiltonian (10.203).

Hopfield's crucial observation of 1982 [Hopf82] is that a system of this type behaves like an **associative memory**, which is addressable by content. In standard computer memory, it is necessary to identify the address of each stored item in order to retrieve its contents. In associative memory, the stored items can be retrieved if one inputs the system with partial information about their contents. It is possible, for instance, to retrieve a name like *Hopfield* starting from a substring (with errors) like *Kopfi*.

Let us now assume that we have a system represented by a Hopfield model. The vectors ξ^a represent the **patterns** that are stored in memory—in other words, the strings we want to retrieve. The system is subjected to a **stimulus**—in other words, it is prepared in an initial configuration $\sigma(0)$. The system then evolves following the asynchronous updating rule we discussed and that we have seen is equivalent to a Monte Carlo dynamics. After a while, the system's state will be represented by one of the model's equilibrium phases, and one will therefore have $m^a = (\xi^a \cdot \langle\sigma\rangle \neq 0)$ for one (and only one) of ξ^a's patterns. The measure of $\langle\sigma_i\rangle$ allows one to reconstruct the pattern. One can expect the chosen pattern to be that which most resembles the initial stimulus $\sigma(0)$—in other words, that for which $[\xi^a \cdot \sigma(0)]$ is largest (in absolute value). In this manner, the evolution of Hopfield's model has allowed us to retrieve the information that was not completely present in the stimulus but was stored in the J_{ij}.

The Hopfield model is one of the simplest cases of application of statistical concepts to the description of a biological system.

Recommended Reading

The theory of linear polymers is presented in a simple fashion in M. Doi, *Introduction to Polymer Physics,* Oxford, UK: Clarendon Press, 1996. One can also usefully consult P.-G. de Gennes, *Scaling Concepts in Polymer Physics,* Ithaca, NY: Cornell University Press, 1979. A discussion of percolation theory can be found in D. Stauffer and A. Aharony, *Introduction to Percolation Theory,* 2nd ed., London: Taylor and Francis, 1992. An introduction to spin-glass theory is provided by K. H. Fischer and J. A. Hertz, *Spin Glasses* (Cambridge, UK: Cambridge University Press, 1991). An important collection of articles on spin glasses is contained in M. Mézard, G. Parisi, and M. Virasoro, *Spin Glass Theory and Beyond,* Singapore: World Scientific, 1987. The theory of spin glasses has recently found several unexpected applications: a very didactic presentation can be found in O. C Martin, R. Monasson, and R. Zecchina, Statistical mechanics methods and phase transitions in optimization problems, *Theoretical Computer Science* **265**, 3–67 (2001), available at arXiv:cond-mat/0104428. One can also read with benefit A. Hartmann and M. Weigt, *Phase Transitions in Combinatorial Optimization Problems* (Weinheim: Wiley-VCH, 2005). A thorough exposition of these applications is contained in M. Mézard and A. Montanari, *Information, Physics, and Computation* (Oxford, UK: Oxford University Press, 2009). The developments of the theory of neural networks based on the Hopfield approach are discussed in D. J. Amit, *Modelling Brain Function: The World of Attractor Neural Networks* (Cambridge, UK: Cambridge University Press, 1989).

Appendices

Appendix A Legendre Transformation

A.1 Legendre Transform

Let us consider a function $f(x,\xi)$ with arguments x and ξ. Let us assume that, instead of x, we want to use the quantity

$$p = \left.\frac{\partial f}{\partial x}\right)_{\xi}, \tag{A.1}$$

as an independent variable.

A priori, we could simply solve equation (A.1) with respect to x, obtaining $x = x(p,\xi)$, and substitute the result in $f(x,\xi)$. In order for this equation to admit a single solution, it is necessary that the second derivative of f, evaluated with respect to x, with ξ kept constant, has a definite sign, and that f therefore is either concave or convex with respect to the variable x. We thus obtain the function $\phi(p,\xi)$, defined by

$$\phi(p,\xi) = f(x(p,\xi),\xi). \tag{A.2}$$

Proceeding in this fashion, we notice, however, that our knowledge of $\phi(p,\xi)$ does *not* (generally) allow us to reconstruct the $f(x,\xi)$. Let us consider, e.g., *two* functions, $f_1(x,\xi)$ and $f_2(x,\xi)$, where

$$f_2(x,\xi) = f_1(x + \psi(\xi),\xi), \tag{A.3}$$

where $\psi(\xi)$ is some function of ξ. The plots of $f_1(x,\xi)$ and $f_2(x,\xi)$, for a fixed value of ξ, differ by a translation parallel to the x axis. We will now proceed as described earlier, by defining $x_1(p,\xi)$ and $x_2(p,\xi)$ by means of

$$p = \left.\frac{\partial f_1(x,\xi)}{\partial x}\right|_{x = x_1(p,\xi)}; \qquad p = \left.\frac{\partial f_2(x,\xi)}{\partial x}\right|_{x = x_2(p,\xi)}. \tag{A.4}$$

It is easy to see that

$$f_1(x_1(p,\xi),\xi) = f_2(x_2(p,\xi),\xi). \tag{A.5}$$

and that, therefore, the two functions f yield the *same* function ϕ! If the $\phi(p,\xi)$ is known, therefore, the $f(x,\xi)$ is not univocally known.

In order to solve this problem, let us consider the plot of $f(x,\xi)$ as a function of x. Let us assume that $f(x,\xi)$, considered as a function of x at fixed ξ, is *concave*. (One would obtain the same results if it were convex instead.) The same curve $y=f(x,\xi)$ can be seen as the locus of points (x, y) that satisfy the condition $y=f(x,\xi)$, or as the envelope of tangents of the form $y = p(x - x_0) + f(x_0,\xi)$. From this second point of view, p identifies the slope's tangent. We can thus unequivocally identify $f(x,\xi)$ if we associate p with the tangent of slope p to the curve $y=f(x,\xi)$, identified, e.g., by its intercept with the y axis. This intercept can be obtained by substituting $x = 0$ in the expression given earlier. We obtain

$$g(p,\xi) = f(x(p,\xi),\xi) - px(p,\xi). \tag{A.6}$$

This expression defines the **Legendre transform** $g(p,\xi)$ of the $f(x,\xi)$ with respect to the variable x. This relation (in which we used the convention most commonly adopted by physicists) is not symmetrical between f and g, and as a result, the **Legendre antitransform** is defined by

$$f(x,\xi) = g(p,\xi) + px, \tag{A.7}$$

where p is the solution of the equation

$$x = -\left.\frac{\partial g}{\partial p}\right)_{\xi}, \tag{A.8}$$

with x and ξ fixed. The signs ought to be kept in mind. Mathematicians use another notation, which is more symmetrical.

A.2 Properties of the Legendre Transformation

The Legendre transformation, which leads from the function $f(x,\xi)$ to its Legendre transform $g(p,\xi)$, has the following properties:

Linearity: If $f = \lambda_1 f_1 + \lambda_2 f_2$, then the Legendre transform g of the f is given by $f = \lambda_1 g_1 + \lambda_2 g_2$, where g_1 and g_2, respectively, are the transforms of their respective f's.

Convexity: If the f is concave (convex) with respect to x, the Legendre transform is convex (concave) with respect to p.

Variational principle: Let us assume that $f(x,\xi)$ is concave. Let us introduce the *three-variable* function $\Phi(x,\xi,p)$, defined by

$$\Phi(x,\xi,p) = f(x,\xi) - px, \tag{A.9}$$

where p is now an independent variable. One then has

$$\Phi(x,\xi,p) \geq g(p,\xi), \qquad \forall x, \tag{A.10}$$

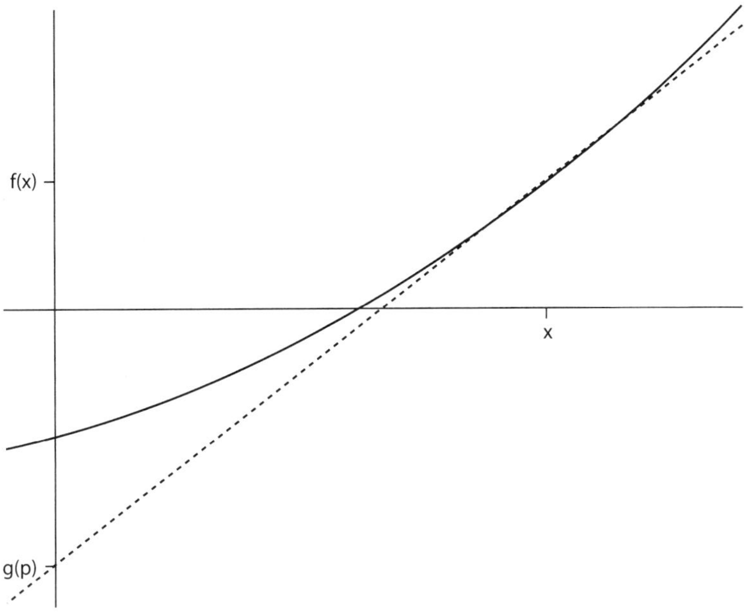

FIGURE A.1. Geometrical definition of the Legendre transform $g(p) = f(x(p)) - px(p)$ as the intercept on the y axis of the tangent in $(x, f(x))$ to $f(x)$. For a concave $f(x)$ function, the curve is always located above the tangent, which corresponds to the variational principle $g(p) \leq f(x) - xp$, $\forall x$.

and the equality holds only for $x = x(p, \xi)$. This property easily follows from the fact that because the f is concave, it is always located above the tangent (see figure A.1).

One should note that if the $f(x, \xi)$ is linear with respect to x within a certain interval $x_1 < x < x_2$—in other words, if one has

$$f(x, \xi) = f(x_1, \xi) + p_0(x - x_1), \qquad \text{for } x_1 < x < x_2, \tag{A.11}$$

then the transform $g(p, \xi)$ exhibits an angular point for $p = p_0$, and the two derivatives (right and left) are equal to $-x_1$ and $-x_2$, respectively. The variational principle of the Legendre transform shows the intimate connection that exists between the Legendre transformation and the theory of **Lagrange multipliers**, which we will discuss in the following section.

A.3 Lagrange Multipliers

Let us suppose that we want to look for the maxima and minima of a function $f(x, \xi)$ of the variables x and ξ. Only, instead of looking for the *absolute* maxima and minima—in other words, within a domain in the plane (x, ξ)—we want to identify them among the points that satisfy a condition of the type

$$\phi(x,\xi) = 0, \tag{A.12}$$

where $\phi(x,\xi)$ is a certain function. The locus of the points that satisfy this condition is a curve, and we want to look for the maximum (or minimum) of f along this curve. Problems of this sort are called **constrained** extrema problems, because one searches for the extremum among the points that satisfy a constraint such as the one expressed by equation (A.12).

Let us assume that we are able to solve equation (A.12) explicitly with respect to ξ, obtaining $\xi = \xi(x)$, such that

$$\phi(x,\xi(x)) = 0, \tag{A. 13}$$

So our problem becomes that of a search for the extremum of a function of x alone, along the curve defined by the constraint

$$\psi(x) = f(x,\xi(x)). \tag{A.14}$$

At the extremum $[x_0, \xi_0 = \xi(x_0)]$, one obviously has $\mathrm{d}\psi/\mathrm{d}x = 0$. We thus obtain

$$0 = \left.\frac{\mathrm{d}\psi}{\mathrm{d}x}\right|_{x_0} = \left.\frac{\partial f}{\partial x}\right|_{(x_0,\xi_0)} + \left.\frac{\partial f}{\partial \xi}\right|_{(x_0,\xi_0)} \left.\frac{\mathrm{d}\xi}{\mathrm{d}x}\right|_{x_0}. \tag{A.15}$$

On the other hand, taking the derivative of equation (A.13), one has

$$\left.\frac{\partial \phi}{\partial x}\right|_{(x_0,\xi_0)} + \left.\frac{\partial \phi}{\partial \xi}\right|_{(x_0,\xi_0)} \left.\frac{\mathrm{d}\xi}{\mathrm{d}x}\right|_{x_0} = 0. \tag{A.16}$$

Let us multiply this equation by an indeterminate number λ, and subtract both sides of (A.16) from equation (A.15). We obtain

$$0 = \left.\left(\frac{\partial f}{\partial x} - \lambda\frac{\partial \phi}{\partial x}\right) + \frac{\mathrm{d}\xi}{\mathrm{d}x}\left(\frac{\partial f}{\partial \xi} - \lambda\frac{\partial \phi}{\partial \xi}\right)\right|_{(x_0,\xi_0)}. \tag{A.17}$$

If we chose λ so that the second member vanishes, this equation tells us that the first term also vanishes at the extremal. In other words, there exists a constant λ such that f's constrained extremum, under the constraint $\phi(x,\xi) = 0$ that we are considering, appears as the absolute extremum of function $\Phi(x,\xi;\lambda) = f(x,\xi) - \lambda\phi(x,\xi)$ with respect to x and ξ. This constant is called the **Lagrange multiplier**.

This allows us to define a strategy for the search for constrained extrema. One searches for the absolute extrema of the $\Phi(x,\xi;\lambda)$ for each value of λ. As λ varies, these extrema trace a curve $[x = x(\lambda), \xi = \xi(\lambda)]$. One then looks for the point (x_0, ξ_0) that satisfies the constraint—in other words, for which $\phi(x_0, \xi_0) = 0$—along this curve.

It is easy to generalize the previous considerations to cases with several variables and define the sufficient conditions for this method to function. In practice, one requires that, in the region we are considering, the vector $(\partial\phi/\partial x_i)$ of the constraint's partial derivatives be linearly independent from vector $(\partial f/\partial x_i)$ of f's partial derivatives. This is equivalent to saying that ϕ's *contour lines*—in other words, the locus of points in which the ϕ assumes a certain value—be transversal to the f's contour lines.

We can now see the relation between Lagrange multipliers and Legendre transforms. We know that given the function $f(x,\xi)$, which we will assume is convex with respect to x, its Legendre transform $g(p,\xi)$ satisfies a variational principle: it is identical to the function $\Phi(x,\xi;p)=f(x,\xi)-px$, at its minimum value with respect to x.

Given a value p_0 of p, let us denote the value of ξ at the extremal $g(p_0,\xi)$ by $\xi_0(p_0)$, where one has

$$\left.\frac{\partial g(p_0,\xi)}{\partial \xi}\right)_p = 0. \tag{A. 18}$$

By explicitly evaluating the derivative of g, we obtain

$$0 = \left[\left.\frac{\partial f}{\partial x}\right)_\xi - p_0\right]\frac{\partial x(p,\xi)}{\partial \xi} + \left.\frac{\partial f}{\partial \xi}\right)_x. \tag{A. 19}$$

The factor in square brackets vanishes since the condition $\partial f/\partial x)_\xi = p_0$ is satisfied. Therefore, g's extremal is also f's extremal.

Summing up, let us consider the point (x,ξ_0) that is $f(x,\xi)$'s extremal with respect to x, with ξ fixed. Let the value of f's partial derivative with respect to x with ξ fixed, calculated in that point, be equal to p_0. The same value ξ_0 of ξ is obtained if one considers $g(p,\xi)$'s extremal with respect to p with ξ fixed (and equal to p_0). At this point, however, because of the variational principle of the Legendre transform, the point (x,ξ_0) is extremal, *both* with respect to x and with respect to ξ, of the function $\Phi(x,\xi;p_0)=f(x,\xi)-p_0x$, which is numerically equal, on the extremal, to the Legendre transform $g(p_0,\xi)$.

Appendix B Saddle Point Method

B.1 Euler Integrals and the Saddle Point Method

It is often necessary to calculate integrals of the form

$$I(N) = \int_a^b \mathrm{d}x \exp[-Nf(x)], \tag{B.1}$$

where the variable N assumes very large and positive values, and where the function $f(x)$ exhibits a minimum on the integration path. These integrals are known as Euler integrals and can be estimated with the **saddle point method**. Let us assume that $f(x)$ exhibits a minimum in x_0, with $a < x_0 < b$. We expand the $f(x)$ into a Taylor series around x_0:

$$f(x) = f(x_0) + \frac{1}{2} f''(x_0)(x - x_0)^2 + \cdots \tag{B.2}$$

The first-order term is absent as a result of the hypothesis that x_0 corresponds to a minimum. In our hypotheses, moreover, $f''(x_0) > 0$. By substituting this expression in the integral (and setting $\Delta x = x - x_0$), we obtain

$$I(N) \approx \exp[-Nf(x_0)] \int_{a-x_0}^{b-x_0} \mathrm{d}(\Delta x) \exp\left[-\frac{1}{2} Nf''(x_0)\Delta x^2\right]. \tag{B.3}$$

The error one introduces by taking the lower integration limit to $-\infty$ and the upper to $+\infty$ is exponentially small in N. We thus obtain

$$\begin{aligned} I(N) &\approx \exp[-Nf(x_0)] \int_{\infty}^{+\infty} \mathrm{d}(\Delta x) \exp\left[-\frac{1}{2} Nf''(x_0)\Delta x^2\right] \\ &= \exp[-Nf(x_0)] \sqrt{\frac{2\pi}{Nf''(x_0)}}. \end{aligned} \tag{B.4}$$

The meaning of this expression is the following:

$$\lim_{N\to\infty} \frac{I(N)}{\exp[-Nf(x_0)]\sqrt{\dfrac{2\pi}{Nf''(x_0)}}} = 1. \tag{B.5}$$

Taking into account the following terms of the Taylor expansion of $f(x)$, and integrating them explicitly, it is possible to obtain the formal series

$$I(N) \approx \exp[-Nf(x_0)]\sqrt{\frac{2\pi}{Nf''(x_0)}}\left[1+\sum_{k=1}^{\infty}\frac{I_k}{N^k}\right]. \tag{B.6}$$

This series is not convergent, however, but is an asymptotic series, in the sense that

$$\frac{I(N)}{\exp[-Nf(x_0)]}\sqrt{\frac{2\pi}{Nf''(x_0)}} = 1+\sum_{k=1}^{r}\frac{I_k}{N^k}+\mathrm{o}(N^{-r}). \tag{B.7}$$

In this expression, $\mathrm{o}(x)$ is the Landau symbol, which has the following interpretation:

$$\lim_{x\to 0}\frac{\mathrm{o}(x)}{x} = 0. \tag{B.8}$$

Although they are not convergent, asymptotic series often represent an excellent approximation, and are preferred to convergent series from a numerical point of view. The method we just outlined is called the saddle point method for the following reason. For simplicity's sake, let us continue to assume that N is positive and that the function $f(x)$ does not present a minimum on the integration path but has an extremal (a point in which the first derivatives vanish) in a point x_0 of the complex plane. If we assume that $f(x)$ is analytic, we can deform the integration path in such a manner as to make it pass through x_0. Then $f(x)$ has the following expansion around x_0:

$$\begin{aligned} f(x) &\simeq f(x_0)+\frac{1}{2}f''(x_0)(x-x_0)^2+\cdots \\ &= f(x_0)+\frac{1}{2}\Re\left[f''(x_0)(x-x_0)^2\right]+\frac{1}{2}\Im\left[f''(x_0)(x-x_0)^2\right]+\cdots \end{aligned} \tag{B.9}$$

The modulus of our integrand can therefore by approximated by

$$\begin{aligned} &\exp\left\{-\frac{N}{2}\Re\left[f''(x_0)\,\Delta x^2\right]\right\} = \\ &\quad\exp\left\{-\frac{N}{2}\left[\Re f''(x_0)\left(\Delta x'^2-\Delta x''^2\right)-2\Im f''(x_0)\,\Delta x'\Delta x''\right]\right\}, \end{aligned} \tag{B.10}$$

where we have denoted the real and imaginary parts of $\Delta x = x - x_0$ by $\Delta x'$ and $\Delta x''$, respectively. The exponential quadratic form can be diagonalized if we locally rotate the integration path by an angle θ with respect to the real axis, where

$$\tan(2\theta) = \frac{\Im f''(x_0)}{\Re f''(x_0)}. \tag{B.11}$$

This equation admits as a solution two directions that are mutually orthogonal. By following one direction, $f(x)$ admits a maximum, while by following the other, it admits a minimum in x_0. It is clear that if we further deform the integration path until we make it reach x_0 along this second direction, we obtain an integral of the same kind as the real integral we studied before, slowly changing factors.

The geometry of the $\exp[-Nf(x)]$ close to x_0 is that of a saddle point—and the integration path is chosen in such a manner as to pass through the saddle by following the maximum slope path. It is easy to convince ourselves that $f(x)$'s imaginary part (and therefore the integrand's phase) does not vary along this path—for this reason, the method is also called the **stationary phase method**.

B.2 The Euler Gamma Function

Let us consider the integral (defined for positive z's)

$$\Gamma(z) = \int_0^\infty \mathrm{d}t\,\mathrm{e}^{-t}t^{z-1}. \tag{B.12}$$

We obviously have $\Gamma(1) = 1$, while an integration by parts shows that

$$z\Gamma(z) = \Gamma(z+1). \tag{B.13}$$

These two relations imply that, if z is a positive integer,

$$\Gamma(z) = (z-1)!. \tag{B.14}$$

We can therefore consider this integral as the generalization of the factorial. The integral converges in all of the right half of the complex plane, and therefore our function can be extended from this region, by means of the functional relation we have just written, to the entire complex plane, except for points $z = 0, -1, -2, \ldots$, where it presents simple poles. In this manner, we have defined **Euler's gamma function**.

For $z = 1/2$, the transformation $t = x^2$ reconnects our integral to the Gaussian integral, and one therefore obtains $\Gamma(1/2) = \sqrt{\pi}$. By applying the functional relation (B.13), we can derive the relation

$$\begin{aligned}\Gamma\left(n+\frac{1}{2}\right) &= \frac{(2n-1)(2n-3)\cdots 1}{2^n}\sqrt{\pi} \\ &= \frac{(2n-1)!!}{2^n}\sqrt{\pi}.\end{aligned} \tag{B.15}$$

In this expression, $n!!$ is the semifactorial, which is equal to $2 \times 4 \times \ldots \times n$ if n is even, and to $3 \times 5 \times \ldots \times n$ if n is odd.

We can now calculate the asymptotic expression $\Gamma(z)$ for large values of z (assumed real—in fact, for large values of $\Re(z)$). By utilizing the saddle point method, we obtain

$$\begin{aligned}\Gamma(z) &= \int_0^\infty \mathrm{d}t \exp[(z-1)\ln t - t] \\ &\approx \exp[(z-1)\ln t_0 - t_0]\int \mathrm{d}(\Delta t)\exp\left(-\frac{z-1}{2t_0^2}\Delta t^2\right),\end{aligned} \tag{B.16}$$

where $t_0 = z - 1$. In this manner, we obtain the well-known **Stirling formula**:

$$\Gamma(z) \approx \exp[(z-1)\ln(z-1)-(z-1)]\sqrt{2\pi(z-1)}. \tag{B.17}$$

B.3 Properties of N-Dimensional Space

We will often have to deal with integrals in N-dimensional space, where N is very large. Since some of this space's geometrical properties are far from intuitive, it is useful to consider them closely.

First let us note that the ratio between the volume of a sphere of radius $R-\delta R$ and that of radius R is given by $(1-\delta R/R)^N$, which, however small δR may be, tends to zero as N increases—in other words, almost all of the sphere's volume can be found near the surface.

The surface of the sphere with R radius is given by $S_N(R) = S_N(1)R^{N-1}$. In order to calculate the unit sphere's surface, we calculate the N-dimensional Gaussian integral in polar coordinates:

$$G_N = \int \prod_{i=1}^{N} \mathrm{d}x_i \exp\left(-\sum_{i=1}^{N} x_i^2\right) = \int_0^\infty \mathrm{d}R S_N(R) \exp(-R^2). \tag{B.18}$$

This integral is the product of N Gaussian integrals and is therefore equal to $\pi^{N/2}$. On the other hand, one has

$$\begin{aligned} \int_0^\infty \mathrm{d}R S_N(R) \exp(-R^2) &= S_N(1)\int_0^\infty \mathrm{d}R R^{N-1}\exp(-R^2) \\ &= \frac{1}{2} S_N(1)\int_0^\infty \mathrm{d}t e^{-t} t^{N/2-1}. \end{aligned} \tag{B.19}$$

In the last integral, we recognize the definition of the gamma function. Therefore,

$$S_N(R) = \frac{2\pi^{N/2}}{\Gamma(N/2)} R^{N-1}. \tag{B.20}$$

We now calculate the distribution of the sphere's projection onto one of the axes—in other words, the density of the points of the sphere whose first coordinate lies (for example) between x and $x+\mathrm{d}x$. This quantity is given by

$$\rho(x_1) = \int \prod_{i=2}^{N} \mathrm{d}x_i \, \theta\left(R^2 - x_1^2 - \sum_{i=1}^{N} x_i^2\right). \tag{B.21}$$

The integral on the right-hand side is the volume of the sphere of radius $\sqrt{R^2 - x_1^2}$ in $(N-1)$–dimensional space. We can therefore write

$$\begin{aligned} \rho(x_1) &= \frac{1}{N-1} S_{N-1}(1)(R^2 - x_1^2)^{\frac{N-1}{2}} \\ &\simeq \exp\left[\frac{N-1}{2}\ln(R^2 - x_1^2)\right]. \end{aligned} \tag{B.22}$$

Because of the large factor $(N-1)$ in the exponential, this function becomes very small as soon as x_1 moves away from zero. By expanding the exponent into a Taylor series around the origin, we obtain

$$\begin{aligned} \rho(x_1) &\simeq \exp\left[\frac{N-1}{2}\ln\left(1-\frac{x_1^2}{R^2}\right)\right] \\ &\simeq \exp\left[-\frac{(N-1)\,x_1^2}{2R^2}\right]. \end{aligned} \tag{B.23}$$

The distribution of x_1 is therefore a Gaussian with zero average and variance equal to $R^2/(N-1)$.

B.4 Integral Representation of the Delta Function

The relation between a function and its Fourier transform can be summarized by the expression

$$\frac{1}{2\pi}\int_{-\infty}^{\infty} \mathrm{d}y \, \exp(\mathrm{i}xy) = \delta(x). \tag{B.24}$$

In effect, one has

$$f(x_0) = \frac{1}{2\pi}\int_{-\infty}^{\infty} \mathrm{d}y \, \tilde{f}(y) \exp(\mathrm{i}x_0 y), \tag{B.25}$$

where $\tilde{f}$ is the Fourier transform of $f(x)$:

$$\tilde{f}(y) = \int_{-\infty}^{\infty} \mathrm{d}x \, \exp(\mathrm{i}xy) f(x). \tag{B.26}$$

Therefore,

$$\begin{aligned} f(x_0) &= \int_{-\infty}^{\infty} \mathrm{d}x \, \delta(x-x_0) f(x) \\ &= \int_{-\infty}^{\infty} \mathrm{d}x \, f(x) \frac{1}{2\pi}\int_{-\infty}^{\infty} \mathrm{d}y \, \exp[\mathrm{i}y(x-x_0)], \end{aligned} \tag{B.27}$$

which is what we wanted to prove.

Appendix C A Probability Refresher

C.1 Events and Probability

Let us consider a set Ω of *elementary events* such that only one of them must in any case occur as the result of a probabilistic experiment. The measurable subsets of Ω we will call *events*. Let us, for instance, consider the throw of a die. The set of elementary events is $\Omega = \{1,2,3,4,5,6\}$; some of the possible events (subsets of Ω) are the *even number* event ($\{2,4,6\}$), the *greater than or equal to 4* event ($\{4,5,6\}$), the *greater than 6* event ($\{\} = \emptyset$), and so on.

A **probability** p is associated with each event—in other words, a real number such that

- $0 \leq p\,(\alpha) \leq 1,\ \forall \alpha \subseteq \Omega$.
- $p(\emptyset) = 0$ and $p(\Omega) = 1$.
- $p(\alpha \cup \beta) = p(\alpha) + p(\beta) - p(\alpha \cap \beta),\ \forall \alpha, \beta \subseteq \Omega$.

More specifically, if α and β are *mutually exclusive events*—in other words, $\alpha \cap \beta = \emptyset$, then one has $p(\alpha \cup \beta) = p(\alpha) + p(\beta)$. This result leads us to the *law of total probability*: given a set of mutually exclusive events $\{\alpha_i\}$ such that $\cup_i \alpha_i = \Omega$, the probability of a generic event β is given by

$$p(\beta) = \sum_i p(\beta \cap \alpha_i). \tag{C.1}$$

With a judicious choice of events $\{\alpha_i\}$, this result can simplify the calculation of the probabilities of a complex event, reducing it to the calculation of the probability of simpler events.

C.2 Random Variables

A **random variable** is a real function defined over Ω that can assume different values, depending on which elementary event takes place. A random variable can be characterized by events (for example, $a \leq f < b$) and by associated probabilities. In many cases, these probabilities are *continuous* functions of the extremes (a, b) of the interval. One can then define the **probability distribution** $P(f)$ of the random variable f:

$$p(a \leq f < b) = \int_a^b P(f) df.$$

Some simple examples are:

- **Uniform distribution:**

$$P(f) = \begin{cases} 1/(b-a), & \text{if } a < f < b, \\ 0, & \text{otherwise.} \end{cases} \tag{C.2}$$

- **Gaussian distribution:**

$$P(f) = \frac{1}{\sqrt{2\pi\sigma^2}} \exp\left[-\frac{(f-f_0)^2}{2\sigma^2}\right]. \tag{C.3}$$

- **The Dirac delta "function":** $\delta(f)$ such that $\int_a^b df\,\delta(f) = 1$ if $a < 0 < b$, and 0 otherwise. It has the following properties:

  1. $\int df\,\delta(f-f_0)F(f) = F(f_0)$.
  2. $\delta(-f) = \delta(f)$.
  3. $\delta(af) = |a|^{-1}\,\delta(f)$.

  Note: One can apply to the delta function all *linear* operations (including, for example, differentiation).

The distribution of the results of the throws of an "honest" (nonbiased) die, for example, can be written

$$P(x) = \frac{1}{6} - \sum_{k=1}^{6} \delta(x-k).$$

C.3 Averages and Moments

The average of a random variable x is defined by

$$\langle x \rangle = \int dx\, x P(x). \tag{C.4}$$

If an x variable is random, its function $f(x)$ is random as well. One then has

$$\langle f(x) \rangle = \int df\, f P(f) = \int dx\, f(x) P(x). \tag{C.5}$$

More specifically, one has the useful relation

$$\langle \delta(x-x_0) \rangle = \int dx\, \delta(x-x_0) P(x) = P(x_0). \tag{C.6}$$

The averages of the powers of a random variable x are called **moments**. One therefore has

$$m_1(x) = \langle x \rangle = \int \mathrm{d}x\, x P(x), \tag{C.7}$$

$$m_2(x) = \langle x^2 \rangle = \int \mathrm{d}x\, x^2 P(x), \tag{C.8}$$

$$\vdots$$

The **variance** of x is defined by

$$\sigma^2 = \langle (x - \langle x \rangle)^2 \rangle = \langle x^2 \rangle - \langle x \rangle^2. \tag{C.9}$$

One obviously has $\sigma^2 \geq 0$.

C.4 Conditional Probability: Independence

Given two events, α and β, the conditional probability that the second will occur, given the first, is defined by

$$p(\beta|\alpha) = \frac{p(\alpha \cap \beta)}{p(\alpha)}. \tag{C.10}$$

More specifically, for the certain event, $\alpha = \Omega$, it coincides with β's probability: $p(\beta \,|\, \Omega) = p(\beta)$.

The probability of the two events' intersection is called **joint probability**. Two events are called **independent** if $p(\alpha \cap \beta) = p(\alpha)p(\beta)$. For independent events, one therefore has

$$p(\beta|\alpha) = p(\beta), \qquad p(\alpha|\beta) = p(\alpha). \tag{C.11}$$

The law of total probability can therefore be reformulated as follows: Given a set of mutually exclusive events $\alpha_i \cap \alpha_j = \emptyset \, (i \neq j)$, such that $\cup_i \alpha_i = \Omega$, the probability of a generic event β is given by

$$p(\beta) = \sum_i p(\beta|\alpha_i)\, p(\alpha_i). \tag{C.12}$$

Given a pair of random variables (x, y), one can consider the corresponding events $\alpha = (a_1 \leq x < b_1)$ and $\beta = (a_2 \leq y < b_2)$. We can therefore define the **joint distribution** analogously to the joint probability:

$$p(\alpha \cap \beta) = \int_{a_1}^{b_1} \mathrm{d}x \int_{a_2}^{b_2} \mathrm{d}y P(x, y). \tag{C.13}$$

If events of this type are independent for any value of the intervals' extremes, the variables are called **independent**, and one has

$$P(x, y) = P_x(x) P_y(y). \tag{C.14}$$

It is easy to see that, for independent variables, one has

$$\langle xy \rangle = \langle x \rangle \langle y \rangle. \tag{C.15}$$

The inverse is not true, but this relation is valid for all moments

$$\langle x^p y^q \rangle = \langle x^p \rangle \langle y^q \rangle, \qquad \forall p, q \text{ integers},$$

if and only if x and y are independent. If it is the case for $p = q = 1$, the variables are called **uncorrelated**.

Let us now consider a certain number N of random variables $x_1, \ldots, x_N$. One has the following properties:

- The *average* of the sum of random variables (*even if not independent*) is equal to the sum of the averages:

$$\left\langle \sum_{i=1}^{N} x_i \right\rangle = \sum_{i=1}^{N} \langle x_i \rangle. \tag{C.16}$$

- The *variance* of the sums of *uncorrelated* variables is equal to the sum of the variances. More particularly, if $\langle x_i \rangle = 0, \forall i$, one has

$$\left\langle \left(\sum_{i=1}^{N} x_i \right)^2 \right\rangle = \sum_{i=1}^{N} \langle x_i^2 \rangle. \tag{C.17}$$

C.5 Generating Function

The **generating function** of variable x's probability distribution is defined by

$$\Gamma(k) = \langle \exp(ikx) \rangle = \int dx \, \exp(ikx) P(x). \tag{C.18}$$

This is called a generating function because of its relation with its moments:

$$\Gamma(k) = \sum_{p=0}^{\infty} \frac{(ik)^p}{p!} \langle x^p \rangle. \tag{C.19}$$

The generating function of the sum of two random *independent* variables is the product of their respective generating functions:

$$\langle \exp[i(x+y)] \rangle = \langle \exp(ix) \rangle \langle \exp(iy) \rangle. \tag{C.20}$$

This allows one to calculate the generating functions of arbitrary linear combinations of independent variables. It is important to observe that the generating function of a Gaussian variable is also a Gaussian:

$$\begin{aligned} \langle \exp(ix) \rangle &= \int dx \, \exp(ikx) \frac{1}{\sqrt{2\pi\sigma^2}} \exp\left[-\frac{(x-x_0)^2}{2\sigma^2} \right] \\ &= \exp\left(ikx_0 - \frac{1}{2} k^2 \sigma^2 \right). \end{aligned} \tag{C.21}$$

C.6 Central Limit Theorem

Let $x_1, \ldots, x_N$, be N random independent variables, identically distributed, with average $\langle x \rangle$ and variance equal to σ^2. The **mean**, as opposed to the average, over the probability distribution, is a *random variable* defined by

$$\bar{x} = \frac{1}{N}\sum_{i=1}^{N} x_i. \tag{C.22}$$

The central limit theorem states that, in the limit $N \to \infty$, the distribution of the mean tends to a Gaussian of average $\langle x \rangle$ and of variance equal to σ^2/N. For the proof, it is sufficient to consider the characteristic functions:

$$\begin{aligned}
\langle \exp(ik\bar{x}) \rangle &= \left\langle \exp\left[ik\left(\frac{1}{N}\sum_{i=1}^{N}\right)\right]\right\rangle = \left\langle \exp\left(\frac{ikx}{N}\right)\right\rangle \\
&= [\Gamma(k/N)]^N \\
&= \left[\exp\left(\frac{ik\langle x\rangle}{N} - \frac{k^2\sigma^2}{2N^2} + \cdots\right)\right]^N \\
&\longrightarrow \exp\left[ik\langle x\rangle - \frac{k^2\sigma^2}{2N} + o(N^{-1})\right].
\end{aligned} \tag{C.23}$$

The theorem can be generalized to nonidentically distributed variables as long as no variable *dominates* the others (in other words, exhibits a variance that is much larger than the others')—it is necessary for each variable's variance to be *finite*. In that case, the mean $\bar{x}$ tends to the average of the averages $\langle x \rangle$, and the variance tends to $1/N$ times the variances' average.

For not strictly independent variables, the theorem remains true if, in the sequence, each variable is correlated with a finite number of neighboring variables. If N is very large, the variable $\bar{x}$ almost certainly assumes a value very close to $\langle x \rangle$.

C.7 Correlations

Let us consider a collection of N random variables $(x_1, \ldots, x_N)$. The correlation between variables x_i and x_j is defined by

$$C_{ij} = \langle x_i x_j \rangle - \langle x_i \rangle \langle x_j \rangle, \tag{C.24}$$

and vanishes when the variables are uncorrelated.

A collection of N variables is called **Gaussian** if the joint probability distribution has the form

$$P(x_1, \cdots, x_N) \propto \exp\left[-\frac{1}{2}\sum_{k\ell} A_{k\ell}\left(x_k - x_k^0\right)\left(x_\ell - x_\ell^0\right)\right]. \tag{C.25}$$

The matrix $\mathsf{A} = (A_{ij})$ must be symmetrical and positive definite. The proportionality constant, which is equal to $(2\pi)^{-N/2}(\det \mathsf{A})^{1/2}$, is defined by the **normalization** condition

$$\int \prod_{i=1}^{N} \mathrm{d}x_i\, P(x_1, \cdots, x_N) = 1. \tag{C.26}$$

Let us prove that the matrix $\mathsf{C} = (C_{ij})$ of the correlations of a Gaussian collection of variables is equal to the inverse of the matrix A, which defines the distribution: $\mathsf{C} = \mathsf{A}^{-1}$. In fact, let us assume for simplicity's sake that for each i, $\langle x_i \rangle = 0$. Let us then define the new integration variables y_k so that $y_k = \sum_\ell Q_{k\ell} x_\ell$, where the symmetrical matrix $\mathsf{Q} = (Q_{k\ell})$ satisfies the relation $\mathsf{Q}^2 = \mathsf{A}$. This transformation's Jacobian is given by $\mathsf{Q}^{-1} = (\det \mathsf{A})^{-1/2}$, and it therefore cancels out with the normalization determinant, while the product $x_i x_j$ is expressed by $\sum_k Q^{-1}_{ik} y_k Q^{-1}_{j\ell} y_\ell$. We thus obtain

$$\begin{aligned} C_{ij} &= \langle x_i x_j \rangle \\ &= \int (\det \mathsf{A})^{1/2} \prod_m \frac{\mathrm{d}x_m}{\sqrt{2\pi}}\, x_i x_j \exp\left(-\frac{1}{2}\sum_{k\ell} A_{k\ell} x_k x_\ell\right) \\ &= \int \prod_p \frac{\mathrm{d}y_p}{\sqrt{2\pi}} \sum_k Q^{-1}_{ik} y_k Q^{-1}_{j\ell} y_\ell \exp\left(-\frac{1}{2}\sum_q y_q^2\right) \\ &= \sum_k Q^{-1}_{ik} Q^{-1}_{j\ell} \delta_{k\ell} = A^{-1}_{ij}, \end{aligned} \tag{C.27}$$

as we wanted to prove.

Appendix D Markov Chains

D.1 Introduction

In the simulation with the Monte Carlo method, we introduce a fictitious dynamical process for the system under consideration, chosen so as to guarantee that the corresponding equilibrium distribution will be the Boltzmann-Gibbs distribution. This process has the following characteristics:

- It takes place at discrete instants in time.
- The system's state at each instant univocally determines the probabilities of the system's states at the next instant.
- The conditional probability that the system is in a certain state at instant $t + 1$, given the state in which it is at instant t, does not depend explicitly on t.

A dynamical process with these properties is called a **Markov chain**. In the following sections, we will provide a general definition of Markov chains and prove (informally) some fundamental theorems. We will then be able to show that it is possible to define a special Markov chain, called the Monte Carlo dynamics, whose stationary distribution is the Boltzmann-Gibbs one. Unless stated otherwise, we will consider only systems with a finite number of states.

D.2 Definitions

Let us consider a system that can be, at each instant $t = 0, 1, 2, \dots$, in one of N states. We will denote by $X_t \in \{1, 2, \dots, N\}$ the state of the system at the instant t. The succession $(X_0, X_1, \dots, X_t, \dots)$ constitutes a random process, which is called a **Markov chain** if the probability that $X_t = a$ depends only on the value of X_{t-1}. In other words, in a Markov chain, once one has determined the system's state at a certain instant, the probabilities of the system's states at the following instant is determined. The dynamics of the process are therefore identified by the collection of all conditional probabilities of X_t given the value of X_{t-1}—in other words, the $W_{ab}(t) = P(X_t = a \mid X_{t-1} = b)$—which are called **transition probabilities.**

This definition does not exclude that the transition probabilities $W_{ab}(t)$ depend explicitly on time. However, we will consider only **stationary** Markov chains, in which these probabilities are independent of t.

Let us denote the probability that $X_t = a$ by $p_a(t)$, where $a \in \{1, 2, \dots, N\}$, and the transition probability from a to b by W_{ba}—in other words, the probability that $X_t = b$, given that $X_{t-1} = a$. We obviously have the following properties:

1. $p_a(t) \geq 0, \forall a, t.$
2. $\sum_a p_a(t) = 1, \forall t.$
3. $W_{ab} \geq 0.$
4. $p_a(t+1) = \sum_b W_{ab} p_b(t).$
5. $\sum_a W_{ab} = 1, \forall b.$

A matrix $\mathsf{W} = (W_{ab})$ that satisfies properties 3 and 5 is called **stochastic**. It follows from this property that

$$p_a(t+n) = \sum_b W^n_{ab} p_b(t), \qquad \forall t, n \geq 0, \tag{D.1}$$

where W^n_{ab} is the matrix element of the n-th power of W.

D.3 Spectral Properties

Let λ now be an eigenvalue of W. One has the following properties:

1. $|\lambda| \leq 1.$
2. 1 is an eigenvalue.
3. If $v_a^{(\lambda)}$ is a right eigenvector of W that belongs to eigenvalue λ, then either $\lambda = 1$ or $\sum_a v_a^{(\lambda)} = 0.$

Proof:

1. Let us assume that $v_a^{(\lambda)}$ is a right eigenvector of W, that belongs to eigenvalue λ. Then,

   $$\sum_b |W_{ab} v_b^{(\lambda)}| = \sum_b W_{ab} |v_b^{(\lambda)}| \geq \left|\sum_b W_{ab} v_b^{(\lambda)}\right| = |\lambda| |v_b^{(\lambda)}|. \tag{D.2}$$

   Taking the sum of this relation with respect to a and taking property 5 into account, we obtain

   $$\sum_b |v_b^{(\lambda)}| \geq |\lambda| \sum_a |v_a^{(\lambda)}|. \tag{D.3}$$

   Therefore, $|\lambda| \leq 1$. Moreover, if $|\lambda| = 1$, one must have $\sum_b W_{ab} |v_b^{(\lambda)}| = |v_a^{(\lambda)}|, \forall a.$

2. Let us consider the vector $u_a^{(1)} = 1, \forall a$. As a result of property 5, one has

$$1 = \sum_a W_{ab} = \sum_a u_a^{(1)} W_{ab} = u_b^{(1)}, \tag{D.4}$$

and therefore $u_a^{(1)}$ is a left eigenvector of W that belongs to eigenvalue 1.

3. Let $v_b^{(\lambda)}$ be a right eigenvalue of W that belongs to eigenvalue λ:

$$\sum_b W_{ab} v_b^{(\lambda)} = \lambda v_a^{(\lambda)}. \tag{D.5}$$

Taking the sum of this relation with respect to a, we obtain

$$\sum_b v_b^{(\lambda)} = \lambda \sum_a v_a^{(\lambda)}, \tag{D.6}$$

and therefore either $\sum_b v_b^{(\lambda)} = 0$ or $\lambda = 1$.

Let us now consider a right eigenvector $v_a^{(\lambda)}$ that belongs to an eigenvalue λ of modulus 1, but different from 1. Because of 3, its components cannot all have the same phase. Let us divide the indices a into subsets K_α such that two indices a and b have the same phase only if they belong to the same subset. Let us consider the vectors

$$w_a^{(\alpha)} = \begin{cases} \left|v_j^{(\lambda)}\right|, & \text{if } a \in K_\alpha, \\ 0, & \text{otherwise.} \end{cases}$$

As a consequence of the relation we proved earlier in the $|\lambda| = 1$ case, the matrix W cannot "mix" the K_α sets. (Otherwise, one would have $\sum_b W_{ab} v_b^{(\lambda)} > v_a^{(\lambda)}$). Therefore, the nonvanishing elements W_{ab} of W, when $a \in K_\alpha$, correspond to indices b that belong to the same set K_α.

D.4 Ergodic Properties

From now on, let us assume that the matrix W is **ergodic**—in other words, that there exists an integer n such that the matrix W^m has all nonvanishing elements for $m > n$. Then, due to the reasoning we just presented, it follows that there exists a single vector (p_a^∞) that satisfies the following conditions:

1. $p_a^\infty > 0, \qquad \forall a.$
2. $\sum_a p_a^\infty = 1.$
3. $\sum_b W_{ab} p_b^\infty = p_a^\infty, \qquad \forall a.$

The third condition expresses the fact that (p_a^∞) is an eigenvector belonging to eigenvalue 1. In our conditions, this eigenvalue is not degenerate, and all other eigenvalues are strictly smaller than 1 in magnitude. Therefore, one can easily show that if (v_a^0) is a vector with nonnegative components, which satisfies $\sum_a v_a^0 = 1$, one has

$$\lim_{n\to\infty} \sum_b W_{ab}^{(n)} v_j^0 = p_a^\infty, \qquad \forall a. \tag{D.8}$$

In other words, given any initial probability distribution (v_a^0), the probability distribution tends to (p_a^∞) for $n \to \infty$.

If the stochastic matrix is not irreducible, it is possible to identify at least two nonempty subsets A and B such that, for whatever integer n, the elements W^n_{ab} of the matrix W^n, where $a \in \mathrm{A}$ and $b \in \mathrm{B}$, all vanish. Then the limit distribution, as $n \to \infty$, is not unique and depends on the initial conditions. In principle, it is possible to divide the states of the process in nonoverlapping subsets K_α, such that the probability that the system is in any state belonging to one such set tends to a limit (depending on the initial conditions) as $n \to \infty$, while the conditional probability that the system is in a state $a \in K_\alpha$, given that it is in K_α, tends to a limit that does not depend on the initial conditions. In other words, the state space separates into different components that are not mutually connected—the probability of remaining in each component is conserved, and therefore depends on the initial conditions, and within each component, one tends to an equilibrium distribution, normalized to the probability of being in that component.

D.5 Convergence to Equilibrium

In this section, I will provide a new, simpler proof of the convergence to equilibrium for an ergodic Markov chain. The proof is due to Kemeny and Snell [Keme60] and can be found on A. Peter Young's website [Youn10].

Let us again consider the evolution equation for the $p_a(t)$:

$$p_a(t+1) = \sum_b W_{ba} p_b(t). \tag{D.9}$$

We know that the matrix W has one eigenvalue equal to 1 (because of the normalization relation), whose left eigenvector's elements are all equal, while the right eigenvector is proportional to the equilibrium distribution p_a^{eq}. We now want to show that this eigenvector dominates when the transformation (D.9) is iterated.

Let us assume that the Markov chain is ergodic and that the number of possible states is finite. There therefore exists (at least) one value of t such that *all* the elements of W^t are greater than zero. In order to simplify the notation, we will from now on denote W^t, where t is this value, with W.

Let us denote W's smallest value with ϵ. We apply W *on the right* to a *row* vector x, whose smallest component is equal to m_0, while its largest is equal to M_0. Let us consider the vector x', obtained by substituting all of x's components, except the smallest, with M_0. Then, the vector's components $x'\mathsf{W}$ have the following expression:

$$w m_0 + (1-w) M_0 = M_0 - w(M_0 - m_0), \tag{D.10}$$

since $\sum_a W_{ab} = 1$, where w is the element of W in the column in question, whose line corresponds to m_0.

Undoubtedly, $w \geq \epsilon$, so that each component is increased by $M_0 - \epsilon(M_0 - m_0)$. On the other hand, since each component of x is increased by the corresponding amount of x', even the largest component of $x\mathsf{W}$, let us call it M_1, must satisfy this inequality—in other words,

$$M_1 \leq M_0 - (M_0 - m_0). \tag{D.11}$$

More precisely, $M_1 \leq M_0$. By applying these same arguments to the vector $-x$, and operating the substitutions $M_0 \to -m_0$, $m_0 \to -M_0$, $M_1 \to -m_1$ (where m_1 is the smallest component of $x\mathsf{W}$), one has

$$-m_1 \leq m_0 - \epsilon(-m_0 + M_0), \tag{D.12}$$

which shows that $m_1 \geq m_0$. We thus obtain

$$M_1 - m_1 \leq (1 - 2\epsilon)(M_0 - m_0). \tag{D.13}$$

By applying this relation to the iterates $x\mathsf{W}^n$, we obtain

$$M_n - m_n \leq (1 - 2\epsilon)^n (M_0 - m_0). \tag{D.14}$$

Therefore, for $n \to \infty$, $x\mathsf{W}^n$ is a vector in which all the components are equal. Let us apply this result to a vector $x^{(k)}$ in which all the components are equal to zero, except for the k-th, which is equal to 1. Let us denote the common value of the resultant vector's components by α_k. Now, $x^{(k)}\mathsf{W}^n$ is the k-th line of W^n. It follows that W^n tends to be a matrix in which all the elements of the k-th line are equal to α_k—in other words,

$$\lim_{n\to\infty} \mathsf{W}^n = \begin{pmatrix} \alpha_1 & \alpha_1 & \cdots & \alpha_1 & \cdots \\ \alpha_2 & \alpha_2 & \cdots & \alpha_2 & \cdots \\ \vdots & \vdots & \ddots & \vdots & \ddots \\ \alpha_k & \alpha_k & \cdots & \alpha_k & \cdots \\ \vdots & \vdots & \ddots & \vdots & \ddots \end{pmatrix}. \tag{D.15}$$

Since the sum of the elements in a column of W (and of W^n) is equal to 1, we have

$$\sum_a \alpha_a = 1. \tag{D.16}$$

Therefore, given a vector x that represents the initial distribution, and as a result of which $\sum_a x_a = 1$, we obtain

$$\lim_{n\to\infty} \mathsf{W}^n x = \alpha, \tag{D.17}$$

where the vector α is given by

$$\alpha = \begin{pmatrix} \alpha_1 \\ \alpha_2 \\ \vdots \\ \alpha_k \\ \vdots \end{pmatrix}. \tag{D.18}$$

Since $\mathsf{W}^n x = \alpha$ for a sufficiently large n, α is a stationary distribution. It is fairly easy to prove that if W is ergodic, there can be only one stationary distribution. In conclusion, we have proven that the distribution converges to the stationary distribution as long as the transition probabilities are stationary and ergodic.

Appendix E Fundamental Physical Constants

Planck constant	h	$6.6260755\ 10^{-34}$ Js
	$\hbar = h/(2/\pi)$	$1.054527\ 10^{-34}$ Js
Boltzmann constant	k_B	$1.384\ 10^{-23}\ JK^{-1}$
Elementary charge	e	$1.60217733\ 10^{-19}$ C
Avogadro constant	N_A	$6.0221367\ 10^{23}\ mol^{-1}$
Speed of light	c	$2.99792458\ 10^{8}\ ms^{-1}$
Magnetic constant	μ_0	$4\pi\ 10^{-7}\ T^2m^3J^{-1}$
		$12.566370614\ 10^{-7}\ T^2m^3J^{-1}$
Dielectric constant	$\epsilon_0 = 1/(\mu_0 c^2)$	$8.854187817\ 10^{-12}\ C^2J^{-1}m^{-1}$
Atomic mass unit		$1.66057\ 10^{-27}$ kg
Bohr radius	a_0	$5.29177\ 10^{-11}$ m
Gas constant	$R = N_A k_B$	$8.3143\ JK^{-1}mol^{-1}$
Standard molar volume	V_{mol}	$22.41383\ m^3kmol^{-1}$
Faraday constant	$F = N_A e$	$9.64846\ 10^{4}\ C\ mol^{-1}$
Electron volt		$1.60217733\ 10^{-19}$ J
		$1.160529\ 10^{4}\ k_B$ K
Standard gravitational acceleration	g	$9.80665\ ms^{-2}$
Von Klitzing constant	$R_K = h/e^2$	$2.581281\ 10^{4}\ \Omega$
Josephson constant	$K_J = 2e/\hbar$	$3.03866\ 10^{15}\ V^{-1}s^{-1}$
Fine structure constant	$1/\alpha = 4\pi\epsilon_0\hbar c/e^2$	137.0359895
Electron mass	m_e	$9.1093897\ 10^{-31}$ kg
Proton mass	m_p	$1.6726231\ 10^{-27}$ kg
Neutron mass	m_n	$1.6749286\ 10^{-27}$ kg
Bohr magneton	$\mu_B = eh/(4\pi m_e)$	$9.2740154\ 10^{-24}\ JT^{-1}$
Nuclear magneton	$\mu_N = eh/(4\pi m_p)$	$5.0507866\ 10^{-27}\ JT^{-1}$

Electron giromagnetic factor	g_e	2.002319304386
Electron magnetic moment	$\mu_e = -(1/2) g_e \mu_B$	$-9.2847701\ 10^{-24}\ JT^{-1}$
Proton magnetic moment	μ_p	$1.41060761\ 10^{-26}\ JT^{-1}$
Gravitational constant	G	$6.6732\ 10^{-11}\ m^3kg^{-1}s^{-2}$

Bibliography

[Alle87] M. P. Allen and D. J. Tildesley, *Computer Simulations of Liquids* (Oxford, UK: Oxford University Press, 1987).

[Amit84] D. J. Amit, *Field Theory, the Renormalization Group, and Critical Phenomena*, 2nd ed. (Singapore: World Scientific, 1984).

[Amit85] D. J. Amit, H. Gutfreund, and H. Sompolinsky, Spin-glass models of neural networks, *Phys. Rev.* **A 32**, 1007–1018 (1985).

[Amit89] D. J. Amit, *Modeling Brain Function: The World of Attractor Neural Networks* (Cambridge, UK: Cambridge University Press, 1989).

[Appe77] K. Appel and W. Haken, Every planar map is four colorable, *Illinois Journal of Mathematics* **21**, 439–567 (1977).

[Ashc76] N. Ashcroft and N. D. Mermin, *Solid State Physics* (New York: Holt, Rinehart and Winston, 1976).

[Bali82] R. Balian, *From Microphysics to Macrophysics: Methods and Applications of Statistical Physics* (Berlin/New York: Springer, 1991).

[Bark76] J. A. Barker and D. Henderson, What is liquid? Understanding the states of matter, *Rev. Mod. Phys.* **48**, 587–671 (1976).

[Bind86] K. Binder, *Monte Carlo Methods in Statistical Mechanics*, 2nd ed. (Berlin: Springer, 1986).

[Bind88] K. Binder and D. W. Heermann, *Monte Carlo Simulation in Statistical Physics* (Berlin: Springer, 1988).

[Binn93] J. J. Binney, N. J. Dowrick, A. J. Fisher, and M.E.J. Newman, *The Theory of Critical Phenomena* (Oxford, UK: Oxford University Press, 1993).

[Bolt95] L. Boltzmann, *Vorlesungen über Gastheorie* (2 vols.) (Leipzig: Barth, 1895–1898).

[Born12] M. Born and T. Von Kármán, On fluctuations in spatial grids, *Physikalische Zeitschrift* **13**, 297–309 (1912); On the theory of specific heat, *Physikalische Zeitschrift* **14**, 15–19 (1912).

[Bose24] S. N. Bose, Plancks Gesetz und Lichtquantenhypothese, *Zeitschrift für Physik* **26**, 178–181 (1924).

[Bouc97] J.-P. Bouchaud and M. Mézard, Universality classes for extreme-value statistics, *J. Phys. A: Math. Gen.* **30**, 7997–8016 (1997).

[Broa57] S. R. Broadbent and J. M. Hammersley, Percolation processes: I. Crystals and mazes, *Proceedings of the Cambridge Philosophical Society* **53**, 629–641 (1957).

[Brus76] S. G. Brush, *The Kind of Motion We Call Heat* (Amsterdam: North-Holland, 1976).

[Call85] H. B. Callen, *Thermodynamics and an Introduction to Thermostatistics* (New York: J. Wiley & Sons, 1985).

[Cara09] C. Carathéodory, Untersuchungen über die Grundlagen der Thermodynamik, *Mathematische Annalen* **67**, 355–386 (1909). English translation in *The Second Law of Thermodynamics*, J. Kestin (ed.) (Stroudsberg, PA: Dowden, Hutchison and Ross, 1976), 229–256.

[Card96] J. Cardy, *Scaling and Renormalization in Statistical Physics* (Cambridge, UK: Cambridge University Press, 1996).

[Carn69] N. F. Carnahan and K. E. Starling, Equation of State for Nonattracting Rigid Spheres, *J. Chem. Phys.* **51**, 635–636 (1969).

[Chah76] C. Chahine and P. Devaux, *Thermodynamique Statistique* (Paris: Dunod, 1976).

[Chai95] P. M. Chaikin and T. C. Lubensky, *Principles of Condensed Matter Physics* (Cambridge, UK: Cambridge University Press, 1995).

[Chan87] D. Chandler, *Introduction to Modern Statistical Mechanics* (Oxford, UK: Oxford University Press, 1987).

[Cloi90] J. des Cloiseaux and J.-F. Jannink, *Polymers in Solution: Their Modelling and Structure* (Oxford, UK: Clarendon Press, 1990).

[Deby12] P. Debye, Zur Theorie der spezischen Wärme, *Ann. Phys. (Leipzig)* **344**, 789–839 (1912).

[Deby23] P. Debye and E. Hückel, The theory of electrolytes. I. Lowering of freezing point and related phenomena, *Physikalische Zeitschrift* **24**, 185–206 (1923).

[Derr81] B. Derrida, Random-energy model: an exactly solvable model of disordered systems, *Phys. Rev.* **B24**, 2613–2626 (1981).

[Doi96] M. Doi, *Introduction to Polymer Physics* (Oxford, UK: Clarendon Press, 1996).

[Domb71] C. Domb and M. Green, eds., *Phase Transitions and Critical Phenomena* (London: Academic Press, 1971–1976).

[Domb78] C. Domb and J. L. Lebowitz, eds., *Phase Transitions and Critical Phenomena* (London: Academic Press, 1978–1994).

[Eins05] A. Einstein, Die von der Molekularkinetische Theorie von Wärme geforderte Bewegung von in den ruhenden Flüssigkeiten suspendierten Teilchen, *Ann. Phys. (Leipzig)* **17**, 549–560 (1905).

[Eins07] A. Einstein, Die Plancksche Theorie der Strahlung und die Theorie der spezischen Wärme, *Ann. Phys. (Leipzig)* **22**, 180–191 (1907).

[Eins25] A. Einstein, Quantentheorie des einatomigen idealen Gases. Zweite Abhandlung, *Sitzungsberichte der Preussischen Akademie der Wissenschaften* **1**, 3–14 (1925).

[Eins56] A. Einstein, *Investigations on the Theory of Brownian Movement* (New York: Dover, 1956), in particular, papers I and II.

[Ferm55] E. Fermi, J. Pasta, and S. Ulam, Studies of Nonlinear Problems, Document LA-1940 (Los Alamos) (May 1955).

[Feyn72] R. P. Feynman, *Statistical Mechanics: A Set of Lectures* (Reading, MA: Benjamin, 1972).

[Fisc77] K. H. Fischer, Spin glasses, *Physica B+C* **86–88** (2), 813–819 (1977).

[Fisc91] K. H. Fischer and J. A. Hertz, *Spin Glasses* (Cambridge, UK: Cambridge University Press, 1991).

[Fish83] M. E. Fisher, Scaling, universality, and renormalization group theory. In *Critical Phenomena, Proceedings of the Summer School Held at the University of Stellenbosch,*

South Africa, January 18–29, 1982, F.J.W. Hahne (ed.), (Berlin: Springer, 1983), pp. 1–139.

[Fors75] D. Forster, *Hydrodynamic Fluctuations, Broken Symmetry, and Correlation Functions* (Reading, MA: Benjamin, 1975).

[Fort72] C. M. Fortuin and P. W. Kasteleyn, On the random-cluster model: I. Introduction and relation to other models, *Physica* (Utrecht) **57**, 536–564 (1972).

[Gall99] G. Gallavotti, *Statistical Mechanics: A Short Treatise* (Berlin: Springer, 1999).

[Gard83] C. W. Gardiner, *Handbook of Stochastic Methods* (Berlin: Springer, 1983).

[Genn79] P.-G. de Gennes, *Scaling Concepts in Polymer Physics* (Ithaca, NY: Cornell University Press, 1979).

[Gibb73] J. W. Gibbs, Graphical methods in the thermodynamics of fluids, *Trans. Conn. Acad. Arts and Sci.* **2**, 309–342 (1873).

[Gibb75] J. W. Gibbs, On the equilibrium of heterogeneous substances, *Trans. Conn. Acad. Arts and Sci.* **III**, 108–248, 343–524 (1875–1876).

[Ginz60] V. L. Ginzburg, *Soviet Physics-Solid State* **2**, 1824 (1960).

[Good85] D. L. Goodstein, *States of Matter* (New York: Dover, 1985).

[Gugg45] E. A. Guggenheim, The principle of corresponding states, *J. Chem. Phys.* **13**, 253–261 (1945).

[Hans86] J.-P. Hansen and I. R. McDonald, *Theory of Simple Liquids*, 2nd ed. (London: Academic Press, 1986).

[Hart05] A. K. Hartmann and M. Weigt, *Phase Transition in Combinatorial Optimization Problems* (Weinheim: Wiley-VCH, 2005).

[Hopf82] J. J. Hopfield, Neural networks and physical systems with emergent collective computational abilities, *Proc. Natl. Acad. Sci. USA* **79**, 2554–2558 (1982).

[Huan87] K. Huang, *Statistical Mechanics* (New York: Wiley, 1987).

[Ito44] K. Itō, Stochastic integral, *Proc. Imp. Acad. Tokyo* **20**, 519 (1944) .

[Kac52] M. Kac and J.-C. Ward, A combinatorial solution of the two-dimensional Ising model, *Phys. Rev* **88**, 1332–1337 (1952).

[Kada66] L. P. Kadanoff, Scaling laws for Ising models near T_c, *Physics* **2**, 263–272 (1966).

[Kada76] L. P. Kadanoff, Notes on Migdal's recursion formula, *Ann. Phys. (N.Y.)* **100**, 359–394 (1976).

[Keme60] J. G. Kemeny and J. L. Snell, *Finite Markov Chains* (Princeton, NJ: Van Nostrand, 1960).

[Kirc45] G. Kirchhoff, Über den Durchgang eines elektrischen Stromes durch eine Ebene, insbesondere durch eine kreisförmige, *Annalen der Physik und Chemie* (also known as *Poggendorffs Annalen*) **64**, 487–514 (1845).

[Knut81] D. E. Knuth, *The Art of Computer Programming*, vol. 2: *Seminumerical Algorithms* (Reading, MA: Addison-Wesley, 1981). Chapter 3, especially section 3.5.

[Kram41] H. A. Kramers and G. H. Wannier, Statistics of the two-dimensional ferromagnet, *Phys. Rev.* **60**, 252–262 (1941).

[Kubo85] R. Kubo, M. Toda, and N. Hashitsume, *Statistical Physics* II, *Nonequilibrium Statistical Mechanics* (Berlin: Springer, 1985), in particular pp. 69ff.

[Land80] L. D. Landau and E. M. Lifshitz, *Statistical Physics*, 3rd ed., part 1 (Oxford, UK: Pergamon Press, 1980).

[Lang08] P. Langevin, Sur la théorie du mouvement Brownien, *C. R. Acad. Sci. Paris* **146**, 530 (1908).

[Lieb98] E. Lieb and J. Yngvason, A guide to entropy and the second law of thermodynamics, *Notices of the American Mathematical Society* **45**, 571–581 (1998). Available at arXiv:cond-mat/9805005.

[Lifs64] S. Lifson, Partition functions of linear-chain molecules, *J. Chem. Phys.* **40**, 3705–3710 (1964).

[Lim90] Y.-K. Lim, *Problems and Solutions in Statistical Mechanics* (Singapore: World Scientific, 1990).

[Lube78] T. C. Lubensky, Thermal and geometrical critical phenomena in random systems. In *Ill-Condensed Matter, Les Houches Session XXXI, 1978*, R. Balian, R. Maynard and G. Toulouse (eds.) (Amsterdam: North-Holland, 1979), pp. 405–475.

[Ma85] S.-K. Ma, *Statistical Mechanics* (Singapore: World Scientific, 1985).

[Mand77] B. Mandelbrot, *Fractals: Form, Chance, and Dimension* (San Francisco: W. H. Freeman and Co., 1977).

[Marc59] P. M. Marcus and H. J. McFee, Velocity distributions in potassium molecular beams. In *Recent Research in Molecular Beams*, I. Estermann (ed.) (New York: Academic Press, 1959), pp. 43ff.

[Mart01] O. C. Martin, R. Monasson, and R. Zecchina, Statistical mechanics methods and phase transitions in optimization problems, *Theoretical Computer Science* **265**, 3–67 (2001).

[Maxw60] J. C. Maxwell, Illustrations of the dynamical theory of gases. Part 1. On the motion and collisions of perfectly elastic spheres, *Phil. Mag.* **XIX**, 19–32 (1860).

[Maxw74] J. C. Maxwell, Van der Waals on the continuity of the gaseous and liquid states, *Nature* **10**, 477–480 (1874).

[Maye40] J. E. Mayer and M. G. Mayer, *Statistical Mechanics* (New York: Wiley, 1940), chapter 13.

[McCu43] W. S. McCulloch and W. Pitts, A logical calculus of the ideas immanent in nervous activity, *Bull. Math. Biol.* **5**, 115–133 (1943).

[Meza09] M. Mézard and A. Montanari, *Information, Physics, and Computation* (Oxford: Oxford University Press, 2009).

[Meza87] M. Mézard, G. Parisi, and M. A. Virasoro, *Spin Glass Theory and Beyond* (Singapore: World Scientific, 1987).

[Migd75] A. A. Migdal, Phase transitions in gauge and spin-lattice systems, *Zh. Eksp. Teor. Fiz. (USSR)* **69**, 1457–1465 (1975). Translated in *Soviet Physics-JETP* **42**, 743–746 (1976).

[Mori65] H. Mori, Transport, collective motion, and Brownian motion, *Prog. Theor. Phys.* **33**, 423–455 (1965).

[Moss99] S. Moss de Oliveira, P.M.C. de Oliveira, and D. Stauffer, *Evolution, Money, War, and Computers* (Stuttgart: Teubner, 1999).

[Naka58] S. Nakajima, On quantum theory of transport phenomena, *Prog. Theor. Phys.* **20**, 948–959 (1958).

[Niem76] T. H. Niemeyer and J. M. J. van Leeuwen, Renormalization for Ising-like spin systems. In *Phase Transitions and Critical Phenomena*, vol. 6, C. Domb and M. S. Green (eds.) (New York: Academic Press, 1976), pp. 425–505.

[Onsa31] L. Onsager, Reciprocal relations in irreversible processes I, *Phys. Rev.* **37**, 405–426 (1931); Reciprocal relations in irreversible processes II, *Phys. Rev.* **38**, 2265–2279 (1931).

[Onsa44] L. Onsager, Crystal statistics. I. A two-dimensional model with an order-disorder transition, *Phys. Rev.* **65**, 117–149 (1944).

[Orns14] L. S. Ornstein and F. Zernicke, The accidental deviations of density and the opalescence in the critical point, *Proc. Acad. Sci. Amsterdam* **17**, 793–806 (1914).

[Orns20] L. S. Ornstein, The Brownian movement, *Proc. Amst. Acad. Sci.* **21**, 96 (1920).

[Peie34] R. Peierls, Remarks on transition temperatures, *Helv. Phys. Acta* **7(2)**, 81–83 (1934).

[Perr08] J. Perrin, La loi de Stokes et le mouvement Brownien, *C. R. Acad. Sci. Paris* **147**, 475 (1908).

[Perr16] J. Perrin, *Atoms* (New York: Van Nostrand, 1916).

[Pipp57] A. B. Pippard, *The Elements of Classical Thermodynamics* (Cambridge, UK: Cambridge University Press, 1957).

[Plis94] M. Plischke and B. Birgersen, *Equilibrium Statistical Physics*, 2nd ed. (Singapore: World Scientific, 1994).

[Pool92] R. Pool, The third branch of science debuts, *Science* **256**, 44–47 (1992).

[Pres92] W. H. Press, S. A. Teukolsky, W. T. Vetterling, and B. P. Flannery, *Numerical Recipes in FORTRAN: The Art of Scientic Computing*, 2nd ed. (Cambridge, UK: Cambridge University Press, 1992).

[Reic80] L. E. Reichl, *A Modern Course in Statistical Physics* (Austin: University of Texas Press, 1980).

[Ruel69] D. Ruelle, *Statistical Mechanics: Rigorous Results* (New York: Benjamin, 1969).

[Sack11] O. Sackur, On the kinetic explanation of Nernst's heat theory, *Ann. Phys. (Leipzig)* **34**, 455–468 (1911).

[Sarm79] G. Sarma, Conformation des polymères en solution. In *Ill-Condensed Matter, Les Houches XXXI, 1978*, R. Balian, R. Maynard, and G. Toulouse (eds.) (Amsterdam: North-Holland, 1979).

[Schu64] T. D. Schulz, D. C. Mattis, and E. H. Lieb, Two-dimensional Ising model as a soluble problem of many fermions, *Rev. Mod. Phys.* **36**, 856–871 (1964).

[Sred06] M. Srednicki, Quantum field theory, available at the author's web page: http://www.physics.ucsb.edu/~mark/qft.html. Retrieved on August 13, 2010.

[Stan71] H. E. Stanley, *Introduction to Phase Transitions and Critical Phenomena* (Oxford, UK: Oxford University Press, 1971).

[Stau92] D. Stauffer and A. Aharony, *Introduction to Percolation Theory*, 2nd ed. (London: Taylor and Francis, 1992).

[Sved06] Th. Svedberg, Über die Eigenbewegung der Teilchen in kolloidalen Lösungen, *Zeit. für Elektrochemie* **12**, 853 (1906).

[Tetr14] H. M. Tetrode, Theoretical determination of the entropy constant of gases and liquids, *Proc. Sect. Sci. Koninklijke Nederlandse Akad. Wet., Ser.* B **17**, 1167–1184 (1914/1915).

[Tisz66] L. Tisza, *Generalized Thermodynamics* (Cambridge, MA: The MIT Press, 1966).

[Tole87] J.-C. Tolédano and P. Tolédano, *The Landau Theory of Phase Transitions: Application to Structural, Incommensurate, Magnetic and Liquid-Crystal Systems* (Singapore: World Scientific, 1987).

[Tolm38] R. C. Tolman, *The Principles of Statistical Mechanics* (Oxford, UK: Oxford University Press, 1938). Reprinted New York: Dover, 1979.

[Toul75] G. Toulouse and P. Pfeuty, Critère de Ginzburg et dimensionnalité caractéristique pour le probleme de la localisation dans les systèmes désordonnés, *C. R. Acad. Sci. Paris, Série B* **280**, 33 (1975).

[Vdov64] N. V. Vdovichenko, A calculation of the partition function for a plane dipole lattice, *Zh. Eksp. Teor. Fiz. (USSR)* **47**, 715–720 (1964). Translated in *Soviet Physics-JETP* **20**, 477–479 (1964).

[Vdov65] N. V. Vdovichenko, Spontaneous magnetization of a plane dipole lattice, *Zh. Eksp. Teor. Fiz. (USSR)* **48**, 526–530 (1965). Translated in *Soviet Physics-JETP* **21**, 350–352 (1965)

[Waal73] J. D. van der Waals, On the continuity of the gaseous and liquid states, doctoral dissertation, Universiteit Leiden (1873).

[Wann66] G. H. Wannier, *Statistical Physics* (New York: J. Wiley & Sons, 1966).

[Wien30] N. Wiener, Generalized harmonic analysis, *Acta Mathematica* **55**, 117–258 (1930).

[Wils72] K. G. Wilson and J. Kogut, The renormalization group and the ϵ-expansion, *Physics Reports* **12**, 75–200 (1972).

[Wils79] K. G. Wilson, Problems in physics with many scales of length, *Sci. Am.* **241**(2), 140–157 (1979).

[Youn10] A. P. Young, Monte Carlo simulations in statistical physics. Available at the author's web page http://physics.ucsc.edu/~peter/219/convergenew.pdf. Retrieved on August 13, 2010.

[Zwan54] R. W. Zwanzig, High-temperature equation of state by a perturbation method. I. Nonpolar gases, *J. Chem. Phys.* **22**, 1420–1426 (1954).

[Zwan60] R. Zwanzig, Ensemble method in the theory of irreversibility, *J. Chem. Phys.* **33**, 1338–1341 (1960).

Index